A First Course in

FLUID MECHANICS

A First Course in
FLUID MECHANICS

S Narasimhan

Universities Press

Universities Press (India) Private Limited

Registered Office
3-6-747/1/A & 3-6-754/1, Himayatnagar,
Hyderabad 500 029 (A.P.), India
e-mail: info@universitiespress.com

Distributed by
Orient Blackswan Private Limited

Registered Office
3-6-752, Himayatnagar, Hyderabad 500 029 (A.P.), India

Other Offices
Bangalore / Bhopal / Bhubaneshwar / Chennai
Ernakulam / Guwahati / Hyderabad / Jaipur / Kolkata
Lucknow / Mumbai / New Delhi / Patna

First published 2006
Reprinted 2009, 2013

ISBN: 978 81 7371 564 8

Set in Times New Roman 10 on 12 by
Krishtel eMaging Solutions Pvt.Ltd., Chennai 600 017

Printed at
Charita Impressions
Hyderabad 500 020

Published by
Universities Press (India) Private Limited
3-6-747/1/A & 3-6-754/1, Himayatnagar
Hyderabad 500 029 (A.P), India

Contents

Preface

Classical hydrodynamics, since the days of its founder, Euler (1703–1787) has largely remained a mathematician's conception of the motion of an ideal fluid – a fluid that is both incompressible and inviscid. The results of such a theoretical analysis of a hypothetical fluid have proved useful and exact in certain types of fluid motion, for example, wave motion. Yet they have little value where problems of fluid resistance are involved. On the other hand, the empirical science of hydraulics grew out of the specific needs of engineers in their search for limited solutions to complex problems connected with irrigation, water supply, water power, river control and the like. The art and science of hydraulics though they had their origin in the ancient civilisations of Rome, Egypt and the Indus Valley, received a great impetus to their growth in the eighteenth and nineteenth centuries. Based as they were on professional experience, observation and limited experimentation, the solutions lacked generality and were valid only within a certain range of values of the variables. With the advent of aeronautics, however, empiricism had fallen by the wayside. In aircraft design, since large factors of safety are not practicable, great precision is required in the description of behaviour of fluid around the aircraft. The subject of fluid mechanics was evolved out of forging the mathematical rigour in analysis with experimental data presented in generalised forms. The development of the subject of mechanics of fluids owes a great deal to Ludwig Prandtl (1875–1953). Prandtl, Von Karman and G. I. Taylor have been among the principal contributors responsible for the blending of rigour in analysis with carefully conducted experimental results to amplify, substantiate and complete the solutions to many intricate fluid flow problems.

The scope of this book is primarily devoted to the application of the laws of Newtonian mechanics to solve complex problems in fluid motion which is of common interest to civil, mechanical, aeronautical and chemical engineers. The book is addressed to undergraduate students who are undergoing the first and perhaps the second course in fluid mechanics in their academic studies.

The first chapter deals with fluid properties and their role in fluid motion. The second is devoted to fluid statics including relative motion and stability of floating bodies. The third chapter presents fluid kinematics, types of particle motion and concepts of stream lines, path and streak lines and the basic equation of continuity. The fourth chapter deals with Euler's equations of motion, the concept of irrotational flow and Bernoulli's energy equation. The subsequent chapter illustrates some typical forms of irrotational flows such as source, sink, vortex, and combinations of these with uniform flow to yield patterns of flow around bodies and the concept of circulation of lift. Chapter 6 presents the application of Bernoulli theorem and its modifications where energy losses occur but can be estimated. The next chapter deals with force–momentum relationships and derives expressions for forces on bodies, forces on flat and curved surfaces. Chapter 8 introduces the all-important property of viscosity and the Navier–Stokes equations of motion, followed by examples of laminar motion with specific boundary conditions finally ending with the treatment of laminar boundary layers and laminar stability. The ninth chapter presents an elementary discussion of turbulence characteristics, the growth of the turbulent boundary layer and expressions for velocity distributions along flat plates and in pipes. The velocity distributions in pipes and open channels are derived and surface resistance factors defined. The tenth chapter is devoted to steady and unsteady flow in pipes and is largely design-oriented. An elementary treatment of unsteady flow in pipes to describe the phenomenon of water hammer and surges is also included. Chapter 11 illustrates the use of the energy and momentum relationships to determine lift and drag forces on submerged bodies of different shapes including aircraft wings. Chapter 12 dwells on open channels with more emphasis on steady uniform and non-uniform flows with only a cursory treatment of unsteady flows. An elementary treatment of compressible flows is presented in Chapter 13 to illustrate the important role of compressibility or variation in density in compressible fluids. Chapter 14 on dimensional analysis and model similitude is included to emphasise the need for guided experimentation, presentation of results in generalised forms and interpretation of results obtained on the model to the prototype.

Acknowledgements

I wish to acknowledge the role of Late Dr. N. Dwarkanath and Dr. C. Natarajan in the preparation of the initial draft of Chapter 12 and Section II of Chapter 10 respectively.

My thanks are also due to the members of my family for their encouragement and unstinted support during the preparation of the book: to my wife Thara for her patience; my son Rajiv and his wife Dr. Kamala for their keen interest; my daughter Rita and her husband Dr. Sunderesh Heragu in whose house in USA, the first draft of the book was prepared.

I am deeply grateful to Ms. Jebah S. David of Universities Press for her valuable editorial assistance and her competence in checking the proofs of the equations.

I would like to record my appreciation of the efforts of Shri. Sridhar of Krishtel eMaging Solutions Pvt. Ltd., for the excellent work of typesetting the text, equations and drawings. I am indebted to Dr. M. Nagendra Kumar for his help in the preparation of the first set of drawings which were later modified and redrawn.

S. Narasimhan
Professor Emeritus, IIT Mumbai.

1 Properties of Fluids

1.1 Definition of a Fluid

The distinction between a solid and a fluid lies in their behaviour under the action of external forces. A solid has a finite tensile strength. A fluid conforms to the shape of its container and can resist compressive forces only when so bounded and the compressive forces are applied to the walls of the container. When subjected to shearing forces, a solid within elastic limits, attains a static equilibrium position. A fluid, on the other hand, suffers continuous deformation under the action of a tangential force, *however small the shear force* may be. Both liquids and gases therefore come under the category of fluids.

1.2 Density

Mass is the quantity of matter in a given volume of substance and density ρ is the mass per unit volume

$$\rho = \frac{M}{\forall} \tag{1.1}$$

where M is the mass of fluid occupying a volume $\forall$. The SI unit of density is kg/cu m. Water has a density of 1000 kg/cu m at 4° C. The specific volume of a fluid, v_s is the volume occupied by a unit mass of fluid. It is also the reciprocal of mass density.

$$v_s = \frac{\forall}{M} = \frac{1}{\rho} \tag{1.2}$$

The second law of Newton, force = (mass × acceleration), shows the importance of mass density in fluid motion. Mass density appears in all problems where an unbalanced force causes acceleration of particles. Thus, unsteady and non-uniform motion invariably depend on the property of density. The role of density is not however excluded from steady or uniform motion, as forces of inertia, viscosity and gravity may be present, though in a state of dynamic equilibrium. Density assumes an important role in all compressible flow systems as it is related to pressure through the equations of state. Subsonic flows (flow velocity less than the speed of sound), where the usual range of changes in velocity do not bring about large changes in pressure, are generally treated as incompressible flows. In supersonic flows, changes in velocity bring about large changes in pressure and density.

1.3 Specific Weight

The specific weight $\gamma = \rho g$ is the weight of the material per given volume. It is the force due to gravity on a unit volume of fluid. This quantity is not an invariable one as mass is, but may vary with both latitude and distance from earth's surface. The specific weight of water at 5° C, at sea level is 9806 N/cu m.

Relative density, or more correctly, specific gravity is the ratio of unit weight of a given fluid to that of water in case of liquids or to that of hydrogen (or air) at standard temperature and pressure in case of gases. If the specific gravity of a fluid S_f is known, then the specific weight of the fluid is $\gamma_f = \gamma_w S_f$, where γ_w is the specific weight of water. For example, mercury with a specific gravity of 13.6 has a specific weight of $13.6 \times 9806 = 133361.6$ N/cu m.

1.4 Pressure

A fluid exerts a pressure normal to a solid boundary or any plane drawn through the fluid. The magnitude of the force exerted on a small area of the boundary or plane may vary from place to place. It is convenient to work in terms of the pressure p of the fluid defined as the force exerted on unit area. If the force exerted on each unit of area of a boundary or plane is the same, the pressure is said to be uniform.

$$\text{Pressure} = \frac{\text{Force exerted}}{\text{Area of boundary}} \quad \text{or} \quad p = \frac{F}{A}$$

If, as is more commonly the case, the pressure changes from point to point, we consider the element of force, s_F normal to a small area s_A surrounding the point under consideration. Then we write,

$$p = \frac{s_F}{s_A}$$

As $s_A \rightarrow 0$, the pressure at a point is

$$p = \underset{s_A \rightarrow 0}{\text{lt}} \frac{s_F}{s_A} = \frac{dF}{dA}$$

In SI units, the unit of pressure is N/sq m known as pascal, Pa.

1.5 Equations of State

In cases of gases, it is necessary to find relationships between pressure, p; density, ρ; specific volume, v_s; temperature, T and bulk modulus, K. A fundamental equation for gases is the law connecting pressure and specific volume,

$$pv_s^n = \text{constant} \tag{1.3}$$

where $n = 1$ for isothermal changes and $n = k = c_p/c_v$ for adiabatic changes, k being the ratio of the specific heats of the gas at constant pressure, c_p and constant volume, c_v respectively. A perfect gas is one that has constant specific heats. The general equation of state for a perfect gas is

$$pv_s = RT \text{ or } p = \rho RT \tag{1.4}$$

where T is the absolute temperature.

Equations (1.3) and (1.4) yield the relationship among the changes in pressure, specific volume and absolute temperature from one set of conditions to another, the conditions indicated by the subscripts

$$\frac{p_2}{p_1} = \left(\frac{v_{s1}}{v_{s2}}\right)^n = \left(\frac{T_2}{T_1}\right)^{n/n-1} \tag{1.5}$$

EXAMPLE 1.1 Air is compressed from an initial volume of $0.6\,\text{m}^3$ at an initial pressure of $120\,\text{kN/m}^2$ to a final volume of $0.3\,\text{m}^3$. Find the final pressure under (a) isothermal conditions and (b) adiabatic conditions. ($k = 1.4$)

(a) Isothermal conditions

$$p_1V_1 = p_2V_2$$

$$p_2 = \frac{p_1V_1}{V_2}$$

$$= 120 \times \frac{0.6}{0.3} = 240\,\text{kN/m}^2$$

(b) Adiabatic conditions

$$p_1V_1^k = p_2V_2^k$$

$$p_2 = p_1\left(\frac{V_1}{V_2}\right)^k$$

$$= 120\left(\frac{0.6}{0.3}\right)^{1.4} = 316.68\,\text{kN/m}^2$$

Temperature ratio

$$\left(\frac{T_1}{T_2}\right) = \left(\frac{p_1}{p_2}\right)^{0.4/1.4}$$

$$= \left(\frac{120}{316.68}\right)^{0.2857} = 0.7579$$

$$T_2 = 1.318\,T_1$$

$$\frac{\rho_2}{\rho_1} = \left(\frac{p_1}{p_2}\right) \times \left(\frac{T_2}{T_1}\right)$$

$$= \left(\frac{120}{316.8}\right) \times 1.319 = 0.5$$

1.6 Bulk Modulus

Gases are more compressible than liquids. For most purposes, liquids may be treated as incompressible under normal i.e., small and gradual changes in pressure. However, if the change in pressure is sudden and very large as in the case of water hammer in pipes due to sudden closure of valves, compressibility of the liquid must be taken into account.

Bulk modulus, K an important property of compressible fluids is of considerable interest in the propagation of elastic waves in a fluid medium. It is defined as the ratio between volumetric stress and volumetric strain.

$$K = -\frac{dp}{(d\text{¥}/\text{¥})} \tag{1.6}$$

The negative sign is to account for the reduction in volume due to an increase in pressure intensity, dp. Again, as mass is conserved, the bulk modulus may be expressed as

$$K = \frac{dp}{(d\rho/\rho)} = \frac{dp}{(d\gamma/\gamma)} \tag{1.7}$$

Bulk modulus is a function of pressure and temperature. The bulk modulus of elasticity, K has the same dimension as that of pressure. The bulk modulus of water at 20° C is 2.2 GPa (G stands for gega, 10^9).

EXAMPLE 1.2 What is the bulk modulus of elasticity of a liquid which is compressed from a volume of 0.0125 cu m at 80 N/sq cm to a volume of 0.0124 cu m at 150 N/sq cm?

$$\text{Change in volume, } d\text{¥} = 0.0124 - 0.0125 = -0.0001 \text{ cu m}$$

$$\text{Change in pressure, } dp = 150 - 80 = 70 \text{ N/sq cm} = 700000 \text{ N/sq m}$$

$$\text{Bulk modulus, } K = -\frac{dp}{d\text{¥}/\text{¥}} = \frac{700000}{0.0001/0.0125}$$

$$= 8.75 \times 10^7 \text{ N/sq m}$$

1.7 Viscosity

Physically, the difference between a solid and a fluid lies in the relative mobility of the molecules in them; each molecule in a solid has a fixed average position though the molecule may undergo random vibratory or rotary motions about the mean position. The molecule in a fluid, on the other hand, by virtue of its mobility accompanied by translation of molecular momentum may translate in addition to the other two types of motion; these random motions follow certain probability laws, hence the distinction between solids and fluids as pointed earlier.

Viscosity is the property of a fluid that enables it to develop resistance to deformation. The viscous resistance offered by liquids is a function of molecular spacing which is in turn a function of temperature. Molecular spacing increases with temperature and hence viscosity of a liquid varies as an inverse function of temperature. In case of gases, it is found that the influence of molecular spacing is less significant than that of molecular activity, and the latter is found to increase greatly with increase in temperature. Hence we find that the viscosity of gas increases with increase in temperature.

Let a certain liquid fill the space between two closely spaced parallel plates as shown in Fig. 1.1, and let one of the plates be moved with a uniform velocity, thus establishing a velocity gradient across the space between the plates. The original shape of the element ABCD is distorted to the shape A′B′CD. Under steady conditions, there is a balance between the external force F and the internal force developed due to viscosity. Newton in his classical experiments was able to show that the shear stress F/A (A being the area on which the force is applied) is proportional to the velocity gradient, du/dy. Newton's law of viscosity may be stated as that the shear stress is proportional to the rate of angular deformation, $d\theta/dt = du/dy$. The proportionality constant μ connecting the shear stress $\tau = F/A$ and du/dy is known as the dynamic coefficient of viscosity.

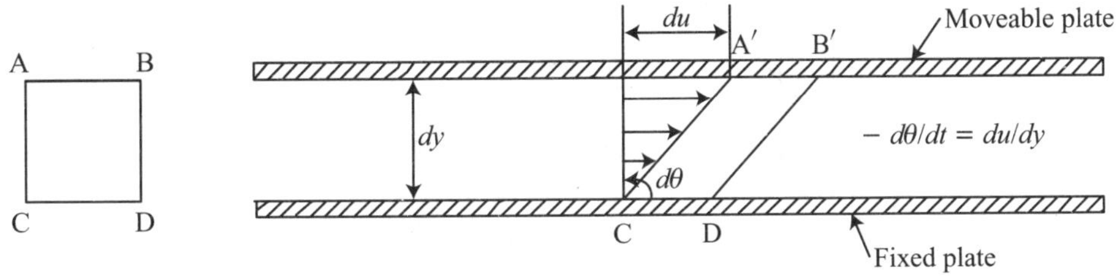

Fig. 1.1 Fluid deformation

$$\tau = \mu\left(\frac{du}{dy}\right) \tag{1.8}$$

Since F has the dimension [MLT^{-2}], A[L^2], u[LT^{-1}] and y[L], the dimension of μ is [ML^{-1}T^{-1}]. μ is practically independent of pressure but is a function of temperature. The SI unit of viscosity is the pascal second (denoted as Pa s).

Example 1.3 A flat plate of area 1.5 sq m is pulled with a velocity of 0.4 m/s, as shown in Fig. 1.1, relative to another parallel plate located at a distance 0.15 mm from it. Find the force required to maintain this velocity, if the viscosity of the fluid filling the space between the plates is known to be 0.1 Pa s.

The velocity gradient between the plates is linear and is obtained as

$$\frac{du}{dy} = \frac{U}{t} = \frac{0.4}{0.00015} = 2666.7$$

The shear stress $\tau = \frac{\mu du}{dy} = 0.1 \times 2666.7 = 266.67$ N/sq m

Force required to maintain the velocity, $F = \tau . A = 266.67 \times 1.5 = 400$ N.

A Newtonian fluid is one for which the value of μ is constant at a given temperature. When the shear stress is dependent on the rate of angular deformation, a non-linear relationship exists between τ and the velocity gradient as illustrated in Fig. 1.2. Such fluids (for example, colloidal suspensions, grouts, gels) are classified as Non-Newtonian fluids. Certain substances known as plastics require a finite initial stress (yield stress) before they behave like fluids. If the deformation thereafter is linear with stress, it is known as a Bingham plastic or ideal plastic. If the relationship is non-linear after the yield stress, then the substance is called real plastic. Non-Newtonian fluids which deform as soon as shear stress is applied may also be classified as pseudo-plastic or dilatant depending on whether the viscosity decreases or increases with rate of shear, as illustrated in Fig. 1.2.

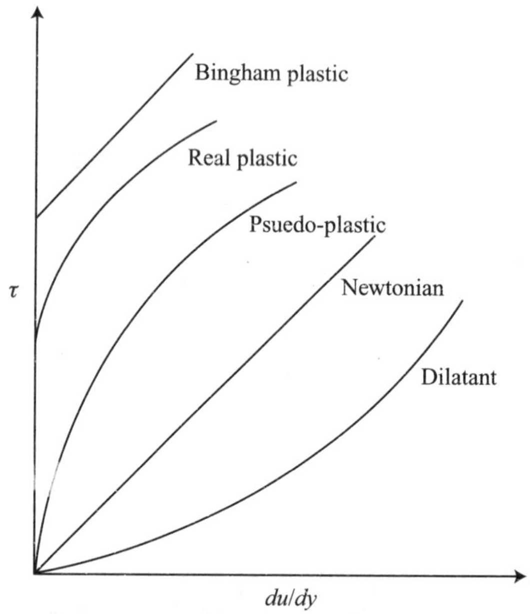

Fig. 1.2 Relationship between shear stress and angular deformation

The ratio μ/ρ having the dimensions of length and time [LT^{-1}] is known as the kinematic viscosity, ν as distinct from μ which also has the dimension [M] in it and hence called the dynamic coefficient of viscosity.

Kinetic Theory of Gases

The kinetic theory of gases postulates that the molecules of a gas are constantly in motion with velocities varying in a random manner. The very nature of the random distribution of these molecular velocities ensures that there is no mean motion of the gas as a whole on this account. Should the gas possess a mean motion with a velocity gradient as illustrated in Fig. 1.3, the random motion of the molecules in a transverse direction causes an exchange of momentum. The molecules in the upper layers possess greater horizontal momentum in relation to those in lower layers. An exchange of molecules from one layer to another tends to equalise the velocities of the upper and lower layers.

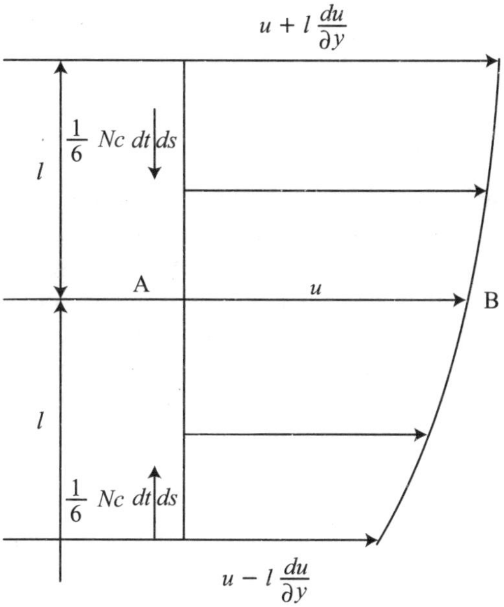

Fig. 1.3 Momentum transfer by molecular exchange

Let us assume that out of N molecules present in a unit volume of gas, equal number of molecules i.e., a third of the molecules $(N/3)$ are directed along the three axes directions. Let half of the molecules moving in the y-direction transverse to the flow in a vertical column of area ds be directed downwards and the other half namely $(N/6)$ upwards. If all the molecules are moving with a speed c, then the number of molecules passing a given plane AB in time dt either upwards or downwards is $(1/6)\, Nc\, dt\, ds$

If the molecule while passing the plane AB has horizontal velocity u, corresponding to a level l units from AB, then the average momentum of each molecule having mass m is $m\ (u - l(\partial u/\partial y))$. du/dy is the velocity gradient in the y-direction. Similarly a molecule passing upwards has an average momentum $m\ [u - l(\partial u/\partial y)]$. The net transport of momentum across the plane AB is therefore 2 m $(l.du/dy)$. Multiplied by the number of molecules passing in any one direction, the transport of momentum per unit area per unit time is given by $(1/3)$ $N.m\ cl\ du/dy$ which should be equal to the shear stress along AB.

$$\tau = \frac{\mu du}{dy} = \left(\frac{1}{3}\right) (N.m)\ c.l.\frac{du}{dy} = \left(\frac{1}{3}\right) \rho cl\frac{du}{dy}$$

since Nm is the total mass per unit volume which is equal to ρ, where l is the mean free path of the molecules. Hence

$$\mu = \left(\frac{1}{3}\right) \rho cl$$

A more accurate hypothesis gives the value of μ as $0.499\rho cl$.

1.8 Surface Tension

Every molecule exerts upon every other neighbouring molecule an electrochemical force known as molecular attraction which gives rise to additional properties of adhesion and cohesion in liquids. At a point P, not on the surface of a liquid, a molecule is attracted equally in all directions. A molecule lying very close to the free surface experiences a lesser force above, as the number of molecules above it are less than those below. The resultant force has to act in a direction normal to the free surface, since an unbalanced force in a direction parallel to the surface must cause continuous deformation, which is the essential property of a fluid. The attraction between neighbouring molecules forms an imaginary film capable of resisting tension at the interface of a liquid and a gas (free surface) or of two immiscible fluids. If a small needle is placed on an undisturbed water surface, it is seen that the needle is supported by the film. It is as if this film is held stretched all over the surface.

In the absence of all external forces, an unconfined volume of liquid would attain a spherical shape due to molecular attraction. Let the sphere be enlarged by some means. The formation of additional free surface involves movement of molecules from the interior to the surface, and work is done against the mutual attraction of the molecules. The work done is proportional to the increase in surface area. Surface tension, σ is defined as the force required to maintain unit length of the surface film in equilibrium. It may be alternately defined as the work done per unit area of the surface. The SI unit of surface tension is accordingly N/m.

One of the most important consequences of surface tension is the pressure difference on the two sides of a curved free surface.

Consider a surface element of sides ds_1 and ds_2 with radii of curvature R_2 and R_1 respectively as shown in Fig. 1.4. The surface tension σ is uniform and the pressure on the two sides of the surface are denoted by p_1 and p_2. Resolving the forces in the vertical direction, we have

$$(p_1 - p_2)ds_1 ds_2 - \sigma ds_2 d\alpha - \sigma ds_1 d\beta = 0$$

The arc lengths ds_1 and ds_2 are respectively equal to $R_1 d\alpha$ and $R_2 d\beta$ On substitution for $d\alpha$ $d\beta$, one obtains

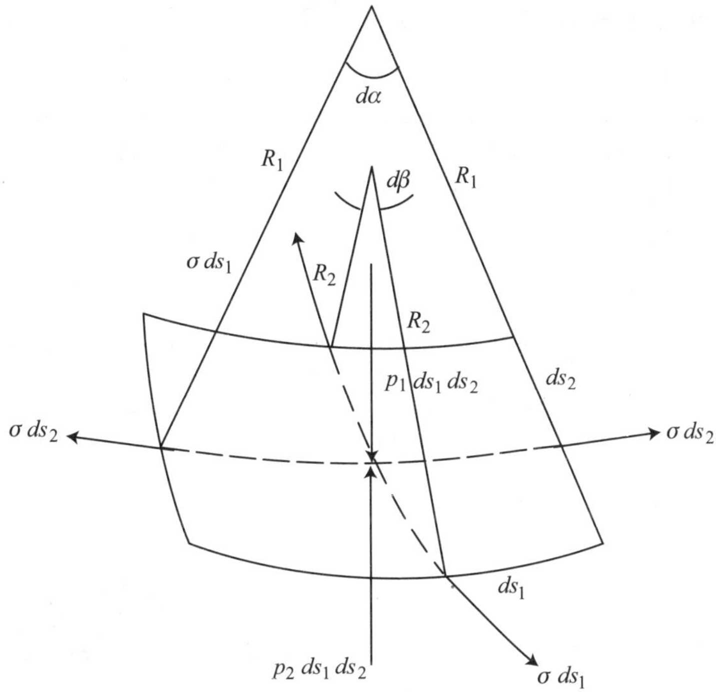

Fig. 1.4 Forces on a surface film

$$p_1 - p_2 = \sigma\left(\frac{1}{R_1} + \frac{1}{R_2}\right) \tag{1.9}$$

If the surface is a part of a sphere of radius R, the above equation simplifies to

$$\Delta p = \frac{2\sigma}{R} \tag{1.10}$$

and for a cylindrical surface (one radius is infinite), the pressure difference is given by σ/R. Soap bubbles have two surfaces and the pressure inside is in excess of the pressure outside by $4\sigma/R$.

The value of the surface tension depends on the nature of substances on different sides of the surface. Along a liquid interface, the molecules of one liquid may attract or repel the molecules of the other. Molecular attraction between molecules of two different substances is called adhesion. When three liquids meet, the surface tensions $\sigma_{12}, \sigma_{13}, \sigma_{23}$ may or may not be in equilibrium Fig. 1.5a.

Under equilibrium conditions, we have

$$\sigma_{13} = \sigma_{12}\cos\alpha + \sigma_{23}\cos\beta$$

An example where equilibrium exists is the case of floating globules of oil or fat on soup. While lubricating oil on a wet street has a tendency to spread continuously without achieving equilibrium. In such a case, the fluids 1, 2, 3 are respectively air,oil and water. For this combination,

$$\sigma_{13} > (\sigma_{12} \cos \alpha + \sigma_{23} \cos \beta)$$

and hence the spreading.

If one of the three substances is a solid, movement is possible only in a direction tangential to the solid surface. Figures 1.5*b* and 1.5*c* illustrate two possible formations of liquid drops. The angle of contact α is determined by the equilibrium of force diagram

$$\cos \alpha = \frac{(\sigma_{13} - \sigma_{23})}{\sigma_{12}}$$

A mercury drop illustrates the condition in Fig. 1.5*b* and a water drop in Fig. 1.5*c*. For a petrol drop, on the other hand, equilibrium does not exist. The meniscus formation at the free surface of a column of liquid in a tube may be discussed in a similar manner as illustrated in Figs. 1.5*d* and *e*. These are the meniscus formations for mercury and water respectively, the adhesion to the solid surface being greater in the case of water but less in the case of mercury than their corresponding neighbouring molecules.

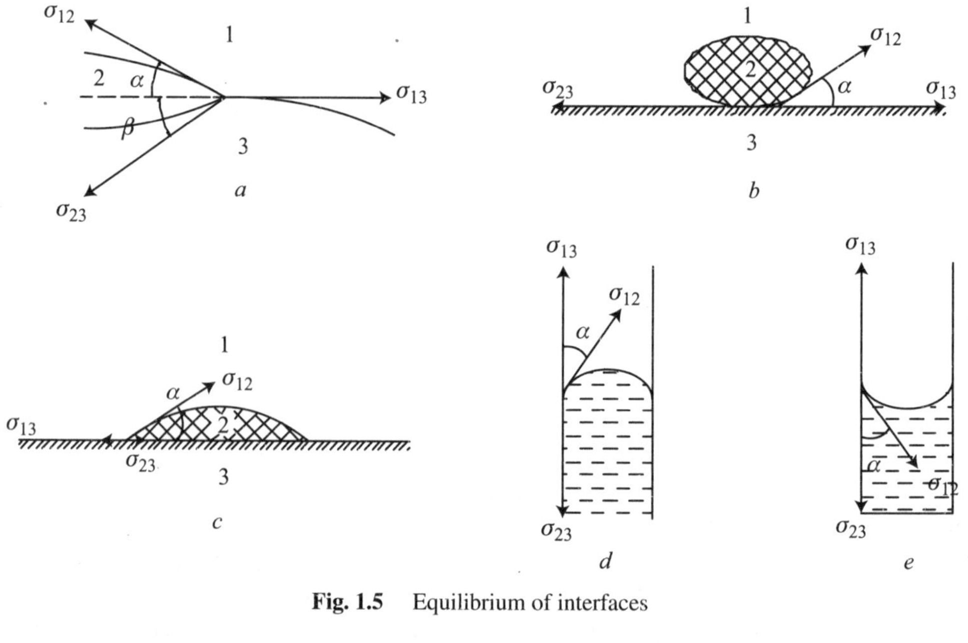

Fig. 1.5 Equilibrium of interfaces

1.9 Capillarity

The capillary rise of liquids in a narrow tube is also of great practical significance (e.g., soil moisture). Let a be the radius of the capillary tube in Fig. (1.6).

The rise in the level of liquid in the tube is found by equating the vertical forces, namely the surface tension forces and the weight of the column of the liquid above the free surface. When

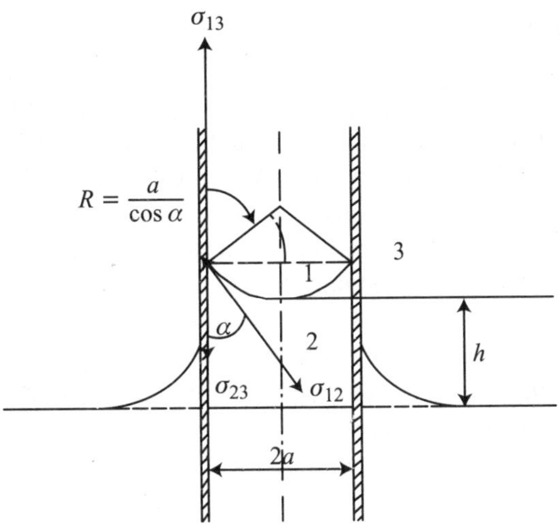

Fig. 1.6 Capillary rise

the rise, h is much larger than the radius of the tube, the meniscus may be approximated by a sphere of radius $a/\cos\alpha$. The surface tension acts over a peripheral length of $2\pi a$.

$$(\sigma_{12}\cos\alpha)\cdot 2\pi a = \gamma\pi a^2 h$$

$$h = \frac{2\sigma\cos\alpha}{\gamma a} \tag{1.11}$$

The role of surface tension is thus seen to have an important bearing whenever there are density discontinuities (interface problems), pressure discontinuities (capillary waves) and when the cross-sectional area of flow is very small (pores in soil).

EXAMPLE 1.4 Calculate the capillary rise/fall in mm in a glass tube of 4 mm diameter when immersed in a container of (a) water; (b) mercury. Surface tensions of water and mercury are 0.0735 N/m and 0.510 N/m respectively. The wetting angle of contact of water with glass is 0° and that for mercury is 130°.

Water

$$h = \frac{2\sigma\cos\theta}{\gamma a}$$

$$= \frac{2\times 0.0735}{9806\times 2\times 10^{-3}} = 0.007495\ \text{m} = 7.495\ \text{mm}$$

Mercury

$$h = \frac{2\times 0.510\times\cos 130^\circ}{13.6\times 9806\times 2\times 10^{-3}} = -0.002458\ \text{m} = -2.458\ \text{mm}$$

The minus sign indicates that the level in the tube is lower than that in the container.

1.10 Vapour Pressure

Molecular activity and molecular attraction were briefly discussed in the sections on viscosity and surface tension respectively. It was also remarked that the molecular momentum has a certain distribution and follows certain probability laws. When we describe the molecular motion by a mean velocity, it is implied that certain molecules move at faster speeds than some, while others move at slower speeds than the mean. At a liquid surface, the transitional momentum of some molecules may be greater than the molecular attractive forces, in which case, these molecules escape from the liquid and form vapour. At a constant temperature, an equilibrium is achieved between the number of molecules leaving the liquid and the number of molecules entering it due to condensation. The vapour molecules exert a partial pressure in the space above the free surface known as vapour pressure. The vapour pressure is dependent on the molecular activity which in turn depends on the temperature. Vapour pressure increases with increase in temperature. When the vapour pressure at a given temperature equals the atmospheric pressure above a liquid surface, the phenomenon of boiling occurs. Thus, the vapour pressure of water equals the atmospheric pressure at 100° C at sea level, at which point boiling occurs. Water may be made to boil at a lower temperature if it is kept in a sealed container and the pressure above the surface is reduced by suction or in the case of water being heated at higher altitudes.

Mercury has a low vapour pressure and hence it is used in barometers. The role of vapour pressure in flow of liquids becomes important when the pressure at certain locations (narrow constrictions) in the flow systems become equal or lower than the vapour pressure at the ambient temperature of the liquid. Then the liquid is vapourised forming a cavity or vapour pocket. These cavities or vapour pockets move in the direction of stream into regions where the local pressure is higher than the vapour pressure, causing the pockets to collapse. Not only does the growth and decay of these cavities affect the efficiency of the flow system but if the cavities collapse near a boundary, it may cause erosion of the boundary surface due to pressure impacts. Cavitation may thus play an important role in the design of body shapes and machinery parts. Cavitation also determines the length of suction pipes of pumps, elevation of siphon systems, degree of pipe constrictions, design of propellors etc.

Vapour pressures of certain liquids in N/m^2 at 20° C are water: 2.348×10^3; mercury: 1.726×10^{-1}; glycerine: 1.373×10^{-2}; benzene: 1.0×10^4; ethyl alcohol: 5.786×10^3.

Vapour pressure expressed in terms of pressure head is $h_v = (p_v/\gamma)$. For example for water at 32° C, $h_v = (4.83 \times 10^3)/9806 = 0.492$ m. Since dissolved gases are released from water at pressures higher than vapour pressure, design pressure heads are not allowed to fall below 2.5 m. A cavitation parameter expressed as $(p_o - p_v)/\frac{1}{2}\rho u^2$ should not exceed a limiting value.

Exercises

1.1 The mass density of keresone is 800 kg/m^3. Find the relative density, specific weight and specific volume.

1.2 Air has a mass density of 1.20 kg/m^3 at 20° C. Find its specific weight and relative density with respect to hydrogen ($\rho = 0.090$ kg/m^3) and water ($\rho = 998$ kg/m^3)

1.3 The velocity distribution between two parallel plates is given by $u = -\frac{1}{2\mu}\left(\frac{dp}{dx}\right)(a^2 - y^2)$. Find the shear stress and velocity at $y = 0$ and at $y = 5$ mm. $a = 20$ mm and $\mu = 8.63$ poise, given $dp/dx = -1500$ N/m.

1.4 A shaft of diameter D rotates in a bearing at N rpm. The clearance between the shaft and the bearing is e. A lubricating oil of dynamic viscosity μ fills the space between the shaft and the bearing wall. If the length of the bearing is L, find an expression for the power required to overcome viscous resistance in the bearing. The velocity profile is assumed to be linear.

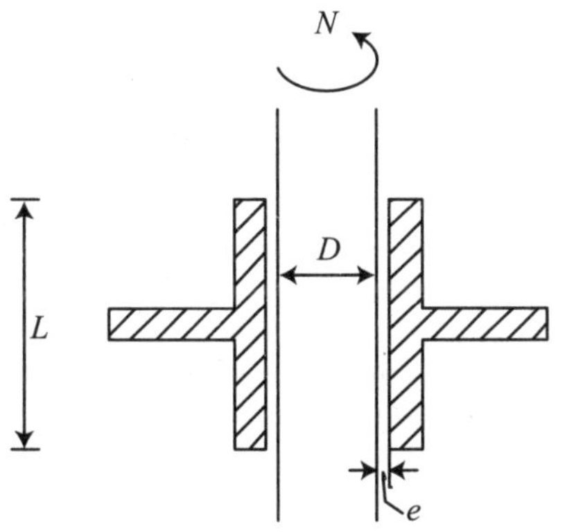

Figure for Problem 1.4

1.5 Find the torque exerted on the lower surface of a disk rotating at N rpm. The gap between the disc and the bottom of the container is e.

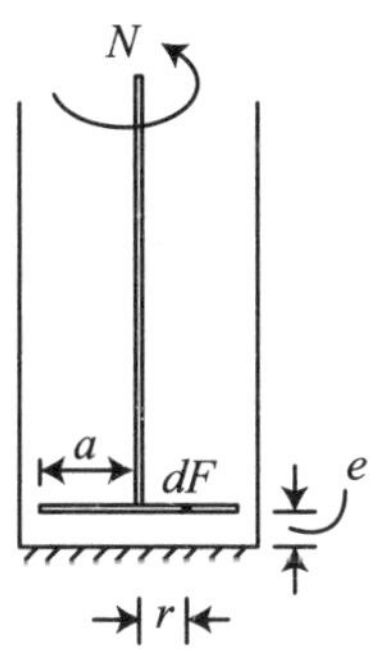

Figure for Problem 1.5

1.6 A cylinder of diameter 100 mm and length 200 mm, weighing 20 N is falling with a uniform velocity, U in an outer cylinder of diameter 100.5 mm. If the kinematic viscosity of the oil in the space between the cylinders is 6.7×10^{-2}m/s^2, find U. $\rho = 880$ kg/m^3

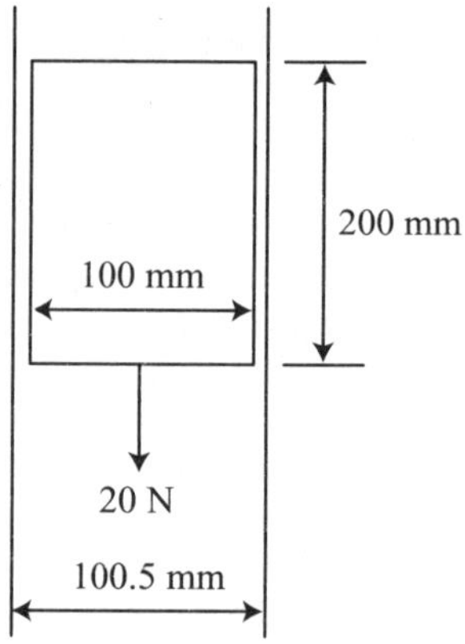

Figure for Problem 1.6

1.7 If in Problem 1.6, the cylinder is falling at 0.5 m/s and is decelerating at the rate of 14 m/s^2, find the weight of the body.

1.8 Water at the nozzle of a Pelton wheel is under a pressure head of 600 m. If the bulk modulus of water is 2.05×10^9 N/m^2, find the mass density of water at the Pelton wheel. What is the error in density if compressibility is ignored.

1.9 Find the mass density of sea water at a depth of 10,000 m. The density of sea water is 1026 kg/m^3. $K = 2.05 \times 10^9$ N/m^2.

1.10 A cylinder with an initial volume of 60 litres has 0.10 kN of gas at an initial pressure of 0.15 kN/m^2. A piston compresses the gas to half its volume (30 litres) isothermally. What is the pressure and density of gas after compression.

1.11 If the gas in Problem 1.10 is compressed adiabatically, what will be the resulting pressure and density. $k = 1.40$.

1.12 An automobile tyre has a capacity of 0.3 m^3. The tyre is filled at a pressure of 2600 kN/m^2 at 25° C. What will be the pressure if the tyre is heated to a temperature of 60° C.

1.13 A volume of 30 litres of alcohol subjected to a pressure of 500 atmospheres at 25° C contracts to 28.8 litres. What is the modulus of elasticity of alcohol and its compressibility. Compressibility is defined as $K_T = -(1/V_0)(\Delta V/\Delta p)$

1.14 A cylinder contains 2 kg of air at 50° C and a pressure of 3 bars. The air is compressed to 8 bars. What is the initial volume. Find the final volume if the compression takes place (a) isothermally; (b) isentropically. ρ_{air} at 50° C = 1.10 kg/m^3 at atmospheric pressure. 1 bar = 10^{-5} pascals; $K = 2.42 \times 10^{-5}$ N/m^2

1.15 Determine the pressure intensity within a bubble of air, 0.25 cm in diameter, at a depth of 1 m below the free surface. $\sigma = 0.073$ N/m

1.16 Compute the power required to convert 0.5 m^3 of water into mist having an average droplet size of 2.5×10^{-5} cm $\sigma = 0.073$ N/m

1.17 Liquids A and B are used in a compound glass manometer. A is lighter than B. The surface tension between the glass and liquid A is 0.050 N/m and that between the glass and liquid B is 0.070 N/m. The surface tension between the liquids is 0.040 N/m. Find the angle at which the liquid interface will meet the wall of the manometer

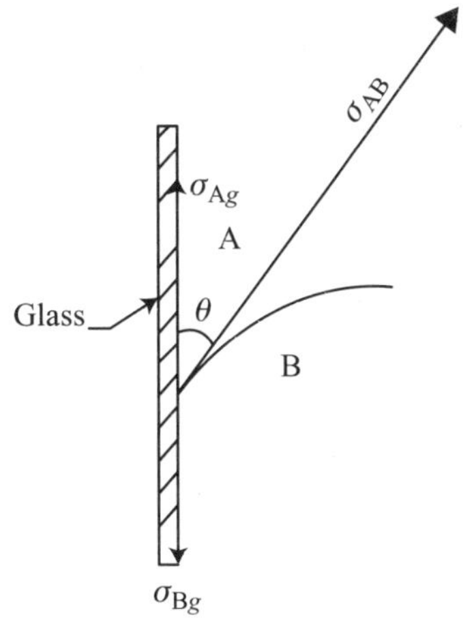

Figure for Problem 1.17

1.18 Two parallel vertical plates placed 2.5 mm apart is immersed in a container containing water. Determine the capillary rise between the plates. $\sigma = 0.073$ N/m

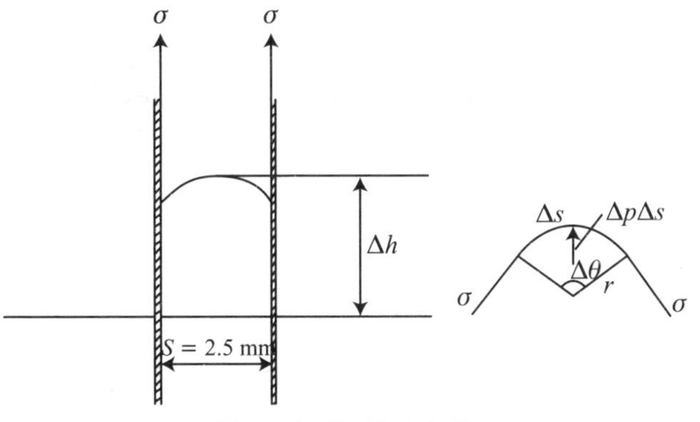

Figure for Problem 1.18

1.19 Water an in an earth embankment rises to a height of 2 m. What is the uniform diameter of the interstices in the earth? $\sigma = 0.073$ N/m

1.20 A tube of 2 mm bore is immersed in an oil of specific gravity 0.8 up to 2 cm below the free surface. If the pressure required to blow a bubble is 188 N/m^2, find the surface tension of the oil.

1.21 If the pressure inside a droplet of water is 200 N/m^2, what is the diameter of the droplet? $\sigma = 0.073$ N/m

1.22 A water bug is sitting on the surface of water by virtue of its surface tension. The bug has six legs and the contact length of each leg with water is 5 mm. What is the maximum weight of the bug above which it would sink. $\sigma = 0.073$ N/m

1.23 Estimate the height to which water at 20° C will rise in a capillary glass tube of 3 mm dia. Contact angle between water and glass is 0°. $\sigma = 0.073$ N/m

2 Fluid Statics

2.1 Introduction

A fluid is considered static if all the particles of the fluid are motionless or moving with the same velocity or there is no relative motion among the particles. As per the definition, a fluid is set in motion by tangential stresses, however small they may be. It is therefore implicit that if the a fluid is at rest, tangential stresses are absent.

2.2 Pressure at a Point

The pressure at a point in a static fluid, with no shear stresses present, is in fact a scalar quantity. In other words, the pressure at any point in a static fluid is the same in all directions. This can be readily proved with reference to Fig. 2.1.

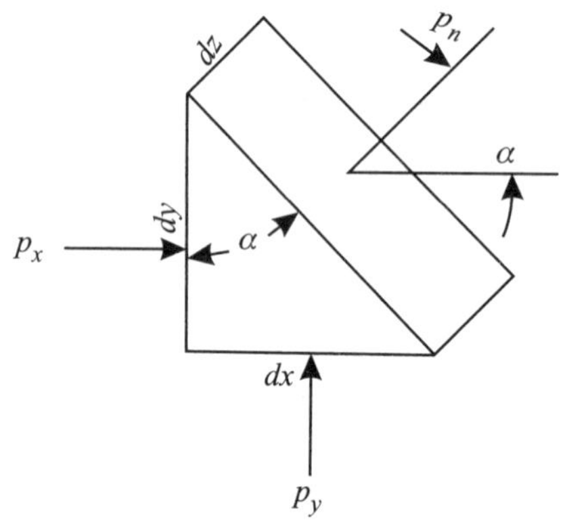

Fig. 2.1 Pressure on wedge faces

Consider a small wedge-shaped fluid element. Let p_x be the pressure acting on the face $dydz$ and p_y be the pressure acting on face $dxdz$ with p_n, the pressure acting on the inclined surface $dsdz$. Resolving the forces in the x-direction

$$-p_x dydz + p_n dsdz \cos\alpha = 0$$

and since $ds \cos\alpha = dy$, we get $p_x = p_n$

Similarly, resolution of the forces in the y-direction will yield

$$-p_y dx dz + p_n ds dz \sin\alpha - \gamma dV$$

where dV is the volume of the wedge.

Again, since $dx = ds \sin\alpha$ and allowing the element to shrink to a point ($dV \to 0$), we get $p_y = p_n$. We can show, by similar arguments, that $p_z = p_n$.

Thus we have $p_x = p_y = p_z = p_n$. The relationships that we have obtained is true for all angles of α and therefore independent of it. Hence the pressure at any point is the same in all directions and is thus a scalar quantity in fluid statics. This is known as Pascal's law. This statement is true for non-viscous or ideal fluids even though they may be in motion since there are no shear stresses in ideal fluids.

2.3 Variation of Pressure within a Static Fluid

Consider a differential element of a static fluid as shown in Fig. 2.2. For an incompressible fluid, the density is constant. Within the small element, we assume that the density remains constant. The pressure variations across the opposite faces are shown in the figure. Since the fluid element is in static equilibrium, the forces in each of the three directions must be equal to zero. The system is only under the action of gravitational forces and tangential stresses are absent.

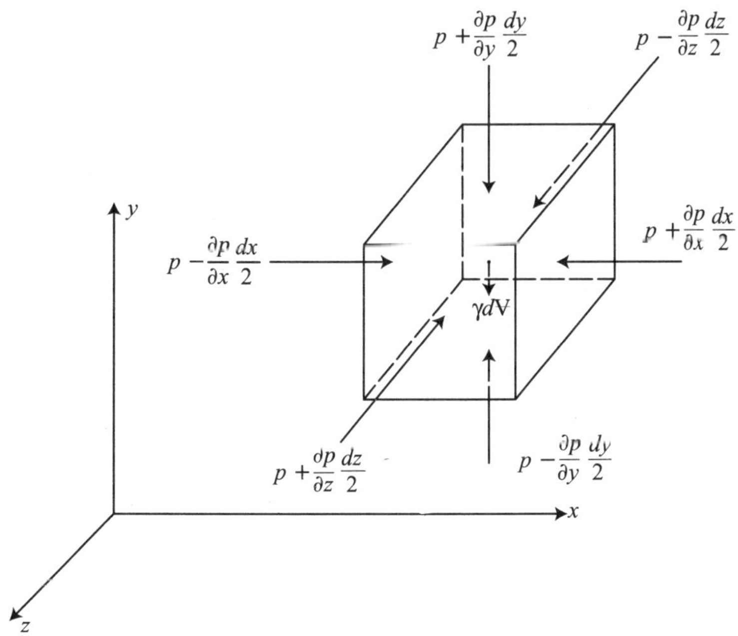

Fig. 2.2 Pressures on a differential element

Resolution in the three directions yield

x-direction:

$$\left(p - \frac{1}{2}\frac{\partial p}{\partial x}dx\right) dy \cdot dz - \left(p + \frac{1}{2}\frac{\partial p}{\partial x} \cdot dx\right) dydz = 0$$

y-direction:

$$\left[p - \frac{1}{2}\left(\frac{\partial p}{\partial y}\right) dy\right] dxdz - \left[p + \frac{1}{2}\left(\frac{\partial p}{\partial y}\right) dy\right] dxdz - \gamma dxdydz = 0$$

z-direction:

$$\left[p - \frac{1}{2}\left(\frac{\partial p}{\partial z}\right) dz\right] dxdy - \left[p + \frac{1}{2}\left(\frac{\partial p}{\partial z}\right) dz\right] dxdy = 0$$

The pressure gradients in the x- and z-directions are zero. The pressure gradient in the vertical direction is given as

$$\frac{\partial p}{\partial y} = -\gamma$$

which we can write as

$$\frac{dp}{dy} = -\gamma \tag{2.1}$$

as there are no changes of pressure in the x- and z-directions. Integrating (2.1) between the limits of the known pressure p_0 at a known value y_0 and the pressure p at a given depth y, we get,

$$p - p_0 = -\gamma(y - y_0)$$

But if y is measured downwards below the free surface ($y_0 = 0$), where $p_0 = p_a$, (the atmospheric pressure), the pressure, p at a depth $y = -h$ is

$$p - p_a = \gamma h \tag{2.2}$$

The pressure $p - p_a$, that is, the pressure above atmosphere is known as the gauge pressure, p_g.

$$p_g = \gamma h \tag{2.3}$$

The pressure p is known as the absolute pressure above an absolute zero. p_a is the local pressure indicated by a barometer. All values of absolute pressure are always positive though gauge pressure below the local atmospheric pressure is negative indicating a certain degree of vacuum. Figure 2.3 shows clearly the difference between gauge pressure and absolute pressure.

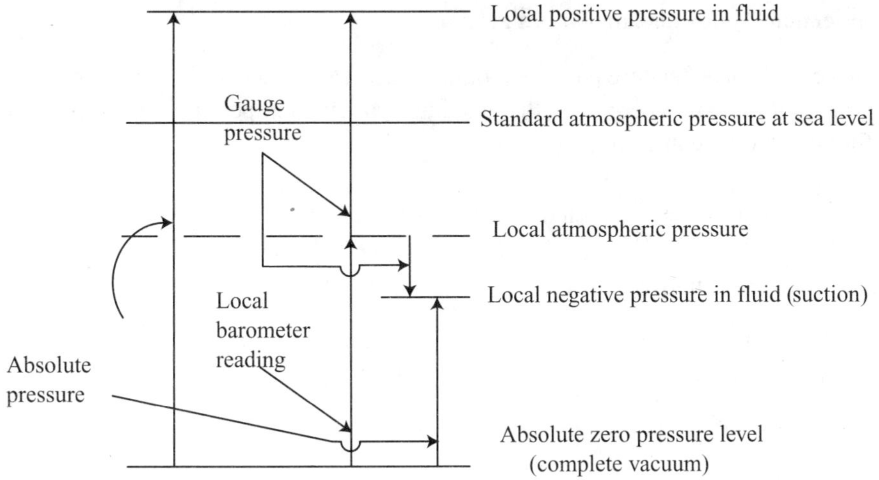

Fig. 2.3 Definition sketch for gauge and absolute pressure

2.4 Pressure Variation in Compressible Fluids with Elevation

2.4.1 Isothermal Condition

If the temperature remains constant with altitude, we have $pV_s = $ constant or $pV_s = RT_0$, where R is the gas constant and T_0 is the absolute temperature. Since $V_s = 1/\rho$, we can rewrite the relationship $p/\rho = pg/\gamma = RT_0$. Introducing this in the basic equation (2.1), we obtain

$$\frac{dp}{dy} = -\gamma = \frac{-pg}{RT_0}$$
$$\frac{dp}{p} = \frac{-g dy}{RT_0} \tag{2.4}$$

Integrating the above equation, we get

$$\ln (p)_{p_1}^{p} = \frac{-g(y)_{y_1}^{y}}{RT_0} \quad \text{or}$$

$$\ln\left(\frac{p}{p_1}\right) = \frac{-g}{RT_0}(y - y_1)$$

$$p = p_1 \exp\left[\frac{-g}{RT_0}(y - y_1)\right] \tag{2.5}$$

If we designate sea level as the reference datum with $y_1 = 0$ and $p_1 = p_a$

$$p = p_a \exp\left[\frac{-gy}{RT_0}\right]$$

2.4.2 Temperature Varies Linearly with Elevation

The temperature in the troposphere has been found to decrease linearly with attitude, y and may be written as $T = T_0 - \beta y$, where T_0 is the absolute sea level temperature and β is a constant. Differentiating the above relationship, we get

$$dy = \frac{-dT}{\beta} \quad \text{and} \quad \gamma = \frac{pg}{RT} \quad \text{and} \quad \frac{dp}{dy} = -\gamma$$

Eliminating dy and γ, we find

$$\frac{dp}{p} = \left(\frac{g}{R\beta}\right)\left(\frac{dT}{T}\right)$$

Integration of the preceding equation yields

$$\ln p = \left(\frac{g}{R\beta}\right)\ln T + C$$

The constant of integration may be evaluated from the sea level conditions: $p = p_a$ and $T = T_0$. Finally we have

$$\ln\frac{p}{p_a} = \left(\frac{g}{R\beta}\right)\ln\left(\frac{T}{T_0}\right) = \ln\left(\frac{T}{T_0}\right)^{g/R\beta}$$

$$\frac{p}{p_a} = \left(\frac{T}{T_0}\right)^{g/\beta R} = \left[\frac{(T_0 - \beta y)}{T_0}\right]^{g/\beta R} \tag{2.6}$$

2.4.3 Adiabatic Process

For an adiabatic process, we have the relationship: pV^k = constant, $p_1 V_1^k = p_2 V_2^k$ and $p = p_1(T_1/T)^{k/1-k}$, where p_1, T_1 are the pressure and temperature at level y_1 respectively.

Differentiating with respect to T

$$dp = -\frac{k}{1-k} p_1 T_1^{(k/1-k)} T^{-(1/1-k)} dT$$

But from $pV = RT$, and $dp = -\gamma dy$, we get $dp = -pgdy/RT$. Hence we have

$$dy = \frac{-RT\left[-\left(\frac{k}{1-k}\right) p_1 T_1^{(k/1-k)} T^{-(1/1-k)}\right]}{gp_1 T_1^{(k/1-k)} T^{-(1/1-k)}} \, dT = \left(\frac{k}{1-k}\right)\frac{RdT}{g}$$

$$\frac{dT}{dy} = -\left[\frac{(k-1)}{k}\right]\left(\frac{g}{R}\right)$$

Integrating the above equation between the limits of altitudes y_1 and y_2 and the corresponding temperatures T_1 and T_2, we obtain

$$y_2 - y_1 = \left(\frac{k}{1-k}\right)\left(\frac{R}{g}\right)(T_2 - T_1) \tag{2.7}$$

Since $T_2/T_1 = (p_2/p_1)^{(k-1)/k}$, we find the relationship between pressure and altitude.

$$y_2 - y_1 = \frac{k}{1-k}\frac{RT_1}{g}\left[\left(\frac{p_2}{p_1}\right)^{k-1/k} - 1\right] \tag{2.8}$$

2.5 Force on Plane Areas Immersed in a Fluid

Suppose a rectangular plane area ABCD is immersed horizontally at a depth h below the free surface. The pressure at any point on the surface is uniform and equal to $p = \gamma h$ and hence the forces on the top and bottom of the plane are equal and equal to $F = pA = \gamma h A$.

Suppose now the plane is placed inclined to the horizontal at an angle θ, the pressure on the plane is obviously not uniform and the pressure at any point on the surface is proportional to the vertical depth of the point below the free surface. We shall take the origin of the coordinate system to be at point O, which is the point of intersection of the extended line AB and the free surface, as shown in Fig. 2.4.

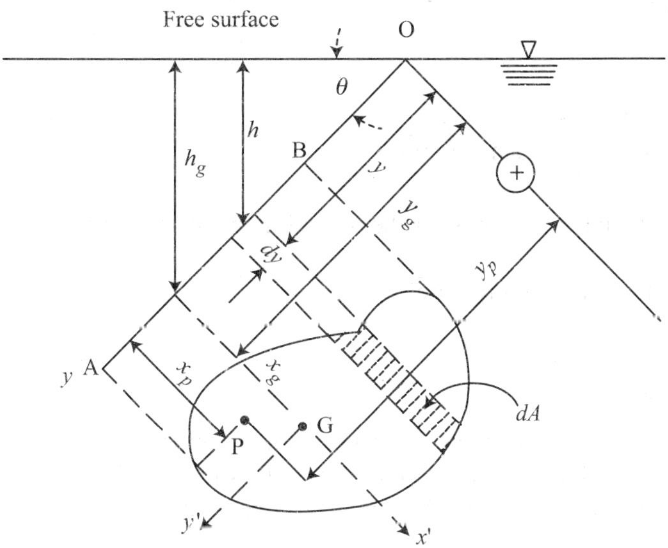

Fig. 2.4 Force on an inclined plate

Consider an elemental strip on AB parallel to the free surface at a distance y from it. Let the width of the strip be dy and the area of the strip be dA located at a depth h below the surface. The pressure on this elemental area is uniform as all points in this area are at a depth h. The force on this element is $\gamma h dA$. To find the force on the entire plate, we integrate over the whole area. Remembering $h = y \sin\theta$, we write

$$F = \int_A \gamma y \sin\theta dA = \gamma \sin\theta \int_A y dA$$

The integral in the preceding equation is the first moment of the area of the plate about the x-axis and is equal to Ay_g, where y_g is the distance of the free surface from the centroid of the plate along AB. Hence we have

$$F = \gamma \sin\theta y_g A = \gamma h_g A = p_g A \tag{2.9}$$

The pressure at the centroid of the plate is $p_g = \gamma h_g$, h_g being the depth of the centroid of the area below the free surface. The magnitude of the resultant force due to fluid pressure on a plane surface is given by the product of the pressure at the centroid and the area of the surface. We may rotate the plate about its centroid without changing the force on the plate. Equation (2.9) gives the magnitude of the force but the centre of pressure or the location of the resultant force is not at the centroid.

We again consider an elementary area dA. Let the total elementary force on the area be denoted by $dF = pdA$ and the total force by F. By definition, the moment produced by F about a point or axis must be equal to the sum of the moments produced by individual forces (forces on individual elementary areas such as dA). Thus we have

$$Fy_p = \int_A pdAy$$

Since $p = \gamma y \sin\theta$

$$Fy_p = \int_A \gamma y^2 \sin\theta dA = \gamma \sin\theta \int_A y^2 dA$$

The expression under the integral stands for the second moment of the area about the trace axis x, denoted by I_{xx}.

The expression for the centre of pressure is

$$y_p = \frac{I_{xx}}{F} = \frac{I_{xx}}{\gamma A y_g} \tag{2.10}$$

It is usually more convenient to express y_p in terms of the moment of inertia about an axis passing through the centroid and parallel to the x-axis. This may be done by using the parallel axes theorem of the moments of inertia, which may be expressed as

$$I_{xx} = I_{x'x'} + Ay_g^2$$

where I_{xg} is the moment of inertia about an axis passing through the centroid of the area and parallel to the x-axis. The substitution yields

$$y_p = \frac{I_{x'x'} + Ay_g^2}{Ay_g} = y_g + \left(\frac{I_{x'x'}}{Ay_g}\right) \tag{2.11}$$

Since I_{xg}, A, y_g are all positive quantities, it follows that the centre of pressure is always below the centroid.

To find the lateral position of the centre of pressure, we take moments about the y-axis as shown in Fig. 2.4.

$$Fx_p = \int_A xy \sin\theta \gamma dA = \gamma \sin\theta \int_A xydA = \gamma \sin\theta I_{xy}$$

where I_{xy} is the product of inertia about the reference axes.

$$x_p = \frac{I_{xy}}{Ay_g} \tag{2.12}$$

If x' and y' are the axes passing through the centroid and parallel to the x- and y-axes, we write by the transfer formula that the products of inertia

$$I_{xy} = I_{x'y'} + Ax_g y_g$$

Substituting the preceding expression in (2.12), we arrive at

$$x_p = x_g + \frac{I_{x'y'}}{Ay_g} \tag{2.13}$$

EXAMPLE 2.1 Consider a triangular plate PQR (with PR = QR) immersed in water, as shown in the figure. The base PQ is horizontal. The plate is making an angle of 60° with the horizontal. All required distances are given. Determine the magnitude and the point of application of the force due to water pressure on the triangular plate.

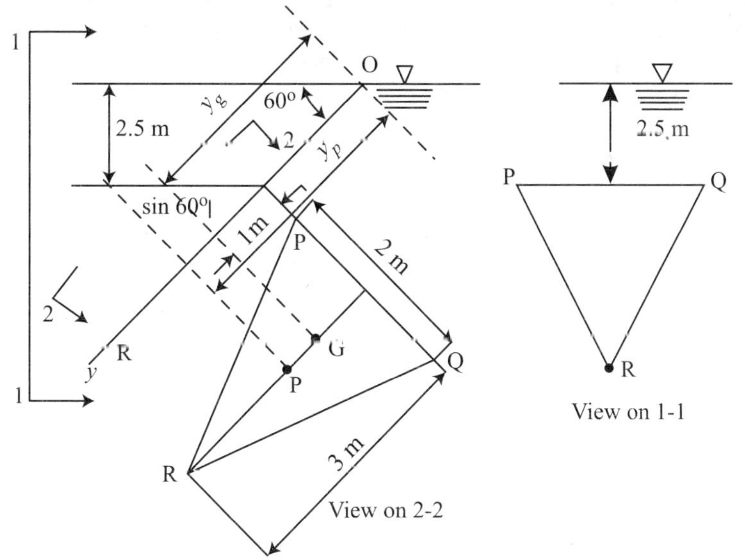

Figure for Example 2.1 (not to scale)

The force on the plate is equal to the product of the pressure at the centroid and the area of the plate. The depth of the centroid below the free surface is

$$2.5 + 1 \sin 60^\circ = 3.366\,\text{m}.$$

The resultant force on the plate is

$$F = \gamma h_g A = 9806 \times 3.366 \times 3 = 99021 \text{ newtons}$$

The y-coordinate of the centre of pressure, measured from O along the y-axis y_p is given by $(I_{x'x'}/Ay_g) + y_g$

$$I_{x'x'} = \frac{bh^3}{36} = \frac{2 \times 3^3}{36} = 1.5$$

where b is the base and h is the altitude of the triangle.

$$y_g = \frac{h_g}{\sin 60^\circ} = \frac{3.366}{\sin 60^\circ} = 3.887\,\text{m} \text{ and } A = 3\,\text{sqm}$$

Hence, we have

$$y_p = \frac{1.5}{3.887 \times 3} + 3.887 = 4.015\,\text{m}$$

$$h_p = y_p \sin 60^\circ = 3.447\,\text{m}$$

Since the triangle is symmetric, the centre of pressure lies on the median with $I_{x'y'} = 0$.

EXAMPLE 2.2 An inclined triangular plate as shown in the sketch is held in water with the base at the surface. $B = 4$ m and $H = 6$ m. The inclination of the plate to the horizontal is 60°. Find the resultant force and centre of pressure. $\gamma = 9810\,\text{N/m}^3$

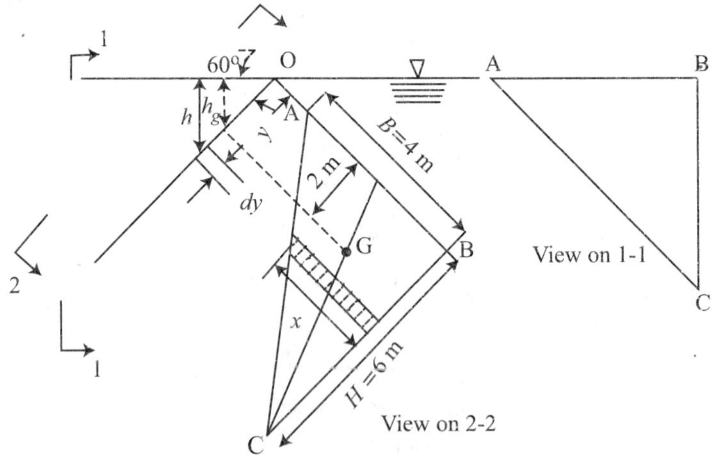

Figure for Example 2.2

Area $= 0.5 \times 6 \times 4 = 12\,\text{m}^2$. The centre of gravity lies at a vertical depth of $h_g = (6/3) \times \sin 60^\circ = 1.732\,\text{m}$.

Hence the hydrostatic force on the plate is

$$\gamma A h_g = 9810 \times 12 \times 1.732 = 203897\,\text{N}$$

$$y_p = y_g + \frac{I_{x'x'}}{A y_g} = y_g + \frac{BH^3/36}{(BH/2)\,y_g} = 2 + \frac{24}{12 \times 2} = 3\,\text{m}$$

$$x_p = x_g + \frac{I_{x'y'}}{A y_g} = x_g + \frac{\left(-B^2H^2/72\right)}{(BH/2)\,y_g} = \frac{4}{3} - \frac{8}{12 \times 2} = 1\,\text{m}$$

The centre of pressure can be found by an alternate method of taking moments of elementary forces about O.

$$\text{Total moment} = \int_0^H \gamma h(x dy)y = \int_0^H \gamma y \sin 60^\circ \left[B\left(1 - \frac{y}{H}\right)\right] y dy = \gamma B \sin 60^\circ \left[\frac{H^3}{12}\right]$$

The location of the centre of pressure along the y-axis is

$$F.y_p = 203897 y_p = \gamma B \sin 60^\circ \frac{H^3}{12} \quad \text{or} \quad y_p = \frac{H}{2} = 3\,\text{m}$$

Similarly we can evaluate the location of the centre of pressure in the x-direction by taking moments of the elementary forces about BC.

$$\int_0^H \gamma h x dy \frac{x}{2} = \int_0^H \frac{1}{2}\gamma y \sin 60^\circ \left[B\left(1 - \frac{y}{H}\right)\right]^2 = \frac{1}{2}\gamma B^2 \sin 60^\circ \left(\frac{H^2}{12}\right)$$

$$F x_p = \frac{1}{2}\gamma B^2 \sin 60^\circ \left(\frac{H^2}{12}\right)$$

$$x_p = \frac{B}{4} = 1\,\text{m}$$

Example 2.3 An elliptical gate covers a circular pipe 4 m in diameter. If the gate is hinged at the top, find the normal force, F required to open the gate when the water depth is 8 m. The gate is open to the atmosphere on the other side.

$$\text{Area of the elliptical gate} = \pi ab = \pi \times 2 \times 2.5 = 5\pi\,\text{m}^2$$

$$\text{Force on the elliptical gate} = \gamma h_g A = 9810 \times 10 \times 5\pi = 1540951\,\text{N}$$

The centre of pressure

$$y_p = \frac{I}{y_g A} + y_g = \frac{1}{4}\frac{\pi a^3 b}{y_g \pi ab} = \frac{1}{4}\frac{a^2}{y_g} + y_g$$

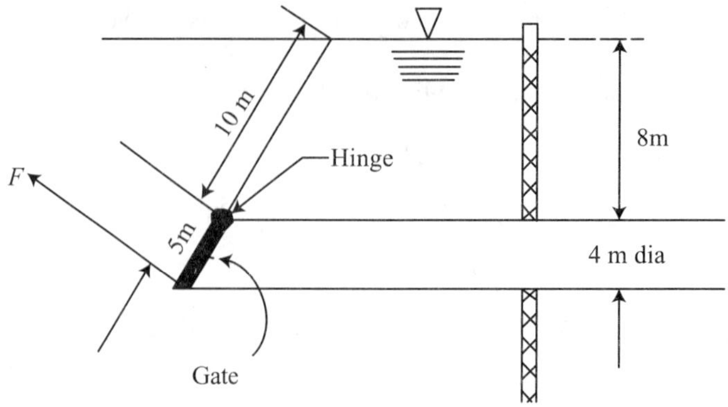

Figure for Example 2.3

y_g, the slant distance from the water surface to the centroid of the plate = 12.5 m

$$y_p = \frac{6.25}{4 \times 12.5} + 12.5 = 0.125 + 12.5 = 12.625$$

from the surface but from the hinge, the centre of pressure is located at a distance of $0.125 + 2.5 = 2.625$ m. Hence taking moments about the hinge

$$F \times 5 = 1540951 \times 2.625 \quad \text{and} \quad F = 809 \text{ kN}$$

2.6 Forces on Curved Surfaces

When the surface is curved, such as shown in Fig. 2.5, the system of forces on elemental areas will not constitute a system of parallel forces and therefore we will not be able to obtain the

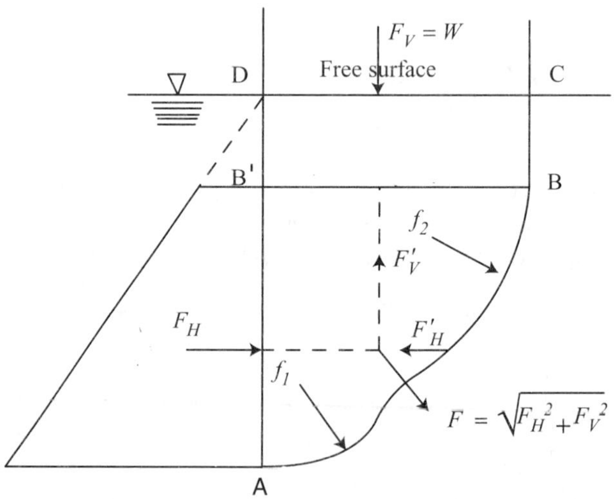

Fig. 2.5 Force on curved surfaces

resultant force by a single integration. Again referring to Fig. 2.5, f_1 and $f_2, \ldots$ are the forces on individual elements and they are normal to the respective elemental areas.

Let F be the resultant force on the curved surface. F_H and F_V are the horizontal and vertical force components respectively of the resultant. We determine these components, F_H and F_V instead of finding F directly. Consider the fluid enclosed between the curved surface AB and its projection on a vertical plane AB$'$. In the x-direction, there are only two forces acting on the volume of the fluid, viz., F_H, the force acting on the vertical projection AB$'$ and the horizontal force F'_H on the curved surface. Since the fluid within the volume AB$'$BA is in equilibrium, it follows that $F' = F_H$. From this we draw the important inference that the horizontal component of the fluid force on a curved surface is equal to the fluid force on the projection of the surface on a vertical plane. Also the line of action of F_H must be the same as that of F'_H. For example, as in Fig. 2.6*a* if the curved surface AB is a half-cylinder, with diameter d and length l and the top of the vertical cylinder is just touching the free surface, its projection on a vertical plane is a rectangle of width d and length l. Hence the horizontal force on the curved surface is equal to the force on the rectangle equal to $\gamma d^2 l/2$ and the line of action of this force is located at a distance $2d/3$ from the free surface.

We shall determine the vertical component of the resultant force by a similar reasoning. Consider a section of unit width of the curved surface. The only two forces that keep the volume of liquid in equilibrium in the vertical direction are the weight, W of the volume of liquid ABCDE (Fig. 2.6*b*) and the vertical component of the force on the curved surface, F_V.

Hence $F_V = W$. The vertical component, F_V of the resultant force on the curved surface is equal to the weight of the volume of liquid enclosed between the curved surface and the free surface of the fluid. The line of action of the vertical component, F_V must be the same as the weight of the fluid and hence it must pass through the centroid of the fluid volume between the boundary of the curved surface and the free surface.

For example as shown in Fig. 2.6*b* if a semi-cylinder of diameter d and length l is lying horizontally with its diameter at a depth h below the free surface, then the vertical component of the force is equal to the weight of the fluid above. This is equal to $\gamma(dlh - l\pi d^2/8)$.

Once we get the horizontal and vertical components of the resultant force, we can determine the resultant force and its direction from first principles.

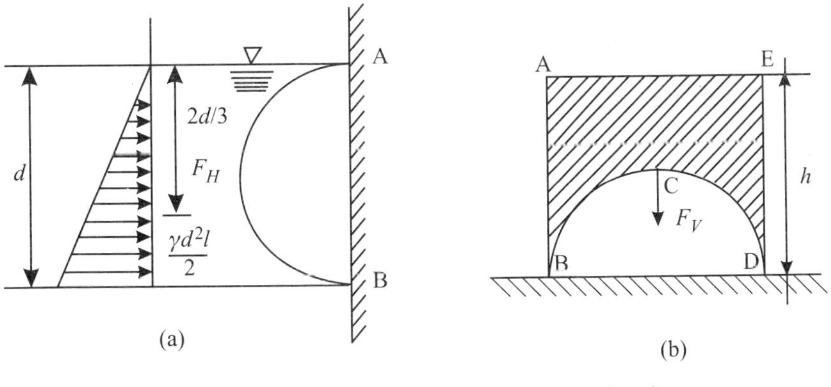

Fig. 2.6 Horizontal and vertical forces on curved surfaces

$$F = (F_H^2 + F_V^2)^{1/2} \text{ and } \theta = \tan^{-1}(F_H/F_V)$$

θ being the angle made by the resultant with the vertical.

EXAMPLE 2.4 Calculate the magnitude and direction of the resultant force on a semi-circular gate as shown in the sketch below. The width of the gate is 3 m.

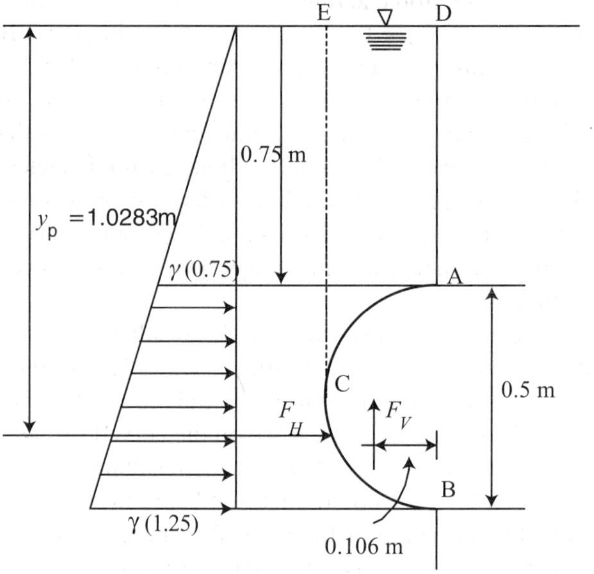

Figure for Example 2.4

The horizontal force on the gate, F_H is the force on the projection of the curved surface on a vertical plane. The area of projection of the half-cylinder is $3 \times 0.5 = 1.5\,\text{m}^2$ and the CG of the area is located at a depth $0.75 + 0.25 = 1.0\,\text{m}$. The magnitude of the horizontal component is

$$F_H = \gamma A h_g = 9806 \times 1 \times 1.5 = 14715\,\text{N}$$

The horizontal force acts at the centre of pressure c_p located at a distance given by the expression in (2.11).

$$y_p = \frac{(bd^3/12)}{bdy_g} + y_g = \frac{(0.5)^2}{12.1} + 1 = 1.0283\,\text{m below the surface}$$

The vertical component of the force consists of two parts: a downward force on the upper arc AC and an upward force on the lower arc CB. These forces correspond to the weight of the liquid in the respective volumes above the surfaces as derived earlier.

Force on CB = weight of liquid in volume ECBAD

Force on AC = weight of liquid in volume ADECA

The resultant weight acting upward is given by the weight of the liquid in volume ACBA $= 9806 \times 3 \times \pi(0.5)^2/8 = 2888.1$ N being the same as the weight of the volume of liquid displaced by the gate form.

The vertical force acts at the centroid of the semi-circle at a distance of $0.212d$ from the diameter.

The resultant force

$$F = (14715^2 + 2888.1^2)^{1/2} = 14990\,\text{N} = 14.99\,\text{KN}$$

And the angle it makes with the horizontal is given by

$$\theta = \tan^{-1}\frac{F_V}{F_H} = 10.9^\circ$$

EXAMPLE 2.5 Find the force per unit width on a circular quadrant as shown in the figure. The water surface is at the top of the gate.

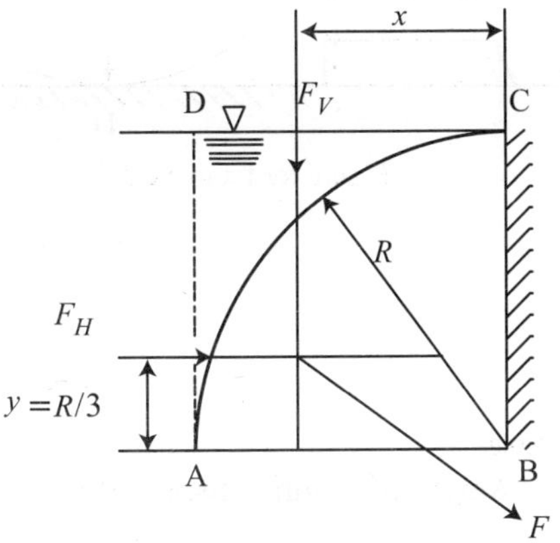

Figure for Example 2.5

$$F_H = \frac{1}{2}\gamma R^2$$

and acts at a distance of $R/3$ above the bottom AB.

The vertical force on the curved surface is equal to the weight of water in the volume ADCA above the curved surface,

$$F_V = \gamma\left(R^2 - \frac{1}{4}\pi R^2\right) = 0.2146\gamma R^2$$

and acts at a distance x_p from the wall BC.

$$0.2146\gamma R^2 x_p = \frac{\gamma R^2 R}{2} - \gamma\left(\frac{\pi R^2}{4}\right)\frac{4R}{3\pi} = \frac{\gamma R^3}{6}$$

$x_p = 0.7766R$ from the wall

$F_{\text{resultant}} = (F_H^2 + F_V^2)^{1/2} = [(0.5\gamma R^2)^2 + (0.2146\gamma R^2)^2]^{1/2} = 0.5441\gamma R^2$ and acts at an angle $\theta = \tan^{-1}(F_H/F_V) = \tan^{-1}(0.5\gamma R^2/0.2146\gamma R^2) = 66.77°$

EXAMPLE 2.6 Find the resultant force per unit width on the half-circle ACD as shown in the figure. The water surface coincides with the top of the circle at A.

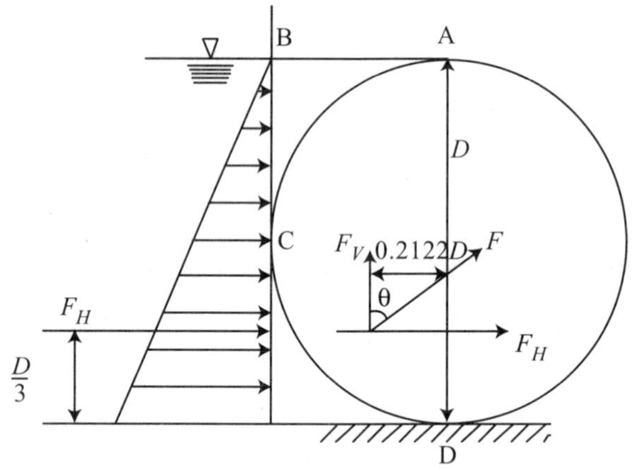

Figure for Example 2.6

Horizontal component

$$F_H = \frac{1}{2}\gamma D^2$$

and it acts at $D/3$ above the base.

Vertical component, F_V = Weight of volume displaced = $\gamma\pi D^2/8$ and acts upward at the CG of the half-circle i.e., at

$$x_g = \frac{2D}{3\pi} = 0.2122D$$

$$F_R = [(0.5\gamma D^2)^2 + (0.3927\gamma D^2)^2]^{1/2} = 0.6358\gamma D^2$$

$$\theta = \tan^{-1}\left(\frac{0.5\gamma D^2}{0.3927\gamma D^2}\right) = 51.85°$$

2.7 Measurement of Pressure

In this section, we shall study some devices used to measure pressure. For measuring purposes, the pressure is frequently expressed as the height of a liquid column, known as pressure head. For example, a pressure head of 760 mm of mercury corresponds to the pressure exerted at the base of a column of mercury 760 mm high. When the pressure is expressed as force per unit area, it is referred to as the pressure intensity. Pressure intensity may be expressed as an equivalent pressure head of any liquid. The pressure intensity and the pressure head are linked by the expression $p = \gamma h$, where γ is the specific weight of the fluid. For two different liquids measuring the same pressure intensity, the relation is

$$p = \gamma_1 h_1 = \gamma_2 h_2$$

and the pressure head for the second liquid is

$$h_2 = \left(\frac{\gamma_1}{\gamma_2}\right) h_1$$

The atmospheric pressure is normally taken as 760 mm of mercury and we may express this in terms of a water column equivalent to

$$h_w = \frac{13.6 \times 9806 \times 760}{9806} = 10336\,\text{mm} = 10.336\,\text{m}$$

The pressure head of 760 mm of mercury is equivalent to a pressure head of 10.336 m of water.

2.7.1 Manometers

Manometers are simple devices to measure pressure. The simplest of the manometers is the piezometer. It is simply a glass tube fitted to the pressure source as shown in Fig. 2.7*a*. The pressure source may be a flow in a pipe or a pressure tank. As the glass tube (piezometer) is open to atmosphere, the liquid will rise to a certain height h inside the tube to balance the pressure at A. The height itself will give the pressure head at A in terms of the fluid flowing in the tube. Piezometers are obviously unsuitable for measuring negative pressures, since air would be sucked into the source through the tube. Also if the positive pressures to be measured are very high, the length of the glass tube required will have to be very large, so much as to be sometimes impracticabie. In such cases, U-tube manometers are used as shown in Fig. 2.7*b* and *c*.

The fluid from the pressure source will enter the U-tube and attain a level B when equilibrium is reached. If the pressure at A is negative, the point B will be below A and if positive, B will be

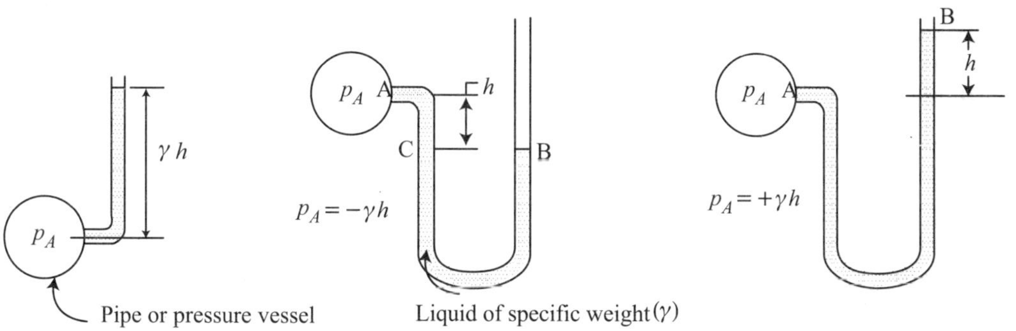

(a) Piezometer (b) U-tube manometer (–ve pressure) (c) U-tube manometer (+ve pressure)

Fig. 2.7 Types of manometers

above A. To get the negative pressure at point A, we start at point B where the pressure is known to be atmospheric. In most problems, we will only be interested in finding the gauge pressure (pressures relative to atmospheric pressure), and therefore we take the atmospheric pressure as the reference pressure set to zero. Because the point C is at the same level as B, the pressure at C is equal to that at B. The point C is itself at a level h below A, and there is continuity of fluid between A and C. Therefore the pressure at A plus the pressure due to the liquid column between A and C must be equal to the pressure at C. That is, $p_A = -\gamma h$ and knowing the specific weight of the fluid, we can determine the pressure at A. When the pressures are large, the tube length has also to be large, sometimes so large as to be impractical. To overcome this difficulty, a second fluid of higher specific weight is used. The advantage is that the rise in the level of higher density is smaller than that of the fluid of lower density for the same pressure difference. Figure 2.8 shows such a manometer. Normally mercury is used as the fluid of the manometer since it has a relatively high specific gravity ($S = 13.6$) and is immiscible with most fluids. An advantage of such manometers which use another fluid to measure the pressure is that they can be used when the pressure source is a gas.

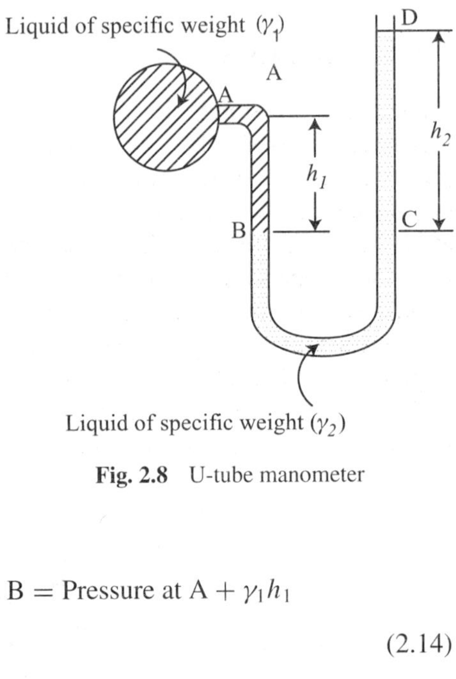

Fig. 2.8 U-tube manometer

In Fig. 2.8, to determine the pressure p_A at the point A, we proceed as follows. Point B is at the interface of fluid 1 and fluid 2. We know that two points along the same horizontal line will have equal pressures (of course, as long as there is continuity in the fluid between the two points). Thus the pressure at C will be equal to the pressure at B. We also know that the pressure at point D which is open to the atmosphere is zero (since we are talking of gauge pressure, i.e., pressures relative to the atmospheric pressure). The obvious procedure to adopt therefore would be to start with pressure at D and add to it the pressure due to the column DC of fluid 2, to get the pressure at C. This is also the pressure at B. From the point B, to reach point A, we need to move up the column of fluid 1 and as we move up the column, the fluid pressure reduces. Thus we write

Pressure at D + $\gamma_2 h_2$ = Pressure at C = Pressure at B = Pressure at A + $\gamma_1 h_1$

$$p_A = \gamma_2 h_2 - \gamma_1 h_1 \tag{2.14}$$

In many practical problems, we will only be interested in the pressure difference between two sources and not in the actual individual pressures. In such a case we use a differential manometer. The differential manometer is very much similar to the U-tube manometer shown in Fig. 2.8, except that a pressure source is connected to each end of the manometer. To get the pressure difference between two sources A and B, $p_B - p_A$, we proceed exactly in the same manner as before. That is, we reach the point E (Fig. 2.9), where we add to the unknown pressure p_A, a pressure $\gamma_1 h_1$.

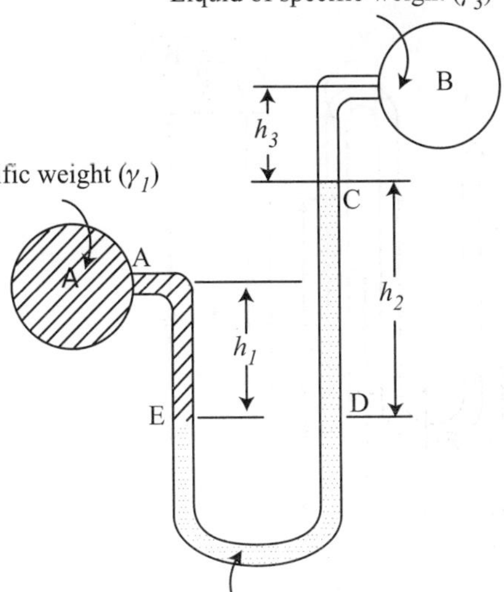

Fig. 2.9 Differential manometer

Since E and D are at the same level and the fluid between them is the same and continuous, the pressure at D is the same as the pressure at E. There are two fluids above D. The pressure at C, p_C is equal to $p_D - \gamma_2 h_2$ as we move up the column. The pressure at B is similarly $p_C - \gamma_3 h_3$. Thus we have

$$p_A + \gamma_1 h_1 - \gamma_2 h_2 - \gamma_3 h_3 = p_B$$

$$p_B - p_A = \gamma_1 h_1 - \gamma_2 h_2 - \gamma_3 h_3 \tag{2.15}$$

Since the specific weights of the three liquids and the measured values of h_1, h_2, h_3 are known, the pressure difference between the two sources can be determined. In all manometer problems, we use the same procedures.

1. Start at one end and write the pressure there in appropriate units if the pressure is known or in an appropriate symbol if the pressure is unknown.
2. Add to this the change in pressure, in the same units, from one meniscus to the next, plus if the next is lower and minus if it is higher. Always keep in mind, the specific weight of the fluid between successive meniscus.
3. Continue this procedure till the desired point is reached. Two general examples of multiple U-tube or multiple liquid manometers are shown in Fig. 2.10.

The pressure at A is given by the general formula given below.

$$\begin{aligned} p_A = {} & p_B + \gamma_{5-6}(z_6 - z_5) + \gamma_{4-5}(z_5 - z_4) + \gamma_{3-4}(z_4 - z_3) \\ & + \gamma_{2-3}(z_3 - z_2) + \gamma_{1-2}(z_2 - z_1) \end{aligned}$$

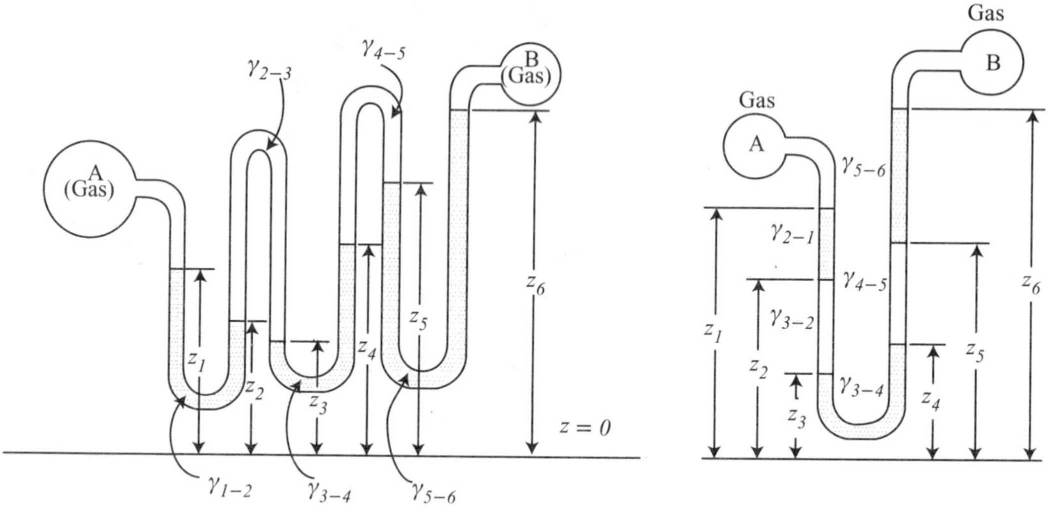

Fig. 2.10 (a) Multiple U-tube manometer; (b) Multiple liquid manometer

or in general

$$p_A = p_B + \Sigma_1^n \quad \gamma_{i \to i+1}(z_{i+1} - z_i)$$

Precision manometers

Inclined manometers In some applications, we wish to measure small pressures cr small differences in pressure. In such cases, we may use the inclined manometer shown in Fig. 2.11. As the name indicates, the limbs of the manometer are inclined to the horizontal.

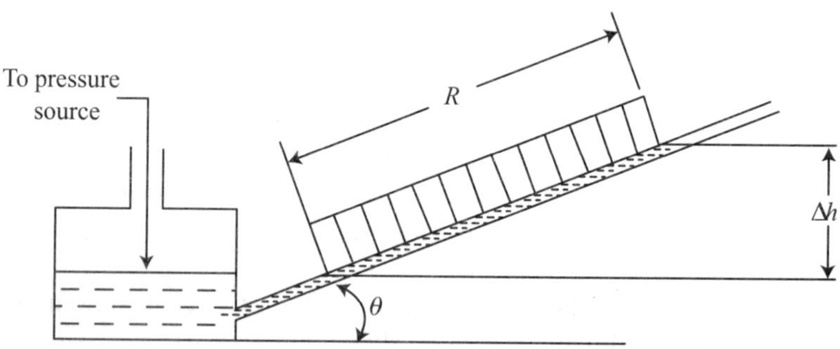

Fig. 2.11 Inclined manometer

As a result, a larger reading or a larger difference in the levels of the manometer fluid in different limbs is produced even for small pressures or pressure differences. The pressure measurement is thus made more accurate. The magnification is readily seen to be $\Delta h / \sin \theta$, where Δh is the value of the pressure head if the tube had been vertical. Thus the magnification may be set at the desired level by adjusting the angle of inclination.

Inverted U-tube manometer An inverted U-tube manometer as shown in Fig. 2.12 may be used to measure small pressures. In the figure illustrated, we can write

$$p_A - \gamma_1 h_1 - \gamma_2 h_2 + \gamma_3 h_3 = p_B$$

or

$$p_A - p_B = \gamma_1 h_1 + \gamma_2 h_2 - \gamma_3 h_3$$

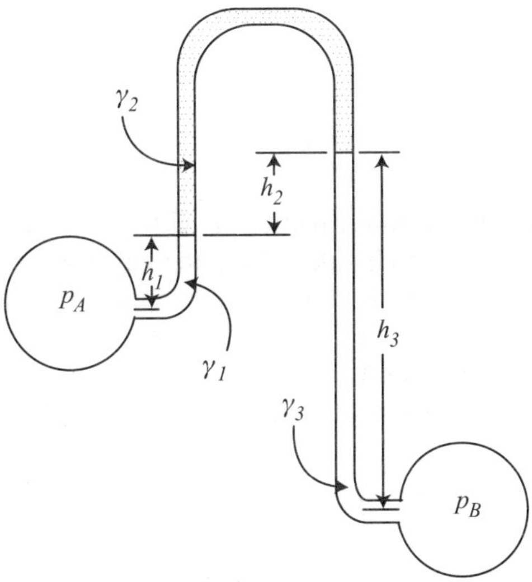

Fig. 2.12 Inverted U-tube manometer

Suppose $\gamma_1 = \gamma_3$ and $h_1 + h_2 = h_3$ as it usually happens between two points along a horizontal pipe, then we have

$$p_A - p_B = (\gamma_2 - \gamma_1) h_2$$

or

$$h_2 = \frac{(p_A - p_B)}{(\gamma_2 - \gamma_1)} \tag{2.16}$$

From the above expression, it becomes clear that the value of the differential head h_2 becomes magnified if the difference in the specific weights of the manometer fluid and the pressure source fluid is kept as low as possible.

Manometers with enlarged ends Manometers with enlarged top ends, as illustrated in Fig. 2.13 may also be used to produce a large gauge difference for a small differential pressure.

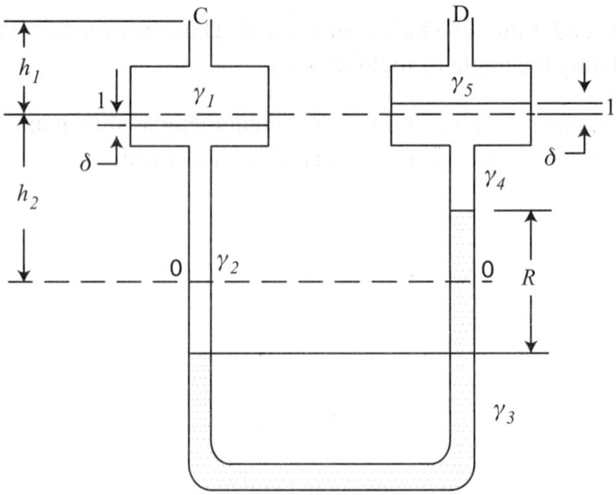

Fig. 2.13 Manometer with enlarged areas

The heavier gauge liquid fills the lower part of the U-tube and the initial level is at 0-0. Then the lighter gauge fluid is added to both sides filling the larger reservoirs up to 1-1. The gas or liquid in the system fills the space above 1-1. When the pressure at C is slightly greater than at D, the gauge fluid moves as shown in the figure and the deflection is measured as R. The volume of liquid moved in each reservoir is $\delta \cdot A = a \cdot R/2$, where A and a are the areas of the reservoir and the U-tube respectively. The manometer equation may be written starting from A.

$$p_C + (h_1 + \delta)\gamma_1 + \left(h_2 - \delta + \frac{R}{2}\right)\gamma_2 - \gamma_3 R - \left(h_2 - \frac{R}{2} + \delta\right)\gamma_4 - (h_1 - \delta)\,\gamma_5 = p_D$$

The specific weights of different fluids are indicated in the figures as well as the depths h_1 and h_2. Further simplification and substitution for $\delta = (a/A)(R/2)$ yields (assuming$\gamma_2 = \gamma_4$ and $\gamma_1 = \gamma_5$)

$$p_C - p_D = R\left[\gamma_3 - \gamma_2\left(1 - \frac{a}{A}\right) - \gamma_1\frac{a}{A}\right] \tag{2.17}$$

The expression inside the bracket is a gauge constant and hence the pressure difference is directly proportional to R.

Several patented micro-manometers are available and they make use of both enlarged areas and inclined tubes.

2.8 Relative Motion

2.8.1 Linear Acceleration

It is important to realise that lines of equal pressure are perpendicular to the line of resultant force acting on the fluid mass. We have already seen that when the fluid mass is subjected to only gravitational force, the lines of equal pressure are horizontal, including the free surface in case of liquids. However there may be situations where the fluid mass may be subjected to other external forces which cause acceleration of the entire mass in directions other than the vertical gravitational direction. In such situations, it should be noted that the entire mass of fluid is subjected to the same force and there is no relative motion of the particles with respect to each other. This is again a case which falls under the ambit of fluid statics. Let us first consider the container shown in Fig. 2.14 subjected to a horizontal acceleration a_x.

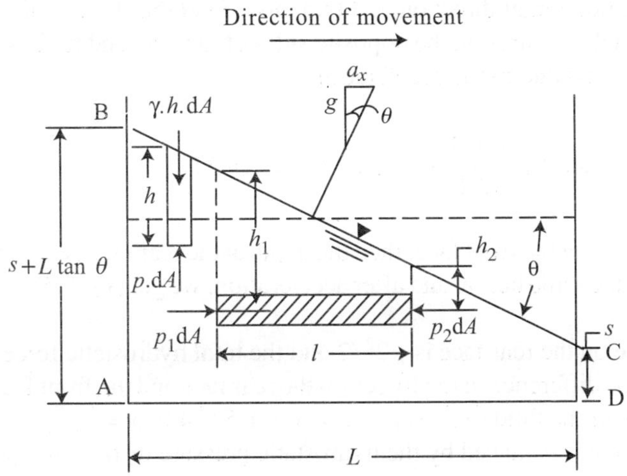

Fig. 2.14 Container under horizontal acceleration

Under the combined action of the horizontal and gravitational accelerations, the liquid mass adjusts itself to move as a solid with no relative motion of the particles with respect to each other. Consider first a vertical column of the liquid. Since there is no additional vertical acceleration, the only vertical forces are due to the weight of the fluid $\gamma h dA$ and the pressure force at the base of the column, pdA. Hence at any depth h below the free surface, the pressure is $p = \gamma h$. Now consider the equilibrium of a horizontal prism of the liquid.

$$p_1 dA - p_2 dA = \rho L dA a_x$$

$$\frac{(p_1 - p_2)}{L} = \gamma \frac{(h_1 - h_2)}{L} = \gamma \frac{a_x}{g} \tag{2.18}$$

where h_1 and h_2 are the depths of the two ends of the prism below the water surface as shown in Fig. 2.14.

The term $(h_1 - h_2)/L$ is the slope of the free surface viz., $\tan\theta$. Hence we have

$$\tan\theta = \frac{a_x}{g} \tag{2.19}$$

It must be remembered that the free surface is perpendicular to the resultant vector $-a_x$ and g. Further lines of equal hydrostatic pressure are lines parallel to the free surface. The difference in the hydrostatic forces on the opposite faces AB and CD (Fig. 2.14) per unit width

$$\frac{\gamma}{2}[(s + L\tan\theta)^2 - s^2]$$

must be equal to the force due to acceleration of the mass of liquid $\gamma/2g\,[2s + L\tan\theta]\,2a_x$. Simplifying we get $\tan\theta = a_x/g$. The hydrostatic forces on the opposite forces counterbalance the force due to acceleration of the tank.

EXAMPLE 2.7 A rectangular box 2 m long, 1 m wide and 3 m deep is half filled with water and is accelerated in the horizontal direction at 4.095 m/s^2. Find the slope of the free surface and its location. Find the total pressures on the opposite sides of the wall and find its relation to the force exerted on the fluid mass due to the acceleration.

$$\text{By (2.19)}, \theta = \tan^{-1}\left(\frac{4.095}{9.81}\right) = \tan^{-1} 0.5 = 26.57^\circ$$

If s is the depth of water at the front face, then the depth at the rear face is $s + 2\tan\theta = s + 1$. Since the volume remains the same before and after acceleration, we get $(s + 0.5) \times 2 \times 1 = 2 \times 1 \times 1.5$. Hence $s = 1$ m.

The total pressure on the rear face is $\gamma 2^2/2$ and the total hydrostatic force on the front face is $\gamma 1^2/2$. Therefore, the difference in the forces on the rear face and the front face is 1.5γ. The force causing acceleration of the fluid mass is $\rho \times 2 \times 1 \times 1.5 \times 4.095 = (\gamma/g) \times 3 \times 4.095 = 1.5\gamma$. It is seen that this force is balanced by the hydrostatic pressure distribution on the opposite sides of the container.

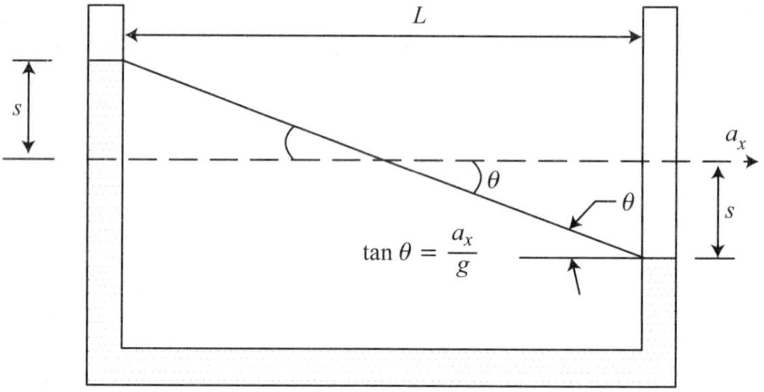

Fig. 2.15 U-tube under horizontal acceleration

The shape of the container is immaterial so long as continuity is maintained in the fluid. For example if the U-tube open at ends shown in Fig. 2.15 is accelerated with the same acceleration, 4.095 m/s^2, the line joining the free surfaces in the two arms of the U-tube will correspond to a slope of 26.57° and $s = (L\tan\theta)/2$, where L is the distance between the two vertical columns and s is the displacement in the water level in the tubes as shown in the sketch.

Vertical acceleration

If the container is subjected to a vertical acceleration, $a_y(a_x = 0)$, the equilibrium of the vertical prism shown in Fig. 2.16 is given by the equation

$$p \cdot dA - \gamma h dA = \left(\gamma h \frac{dA}{g}\right) a_y$$

$$p = \gamma h \left(1 + \frac{a_y}{g}\right)$$

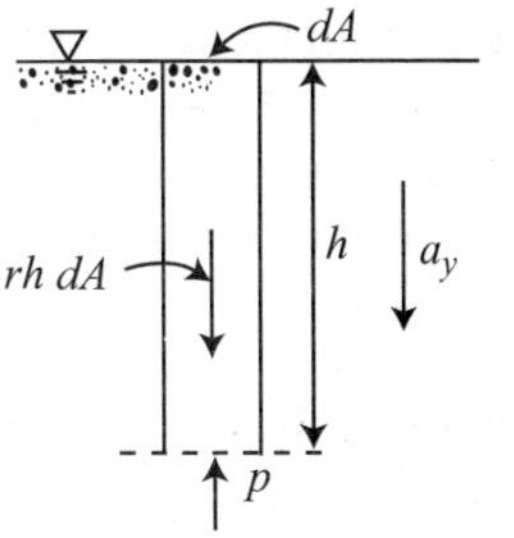

Fig. 2.16 Vertical acceleration

If the acceleration is vertically upwards, the pressure at a point located at a depth h below the free surface increases by a factor of a_y/g above the normal hydrostatic pressure. If a_y is vertically downwards, the pressure decreases by a factor a_y/g. For example if $a_y = -g$, then the pressure for a freely falling liquid mass is zero everywhere.

Acceleration along an inclined plane

If a container containing liquid is accelerated up a slope inclined at an angle α to the horizontal as shown in Fig. 2.17, we can resolve the acceleration a into two components $a \cos \alpha$ in the horizontal direction and $a \sin \alpha$ in the vertical direction. We have to add this vertical component to the gravitational component, g. Then (2.19) is modified to

$$\tan \theta = \frac{a \cos \alpha}{(g + a \sin \alpha)} \tag{2.20}$$

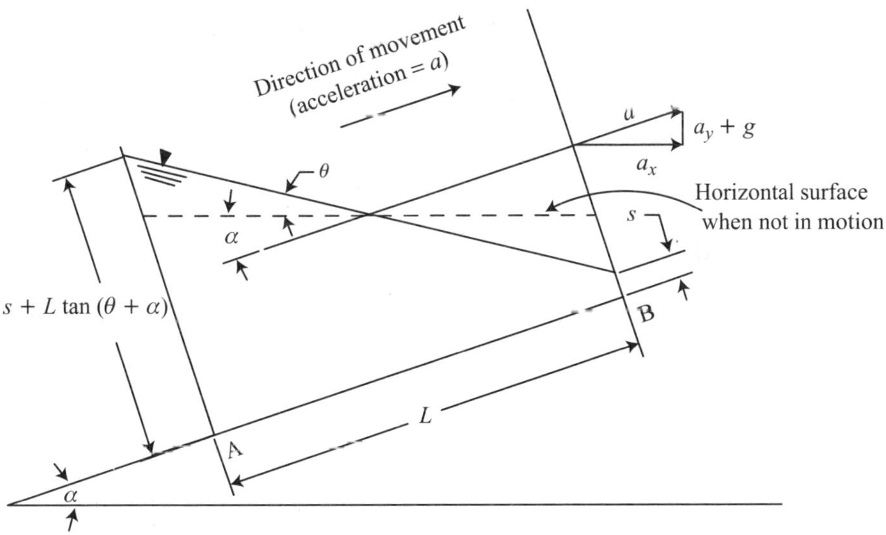

Fig. 2.17 Acceleration along an inclined plane

where θ is the angle with respect to the horizontal. The free surface makes an angle $\theta + \alpha$ with the bottom of the container or with the inclined plane. The free surface is always perpendicular to the resultant vector of $-a$ and g. If s is the depth of the liquid at the front face, the depth at the rear is $s + L\tan(\theta + \alpha)$. The pressure at any point can be determined by finding the vertical depth of the point below the free surface. For example, the pressure at the rear bottom edge, A is given as $\gamma[s + L\tan(\theta + \alpha)]\sin\alpha$ and at the leading bottom edge, B the pressure is $\gamma s\sin\alpha$.

EXAMPLE 2.8 As an illustration, let us consider a 2 m cube container half filled with liquid of specific weight, γ moving up a slope $\alpha = 30°$ with an acceleration of 2.45 m/s^2; then $a_x = 2.45\cos 30° = 2.12$ m/s^2 and $a_y = 2.45\sin 30° = 1.225$ m/ s^2. The net vertical acceleration is $g + a_y = 9.80 + 1.225 = 11.025$ m/s^2. Hence the slope of the free surface with respect to the horizontal is

$$\tan\theta = \frac{a_x}{(g + a_y)} = \frac{2.12}{11.025} = 0.192$$

or $\theta = 10.88°$. The angle that the free surface makes with the bottom of the container is therefore $30 + 10.88 = 40.88°$. We can determine the wetted length s at the front by equating the volumes before and after the acceleration.

$$2 \times 2[s + s + 2\tan(\theta + \alpha)]/2 = 2 \times 2 \times 1$$

$$s + (\tan 40.88) = 1 \quad \text{or} \quad s = 0.135 \text{ m}$$

The wetted length at the rear end is $0.135 + 2\tan 40.88 = 1.866$ m. The pressure at A is equal to $\gamma 1.866\sin 60° = 1.616\gamma$ and at B, the pressure is $0.135\gamma\sin 60° = 0.117\gamma$.

If a_s is directed downwards along a downward slope, the above equation is again modified to read

$$\tan\theta = \frac{a_s\cos\alpha}{(g - a_s\sin\alpha)} \tag{2.21}$$

If the container is moving down the slope due to its own weight, the horizontal acceleration is zero and hence $\theta = 0$ with the result that the free surface is parallel to the slope.

2.8.2 Rotating Cylinder

Another equally important practical problem that is related to a fluid mass moving as a solid body under an external force is that of uniform rotation of a container containing liquid about a vertical axis. This is often called a forced vortex motion in which all the particles move at the same angular velocity, ω and the liquid in the container moves as a solid body some time after the initial rotation. There is, therefore, no tangential stress anywhere in the liquid. The only acceleration is due to the centrifugal force directed normally outwards along a radial direction. The pressure distribution should counteract this force to maintain equilibrium and the net pressure on any fluid element is directed radially inwards. In the vertical direction, the only force acting on the liquid is the gravitational force. Hence the pressure distribution is hydrostatic as the force due to pressure at the bottom of a vertical column must balance the weight of the fluid above. Hence the pressure at any point is given by $p = \gamma h$, where h is the vertical distance between the point and the free surface.

The equation of motion in the radial direction which is also the horizontal direction is obtained with reference to Fig. 2.18.

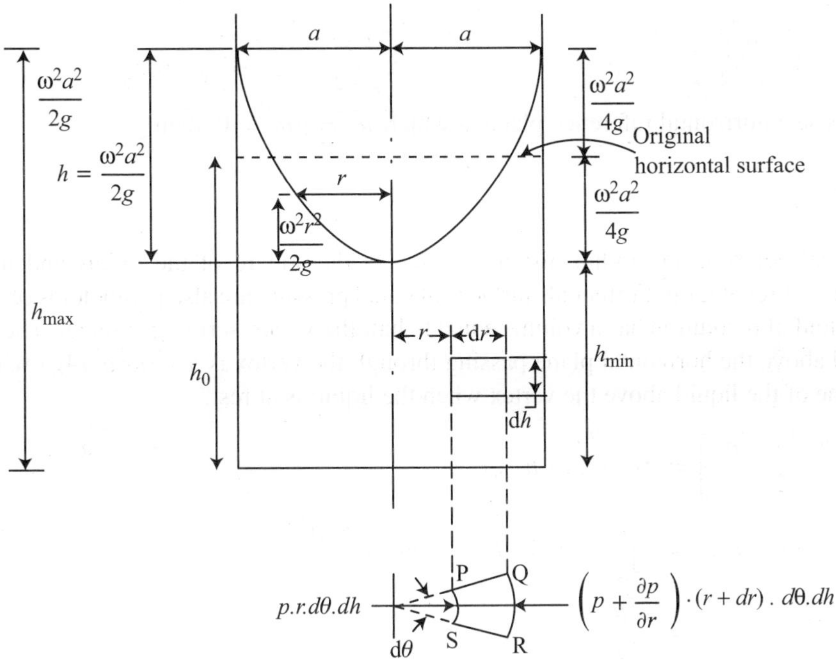

Fig. 2.18 Rotating cylinder

Consider a radial element PQRS as shown in the figure. The forces keeping the segment in equilibrium are

$$p r d\theta dh - \left(p + \frac{\partial p}{\partial r}\right)(r + dr) d\theta dh = \rho dh \left(r + \frac{dr}{2}\right) d\theta(-\omega^2 r)$$

Neglecting small quantities (cube of small quantities disappear faster than square of small quantities), we obtain

$$\frac{dp}{dr} = \frac{\gamma}{g}\omega^2 r$$

Integration yields

$$p = \frac{\gamma \omega^2 r^2}{2g} + \text{constant}$$

The value of the constant is obtained in terms of the pressure p_0 at the axis of rotation where $r = 0$. Finally we have

$$p = p_0 + \frac{\gamma \omega^2 r^2}{2g} \tag{2.22}$$

or in terms of pressure head

$$h = h_0 + \frac{\omega^2 r^2}{2g}$$

If we choose a horizontal reference plane at which $p_0 = \gamma h_0 = 0$, then

$$h = \frac{\omega^2 r^2}{2g} \tag{2.23}$$

The vertical depth at any radial distance varies as the square of the radius and it describes paraboloids of revolution. Further all surfaces of equal pressure are also paraboloids of revolution. A paraboloid of revolution has a volume equal to half the circumscribing cylinder. The volume of the liquid above the horizontal plane passing through the vertex is $\pi a^2(\omega^2 a^2/4g)$ which is also the volume of the liquid above the vertex when the liquid is at rest.

$$\frac{1}{2}\left[\frac{\pi a^2 \cdot \omega^2 a^2}{2g}\right] = \pi a^2 (h_0 - h_{\text{min}})$$

$$h_0 - h_{\text{min}} = \frac{\omega^2 a^2}{4g} \quad \text{and} \quad h_{\text{max}} - h_0 = \frac{\omega^2 a^2}{4g}$$

$$h_{\text{max}} - h_{\text{min}} = \frac{\omega^2 a^2}{2g}$$

Therefore, the height of liquid rising at the sides above the initial level is equal to the depth of liquid falling in the central region.

Example 2.9 A cylinder of diameter 1 m and height 1.4 m, filled with water to a depth of 0.75 m is rotated about its central vertical axis with a constant speed of 90 rpm. Determine (a) the depths of liquid at the walls and at the centre, (b) the total pressure on the cylindrical walls, the total pressure on the bottom of the vessel, (c) the rpm required for the water level to just reach the top of the vessel, (d) the rpm required to attain a zero depth at the centre and the amount of water spilled over the brim of the container.

$$\omega = \frac{2\pi N}{60} = 3\pi \,\text{rad/s}$$

$$h_c = \frac{\omega^2 a^2}{4g} = \frac{9\pi^2 \times 0.5^2}{4 \times 9.80} = 0.5665 \text{ m}$$

below the original level and the depth at the wall is also above the original level at the wall by the same amount or the total depth at the wall is $0.75 + 0.5665 = 1.3165$ m and the depth at the centre is $0.75 - 0.5665 = 0.1835$ m

The force on the outer wall of the cylinder is the force on the vertical projection of the curved surface equal to $(\gamma h_w^2/2) \times d = 0.866\gamma$. The total pressure on the bottom will remain the same as the one when the liquid is at rest at a uniform depth of $h = 0.70$ m which is easily estimated as $\pi a^2(\gamma h) = 0.55\gamma$

If the wall depth should reach the level of the top of the vessel, the rise in the level above the original depth is $1.4 - 0.75 = 0.65$ which should equal $\omega^2 a^2/4g$.

$$\text{Hence } \omega = \frac{2\pi N}{60} = \left(\frac{0.65 \times 4 \times 9.80}{0.25}\right)^{1/2} = 10.096 \text{ or } N = 96.40 \text{ rpm}$$

The depth at the centre is $0.75 - 0.65 = 0.10$ m. Any further increase in the speed rotation will cause further decrease in the level at the centre and overflow of the liquid over the top of the container.

If the depth at the centre has to become zero, the depth at the wall being 1.40 m, we equate $\omega^2 a^2/2g = 1.40$ and $N = (60/2\pi)(1.40 \times 2 \times 9.8/0.25)^{1/2} = 100.05$ rpm. The amount of overflow is $\pi(0.5)^2 \times (0.75 - 0.70) = 0.0393$ cu m.

EXAMPLE 2.10 The U-tube shown in the figure is rotated at 50 rpm about a vertical axis passing through C. Find the difference in level in the two limbs of the tube and the original level in the tubes. The distances from the vertical axis of the two tubes are shown in the figure (30 and 15 cms).

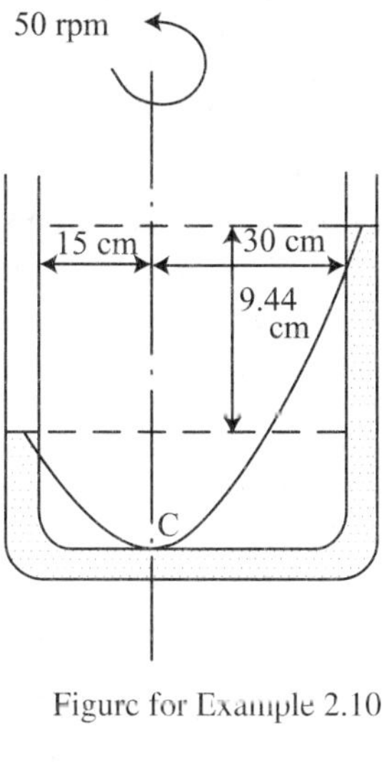

Figure for Example 2.10

The level of liquid in the tube above the vertex of the paraboloid which lies on the axis of rotation is given by (2.21). The difference in levels in the two tubes at different radii is given by

$$\Delta h = \left(\frac{\omega^2}{2g}\right)(r_1^2 - r_2^2) = \frac{(2\pi \times 50/60)^2(30^2 - 15^2)}{2 \times 980} = 9.44 \text{ cms}$$

The original height of the liquid column is obtained as equal to $\omega^2 r_1^2/4g = 6.29$ cms

2.9 Floating Bodies

2.9.1 Buoyant Force and Centre of Buoyancy

The buoyant force on a submerged or floating body is defined as the vertical upward force exerted by a static fluid on the body. The buoyant force, F_B on a submerged body is the difference between the upward pressure force acting on the underside of the body and the vertical downward force exerted by the pressure distribution on the upper side of the body. This has already been shown to be equal to the weight of the volume, $\forall$ displaced by the body: $F_B = \gamma \forall$.

This is also true for floating bodies except that the volume now is the volume displaced by that portion of the body which is below the liquid surface. Since the body is in static equilibrium, it is evident that the buoyant force is equal to the weight of the fluid and their lines of action also coincide. A floating body therefore displaces a certain volume of fluid whose weight is equal to the weight of the body. The centre of gravity of the displaced volume of liquid is known as the centre of buoyancy.

The relative position of the centre of buoyancy with respect to the centre of gravity of the body is very important to determine the stability of the body when the body is subjected to rotation about its axes. If the centre of gravity is below the centre of buoyancy, the body is stable. If the centre of gravity is above the centre of buoyancy, the body is stable subject to certain conditions which shall be discussed further in the next section. Some objects like a sphere are in neutral equilibrium since the relative positions of the two points do not change for all angles of tilt.

2.9.2 Metacentric Height

Consider a general non-prismatic body of cross-section shown in Fig. 2.19. The original position of the body is indicated by PQRSP.

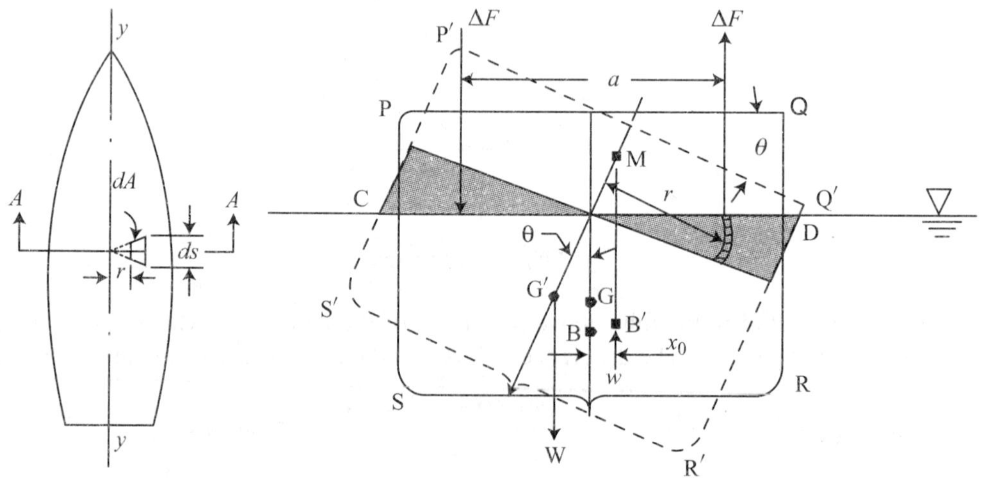

Fig. 2.19 Metacentric height

When the body is subjected to a small angle of rotation, θ, the centre of buoyancy shifts to the side where greater submergence of the body takes place. In the figure, the clockwise rotation has shifted the centre of buoyancy to the right. The new centre of buoyancy is located at B′ which is the centroid of the displaced volume CDR′S′C. The buoyant force acts vertically upwards through B′ and the weight of the body acts vertically downwards through the new position of the centre of gravity G′. The two forces now form a couple. If the vertical through B′ intersects the axis of the body at a point M above the centre of gravity, G, then the couple tries to restore the body to its original position. This is known as the righting couple and the body is stable for small angles of tilt. If on the other hand, the point M lies below G, then the couple acts in the same direction as the tilting moment and the body will tilt further, leading to the instability of the floating body and the couple is known as the overturning couple. If M coincides with G, the body is in neutral equilibrium, since there would be no moment acting on the body. The point M is called the metacentre. The metacentre is defined as the point at which the vertical through the centre of buoyancy intersects the vertical centre line of the floating body's section after a small angle of heel. The object may be seen as rotating about the point M which may be regarded as the instantaneous centre of rotation. Generally the metacentre lies close to the free surface and the distance GM (for the limit of zero angle of heel) is known as the metacentric height. A certain minimum value of GM is often specified for ships to ensure stability and this value varies depending on the type of vessel. It generally lies in the range of 0.3 to 1.25 m.

2.9.3 Analytical Method of Determining Metacentric Height

In Fig. 2.19, the shift in the centre of buoyancy BB′ designated as x_0 is found by considering two wedges, one on the right which increases the upward buoyant force component on the right and the second one on the left which has decreased the buoyancy as the wedge has come above the free surface. These two wedge forces, ΔF must be equal since the weight of the body and hence the buoyant force remain unchanged. If the distance between the centroid of these wedges is a, then the shift in the centre of buoyancy is given by

$$Wx_0 = \Delta F \cdot a \tag{2.24}$$

The value of the couple on the right-hand side of the equation may be found by integrating moments about O of the elementary forces in the wedges

$$\Delta F \cdot a = \gamma \int (rd\theta dr ds)r = \gamma\theta \int_A r^2 dA = \gamma\theta I \tag{2.25}$$

where ds is the distance along the length of the body $drds = dA$ and I is the moment of inertia of the area about the axis yy. Reverting to (2.24)

$$Wx_0 = \gamma \forall x_0 = \gamma\theta I$$

$\forall$ being the total volume displaced. From the geometry of the figure, we get, with the angle of tilt θ being very small

$$\text{MB} \sin\theta = \text{MB} \tan\theta = \text{MB}\theta = x_0 \quad \text{or} \quad \text{MB} = \frac{x_0}{\theta} = \frac{I}{\forall}$$

It follows that the expression for the metacentric height is

$$\text{MG} = \text{MB} - \text{GB} \quad \text{or} \quad \text{MG} = \text{MB} + \text{GB}$$

The minus sign applies if G is above B and the plus sign is operative if G is below B.

It is important to note that I is the moment of inertia of the horizontal section of the body at the water line. Usually the sides of a vessel are vertical at this section, so that I may be taken as the moment of inertia of the deck about a longitudinal axis. For the plan of the vessel shown in Fig. 2.19, the moment of inertia about the longitudinal axis, yy is got by dividing the section into small horizontal rectangles and adding together their moments of inertia. The common practice is to describe the moment of inertia of the deck of the body as a function of the circumscribing rectangle. In such a case, if the breadth of the rectangle is b and the length L, the moment of inertia is given as $I = kLb^3/12$, where k is a coefficient depending on the shape of the ship. If the deck is rectangular, $k = 1$. The stability of the body for pitch is also similarly determined and the moment of inertia is taken about the xx axis.

EXAMPLE 2.11 A cylindrical buoy 2 m in diameter and 1.2 m long, weighing 9806 N is floating in water. Find the centre of buoyancy and the metacentric height.

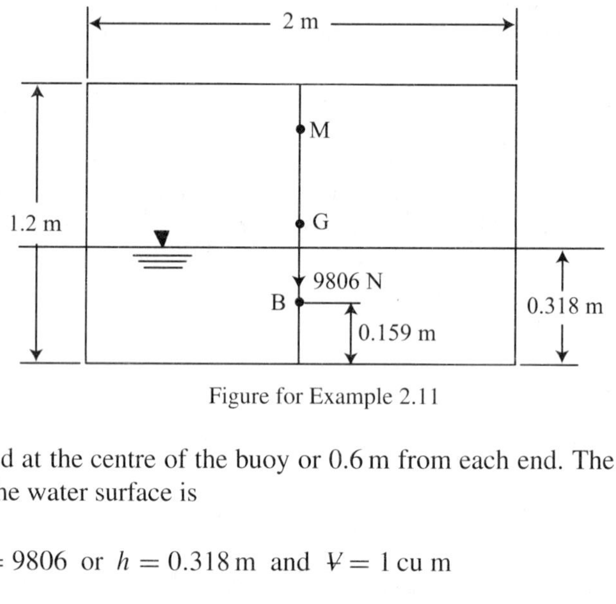

Figure for Example 2.11

The CG is located at the centre of the buoy or 0.6 m from each end. The submerged depth h of the body below the water surface is

$$\gamma\left(\pi\frac{d^2}{4}\right)h = 9806 \quad \text{or} \quad h = 0.318\,\text{m} \quad \text{and} \quad \forall = 1\,\text{cu m}$$

and the centre of buoyancy lies at a depth of 0.159 m below the water surface.

$$\text{BM} = \frac{I}{V} = \frac{(\pi d^4/64)}{V} = 0.785\,\text{m}$$

$$\text{BG} = 0.6 - 0.159 = 0.441\,\text{m}$$

$$\text{GM} = 0.785 - 0.441 = 0.344\,\text{m}$$

Since M is above G, the body is stable.

EXAMPLE 2.12 A cylindrical buoy is 1.5 m in diameter and 2 m high, of mass 700 kg with its centre of gravity 0.8 m above the base. Is the buoy stable when it is floating with its axis vertical in sea water? If the buoy is anchored to the bottom by a chain attached to it, find the tension required in the chain in order to make the buoy just float in a stable condition. P (sea water) $= 1020\,\text{kg/m}^3$

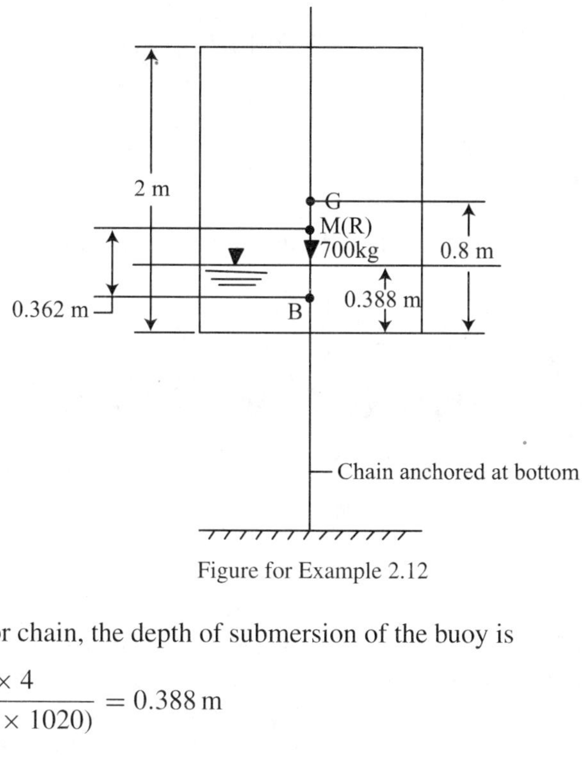

Figure for Example 2.12

Without the anchor chain, the depth of submersion of the buoy is

$$h = \frac{700 \times 4}{(\pi \times 1.5^2 \times 1020)} = 0.388\,\text{m}$$

Then,

$$\text{BG} = 0.8 - 0.388/2 = 0.606\,\text{m}$$

$$\text{BM} = \frac{I}{V} = \frac{(\pi \times 1.5^4/64)}{(\pi \times 1.5^2/4 \times 0.388)} 0.362\,\text{m}$$

Since M lies below G, the body is unstable.

When the anchor chain exerts a tension T on the body, the submerged depth has to be found again.

$$700 \times 9.81 + T = 1020 \times 9.81 \times \pi \times 1.5^2/4h_1 = 1802.49 \times 9.81h_1$$

or

$$T = (1802.49h_1 - 700) \times 9.81$$

Let R be the location of the resultant of the two forces W and T

Then we have

$$700 \times 9.81 \times \text{RG} = T \times (0.8 - \text{RG})$$

Eliminating T from the above two relations, we get

$$T = (1802.49h_1 - 700) \times 9.81 = 700 \times 9.81 \times \left[\frac{\text{RG}}{(0.8 - \text{RG})}\right]$$

Simplifying,

$$\text{RG} = 0.8 - 0.311/h_1$$

If M has to coincide with R, for minimum stability condition, we have the relation

$$0.8 - \text{RG} - h_1/2 = \text{BM} = \frac{I}{V} = \frac{(\pi \times 15^4/64)}{(h_1\pi \times 1.5^2/4)} = \frac{0.141}{h_1}$$

$$\text{RG} = 0.8 - 0.141/h_1 - h_1/2 = 0.8 - 0.311/h_1$$

Therefore

$$h_1 = 0.583\text{ m}$$

$$T = (1802.49 \times 0.583 - 700) \times 9.81 = 3440\text{ N}$$

EXAMPLE 2.13 A diamond-shaped pontoon 8 m long, 2 m in height and 3 m at the widest part weighs 6 tonnes. The centre of gravity is located at a distance of 0.75 m above the bottom. It floats in fresh water and has a central lifting device to hoist weight. Find the stability of the body both for roll and pitch. What is the maximum height to which a weight of 500 kg can be lifted without upsetting the stability.

The moment of inertia of the triangular plane about its base is $bd^3/12$, b being the base width and d its altitude.

$$I_{yy} = 2 \times 8 \times 1.5^3/12 = 4.5\text{ m}^4 \text{ and } I_{xx} = 2 \times 3 \times 4^3/12 = 32\text{ m}^4$$

Area $= 2 \times 3 \times 4/2 = 12\text{ m}^2$

Submerged depth: $12 \times h \times 1000 = 6000$ or $h = 0.5$ m; $V = 12 \times 0.5 = 6\text{ m}^3$

The centre of buoyancy is located at a distance of 0.25 m from the bottom.

Hence BG = 0.5 m

$$\text{BM(for roll)} = \frac{I_{yy}}{V} = \frac{4.5}{6} = 0.75\text{ m}$$

Metacentric height GM = 0.5 m

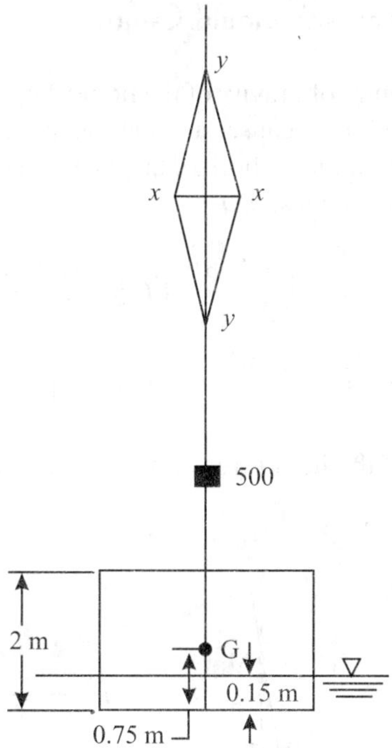

Figure for Example 2.13

Since BM > BG, the body is stable for small angles of roll.

$$\text{BM(for pitch)} = \frac{I_{xx}}{V} = \frac{32}{6} = 5.33\ \text{m}$$

and again M is above G and stable.

Let y be the distance above the bottom of the weight of 500 kg for which the body is just stable or M coincides with G or BM = BG.

The submerged depth now is : $12 \times h_1 \times 1000 = 6500$ or $h_1 = 0.542$ m

And the centre of buoyancy is located at 0.271 m above the bottom.

To find the location of the new centroid y_1 above the bottom, we take moments of the two weights about the bottom plane.

$$500y + 6000 \times 0.75 = 6500y_1 \quad \text{or} \quad y_1 = 0.077y + 0.692$$

$$\text{BM} = 4.5/12 \times 0.542 = 0.692\ \text{m}$$

and we equate

$$\text{BG} = \text{BM} \quad \text{or} \quad 0.077y + 0.692 - 0.271 = 0.692$$

$$\text{or} \quad y = \frac{0.271}{0.077} = 3.52\ \text{m}$$

2.10 Stability of Floating Bodies with Liquid Cargo

When the floating body heels, the centre of gravity of the liquid cargo is displaced in the direction of tilt. This movement of CG of liquid mass causes a reduction in the righting moment. As shown in Fig. 2.20, the surface of the liquid remains horizontal. The reduction in the righting moment may be estimated by considering the wedges, a_0a_1.

$$\text{Weight of wedge of liquid } a_0a_1 = \gamma\left(\frac{1}{2}\frac{B}{2}\frac{B\theta}{2}\right)L = \frac{\gamma L B^2\theta}{8}$$

$$\text{Reducing moment of the wedge} = \frac{\gamma L B^2\theta}{8} \times \frac{2}{3}B = \gamma\theta\left(\frac{LB^3}{12}\right) = \gamma\theta I_L$$

where I_L is the moment of inertia of the liquid free surface about the longitudinal axis.

$$\text{Net restoring moment} = W.\text{GM}.\theta - \gamma\theta I_L.$$

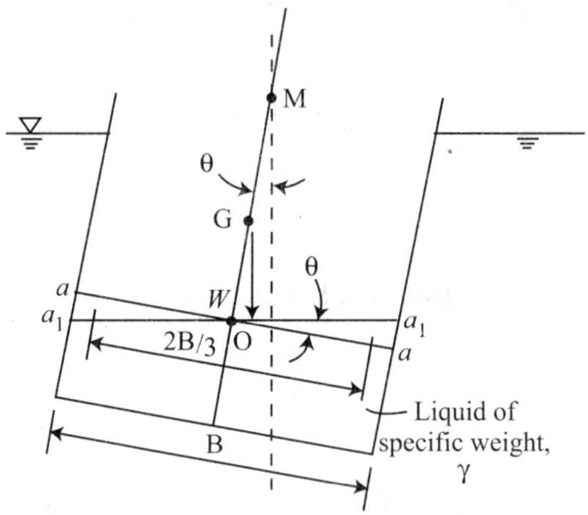

Fig. 2.20 Floating body with liquid cargo

Exercises

2.1 Calculate the pressures at A, B, C, and D in the container shown in the figure.

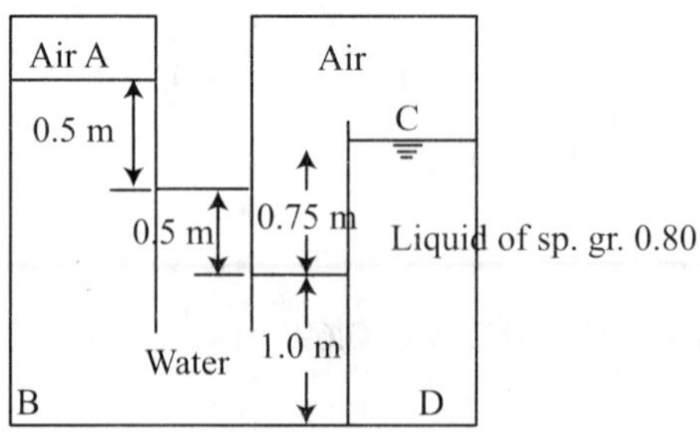

Figure for Problem 2.1

2.2 Determine the levels in the piezometers A and B shown in the figure. What are the pressures at C and D?

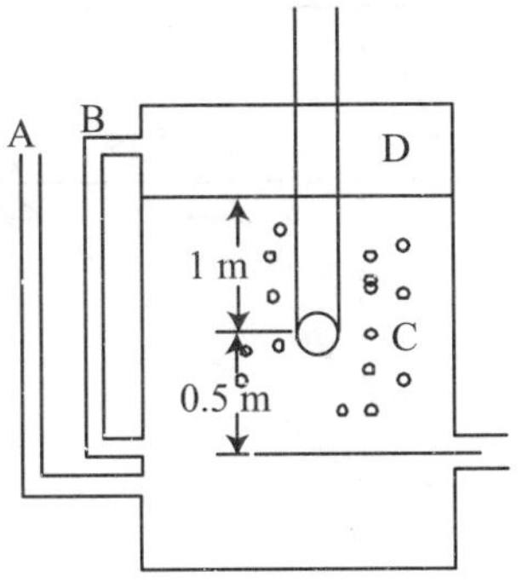

Figure for Problem 2.2

2.3 A tank is divided into two parts and on each side there is a different fluid. What are the resultant horizontal and vertical forces on the stopper. Determine the moment of all forces at the mid-point of the central wall as shown in the figure.

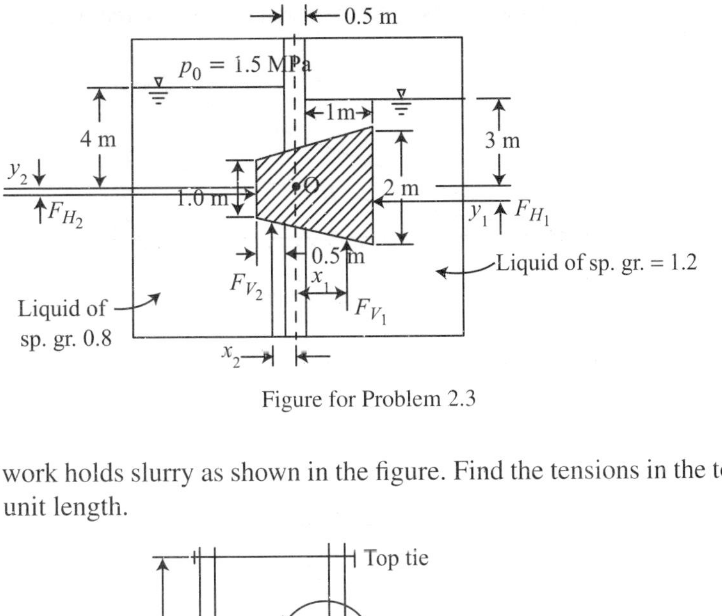

Figure for Problem 2.3

2.4 A form work holds slurry as shown in the figure. Find the tensions in the top and bottom ties per unit length.

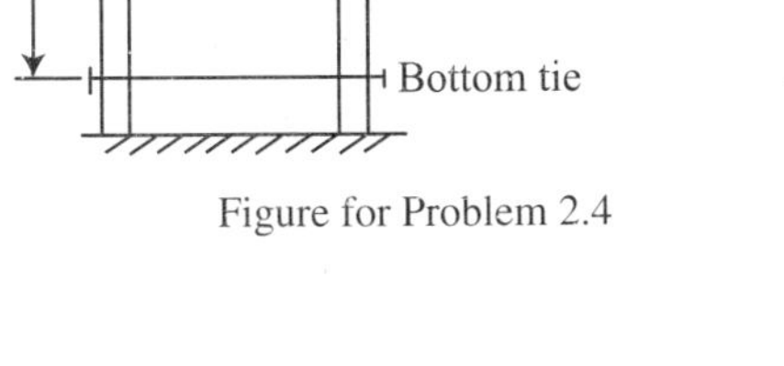

Figure for Problem 2.4

2.5 A gate 2 m wide has a lifting device at A. What is the force P required to lift the gate.

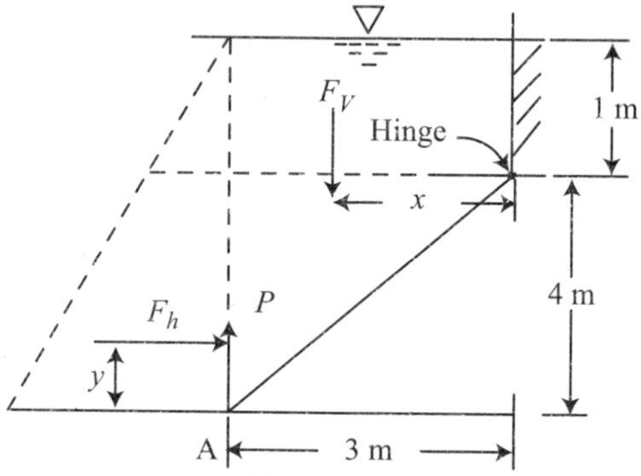

Figure for Problem 2.5

2.6 A cone of base diameter D and height H is filled with water and is lying on its face as shown in the figure. Find the pressure on the curved surface. There is a hole at the top-most point of the base, A.
Hint: If the base is removed all the water will flow out. Hence, the force on the curved surface must be equal and opposite to the force exerted on the base.

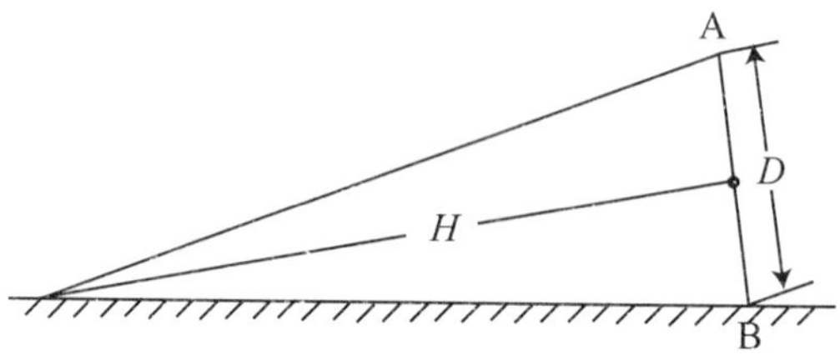

Figure for Problem 2.6

2.7 A radial gate with radius equal to 5.0 m and length 10 m holds water on a spillway as shown in the figure. Find the magnitude and direction of the resultant force on the gate.

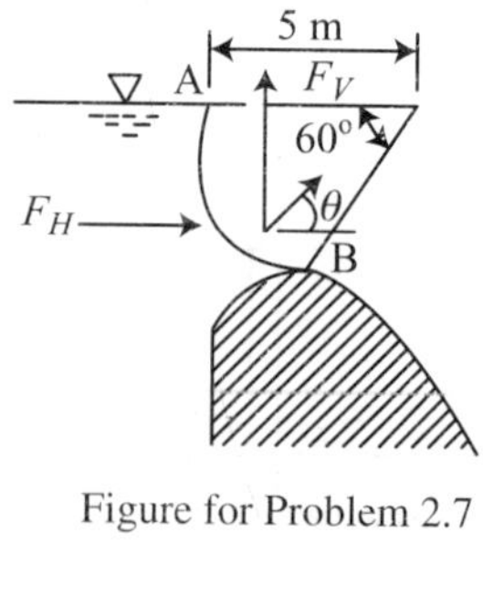

Figure for Problem 2.7

2.8 A cylindrical gate 4 m long and 2 m in diameter holds water to a depth of 1 m on one side. Find the horizontal and vertical forces on the gate. What should the weight of the gate be to keep it in place if the cylinder is supported by a fulcrum at A as shown in the figure?

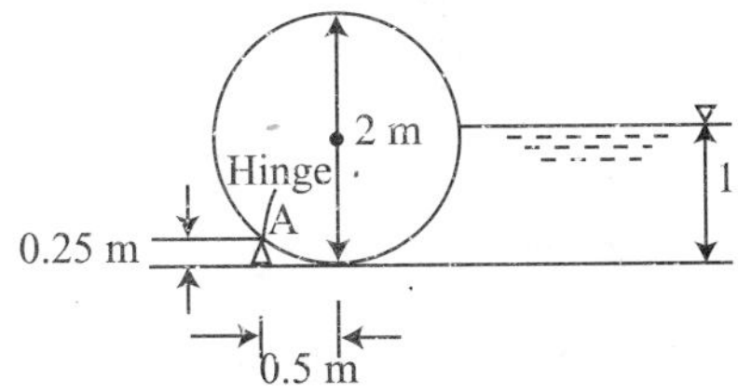

Figure for Problem 2.8

2.9 A cylindrical weir as illustrated in the figure is 5 m long and 3 m in diameter. Determine the magnitude, location and direction of the horizontal and vertical forces.

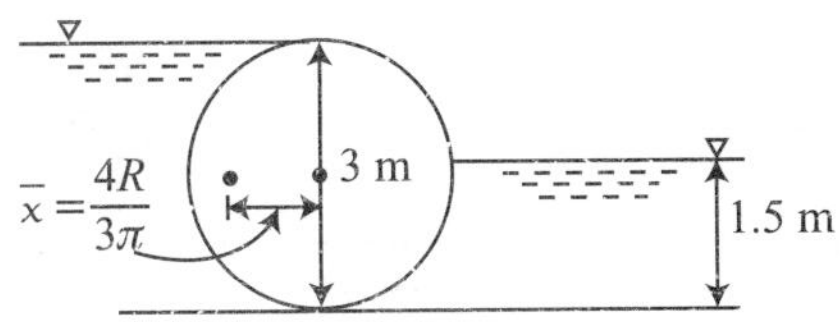

Figure for Problem 2.9

2.10 A trapezoidal plate with the smaller side parallel to the water surface is immersed in water as shown in the figure. Find the horizontal and vertical forces and their centres of pressure. Find also the direction and magnitude of the resultant force.

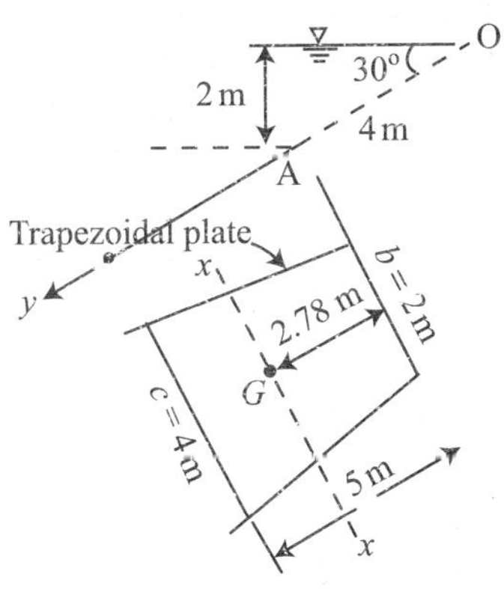

$$I_{xx} = \frac{(b^2 + 4bc + c^2)\,h^3}{36\,(b + c)}$$

Figure for Problem 2.10

2.11 A circular gate 1 m in dia. is located in the tank wall inclined at an angle of 60° to the horizontal. The centre of the gate is at a vertical depth of 1.5 m below the water surface. Find the resultant force and centre of pressure.

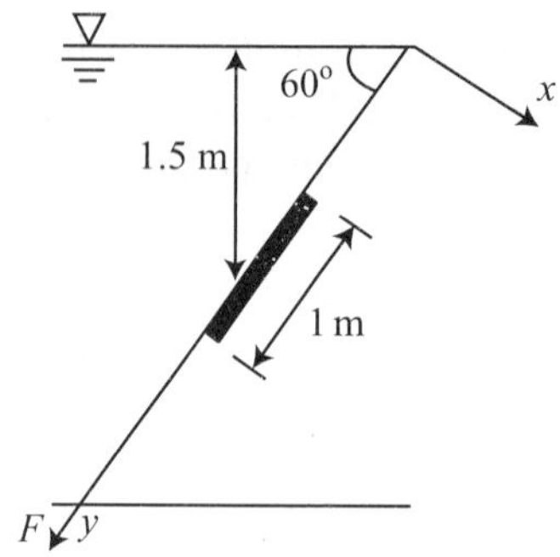

Figure for Problem 2.11

2.12 (a) If a tank containing oil of specific gravity 0.8 is accelerated horizontally at 3 m/s^2, find the surface slope and pressure intensities at points A and B as shown in the figure. (b) What acceleration is necessary to just expose point A and find the depth at B when this happens.

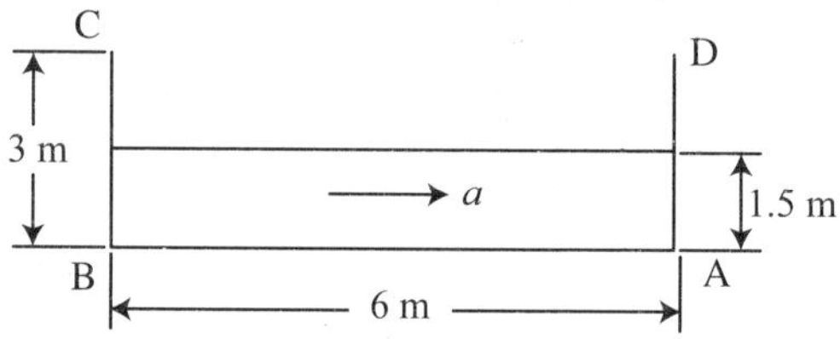

Figure for Problem 2.12

2.13 If the tank in Problem 2.11 is initially moving at a uniform speed of 45 kph and is brought to rest in a distance of 30 m, find the depths at A and B

2.14 A tank measuring 3 m long, 2 m wide and 3 m deep is accelerated at 3 m/s^2 up a slope of 15°, find the surface slope and the depths at the front and rear ends, if the water depth is initially, 1 m.

2.15 A cylindrical tank 2 m in diameter and 4.0 m high is half-filled with water and is rotated at 60 rpm. Find (a) the depths at the axis and outer wall. Determine the rpm required in each of the following cases.

(b) The depth of water at the outer wall is at the brim of the tank. Also find the corresponding depth at the centre.
(c) The depth at the centre is 1.0 m. Find also the corresponding depth at the brink.
(d) The amount of spill is equal to 50% of the original volume.

2.16 Mercury fills a thin tube 0.5 cm in diameter and 0.5 m long and is closed at the outer end. If the tube is rotated at a constant speed at 120 rpm in a vertical plane, what is the maximum force exerted on the closed end.
Hint: Use the equation $\partial(p + \gamma z)/\partial r = a_n/g = \omega^2 r/g$

2.17 If one swings a glass of water held at an arm's length of 0.75 m through a vertical circle at sufficient speed, water will not spill. Determine this speed.

2.18 If the pressure at point B in the figure is 1.5 bar, what is the pressure at point A. The pressures are indicated by the manometer. In which direction does the flow take place?

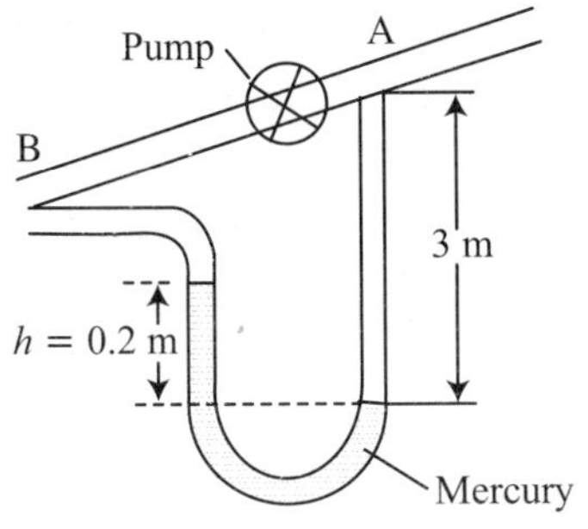

Figure for Problem 2.18

2.19 Find the difference in pressure in the two tanks A and B as indicated in the figure by the differential manometer.

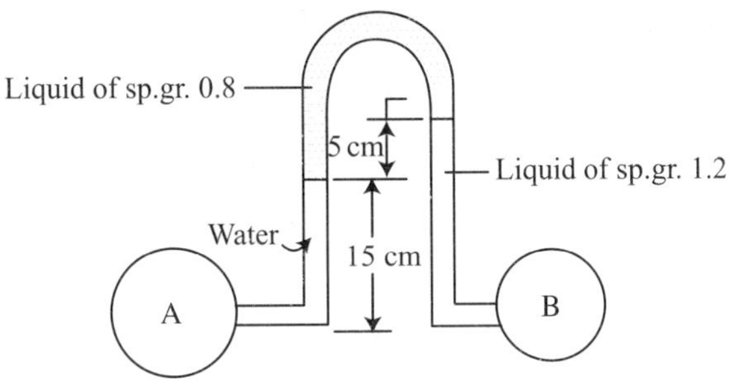

Figure for Problem 2.19

2.20 A cylindrical block of wood 2 m in diameter and 1.75 m long has a specific weight of 8000 N/m^3. Will it float in a stable condition with its axis vertical? (b) If the dia. is 1 m, is the block stable?

2.21 A barge 6 m wide and 15 m long is loaded with containers on the top as shown in the sketch given on the next page. Assume the CG to lie on the top of the barge on the centre line. If the total weight of the barge and its cargo weigh 1800 kN, find the stability of the barge.

2.22 A hollow cylinder open at both ends has an internal diameter of 25 cm and a wall thickness of 10 cms and a length of 100 cms. If the weight of the cylinder is 600 N, find the stability of the floating cylinder.

2.23 A box-like barge, 10 m long and 5 m wide, 4 m in height when loaded has a draft of 3 m. The barge rolls to an angle of 15°. (a) What is the righting moment if the box is solid. (b) What is the righting moment if the load is partially occupying a volume 5 m × 5 m × 1 m in water. (c) If the box is divided into two equal compartments, how is the result modified?

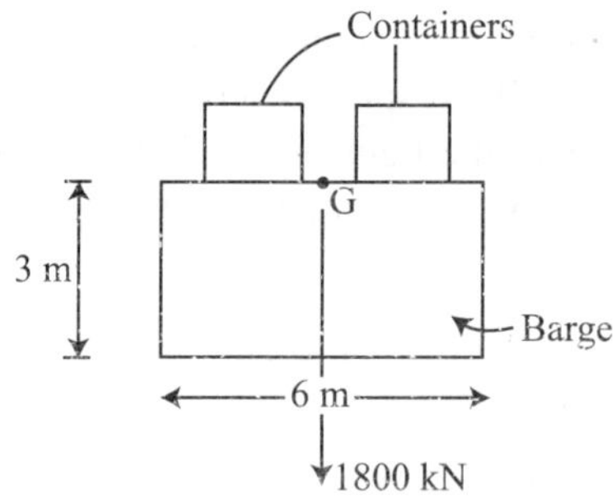

Figure for Problem 2.21

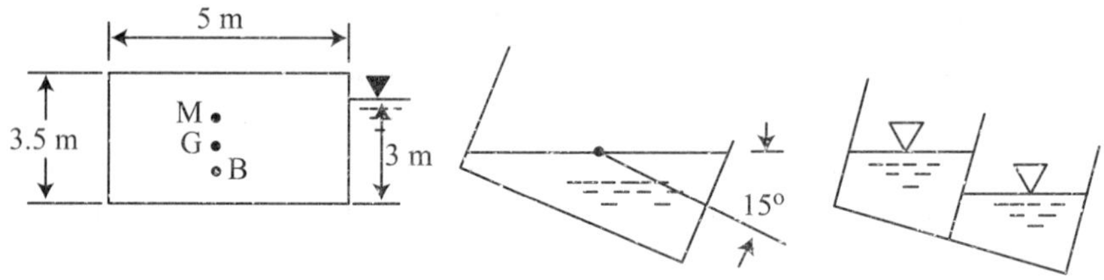

Figure for Problem 2.23

2.24 A wooden cone of base dia. 1 m and height 1.5 m and of specific gravity 0.6 floats in water. Is it stable?

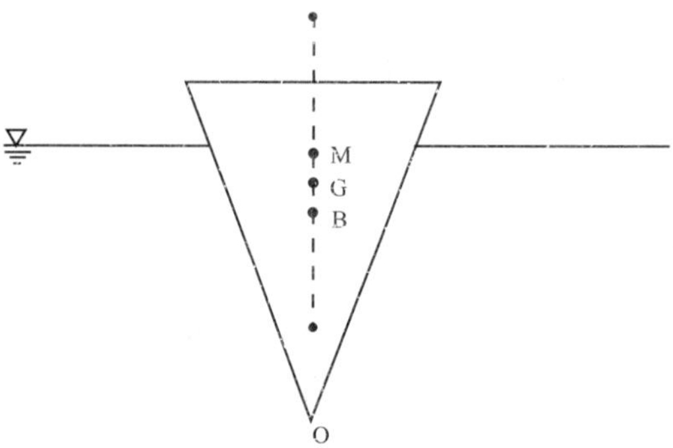

Figure for Problem 2.24

3 Kinematics of Fluid Flow

3.1 Flow Variables and Classification

Apart from the physical properties of the fluid, the principles of fluid mechanics are based on the law of conservation of mass, Newton's laws of motion, the laws of thermodynamics and certain boundary conditions that the flow must satisfy. The description of the motion of fluid particles under various fields forms the essence of fluid mechanics. The rates of displacement of fluid particles (or their velocities) is termed kinematics of fluid flow whereas dynamics of fluid flow involves particle acceleration in a given pressure and/or other force fields.

It is necessary to realise at this stage that, in general, the flow parameters such as direction and magnitude of the velocity vector, pressure, density, and temperature vary not only from point to point but also from one instant to another at the same point. In other words, these variables are functions of coordinates of space and of time as well. The fluid properties (e.g., density) which are dependent on pressure and temperature would also vary with position and time. Thus if we denote the components of the velocity vector as u, v, w, in the three coordinate axes x, y, z and pressure, density and temperature as p, ρ, T respectively at any point A and at any given instant, these variables are all functions of x, y, z, and t.

$$u, v, w, p, \rho, T = f_{1,2,3,4,5,6}(x, y, z, t)$$

where the subscripts on the right denote different functional relationships of the variables on the left. If the rate of change of u is to be described, it should be done with respect to each of the independent variables x, y, z, t. For example, the rate of change of u with respect to the distance x along the x-axis is given by the partial derivative $\partial u/\partial x$. Similarly $\partial u/\partial t$ indicates the rate of change of u in relation to time at the given point.

We classify the flow systems as

1. uniform and non-uniform
2. steady and unsteady

The first classification is based on whether or not the flow is accelerating with respect to space. The second classifies time-dependent flow as unsteady motion and those constant for all periods of time as steady.

We can have different combinations of the above, such as steady uniform or steady non-uniform, unsteady uniform, unsteady non-uniform flows. There are other classifications also, such as laminar and turbulent flow, compressible and incompressible flow, subsonic and supersonic flow, reversible and irreversible flow, and real and ideal fluid flows.

A steady uniform flow in a pipe or channel of constant shape and cross-section is shown in Fig. 3.1*a*. In this case $\partial u/\partial x = 0$ but due to the boundary condition that fluid velocity at the boundary must be the same as at the stationary boundary, $u = 0$ at the stationary boundary and the velocity increases as one moves away from the boundary reaching a maximum mid-way between the boundaries. Therefore, in this example $\partial u/\partial y$ (or $\partial u/\partial r$) exists.

An example of a steady non-uniform flow is found in converging or diverging boundaries (pipe or channel) with constant discharge: accelerations in the direction of flow exist (or $\partial u/\partial x$ exists) as shown in Fig. 3.1*b*. Examples of unsteady flow may be found in the tidal flow in estuaries, wave motion, surges in open channels and water hammer in conduits.

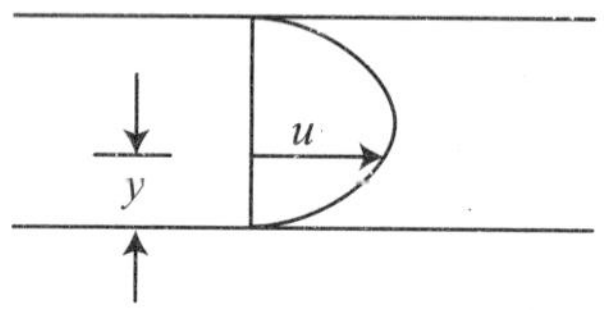

Steady uniform flow

a

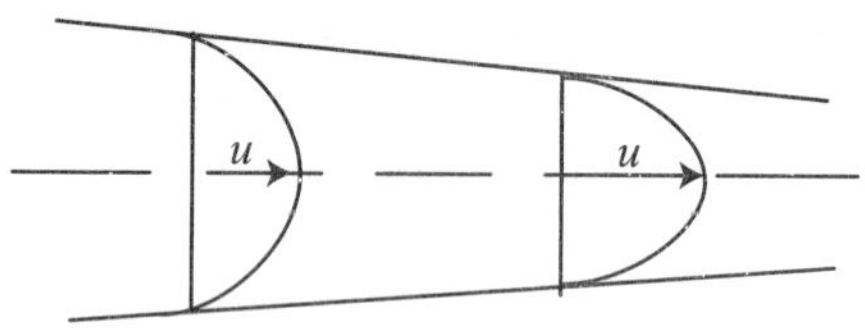

Steady non-uniform flow

b

Fig. 3.1 Spatial variations of velocity

3.2 Continuum

In Chapter 2, mention was made about the random motion of molecules of fluids. Any liquid – liquid or gas – possesses molecular structures (conglomeration of molecules or atoms) peculiar to itself and the molecular motion, dependent on the structure, takes place continuously through space. The details of such motion become of importance only when the scale of motion is small and pressure is low (the gas is rarified). In most applications in fluid mechanics, the physical dimensions of flow are extremely large compared to the size of molecules and the molecular path. In such cases, we assume that a fluid may be very finely divided, yet with the division not reaching the lower limit of the molecular scale. Based on these assumptions, changes in flow parameters such as velocity, pressure, and temperature are regarded as continuous. Thus, when we define density of a fluid at a given point, we must specify the lower limit to which the volume of the element has shrunk.

Let $\delta V\!\!\!\!-$ be an arbitrary volume around the given point and δm the mass of material in it. The density is then defined as

$$\rho = \operatorname*{lt}_{\delta V\!\!\!\!- \to (\delta V\!\!\!\!-)_m} \left(\frac{\delta m}{\delta V\!\!\!\!-} \right)$$

where $(\delta \forall)_m$ is the smallest volume around the point for which statistical averages may be obtained. It is in this sense that we say that flow parameters are spatially continuous and the expression 'continuum' is used to describe the concept.

3.3 Kinematics of Fluid Flow

Kinematics describes the motion of fluid particles with respect to time and space. The description is complete if the magnitude and direction of velocity could be defined at every point at any given instant of time. There are two fundamentally different systems of reference used to describe fluid motion. The first system, named after Euler, describes the velocity vector, V at every point as a function of time. Here we fix the point of reference and study the motion of fluid particles moving past it at different instants of time. The second system named after Lagrange follows a fluid particle as it travels through space and describes changes in its motion. In this book, we shall follow the Eulerian system exclusively.

3.4 Stream Lines, Path Lines and Streak Lines

Let us assume that at a given instant of time, the direction of velocity at every point in the flow field is known. Then it would be possible to draw a series of curves which are tangential to the velocity vectors everywhere. Such imaginary lines to which the velocity vectors are tangential everywhere at a given instant of time are called stream lines. These stream lines give a picture of the direction of flow at that instant and illustrate the flow pattern as shown in Fig. 3.2.

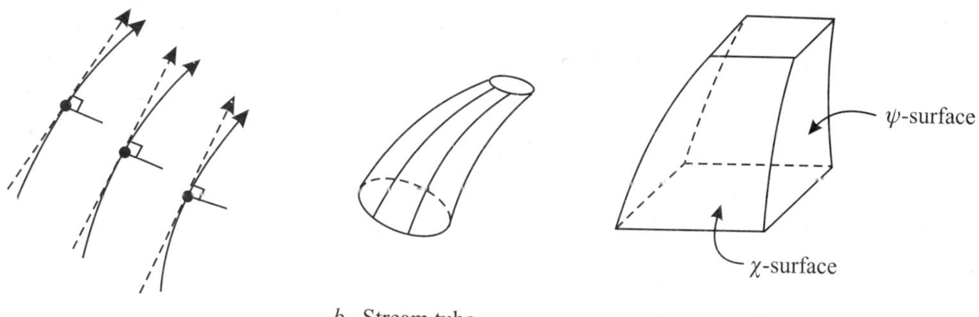

a Stream lines *b* Stream tube *c* Stream surface

Fig. 3.2 Stream lines, stream tubes, stream surfaces

It is easily seen that there is no flow normal to a stream line since the velocity vector is always tangential to it. A fluid volume enclosed by a bunch of stream lines (the bounding surface is called the stream surface) is called a stream tube. If the velocity field is independent of time, the flow pattern represented by the stream lines is a steady one, and combined with the equation of continuity (presented in the next section), the stream line spacing may be used to describe velocity changes.

In addition to the stream line, two other useful definitions are in vogue. A path line is the locus of movement of a fluid particle under observation as it travels in space in relation to time. If a long time exposure is made of a visible fluid particle, the trace left by the particle is the path line. A short time exposure of a number of fluid particles made visible indicates an instantaneous picture of stream lines. At the location of the particle, the tangents to the path and the stream line coincide but may deviate at other locations. It is only in the case of a steady flow that path lines merge with stream lines. In an unsteady flow system, if the stream lines are known at successive instants of time, the path line for any fluid particle may be constructed as explained later.

A streak line connects all particles that have passed successively through a given point in the flow field. Such streak lines are obtained when a dye or tracer material is injected at a fixed point. Again, if the flow is steady, the streak line merges with the stream line and the path line. The construction procedure for drawing the streak lines are explained in the next section. Figure 3.3 illustrates the instantaneous stream line pattern, a path line and a streak line for a symmetric body moving through a fluid.

3.5 Construction of Path and Streak Lines

The construction of path and streak lines is achieved by first drawing the stream lines at different intervals of time. For example, if a strut is moving steadily at a velocity V through a stagnant fluid, the stream lines for the unsteady flow are as illustrated in Fig. 3.3*a*. Let it be required to find the path line of a fluid particle P. The particle will initially move along the tangent to the stream line passing through P. In an interval of time Δt, the particle will move through a distance $V_1 \Delta t$, where V_1 is the velocity assumed constant over the small interval of time Δt. The velocity V_1 may be obtained by the spacing of the stream lines. At the end of the interval Δt, the body has also moved through a distance $V \Delta t$. The position of the body and the stream lines at that instant are shown in dotted lines in Fig. 3.3*b*. The particle having arrived at P′ acquires the direction of the tangent to the stream line passing through P′ and its local velocity will be V_2. The distance P′P″ is covered by the particle in the second interval of time. During this second interval of time, the body would have translated to a new position shown by the chain of dotted lines. The subsequent travel of the particle in the next interval of time would be dictated by the velocity prevailing at P″. Proceeding in the same manner, step by step, the successive positions of the fluid particle may be found over a given interval of time.

To obtain a streak line at any time, say $t = 3\Delta t$, we have to find the end points of the path lines of all the fluid particles which have passed through a given point B at that time. Let B_3 be a particle which was earlier at B when $t = 0$ and at that time it was located on the innermost stream line as shown in Fig. 3.3*a*. Then the path line described by it is shown in Fig. 3.3*c* over three successive intervals of time, finally arriving at point 3. Similarly B_2 was another particle at time $t = \Delta t$ but located on another stream line and its path line over the interval $2\Delta t$ is shown in the figure with its final position at point 2. The particle B_1 has moved tangentially to the stream line passing through B and has arrived at point 1 over an interval of time Δt. The streak line is obtained by joining the terminal positions of the path lines of particles which have earlier passed through B.

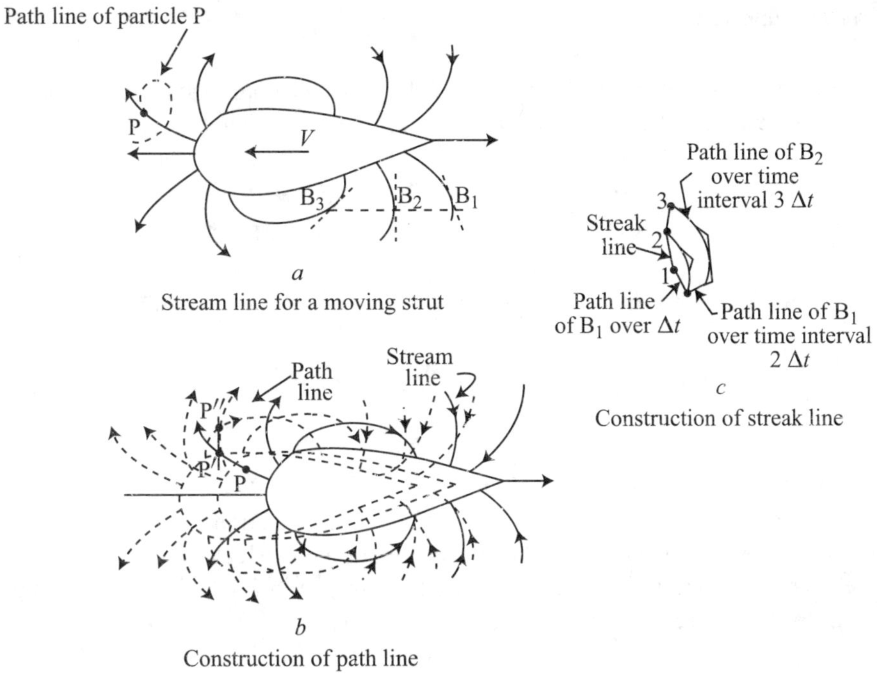

Fig. 3.3 Construction of path lines and streak lines

3.6 Equations for Stream and Path Lines

Let ABC be a stream line at time t_0 as shown in Fig. 3.4. The tangent at B denotes the velocity vector whose components are u, v, w in the three axes directions. The elementary distance ds along the stream line at B has the components dx, dy, and dz along the axes directions. Comparing the similar triangles formed by the velocity and distance diagrams, the equation for the streamline at $t = t_0$ may be written as

$$\frac{dx}{u(x, y, z, t_0)} = \frac{dy}{v(x, y, z, t_0)} = \frac{dz}{w(x, y, z, t_0)} \tag{3.1}$$

In a two-dimensional steady flow (there are no changes in the z components of the velocity and the flow conditions at all points in the flow field is independent of time), the equation for the stream line becomes

$$udy - vdx = 0 \tag{3.2}$$

The equation for a path line should involve the time elapsed since the particle moved from its initial position (x_0, y_0, z_0, t_0) to its new position (x, y, z, t). The equation for a path line is written as

$$\frac{dx}{u(x, y, z, t)} = \frac{dy}{v(x, y, z, t)} = \frac{dz}{w(x, y, z, t)} = dt \tag{3.3}$$

3.7 Stream Function

Equation (3.2) becomes an exact differential if the velocity components can be expressed as derivatives of a scalar quantity, described as stream function, ψ. Thus if

$$u = \frac{\partial \psi}{\partial y} \quad \text{and} \quad v = -\frac{\partial \psi}{\partial x} \tag{3.4}$$

are substituted in (3.2), we get

$$\left(\frac{\partial \psi}{\partial x}\right) dx + \left(\frac{\partial \psi}{\partial y}\right) dy = d\psi = 0$$

or on integration, it is seen that ψ = constant.

Along a stream line, therefore, the value of the stream function is constant. We shall now attempt to give a physical meaning for the stream function. Consider the discharge or the rate of flow between any two stream lines ABC and PQR as shown in Fig. 3.4 along which the values of the stream functions are ψ_1 and ψ_2 respectively.

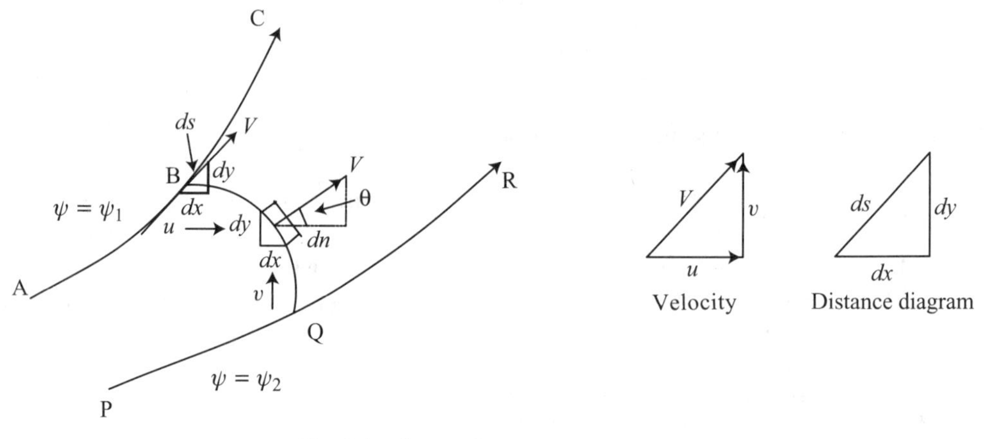

Fig. 3.4 Stream function and discharge

The discharge is obtained as a product of the velocity and the area normal to the velocity vector. The rate of flow between the stream lines is

$$q = \int V\,dn$$

Resolving V and dn into their components along the x- and y-directions, the discharge per unit width may be obtained as

$$
\begin{aligned}
q &= \int (V\cos\theta)(dn\sin\theta) + (V\sin\theta)(-dn\cos\theta) \\
&= \int udy - vdx \\
&= \int \left[\left(\frac{\partial\psi}{\partial y}\right)dy + \left(\frac{\partial\psi}{\partial x}\right)dx\right] \\
&= \int d\psi = \psi_2 - \psi_1
\end{aligned}
$$

The difference between two adjacent stream functions gives the volume rate of flow between them.

In three-dimensional flow systems, one must define two sets of stream surfaces $\psi =$ constant and $\chi =$ constant (Fig. 3.2*c*) which are mutually at right angles. We shall not discuss the general treatment of such three-dimensional flows but confine our attention to a particular case of axi-symmetric flow. In such flow systems, the stream surfaces are concentric surfaces symmetrical about the axis of symmetry. The discharge in the annular region between two stream surfaces whose stream functions differ by $d\psi$ (Fig. 3.5) is expressed as

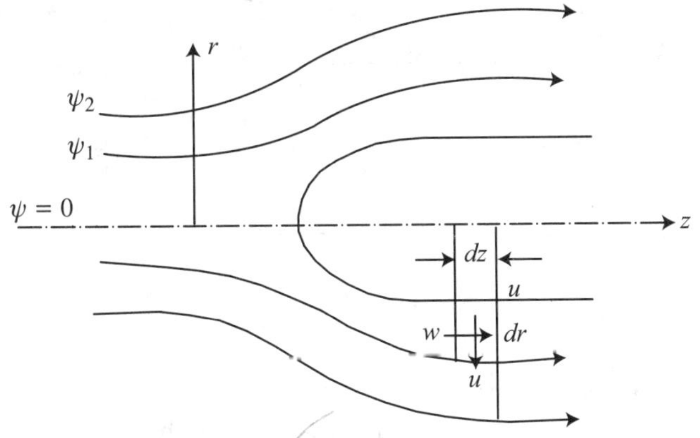

Fig. 3.5 Stream function for an axi-symmetric flow

$$2\pi d\psi = 2\pi r dr w \quad \text{or} \quad w = \left(\frac{1}{r}\right)\left(\frac{\partial\psi}{\partial r}\right) \tag{3.5a}$$

and

$$2\pi d\psi = -2\pi r dz u \quad \text{or} \quad u = -\left(\frac{1}{r}\right)\left(\frac{\partial\psi}{\partial z}\right) \tag{3.5b}$$

with w as axial velocity and u as radial velocity.

If follows that

$$d\psi = -urdz + wrdr$$

$$Q = 2\pi(\psi_2 - \psi_1)$$

3.8 Elements of Particle Motion

The motion of fluid particles may be classified into four basic types known as translation, linear deformation, angular deformation, and rotation. In reality, a fluid particle may undergo one or more of these motions at the same time.

Translation Translation merely implies that all fluid particles in an element ABCD, shown in Fig. 3.6*a*, suffer the same displacement in the three coordinate axes directions. There is no relative motion of the particles and there is no change in size, shape and orientation of the element. Examples of pure translation may be found in a parallel uniform flow.

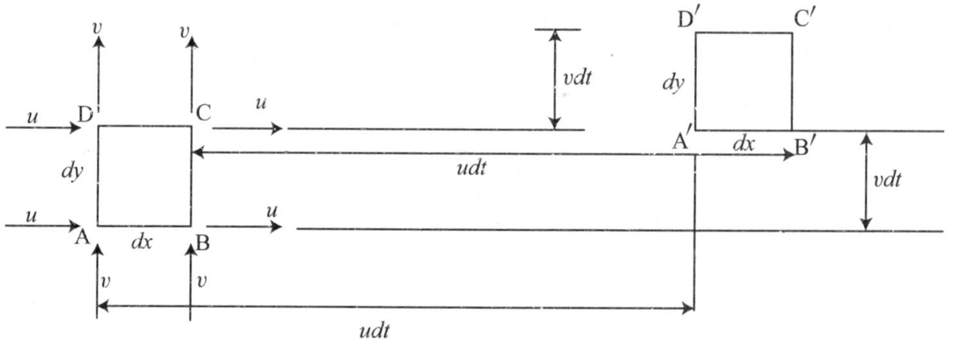

Fig. 3.6*a* Translation

Linear deformation In this kind of motion, the fluid element may undergo linear stretching or shrinkage when the fluid particles are relatively displaced but the displacement occurs along the same linear direction. The fluid particles lying on a straight line AB as shown in Fig. 3.6*b* will continue to remain on the same straight line but the point B has moved relatively to A along the line AB. Similarly D has moved vertically with respect to A and C has moved vertically with respect to B but DC remains parallel to AB. The relative motion of B relative to A is $(\partial u/\partial x) \cdot dx \cdot dt$ and that of D relative to A is $(\partial v/\partial y) \cdot dy \cdot dt$.

In Fig. 3.6*b*, the element has been shown as translated to a new position with a linear deformation to give a clearer picture of the relative positions of the four corners of the element. The rates of linear deformations in the three directions per unit length of the element are therefore

$$a = \frac{\partial u}{\partial x}; \quad b = \frac{\partial v}{\partial y}; \quad c = \frac{\partial w}{\partial z} \qquad (3.6)$$

Pure linear deformations are possible in ideal fluid flow systems but approximate forms may be obtained in rapidly accelerated flows.

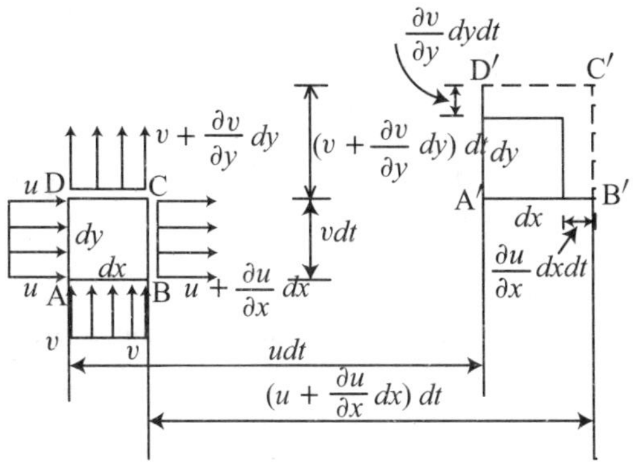

Fig. 3.6*b* Linear deformation

Angular deformation When the transverse velocity component changes along any side of the element, we have two other forms of particle deformation. A typical angular deformation is illustrated in Fig. 3.6*c* indicating the relative displacement of B and D with respect to A.

The angular displacement of side AB may be expressed as $d\alpha = (\partial v/\partial x)dt$ and that of AD as $-d\beta = (\partial u/\partial y)dt$, if the usual convention of measuring anti-clockwise movement as positive is followed. The average angular displacement of the element represented by the median line or a diagonal line is $(d\alpha - d\beta)$. The rate of angular deformation in the xy-plane is

$$\frac{(d\alpha - d\beta)}{dt} = \left(\frac{\partial v}{\partial x} + \frac{\partial u}{\partial y}\right)$$

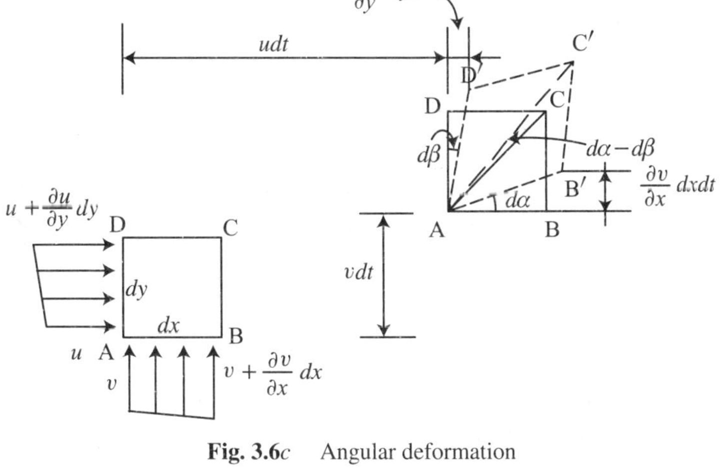

Fig. 3.6*c* Angular deformation

A similar procedure may be followed for slices of fluid volume taken in the zx- and yz-planes. It may be shown that the rates of angular deformations in the three planes are

$$
\begin{aligned}
f &= \frac{\partial w}{\partial y} + \frac{\partial v}{\partial z} \text{ in the } yz\text{-plane} \\
g &= \frac{\partial u}{\partial z} + \frac{\partial w}{\partial x} \text{ in the } zx\text{-plane} \\
h &= \frac{\partial v}{\partial x} + \frac{\partial u}{\partial y} \text{ in the } xy\text{-plane}
\end{aligned} \tag{3.7}
$$

In the above discussion, we observed that the angles at the vertices of the element were also subjected to distortions.

Rotation When these angles remain virtually unchanged, it leads to the rotation of the element. Obviously this is possible when the angular displacements of two adjacent sides are of the same sign as shown in Fig. 3.6*d*.

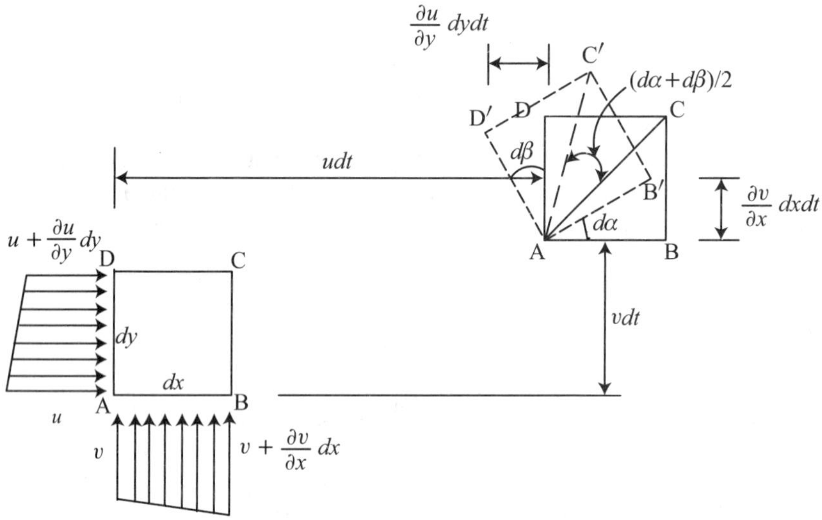

Fig. 3.6*d* Rotation

The average rate of rotation of a diagonal or a median line of the element is the same as the average of the rotation of the adjacent sides. Hence we have

$$\omega = \frac{d\theta}{dt} = \frac{1}{2}\frac{(d\alpha + d\beta)}{dt} = \frac{1}{2}\left(\frac{\partial v}{\partial x} - \frac{\partial u}{\partial y}\right)$$

The rates of rotation in the three different planes are

$$\omega_{yz} = \frac{1}{2}\left(\frac{\partial w}{\partial y} - \frac{\partial v}{\partial z}\right) \text{ in the } yz\text{-plane (about the } x\text{-axis)}$$

$$\omega_{zx} = \frac{1}{2}\left(\frac{\partial u}{\partial z} - \frac{\partial w}{\partial x}\right) \text{ in the } zx\text{-plane (about the } y\text{-axis)} \tag{3.8}$$

$$\omega_{xy} = \frac{1}{2}\left(\frac{\partial v}{\partial x} - \frac{\partial u}{\partial y}\right) \text{ in the } xy\text{-plane (about the } z\text{-axis)}$$

It is generally useful to define vorticities ξ, η, ζ in the three different planes such that they are twice the angular speeds.

$$\begin{aligned} \xi &= 2\omega_{yz} = \frac{\partial w}{\partial y} - \frac{\partial v}{\partial z} \\ \zeta &= 2\omega_{zx} = \frac{\partial u}{\partial z} - \frac{\partial w}{\partial x} \\ \eta &= 2\omega_{xy} = \frac{\partial v}{\partial x} - \frac{\partial u}{\partial y} \end{aligned} \tag{3.9}$$

3.9 Circulation

The line integral of the tangential component of the velocity vector, at any given instant of time, around a closed path is known as circulation. With reference to Fig. 3.7*a*, the circulation, Γ becomes

$$\Gamma = \oint V_s ds \tag{3.10}$$

where V_s is the tangential component of V along the length ds of the curve at any point P. In terms of velocity components u, v and w along the projected length of ds in the x, y, z directions, the circulation may be written as

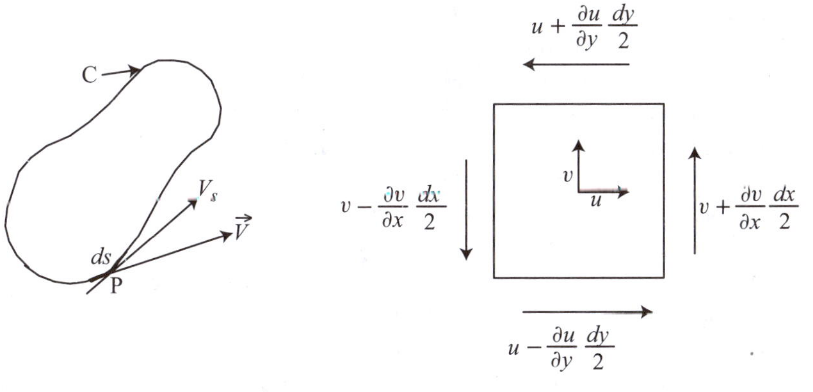

a Circulation around a closed contour *b* Circulation around a rectangle

Fig. 3.7 Circulation

$$\Gamma = \oint (udx + vdy + wdz) \tag{3.11}$$

If the closed curve is chosen as an elementary rectangle (Fig. 3, 7*b*), the circulation is

$$d\Gamma = \left(u - \frac{\partial u}{\partial y} \cdot \frac{dy}{2}\right) dx + \left(v + \frac{\partial v}{\partial x} \cdot \frac{dx}{2}\right) dy - \left(u + \frac{\partial u}{\partial y} \cdot \frac{dy}{2}\right) dx - \left(v - \frac{\partial v}{\partial x} \cdot \frac{dx}{2}\right) dy$$

Each term on the right-hand side is a product of the length of the curve and the velocity along it. The addition of the terms for the four sides of the rectangle represents integration around the closed path and on simplification yields the elementary circulation around the elementary rectangle

$$d\Gamma = \left(\frac{\partial v}{\partial x} - \frac{\partial u}{\partial y}\right) dxdy$$

The circulation per unit area in a given plane is equal to the component of vorticity about a normal to the plane.

$$\Gamma = \left(\frac{\partial v}{\partial x} - \frac{\partial u}{\partial y}\right) \tag{3.12}$$

3.10 Equation of Continuity

Consider Fig. 3.8 which shows the flow through a stream tube or a three-dimensional duct bounded by stream surfaces.

We have already defined a stream line as one to which the velocity vector is totally tangential to it. This would also imply that there is no flow perpendicular to the stream line. A solid boundary is also a stream surface since there cannot be any flow normal to the boundary. Therefore, the discharge within any stream tube is contained within it along the tube length and cannot flow out of the tube normal to it. If V is the velocity normal to the elementary area dA of the stream tube, the discharge is $V \cdot dA$ and the rate of mass flow in the elementary stream tube is $\rho \cdot V \cdot dA$ and this must remain constant as per the law of conservation of mass. For an incompressible flow, ρ is constant and hence $dq = V \cdot dA$ remains constant. The velocity is inversely proportional to the area. In the case of compressible flows, the density may change depending on the local pressure and the mass rate of flow remains constant. If the density, velocity and areas at the two cross-sections of the stream tube shown in Fig. 3.8 are ρ_1, ρ_2; V_1, V_2; dA_1, dA_2 respectively, the equation of continuity is expressed as

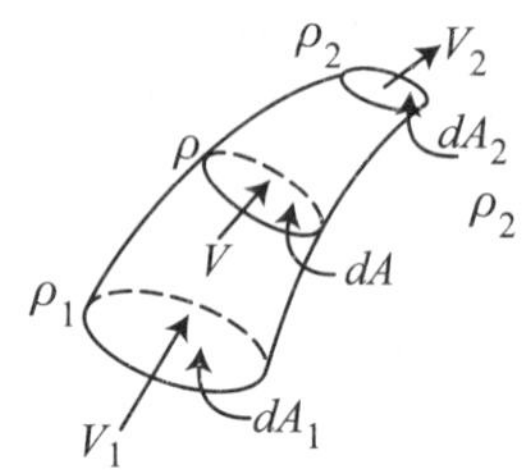

Fig. 3.8 Continuity in discharge

$$\rho_1 V_1 dA_1 = \rho_2 V_2 dA_2$$

The equation specifies that mass inflow (into the tube) must be equal to the mass outflow (out of the tube). A convenient form of the equation of continuity for finite areas such as flows bounded by solid boundaries of pipes or channel walls for incompressible flows is

$$V_1 A_1 = V_2 A_2$$

We can also obtain the equation of continuity in the differential form. Consider the two-dimensional element of sides dx and dy and unit depth in the z-direction shown in Fig. 3.9. We will assume here in this two-dimensional treatment that there are no changes in the z-direction: $\partial u/\partial z = \partial v/\partial z = 0$ and w is either equal to zero or remains constant throughout the flow field.

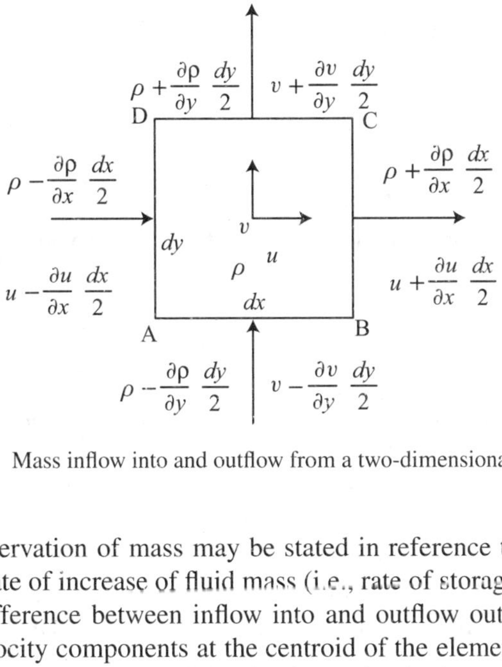

Fig. 3.9 Mass inflow into and outflow from a two-dimensional element

The principle of conservation of mass may be stated in reference to an elementary volume shown in Fig. 3.9. The rate of increase of fluid mass (i.e., rate of storage) within a given volume must be equal to the difference between inflow into and outflow out of that volume. Let the density be ρ and the velocity components at the centroid of the element be denoted by u and v respectively. The corresponding values of the density and velocity components at each pair of opposite faces are shown in the figure. The mass inflow into the element occurs across the faces AD and AB and the mass outflow occurs across the opposite faces AB and CD. The difference between the mass outflow and mass inflow is therefore

$$\left(\rho + \frac{\partial \rho}{\partial x} \cdot \frac{dx}{2}\right)\left(u + \frac{\partial u}{\partial x} \cdot \frac{dx}{2}\right) \cdot dy \cdot 1 + \left(\rho + \frac{\partial \rho}{\partial y} \cdot \frac{dy}{2}\right)\left(v + \frac{\partial v}{\partial y} \cdot \frac{dy}{2}\right) \cdot dx \cdot 1$$

$$- \left(\rho - \frac{\partial \rho}{\partial x} \cdot \frac{dx}{2}\right)\left(u - \frac{\partial u}{\partial x} \cdot \frac{dx}{2}\right) \cdot dy \cdot 1 - \left(\rho - \frac{\partial \rho}{\partial y} \cdot \frac{dy}{2}\right)\left(v - \frac{\partial v}{\partial y} \cdot \frac{dy}{2}\right) \cdot dx \cdot 1$$

On simplification, the difference becomes

$$\left[u\left(\frac{\partial\rho}{\partial x}\right)+v\left(\frac{\partial\rho}{\partial y}\right)+\rho\left(\frac{\partial u}{\partial x}\right)+\rho\left(\frac{\partial v}{\partial y}\right)\right]dx\cdot dy\cdot 1=\left[\partial\frac{(\rho u)}{\partial x}+\partial\frac{(\rho v)}{\partial y}\right]dx\cdot dy$$

The difference must be equal to the mass stored in the element in time dt

$$\text{i.e., }\left[\rho dx\cdot dy\cdot 1-\left(\rho+\frac{\partial\rho}{\partial t}\right)dx\cdot dy\cdot 1\right]=-\left(\frac{\partial\rho}{\partial t}\right)dx\cdot dy\cdot 1$$

Equating the two expressions and dividing throughout by $dx\cdot dy$ and transposing all the terms to one side, we get

$$\frac{\partial\rho}{\partial t}+\frac{\partial(\rho u)}{\partial x}+\frac{\partial(\rho v)}{\partial y}=0 \tag{3.13}$$

If the flow is steady, all changes with respect to time are zero and hence

$$\frac{\partial(\rho u)}{\partial x}+\frac{\partial(\rho v)}{\partial y}=0 \tag{3.14}$$

Further if the flow is incompressible, the density remains constant, yielding

$$\frac{\partial u}{\partial x}+\frac{\partial v}{\partial y}=0 \tag{3.15}$$

Equation (3.13) should be satisfied by all two-dimensional unsteady, compressible flows and (3.14) must be satisfied by all steady compressible flows and (3.15) must be satisfied by all steady, incompressible flows.

Extending the argument to a three-dimensional steady, incompressible flow, the equation of continuity becomes

$$\frac{\partial u}{\partial x}+\frac{\partial v}{\partial y}+\frac{\partial w}{\partial z}=0 \tag{3.16}$$

In vector notation, the above equation can be written as

$$\text{div }\mathbf{V}=0 \text{ or } \nabla\cdot\mathbf{V}=0 \tag{3.17}$$

The equation of continuity in polar coordinates is derived with reference to Fig. 3.10.

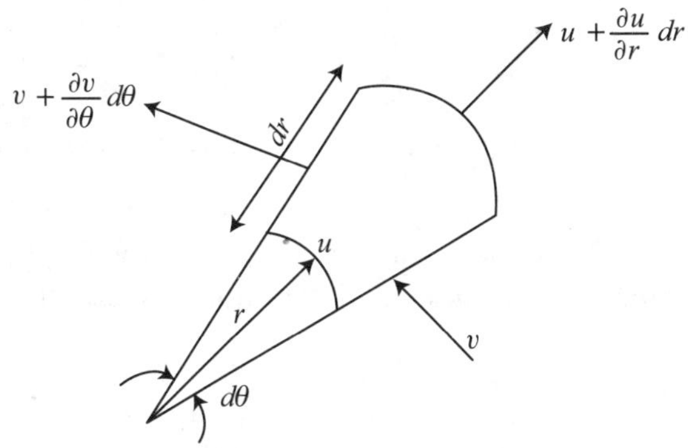

Fig. 3.10 Equation of continuity in polar coordinates

The velocities shown are the average velocity across each face.

$$\text{Inflow} = u r d\theta + v dr$$

$$\text{Outflow} = \left[u + \left(\frac{\partial u}{\partial r}\right) dr\right](r + dr)d\theta + \left[v + \left(\frac{\partial v}{\partial \theta}\right) d\theta\right] dr$$

Equating the inflow and outflow and simplifying after neglecting the term containing the product of three small quantities, we get

$$\frac{\partial u}{\partial r} + \frac{u}{r} + \left(\frac{1}{r}\right)\frac{\partial v}{\partial \theta} = 0 \tag{3.18}$$

EXAMPLE 3.1 In a two-dimensional flow, if $u = 2xy$ and $v = x^2 - y^2$, does the flow satisfy the equation of continuity?

$\partial u/\partial x = 2y$ and $\partial v/\partial y = -2y$ and hence $\partial u/\partial x + \partial v/\partial y = 0$ and the equation of continuity is satisfied.

EXAMPLE 3.2 Does the following velocity field satisfy the equation of continuity for an incompressible fluid?

$$u = ax^2y;\ v = by^2z;\ w = -yz(2ax + bz)$$

$\partial u/\partial x = 2axy;\ \partial v/\partial y = 2byz;\ \partial w/\partial z = -2axy - 2byz$ and therefore the equation of continuity is satisfied.

EXAMPLE 3.3 Given $u = (x^3yz)/3;\ v = (xy^3z)/3$, find w to satisfy the equation of continuity.

$$\frac{\partial u}{\partial x} = x^2yz;\ \frac{\partial v}{\partial y} = xy^2z;\ \frac{\partial w}{\partial z} = -xyz(x + y)$$

Integrating we get

$$w = -xy(x+y)\frac{z^2}{2} + \text{ an integration constant which can be a function of } f(x, y)$$

EXAMPLE 3.4 The rates of linear deformations $\partial u/\partial x$, $\partial v/\partial y$, $\partial w/\partial z$ have been calculated in the above examples. Find the rates of angular deformation and rotations in the above flow fields.

Velocity field in Example 3.1

1. $\partial u/\partial y = 2x$; $\partial v/\partial x = 2x$; hence $h(xy\text{-plane}) = 4x$
2. Rate of rotation $\omega_{xy}(xy\text{-plane}) = (1/2)(2x - 2x) = 0$ and vorticity $= 0$

Velocity field in Example 3.2

1. $\partial u/\partial y = ax^2$; $\partial v/\partial x = 0$; $h = ax^2$; $\omega_{xy} = -1/2ax^2$
2. $\partial u/\partial z = 0$; $\partial w/\partial x = -2ayz$; $g = -2ayz$; $\omega_{zx} = ayz$
3. $\partial w/\partial y = -z(2ax + bz)$; $\partial v/\partial z = by^2$; $f = -2axz - bz^2 + by^2$; $\omega_{yz} = -1/2(2axz + bz^2 + by^2)$

Velocity field in Example 3.3
$f(x \cdot y)$ in the expression for w is assumed as constant.

1. $\partial u/\partial y = x^3z/3$; $\partial v/\partial x = y^3z/3$; $h = z(x^3 + y^3)/3$; $\omega_{xy} = (1/6)\, z(y^3 - x^3)$
2. $\partial u/\partial z = x^3y/3$; $\partial w/\partial x = -xyz^2$; $g = x^3y/3 - xyz^2$; $\omega_{zx} = (1/2)xy\left(\frac{x^2}{3} + z^2\right)$
3. $\partial w/\partial y = -xz^2\,(x/2 + y)$; $\partial v/\partial z = xy^3/3$; $f = -xz^2\,(x/2 + y) + xy^3/3$; $\omega_{yz} = 1/2\left[-xz^2\,(x/2 + y) - xy^3/3\right]$

EXAMPLE 3.5 Evaluate the circulation about a closed square $x = -2$, $x = +2$, $y = +2$, $y = -2$ in a two-dimensional flow field having the velocity components $u = x^2 + y^2$, $v = -2xy$.

Circulation

$$\Gamma = \int_{-2}^{+2} (x^2 + y^2)dx + \int_{-2}^{+2} (-2xy)dy + \int_{+2}^{-2} (x^2 + y^2)dx + \int_{+2}^{-2} (-2xy)dy$$

For the first integral parallel to the x-axis, y remains constant at $y = -2$ and x varies between -2 and $+2$ and the second integral parallel to the y-axis, x remains constant at $x = +2$ and y varies between -2 and $+2$ and so on for the remaining two sides.

$$\Gamma = \left[\frac{x^3}{3} + y^2x\right]_{-2}^{+2} - \left[xy^2\right]_{-2}^{+2} + \left[\frac{x^3}{3} + y^2x\right]_{+2}^{-2} - \left[xy^2\right]_{+2}^{-2} = 0$$

Note that the velocity components satisfy the condition of irrotationality, $\partial u/\partial y - \partial v/\partial x = 0$. The circulation in an irrotational flow field around a closed path not containing any singularities within it, is always zero.

EXAMPLE 3.6 A free vortex has circular stream lines and the velocity is tangential to the circle at any point . There is no radial velocity. If the tangential velocity v_θ is given by C/r at any radius r and C is a constant, find the circulation around a closed circle.

The circulation, $\Gamma = \int_c V \cdot ds = \int_c v_\theta r d\theta = \int_c (C/r) r d\theta = \int_c C d\theta = 2\pi C$

It is important to note that the circulation is independent of radius and is constant for any closed path around the origin. Though the flow field is also irrotational, there is a singular point at the origin since the velocity becomes infinite for $r = 0$.

A forced vortex given by the velocity distribution $v_\theta = Cr$ has the circulation as

$$\Gamma = \int_c Crr d\theta = \int_c Cr^2 d\theta = 2\pi Cr^2$$

EXERCISES

3.1 Does the flow field $u = V(x^3 + xy^2)$ and $v = V(y^3 + xy^2)$ and $w = 0$ satisfy continuity?

3.2 If $u = xt + xy$ and $v = xt^2 - yt$, does it satisfy continuity?

3.3 Show that the velocity distribution $u = -x/(x^2 + y^2)$; $v = -y/(x^2 + y^2)$ satisfy continuity. Find the values of the linear deformation, angular deformation and rotation.

3.4 If $u = C_1 y - C_2 x$ and $v = C_2 y - C_3 x$, find Ψ.

3.5 If $u = -y/b^2$ and $v = x/a^2$, show that $x^2/a^2 + y^2/b^2 = 1$ is a stream line.

3.6 Find the circulation around a closed rectangle $x = -2$ and $x = +2$ and $y = -3$ and $y = +3$ in a two-dimensional field with $u = x + y$ and $v = x + y$.

3.7 Under what conditions does the following velocity field satisfy continuity.

(a) $u = a_1 x + b_1 y + c_1 z$; $v = a_2 x - b_2 y + c_2 z$; $w = a_3 x + b_3 y + c_3 z$

(b) $u = ax^2 + by^2 + cz^2$; $v = -dxy - eyz - fzx$; find w.

3.8 Test the following flow fields for continuity.

(a) $u = \dfrac{U(y^2 - x^2)}{(x^2 + y^2)^2}$; $v = \dfrac{-2Uxy}{(x^2 + y^2)^2}$

(b) $u = kxyzt$; $v = -kxyzt^2$; $w = k\left(\dfrac{z^2}{2}\right)(xt^2 - yt)$

(c) $u_r = U\cos\theta + \dfrac{q}{2\pi r}$; $v_\theta = -U\sin\theta$

(d) $u_r = U\cos\theta - \dfrac{\mu\cos\theta}{r^2}$; $v_\theta = -U\sin\theta - \dfrac{\mu\sin\theta}{r^2}$

4 Dynamics of Flow

4.1 Acceleration

From Newton's second law of motion, we know that the rate of change of momentum of a fluid element is proportional to the force acting on it and takes place in the direction of the force. Stated more explicitly, the force is equal to mass times the acceleration in the direction of the force. We shall now study the application of Newton's second law to fluid motion. We shall first take the simple case of an element accelerating along a stream line.

Figure 4.1*a* shows two adjacent curved and converging stream lines. The velocity vector v_s is tangential to the stream line and the acceleration along the stream line is given by $a_s = dv_s/dt$.

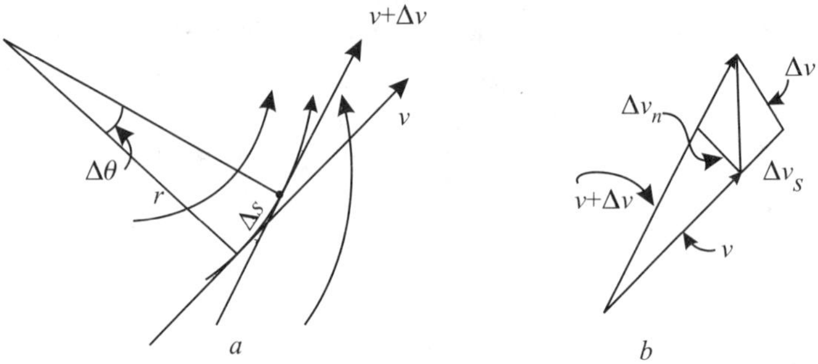

Fig. 4.1 Velocity variation along a stream line

It must be realised that unlike the velocity vector, the acceleration vector has no specific orientation with respect to the stream line. The acceleration vector can have both tangential and normal components. The tangential component of acceleration causes change in magnitude of the velocity whereas the normal component causes change in direction. If the stream line spacings are constant, then there is no acceleration in the direction of flow and $a_s = 0$. If the stream lines are converging, there is acceleration but if they diverge, deceleration takes place. If the stream lines are straight, the normal component $a_n = 0$. If they are curved, there is a centrifugal or centripetal acceleration, a_n.

4.2 Tangential and Normal Components of Accelerations

We will confine ourselves to the treatment of steady flows. Let the fluid element shown in Fig. 4.1*a* move through a distance ds in time dt and the velocity changes slightly in both magnitude and direction. The total change in velocity dv_s is given by $dv_s = (\partial v_s/\partial s)ds$ and $ds = v_s \cdot dt$ which yields

$$\frac{dv_s}{dt} = a_s = v_s\left(\frac{\partial v_s}{\partial s}\right) = \frac{1}{2}\left(\frac{\partial v_s^2}{\partial s}\right) \tag{4.1}$$

Though the velocity vector is always tangential to the stream line, at the end of the travel distance ds, a small component dv_n in the initial direction must be present in order to render the change in direction. This is shown in the vector diagram 4.1*b*. The rate of change of v_n (i.e., dv_n) along ds is $dv_n/ds = (\partial v_n/\partial s)ds = (\partial v_n/\partial s)v_s dt$ or

$$\frac{dv_n}{dt} = v_s\left(\frac{\partial v_n}{\partial s}\right)$$

The similarity between the velocity and space triangles shown in Fig. 4.1*b* give the relation $dv_n/ds = v_s/r$ or substituting in the previous relationship

$$dv_n = \frac{v_s^2}{r} \tag{4.2}$$

which is the well-known expression for centrifugal acceleration.

4.3 Forces

The forces causing fluid acceleration may be classified into two broad groups. The first refers to the distant acting forces commonly known as body forces. Here the driving force does not establish physical contact with the fluid element but exerts its influence through a force field. Examples of these distant acting forces are the gravitational pull of the earth, moon, sun and the planets. A magnet, likewise, establishes a force field around it, while the electrostatic field is yet another example of a force of attraction or repulsion without contact. The components of these body forces per unit mass of fluid in the three coordinate directions are usually designated as X, Y, Z. In this section, we shall mainly restrict ourselves to gravitational forces with the result $X = 0$, $Z = 0$, $Y = -g$, the minus sign attributed to the fact that the gravitational acceleration always acts downwards.

The secondary category of forces called contact forces may be further divided into two classes: (a) normal forces acting perpendicular to a given area and (b) tangential or shear forces acting along the plane of the area. Again in this chapter we shall deal with ideal fluid motion which has no viscosity and hence all shear stresses are zero. The normal stresses are due to pressures acting on the surface. We shall take into account the shear stresses in the chapters on laminar and turbulent flows.

4.4 Equation of Motion along a Stream Line

Consider the fluid element shown in Fig. 4.2 subjected to pressure forces on the top and bottom faces of the element. Shear forces along the length of the element is not considered as the fluid is ideal and forces normal to the length ds of the element have no component in the direction of the stream line.

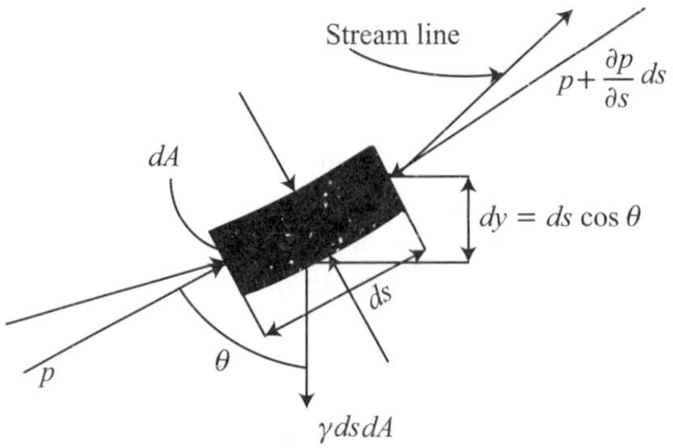

Fig. 4.2 Forces along a stream line

On the bottom face, the force is $p \cdot dA$ and on the top face, the force is $(p + \partial p/\partial s \cdot ds)dA$ giving the net resultant pressure force acting on the element in the direction of the tangent to the stream line as $(\partial p/\partial s) \cdot ds \cdot dA$. The weight of the element acting through the centroid of the element is $\gamma ds \cdot dA$ and its component in the direction of the stream line is $\gamma ds \cdot dA \cos\theta$. Therefore, the force = mass × acceleration equation along the stream line becomes

$$p \cdot dA - \left(p + \frac{\partial p}{\partial s}ds\right)dA - \gamma ds \cdot dA \cdot \cos\theta = \rho ds \cdot dAa_s$$

where a_s is the acceleration of the fluid element. Dividing throughout by the mass of the fluid element $\rho ds \cdot dA$ and simplifying

$$\left(\frac{1}{\rho}\right)\left(\frac{\partial p}{\partial s}\right) + g\cos\theta + a_s = 0 \tag{4.3}$$

From Fig. 4.2, the increase in the vertical elevation between the bottom and upper faces of the element, dy is related to ds as $dy/ds = \cos\theta$. Substituting the above result in (4.3), and also replacing $a_s = dv_s/dt = 1/2\partial v_s^2/\partial s$ from (4.1), we get

$$\left(\frac{1}{\rho}\right)\left(\frac{\partial p}{\partial s}\right) + g\left(\frac{\partial y}{\partial s}\right) + \frac{1}{2}\frac{\partial v_s^2}{\partial s} = 0$$

$$\frac{\partial\left(p/\rho + gy + 1/2v_s^2\right)}{\partial s} = 0 \tag{4.4}$$

The integration of this equation can be carried out with respect to s (along the stream line) only leading to the equation

$$\frac{p}{\rho} + gy + \frac{1}{2}v_s^2 = \text{constant} \tag{4.5}$$

for an incompressible fluid with ρ as constant.

This is the well-known Bernoulli equation along a stream line. The constant may vary from stream line to stream line. Other assumptions made in the above derivation are that the fluid is inviscid and incompressible and the flow is steady. If the fluid is compressible, the equation is modified to take care of the interdependence of pressure and density.

$$\int \frac{dp}{\rho} + gy + \frac{1}{2}v_s^2 = \text{constant} \tag{4.6}$$

A more convenient form of (4.5) commonly used by hydraulic engineers is obtained by dividing throughout by g and with $\gamma = \rho g$

$$\frac{p}{\gamma} + y + \frac{v_s^2}{2g} = \text{constant} \tag{4.7}$$

The term p/γ is called the pressure head, y is the potential energy head and denotes the energy acquired by the fluid element due to its relative position with respect to a datum and the term $v_s^2/2g$ is the kinetic or velocity head. The two terms $(p/\gamma + y)$ together is known as the piezometric head.

Unsteady flow

For unsteady flows, we must introduce local accelerations in addition to spatial accelerations. Accordingly, (4.1) and (4.2) are written as

$$a_s = \frac{dv_s}{dt} - \frac{\partial v_s}{\partial t} + \frac{1}{2}\frac{\partial v_s^2}{\partial s} = \frac{-\partial\,(p/\rho + gy)}{\partial s} \tag{4.8}$$

$$a_n = \frac{dv_n}{dt} = \frac{\partial v_n}{\partial t} + \frac{v_s^2}{r} = \frac{-\partial\,(p/\rho + gy)}{\partial n} \tag{4.9}$$

Integration of the above equations including the local acceleration terms is difficult unless certain conditions are satisfied by the flow field.

Irrotational steady flow: Bernoulli equation

If we add the term $-\partial(v_s^2/2)/\partial n$ to both sides of (4.9), for a steady flow with $\partial v_n/\partial t = 0$ and remembering $\partial v_n/\partial s = v_s/r$ we get,

$$\frac{v_s^2}{r} - \frac{\partial\left(v_s^2/2\right)}{\partial n} = v_s\frac{\partial v_n}{\partial s} - v_s\frac{\partial v_s}{\partial n} = \frac{-\partial\left(p/\rho + gy + v_s^2/2\right)}{\partial n}$$

$$v_s\left(\frac{\partial v_n}{\partial s}-\frac{\partial v_s}{\partial n}\right)=\frac{-\partial\left(p/\rho+gy+v_s^2/2\right)}{\partial n}$$

If the expression on the left-hand side viz., $(\partial v_n/\partial s - \partial v_s/\partial n) = 0$, then the right-hand side is a function of n only and can be integrated to yield along the n-direction

$$\frac{p}{\rho}+gy+\frac{v_s^2}{2}=\text{constant} \tag{4.10}$$

which is the same expression as (4.5). The condition

$$\frac{\partial v_n}{\partial s}-\frac{\partial v_s}{\partial n}=0$$

is easily recognised as the condition of irrotationality in the sn-plane as discussed in Chapter 3. Hence we conclude that if the flow is irrotational, the Bernoulli equation can be applied to any part of the flow field since it is satisfied not only along the stream line but also normal to it.

4.5 Equation of Motion in the xy-Coordinates

We had derived the Bernoulli's theorem using the sn-plane consisting a system of stream lines and normals to the stream lines. We shall derive the same using the xy-coordinate system first for a two-dimensional flow and later extend it to three-dimensional motion. Consider the two-dimensional element having sides dx and dy and unit depth perpendicular to the plane of the paper. We shall further assume that the velocity w in the z-direction is either zero or constant. There is no change of the variables u, v, and p with respect to z. The two-dimensional element is shown in Fig. 4.3 with the forces acting on the element.

By the rules of calculus, the full derivative of any variable, such as u and v, which are both functions of x, y, and t is given as

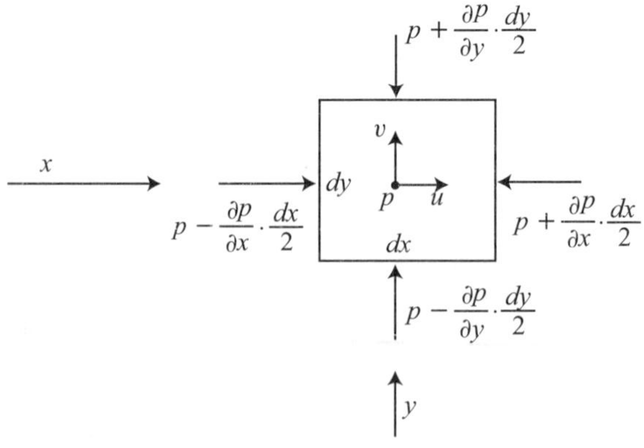

Fig. 4.3 Pressure field in a two-dimensional element

$$\frac{du}{dt} = \frac{\partial u}{\partial t} + \left(\frac{\partial u}{\partial x}\right)\left(\frac{dx}{dt}\right) + \left(\frac{\partial u}{\partial y}\right)\left(\frac{dy}{dt}\right) = \frac{\partial u}{\partial t} + u \cdot \frac{\partial u}{\partial x} + v \cdot \frac{\partial u}{\partial y}$$

$$\frac{dv}{dt} = \frac{\partial v}{\partial t} + \left(\frac{\partial v}{\partial x}\right)\left(\frac{dx}{dt}\right) + \left(\frac{\partial v}{\partial y}\right)\left(\frac{dy}{dt}\right) = \frac{\partial v}{\partial t} + u \cdot \frac{\partial v}{\partial x} + v \cdot \frac{\partial v}{\partial y}$$

The Newtonian relationship: mass $\times$ acceleration $=$ force component is written in the two directions as

$$\rho dxdy \left(\frac{\partial u}{\partial t} + u \cdot \frac{\partial u}{\partial x} + v \cdot \frac{\partial u}{\partial y}\right) = X \cdot \rho \cdot dx \cdot dy \cdot 1 - \left(\frac{\partial p}{\partial x}\right) \cdot dx \cdot dy \cdot 1$$

$$(\rho \cdot dx \cdot dy) \left(\frac{\partial v}{\partial t} + u \cdot \frac{\partial v}{\partial x} + v \cdot \frac{\partial v}{\partial y}\right) = Y \cdot \rho \cdot dx \cdot dy \cdot 1 - \left(\frac{\partial p}{\partial y}\right) \cdot dx \cdot dy \cdot 1$$

If the flow field is subjected to the gravitational field only, then the body forces per unit mass $X = 0$ and $Y = -g$. Dividing throughout by $\rho \cdot dx \cdot dy$, we obtain

$$\frac{\partial u}{\partial t} + u \cdot \frac{\partial u}{\partial x} + v \cdot \frac{\partial u}{\partial y} = -\left(\frac{1}{\rho}\right) \cdot \left(\frac{\partial p}{\partial x}\right) \tag{4.11}$$

$$\frac{\partial v}{\partial t} + u \cdot \frac{\partial v}{\partial x} + v \cdot \frac{\partial v}{\partial y} = -g - \left(\frac{1}{\rho}\right)\left(\frac{\partial p}{\partial y}\right) \tag{4.12}$$

To the left-hand side of the first equation in the x-direction we add and subtract the term $v\partial v/\partial x$ to yield

$$\frac{\partial u}{\partial t} + u \cdot \frac{\partial u}{\partial x} + v \cdot \frac{\partial v}{\partial x} + v\left(\frac{\partial u}{\partial y} - \frac{\partial v}{\partial x}\right) = -\left(\frac{1}{\rho}\right) \cdot \left(\frac{\partial p}{\partial x}\right)$$

If the flow is irrotational, $(\partial u/\partial y - \partial v/\partial x) = 0$ and further we can write $u\partial u/\partial x = \partial(u^2/2)/\partial x$ and $v\partial v/\partial x = \partial(v^2/2)/\partial x$. With these modifications, the equation in the x-direction becomes

$$\frac{\partial u}{\partial t} + \frac{\partial\left[(u^2 + v^2)/2\right]}{\partial x} + \left(\frac{1}{\rho}\right)\left(\frac{\partial p}{\partial x}\right) = 0$$

For a steady flow, we have $\partial u/\partial t = 0$ and $(u^2 + v^2) = V^2$, where V is the resultant velocity. The equation in the x-direction is

$$\frac{\partial\left(V^2/2 + p/\rho\right)}{\partial x} = 0$$

which on integration w.r.t. x yields

$$\frac{V^2}{2} + \frac{p}{\rho} = f(y) \tag{4.13a}$$

We carry out a similar manipulation with the equation of motion in the y-direction. We now add and subtract the term $u \cdot \partial u/\partial y$ to the left-hand side.

$$\frac{\partial v}{\partial t} + u \cdot \left(\frac{\partial v}{\partial x} - \frac{\partial u}{\partial y}\right) + \frac{\partial\left(v^2/2\right)}{\partial y} + \frac{\partial\left(u^2/2\right)}{\partial y} = -g - \left(\frac{1}{\rho}\right)\left(\frac{\partial p}{\partial y}\right)$$

Again for steady flow $\partial v/\partial t = 0$ and for irrotational flow $(\partial v/\partial x - \partial u/\partial y) = 0$. Integration w.r.t. y yields

$$\frac{V^2}{2} + \frac{p}{\rho} = -gy + f(x) \tag{4.13b}$$

Comparing (4.13a) and (4.13b), we conclude $f(x) = \text{const.}$ and $f(y) = -gy$. Finally we arrive at the Bernoulli equation for steady, irrotational, incompressible and inviscid flow

$$\frac{V^2}{2} + \frac{p}{\rho} + gy = \text{constant} \tag{4.14}$$

$$\text{or} \quad \frac{V^2}{2g} + \frac{p}{\gamma} + y = \text{constant} \tag{4.15}$$

If the fluid is compressible and density is dependent on pressure only, (4.14) may be modified as

$$\frac{V^2}{2} + \int \frac{dp}{\rho} + gy = \text{constant} \tag{4.16}$$

4.6 Velocity Potential

The conditions of irrotationality have been described in Chapter 3. These are

$$\frac{\partial u}{\partial y} - \frac{\partial v}{\partial x} = 0; \quad \frac{\partial v}{\partial z} - \frac{\partial w}{\partial y} = 0; \quad \frac{\partial w}{\partial x} - \frac{\partial u}{\partial z} = 0$$

The specified conditions mean that the rotation of any element should be zero over the entire flow field. In other words, vorticity should be absent in the flow field. Such type of flows have been studied extensively. The conditional of irrotationality allows us to assume the existence of a velocity potential, Φ such that its derivatives in a specific direction gives the velocity in that direction. For example in the Cartesian coordinate system, we write

$$u = \frac{\partial \Phi}{\partial x}; \quad v = \frac{\partial \Phi}{\partial y}; \quad w = \frac{\partial \Phi}{\partial z} \tag{4.17}$$

or in the sn-coordinates

$$v_s = \frac{\partial \Phi}{\partial s}; \quad v_n = \frac{\partial \Phi}{\partial n} \tag{4.18}$$

The physical meaning of Φ is not as clear as the stream function Ψ. However, certain analogous comparisons may be made with other physical quantities where similar relationships exist.

1. It is common knowledge that whenever there is a difference in voltage between two points in an electrically conducting medium, a current flows following the law

$$I = K_1 \frac{dE}{ds}$$

where K_1 is the coefficient of electrical conductivity. The current is analogous to the velocity and voltage, E (or electric potential) is compared to the velocity potential, Φ.

2. The quantity of heat, H between two points depends on the difference of temperature between the points

$$H = K_2 \frac{d\theta}{ds}$$

Heat flow is analogous to velocity and temperature to velocity potential.

3. In a porous media, the velocity of percolation is dependent on the difference between two points.

$$V = K_3 \frac{dh}{ds}$$

where K_3 is the coefficient of permeability of the soil and ground, water depth being compared to the velocity potential.

In all these we see that the vector quantities of current, heat, and velocity of percolation are expressed as gradients of scalar quantities E, θ, h.

4.7 Relation between Velocity Potential and Stream Function

In two-dimensional irrotational flow systems, the stream function, Ψ and the velocity potential Φ are related thus

$$u = \frac{\partial \Phi}{\partial x} = \frac{\partial \Psi}{\partial y} \tag{4.19}$$

$$v = \frac{\partial \Phi}{\partial y} = -\frac{\partial \Psi}{\partial x} \tag{4.20}$$

The equation to the stream line at any instant of time is given by

$$d\Psi = u \cdot dy - v \cdot dx = 0$$

and the slope of the stream line at any point is

$$\frac{dy}{dx} = \frac{v}{u} \tag{4.21}$$

The full differential of the velocity potential, $\Phi(x, y)$ is

$$d\Phi = \left(\frac{\partial \Phi}{\partial x}\right) dx + \left(\frac{\partial \Phi}{\partial y}\right) dy = u \cdot dx + v \cdot dy \tag{4.22}$$

Along lines of constant velocity potential

$$d\Phi = 0 = u \cdot dx + v \cdot dy \tag{4.23}$$

The slope of such lines at any point is

$$\frac{dy}{dx} = -\frac{u}{v} \tag{4.24}$$

The product of slopes of the two lines Ψ = constant and Φ = constant, yield a value of -1. Hence, we conclude that the lines Φ = const. and Ψ = const. intersect each other always at right angles. This property of orthogonality of the two sets of lines is put to use in drawing the flow net for given boundary conditions.

4.8 Flow Nets

The flow net is often used to find the pattern of stream lines in two-dimensional motion. The stream lines and potential lines in plane irrotational motion will form a network approaching perfect squares as the mesh size denoting the increments in $d\Phi$ and $d\Psi$ is reduced to a very fine size. However, there is a practical limitation in choosing a very large number of Φ and Ψ lines. When the increments chosen are large, the meshes formed out of fewer $\Phi - \Psi$ lines may deviate considerably from perfect squares, although the orthogonal characteristics is still preserved between the two families.

To construct a flow net for given boundary conditions, first sketch a chosen number of stream lines guided by physical intuition as to their spacing and change in slopes. Next draw in lines intersecting these stream lines at right angles at every point of intersection, and at the same time ensuring that a reasonable semblance to a square is achieved for every mesh. In case of a highly distorted mesh, its median lines should be equal and they should intersect at right angles. The first attempt is rarely successful in meeting all these conditions. Successive adjustments are required in the spacing of stream lines and potential lines, and with a liberal use of pencil and eraser, the final spacings of the two sets of lines mutually at right angles to each other are evolved with the meshes resembling approximate squares. At the corners of the boundary, it may be necessary to sub-divide the mesh size to obtain a better flow net.

In addition to the increments in Φ and Ψ being equal, the spacings between the velocity potential lines (Δs) and between the velocity potential lines (Δn) are also equal. The magnitude of velocity at the centroid of any mesh may be found approximately or in reference to the mesh size Δn_0 and Δs_0, where the magnitude of velocity V_0 is known.

$$\frac{V}{V_0} = \frac{\Delta s_0}{\Delta s} = \frac{\Delta n_0}{\Delta n} \tag{4.25}$$

If the force field is only due to gravitational attraction and the two-dimensional flow plane is horizontal, the Bernoulli equation becomes

$$\frac{p}{\rho} + \frac{V^2}{2} = \text{constant}$$

or if the reference pressure is p_0, where the velocity is V_0, the constant is

$$\frac{p}{\rho} + \frac{V^2}{2} = \frac{p_0}{\rho} + \frac{V_0^2}{2}$$

$$\frac{(p - p_0)}{1/2\rho V_0^2} = 1 - \left(\frac{V}{V_0}\right)^2$$

The change in pressure may be expressed in terms of the spacings of the $\Phi - \Psi$ lines.

$$\frac{(p - p_0)}{1/2\rho V_0^2} = 1 - \left(\frac{\Delta s_0}{\Delta s}\right)^2 = 1 - \left(\frac{\Delta n_0}{\Delta n}\right)^2 \quad (4.26)$$

An example of the flow net into a side aperture is shown in Fig. 4.4. Since the flow is approximately radial, the initial stream lines are guided by radial lines shown dotted in the figure and potential lines are circles at distances far removed from the aperture. The boundary

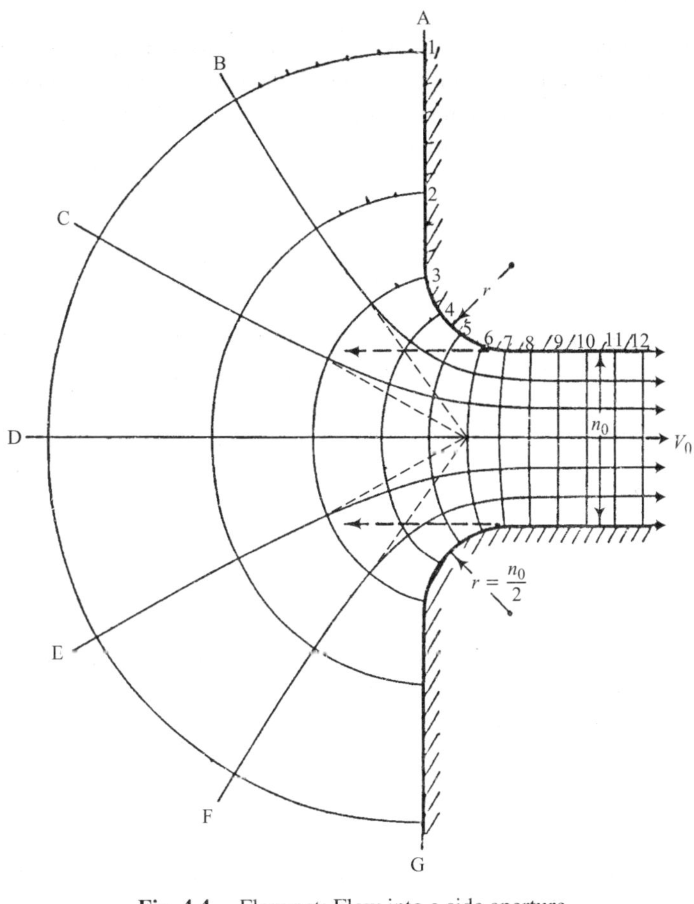

Fig. 4.4 Flow net: Flow into a side aperture

pressures are calculated by noting the intercepts made by the successive potential lines with the boundary.

4.9 Laplace's Equation

The equation of continuity for an incompressible fluid is

$$\frac{\partial u}{\partial x}+\frac{\partial v}{\partial y}+\frac{\partial w}{\partial z}=0$$

and if the flow is irrotational, a velocity potential exists such that $u = \partial\Phi/\partial x$; $v = \partial\Phi/\partial y$; $w = \partial\Phi/\partial z$. Substituting these values for u, v, and w in the equation of continuity, we obtain

$$\frac{\partial^2\Phi}{\partial x^2}+\frac{\partial^2\Phi}{\partial y^2}+\frac{\partial^2\Phi}{\partial z^2}=0 \tag{4.27}$$

which is known as the Laplace's equation. All irrotational problems are therefore solutions of the above second order partial differential equation. For a given boundary geometry, one seeks a solution of the Laplace's equation which simultaneously satisfies the boundary conditions.

In a two-dimensional flow, changes with respect to the z-direction are absent and the Laplace's equation becomes

$$\frac{\partial^2\Phi}{\partial x^2}+\frac{\partial^2\Phi}{\partial y^2}=0 \tag{4.28}$$

The stream function Ψ satisfies the equation of continuity and the velocity components are expressed as derivatives of the stream function $u = \partial\Psi/\partial y$ and $v = -\partial\Psi/\partial x$. Should the flow be irrotational, we have $\partial u/\partial y - \partial v/\partial x = 0$. Substituting the expressions for u and v in the condition for irrotationality, the resulting equation is again the Laplace's equation in the variable $\Psi(x, y)$.

$$\frac{\partial^2\Psi}{\partial x^2}+\frac{\partial^2\Psi}{\partial y^2}=0 \tag{4.29}$$

Both Φ and Ψ satisfy Laplace's equation in two-dimensional irrotational flows. A similar simple relationship in Ψ, however, does not exist for three-dimensional flows for the reason outlined in Chapter 3. In order to familiarise the student with the use of the velocity potential and stream function for solving irrotational flow problems, a few simple cases have been illustrated in Chapter 5.

Laplace's equation in polar coordinates

The radial and tangential components of velocity in polar coordinates may be obtained by taking the derivatives of Φ in these directions.

$$u_r = \frac{\partial \Phi}{\partial r}; \quad v_\theta = \frac{\partial \Phi}{\partial s} = \frac{1}{r}\frac{\partial \Phi}{\partial \theta} \tag{4.30}$$

This satisfies the condition of irrotationality in polar coordinates

$$\frac{\partial v_\theta}{\partial r} + \frac{v_\theta}{r} - \frac{1}{r}\frac{\partial u_r}{\partial \theta} = 0 \tag{4.31}$$

Substituting the expressions for the radial and tangential velocities in the equation of continuity, $\partial u_r/\partial r + u_r/r + \partial v_\theta/r\partial\theta = 0$, we get

$$\frac{\partial^2 \Phi}{\partial r^2} + \left(\frac{1}{r}\right)\frac{\partial \Phi}{\partial r} + \left(\frac{1}{r^2}\right)\frac{\partial^2 \Phi}{\partial \theta^2} \tag{4.32}$$

Examples of irrotational flow are presented in Chapter 5.

Exercises

4.1 If $u = 2x + 3y + z$; $v = 3x - 3y + 2z$; $w = x + y + z$, find the acceleration at location $x = y = z = 1$ m.

4.2 In Exercise 4.1, neglect all body forces and find the pressure at the given location.

4.3 Two plates each of diameter D move towards each other at a uniform velocity V. When the spacing between them is h, what is the radial acceleration at the edge of the disc?

4.4 A stream function ψ is given as

$$\psi = x^2 + 2xy + 4t^2 y$$

When $t = 4$ seconds, what is the flow rate across the semi-circular plate AB shown in the figure? Find also the flow across the x-axis from A(0,0) to C(5,0)

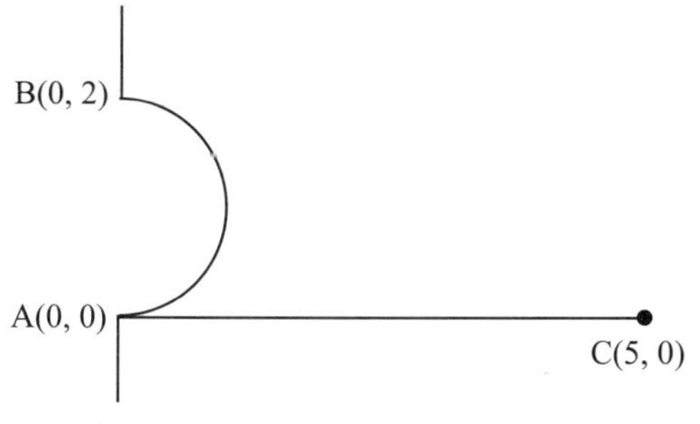

Figure for Problem 4.4

4.5 The velocity field is given as $u = 2x - 6$; $v = 4 - 2y$; $z = t^2$.
Determine the equation for the path taken by a particle which was at $x = 4$, $y = 1$ at $t = 0$.

4.6 Flow through a converging nozzle may be approximated by a one-dimensional velocity parameter $V = V(x)$. If the velocity varies linearly from $V = V_0$ at the entrance to $V = 3V_0$ at the exit in a length L of the nozzle: $V(x) = V_0(1 + 2x/L)$.
Find the acceleration at a distance x from the entrance.

4.7 For the following values of ϕ, find the respective stream functions

(a) $\phi = y + x^2 - y^2$
(b) $\phi = \frac{C}{r^2} \cos 2\theta$

4.8 Find the value of ϕ for the following values of ψ

(a) $\psi = Cr^2 \sin 2\theta$
(b) $\psi = \left(r - \frac{2}{r}\right) \sin \theta$

5 Examples of Irrotational Flow of Ideal Fluids

5.1 Uniform Flows

(a) Let us first consider an uniform flow of constant velocity, U parallel to the x-axis, Then

$$u = U = \frac{\partial \Phi}{\partial x} = \frac{\partial \Psi}{\partial y}$$
$$v = 0 = \frac{\partial \Phi}{\partial y} = -\frac{\partial \Psi}{\partial x}$$

Integrating the above equations, one obtains

$$\Phi = Ux \quad \text{and} \quad \Psi = Uy \tag{5.1}$$

the integration constants disappearing on the assumption that $\Phi = 0$ and $\Psi = 0$ at the origin. The flow diagram is shown in Fig 5.1*a*.

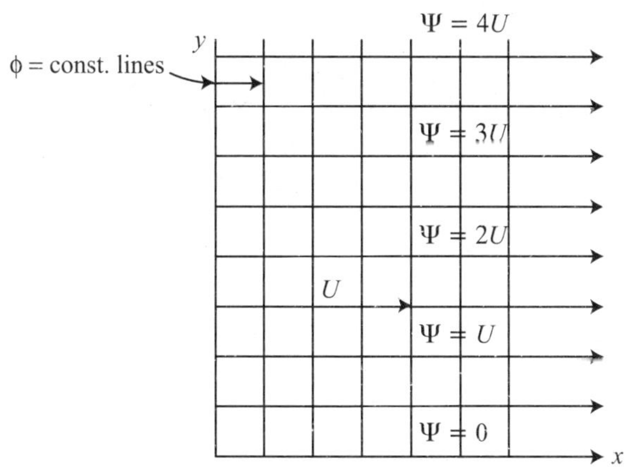

Fig. 5.1*a* Uniform flow parallel to the x-axis

(b) Let the uniform flow now be parallel to the y-axis and its velocity denoted by V. Then we have

$$u = \frac{\partial \Phi}{\partial x} = \frac{\partial \Psi}{\partial y} = 0$$

$$v = V = \frac{\partial \Phi}{\partial y} = -\frac{\partial \Psi}{\partial x}$$

Again on integration along with the assumption that $\Phi = \Psi = 0$ at the origin, we get

$$\Phi = Vy \quad \text{and} \quad \Psi = -Vx \tag{5.2}$$

The flow pattern is similar to that shown in Fig. 5.1*a* except that the Φ- and Ψ-lines are interchanged.

(c) If the flow is inclined to the x-axis at an angle and the uniform velocity in this direction is V_0, then its components in the x- and y-directions are

$$U = V_0 \cos\theta = \frac{\partial \Phi}{\partial x} = \frac{\partial \Psi}{\partial y} \quad \text{and}$$

$$V = V_0 \sin\theta = \frac{\partial \Phi}{\partial y} = -\frac{\partial \Psi}{\partial x}$$

which on integration yield

$$\Phi = Ux + Vy \quad \text{and} \quad \Psi = -Vx + Uy \tag{5.3}$$

The flow net for the uniform flow is shown in Fig. 5.1*b*

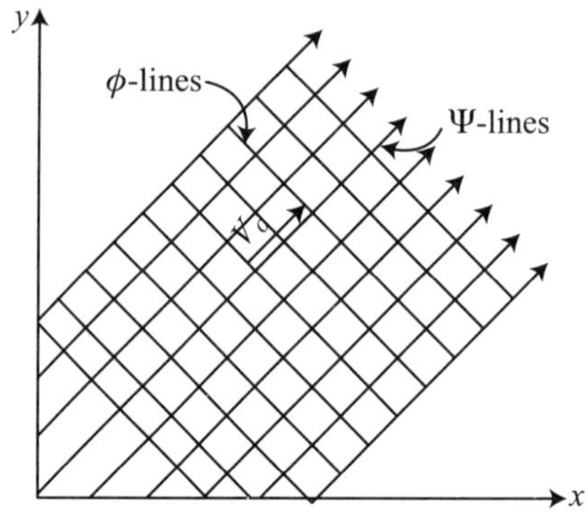

Fig. 5.1*b* Uniform flow inclined to the x-axis

It is seen that the values of Φ and Ψ representing the flow in (c) may be obtained by algebraic additions of those in (a) and (b). This is inherently the advantage of using the scalar quantities Φ and Ψ. Not only does a single variable replace the components of velocity in three directions but also aids in getting solutions to flow problems resulting in a combination of two or more elementary potential flow systems. Superposed solutions are easily obtained by mere algebraic addition of the individual solutions of Laplace's equation. If Φ_1 and Φ_2 are solutions of the Laplace's equation, then we have

$$\nabla^2 \Phi_1 = 0$$

$$\nabla^2 \Phi_2 = 0$$

Now to show that $\Phi_1 \pm \Phi_2$ is also a solution, we write the Laplacian of $\Phi_1 \pm \Phi_2$

$$\frac{\partial^2(\Phi_1 \pm \Phi_2)}{\partial x^2} + \frac{\partial^2(\Phi_1 \pm \Phi_2)}{\partial y^2} = 0$$

$$\left(\frac{\partial^2 \Phi_1}{\partial x^2} + \frac{\partial^2 \Phi_1}{\partial y^2}\right) \pm \left(\frac{\partial^2 \Phi_2}{\partial x^2} + \frac{\partial^2 \Phi_2}{\partial y^2}\right) = 0 + 0$$

by virtue of the fact that both Φ_1 and Φ_2 are solutions of the Laplace's equation. It may be realized that such a simple algebraic procedure is not possible if vector quantities are involved in the solution. After superposition of two or more solutions, the velocity component in any direction may be obtained by the gradients of the resultant velocity potential in that direction.

5.2 Plane Potential Flow in Polar Coordinates

Source

Fluid may emanate from a point (called the point source) radiating outwards in all directions or from a line (known as the line source), the flow being radial and in a plane normal to the source. Depending on the type of flow, they are classified as three-dimensional sources or two-dimensional sources. A practical form of a source is a spring originating from underground sources. The mathematical form of a source originates from a point whereas a spring has finite dimensions. The former has a singular point at the source or origin at which the velocity is infinite since the area is zero. However, away from the origin, the flow field is regular. If Q is the discharge emanating from the point source, the radial velocity at any spherical surface of radius R from the origin is (for a three-dimensional source)

$$u = \frac{\partial \Phi}{\partial R} = \frac{Q}{4\pi R^2}$$

Integration of the preceding equation yields

$$\Phi = -\frac{Q}{4\pi R} = -\frac{M}{R}$$

where $M = Q/4\pi$ is called the strength of the source.

In a two-dimensional source, the radial velocity is expressed in terms of the discharge q (discharge per unit length)

$u_r = \partial\Phi/\partial r = q/2\pi r$ which on integration yields the expression for the velocity potential.

$$\Phi = \left(\frac{q}{2\pi}\right) \ln r = m \ln r \quad \text{with} \quad m = \frac{q}{2\pi} \tag{5.4}$$

Since $u_r = (1/r)(\partial\Psi/\partial\theta) = q/2\pi r$, the value of the stream function is

$$\Psi = m\theta \tag{5.5}$$

where m is the strength of the two-dimensional source. Lines of constant Φ are circles and lines of constant Ψ are radial stream lines. The flow net for a two-dimensional source is shown in Fig. 5.2.

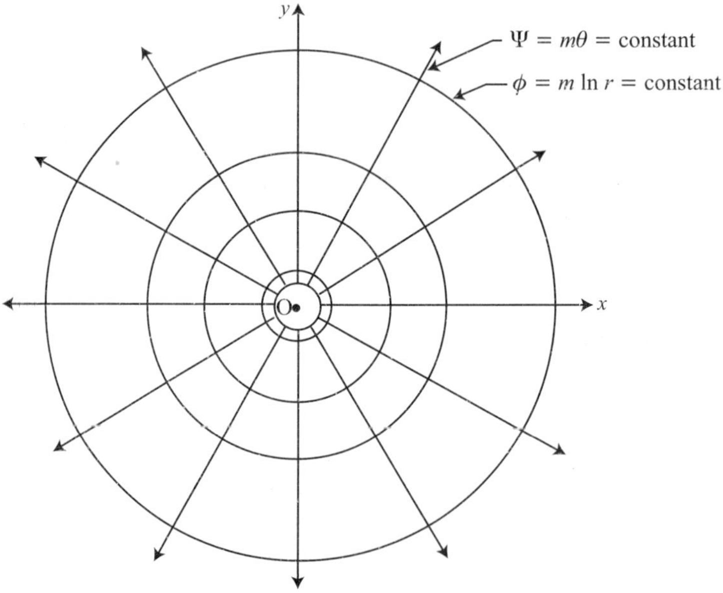

Fig. 5.2 Two-dimensional source

Sink

When the fluid flows radially inward towards a point or a line, we have a sink. The practical illustrations may be seen in a wash basin or ground water flow into a well. The derivations of the expressions for Φ and Ψ are similar to those outlined for the source except that a change in sign occurs due to the flow being radially inward.

$$\Phi = -m \ln r \quad \text{and} \quad \Psi = m\theta \tag{5.6}$$

Vortex

The formation of a vortex is a common phenomena which we can observe even in our wash basin. A whirlpool can form easily in streams and rivers and during a storm, dust and other particles get sucked up by tornadoes into the eye of the storm. The fluid particles in a free vortex move around in circular orbits about a central axis, or in other words, the stream lines are circles with radial lines as lines of constant velocity potential. The tangential veloctiy is inversely proportional to the radial distance from the centre of the vortex and the radial velocity component is absent. We shall consider anti-clockwise motion as positive. For a free vortex, we have

$$u_r = \frac{\partial \Phi}{\partial r} = \left(\frac{1}{r}\right)\left(\frac{\partial \Psi}{\partial \theta}\right) = 0 \quad \text{and} \quad v_\theta = \left(\frac{1}{r}\right)\left(\frac{\partial \Phi}{\partial \theta}\right) = -\frac{\partial \Psi}{\partial r} = \frac{C}{r}.$$

Hence

$$\Phi = C\theta \quad \text{and} \quad \Psi = -C \ln r \tag{5.7}$$

The origin located at the centre or eye of the vortex is a singular point. Consider the circulation around a closed path enclosing the origin, and for the sake of convenience, let the path be considered a circle of radius r_1 as in Fig. 5.3.

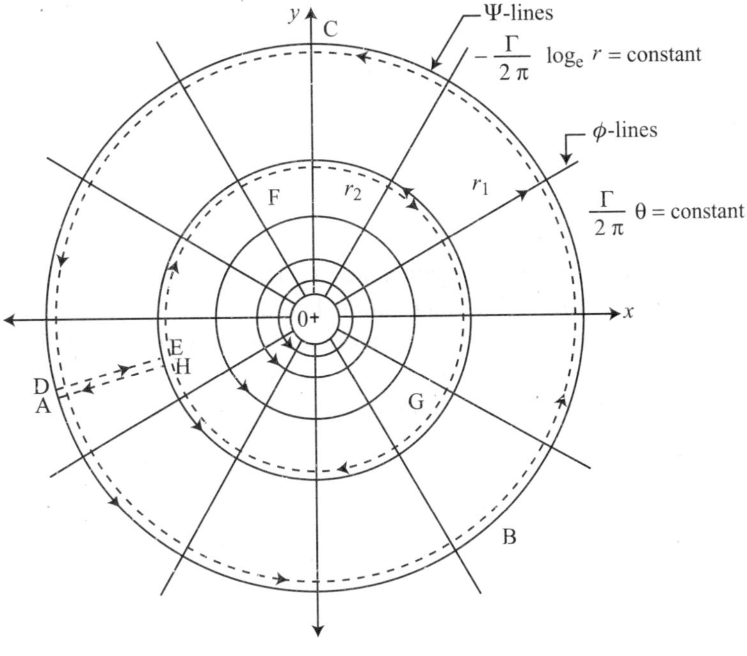

Fig. 5.3 Circulation in a vortex

$$\Gamma = \int_c v_\theta . ds = \int_0^{2\pi} v_\theta r_1 d\theta = \int_0^{2\pi} \left(\frac{C}{r_1}\right) r_1 d\theta = 2\pi C \tag{5.8}$$

The circulation is independent of the radius of the path chosen. The constant C may now be defined as $C = \Gamma/2\pi$, where Γ is the circulation around any closed path including the eye of the vortex.

A region which does not include the origin is regular and hence the circulation should vanish since the motion is irrotational. This may be demonstrated by considering the region between two concentric circles connected by the cut shown in dotted lines ABCDEFGH in Fig. 5.3

$$\int v_\theta . ds = \int_{\text{ABCD}} v_\theta r_1 d\theta + \int_{\text{DE}} v_\theta r dr + \int_{\text{EFGH}} v_\theta r_2 d\theta + \int_{\text{HA}} v_\theta r dr$$

The integral on the left-hand side of the above equation is around the path ABCDEFGHA and the first integral on the right-hand side is around ABCD, the second along DE, the third around

EFGH, the fourth along HA. The sum of the integrals along DE and HA vanish since the traverses are in the opposite direction. Hence we get

$$\int_0^{2\pi}\left(\frac{C}{r_1}\right)r_1 d\theta - \int_0^{2\pi}\left(\frac{C}{r_2}\right)r_2 d\theta = 2\pi C - 2\pi C = 0$$

At large distances from the origin, the peripheral velocity due to the vortex tends to be negligible. Let p_0 be the pressure intensity at an infinite distance from the origin. The pressure distribution in the vortex field is then found by the application of the Bernoulli theorem.

$$p + \frac{1}{2}\rho v_\theta^2 = \text{Constant} = p_0$$

where p is the pressure at any radius r. Substituting for the tangential velocity component $v_\theta = (\Gamma/2\pi r)$, we get

$$p_0 - p = \frac{1}{2}\rho\left(\frac{\Gamma}{2\pi r}\right)^2 = \left(\frac{\rho\Gamma^2}{8\pi^2}\right)\left(\frac{1}{r^2}\right)$$

The pressure at the centre is therefore negative infinity. However, in nature such a singularity at the origin is obviated by the formation of a rotational core.

In the rotational core, the motion of the fluid particles is similar to the solid body rotation, and the tangential velocity is given by $v_\theta = C_1 r$, where C_1 is now to be interpreted as the angular speed, ω. The pressure distribution in the rotational core is obtained by the expression

$$p - p_0 = \frac{1}{2}\rho C_1^2 r^2$$

where p_0 is the pressure at the centre of the vortex. The distribution of velocity and pressure in the combined vortex known as the Rankine combined vortex is shown in Fig. 5.4.

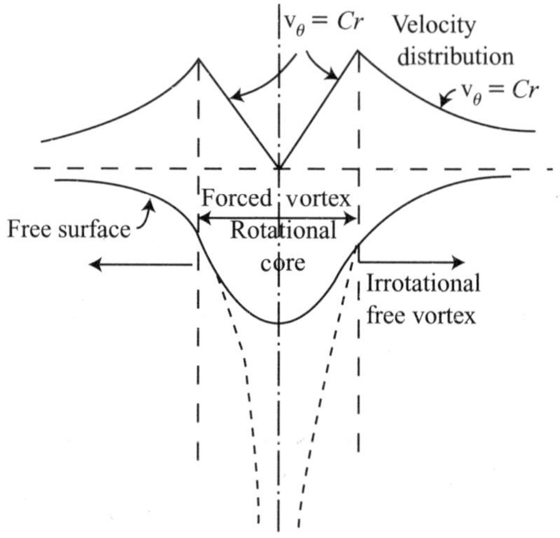

Fig. 5.4 Rankine combined vortex

5.3 Flow Around Corners

Let us now consider a family of general solution to the Laplace's equation:

$$\Phi = Ar^n \cos n\theta$$

By substitution, it may be shown that the preceding equation satisfies Laplace's equation. The velocity components are

$$u_r = \frac{\partial \Phi}{\partial r} = \left(\frac{1}{r}\right)\left(\frac{\partial \Psi}{\partial \theta}\right) = nAr^{n-1} \cos n\theta$$

$$v_\theta = \left(\frac{1}{r}\right)\left(\frac{\partial \Phi}{\partial \theta}\right) = -\frac{\partial \Psi}{\partial r} = -Anr^{n-1} \sin n\theta$$

We can evaluate $d\Psi$ by writing the full differential

$$d\Psi = \left(\frac{\partial \Psi}{\partial r}\right) \cdot dr + \left(\frac{\partial \Psi}{\partial \theta}\right) \cdot d\theta$$

and substituting for the values of $\partial\Psi/\partial r$ and $\partial\Psi/\partial\theta$

$$d\Psi = Anr^{n-1} \sin n\theta.dr + Anr^n \cos n\theta$$

which on integration yields

$$\Psi = \int_{0,0}^{r,o} Anr^{n-1} \sin n\theta.dr + \int_{r,o}^{r,\theta} Anr^n \cos n\theta.d\theta$$

$$\Psi = Ar^n \sin n\theta$$

For every value of n there are two values of θ for which $\Psi = 0$. Since a stream line may be replaced by a boundary, we are in a position to obtain solutions for flow around corners as shown in Fig. 5.5.

The following table illustrates the flow pattern for a few typical positive values of n and real values of A.

n	$\Psi = 0$	Remarks	Flow pattern illustrated in figure
1/2	$\theta = 0$ and $\theta = 2\pi$	Flow around a plate	5.5*a*
4/5	$\theta = 0$ and $\theta = 5\pi/4$	Flow around a 225° bend	5.5*b*
1	$\theta = 0$ and $\theta = \pi$	Flow past a flat plate	
2	$\theta = 0$ and $\theta = \pi/2$	Flow against a normal plate or Flow around a 90° bend	5.5*c*
4	$\theta = 0$ and $\theta = \pi/4$	Flow around a 45° bend	5.5*d*

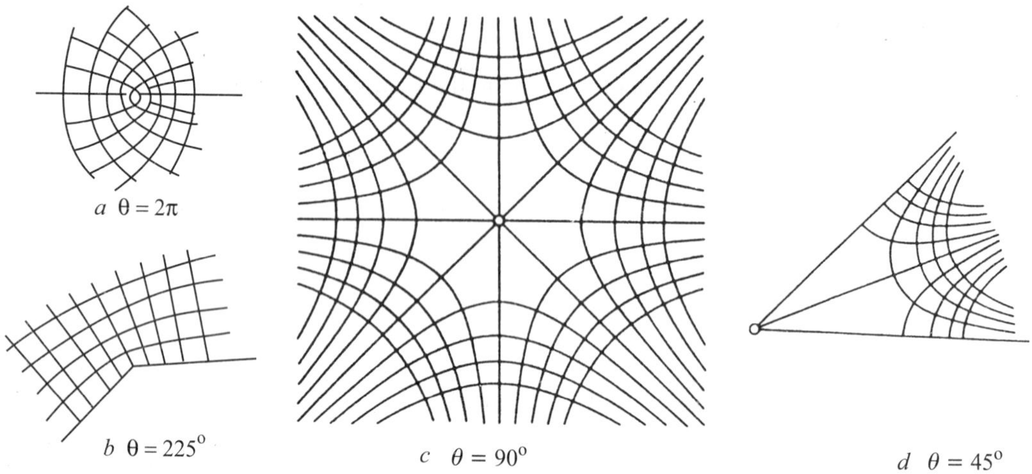

Fig. 5.5 Flow around corners

5.4 Combination of Two or More Flows

It has been shown that if Φ_1 and Φ_2 are solutions of the Laplace's equation, then $\Phi_1 + \Phi_2$ are also solutions. An infinite number of superposed solutions may be thus obtained by a combination of the preceding elementary solutions.

5.4.1 Source + Sink

Let a source and a sink of equal strength be situated at equal distances from the origin along the x-axis. The resultant velocity potential, Φ_P at any point P shown in Fig. 5.6 may be obtained as an algebraic sum of the potentials at that point due to the source and sink individually.

The resultant velocity potential is

$$\Phi_P = \Phi_{\text{source}} + \Phi_{\text{sink}} = m \ln r_2 - m \ln r_1 = m \ln \frac{r_2}{r_1}$$

And the stream function for the combination is

$$\Psi_P = m(\theta_2 - \theta_1) = -m\delta$$

The direction and magnitude of the velocity at P may be obtained by a vectorial combination of the velocities at P due to the source and sink. All discharge emanating from the source should eventually find its way into the sink. The Φ- and Ψ-lines are arcs of circles with their centres lying on the x- and y-axis respectively as shown in Fig. 5.6.

5.4.2 Doublet

We shall consider a particular case where the source and sink of equal strength approach each other till they finally merge at the origin in such a manner that the product of their strengths and

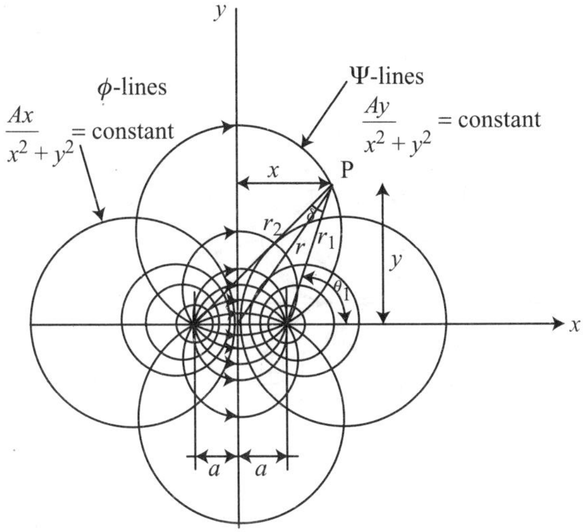

Fig. 5.6 Source and sink

the distance between them is always a constant. i.e., $m.2a = \mu$, a constant. Such a combination is known as a doublet. We have with reference to Fig. 5.6

$$-\tan\delta = \tan(\theta_2 - \theta_1) = \frac{(\tan\theta_2 - \tan\theta_1)}{(1 + \tan\theta_2\tan\theta_1)}$$

In the Cartesian coordinate system, we write

$$-\tan\delta = \frac{y/(x+a) - y/(x-a)}{(1+y^2)/(x^2-a^2)} = \frac{-2y\cdot a}{(x^2+y^2-a^2)}$$

As the source and sink approach each other in the manner described above, $m = \mu/2a$ and $r_1 = r_2 = r$ and as $\theta \to 0$ and $\tan\delta = \delta$ and further as a becomes small, $a^2 \sim 0$. Hence in the limit as a tends to zero

$$\Psi_P = -m\delta = -m\tan^{-1}\left[2y.a/(x^2+y^2)\right] = -\left(\frac{\mu}{2a}\right)\left[2ya/(x^2+y^2)\right] = -\frac{\mu y}{r^2}$$

$\Psi_P = -(\mu\sin\theta)/r$ and Φ can be derived from the relationships, $u_r = \partial\Phi\partial r = (1/r)\partial\Psi/\partial\theta$ and $v_\theta = (1/r)\partial\Phi/\partial\theta = -\partial\Psi/\partial r$

$$\Phi = \mu\cos\theta/r \quad \text{and} \quad \Psi = -\mu\sin\theta/r$$

Note that the doublet has an orientation (direction of the line of movement of source and sink till they meet). In the example we have illustrated in Fig. 5.6, the source and sink were lying on the axis but in general they could lie on any line inclined to the x-axis. The orientation of the doublet is the same as the orientation of the line joining the source and sink. The flow net for a doublet oriented along the x-axis is illustrated in Fig. 5.7.

The flow lines and potential line are circles tangential to the x- and y-axis respectively at the origin. The expression for the doublet can be obtained from the general solution $\Phi = Ar^n\cos n\theta$ for a value of $n = -1$.

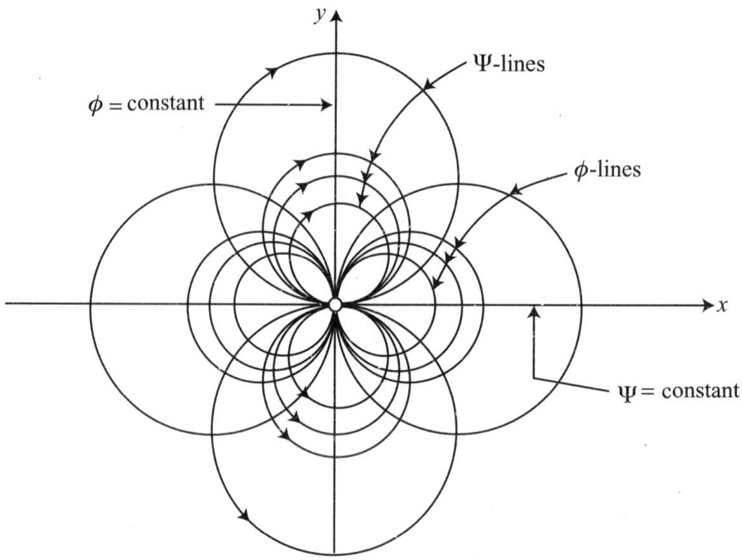

Fig. 5.7 Two-dimensional doublet oriented in the x-axis direction

Source in uniform flow : Rankine half-body

A source of strength m is situated at the origin and a uniform velocity, U parallel to the x-axis is superposed on the source field as shown in Fig. 5.8.

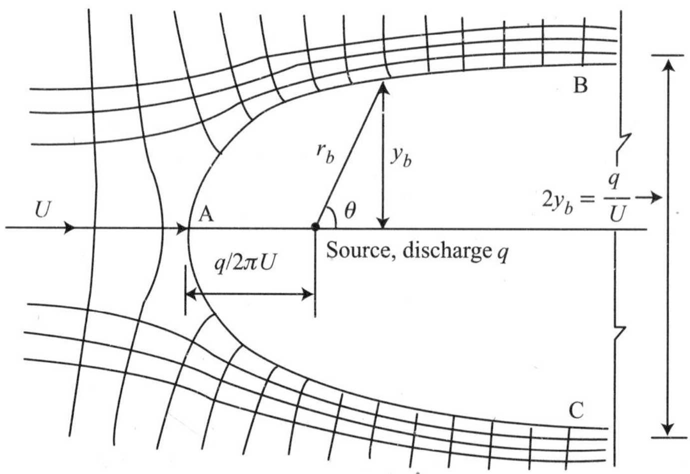

Fig. 5.8 Rankine half-body

The combined velocity and stream functions are obtained by the algebraic sums of the two fields.

$$\Phi = Ur\cos\theta + m\ln r$$

$$\Psi = Ur \sin\theta + m\theta$$

The velocity components u_r and v_θ are given as

$$u_r = \frac{\partial \Phi}{\partial r} = U \cos\theta + \frac{q}{2\pi r}$$

$$v_\theta = \left(\frac{1}{r}\right)\frac{\partial \Phi}{\partial \theta} = -U \sin\theta$$

Obviously a point of stagnation occurs at $\theta = \pi$ and $r = q/2\pi U$. An equation may be obtained for the line $\Psi = q/2$ that passes through the stagnation point.

$$r_b = (q/2U \sin\theta)(1 - \theta/\pi)$$

The stream line divides the flow into two regions. The flow from the source is confined to the semi-infinite space BAC, and the flow due to the uniform flow remains external to this space. We may as well replace the stream line by a solid boundary, profile of which is defined by the foregoing equation for r_b. The combination of a source and uniform flow will then represent the flow about a semi-infinite body called the Rankine half-body. The width of the body at any point is

$$2y_b = 2r_b \sin\theta = \left(\frac{q}{U}\right)(1 - \theta/\pi)$$

and as θ approaches zero for large values of r_b, the final width tends asymptotically towards (q/U). The pressure distribution on the surface of the half-body may be found by the application of the Bernoulli theorem. If p_0 is the stagnation pressure at A, the pressure p at any point (r, θ) is found by the expression

$$p_0 - p = \frac{1}{2}\rho\left(u_r^2 + v_\theta^2\right) = \frac{1}{2}\rho\left(U^2 + (Uq\cos\theta)/\pi r_b + q^2/4\pi^2 r_b^2\right)$$

Substituting for r_b, we get

$$p_0 - p = \frac{1}{2}\left(U^2 - (U^2 \sin^2\theta)/\theta + (\sin^2\theta)/\theta^2\right)$$

$$(p_0 - p)/\frac{1}{2}\rho U^2 = \left\{1 - (\sin 2\theta)/\theta + (\sin^2\theta)/\theta^2\right\}$$

5.4.3 Source, Sink, and Uniform Flow

The velocity potential and stream function for the combination of source and sink of equal strength placed at equal distances from the origin in a uniform stream (as shown in Fig. 5.9) are

$$\Phi = Ur\cos\theta + m\ln r_2 - m\ln r_1$$

$$\Psi = Ur\sin\theta + m\theta_2 - m\theta_1$$

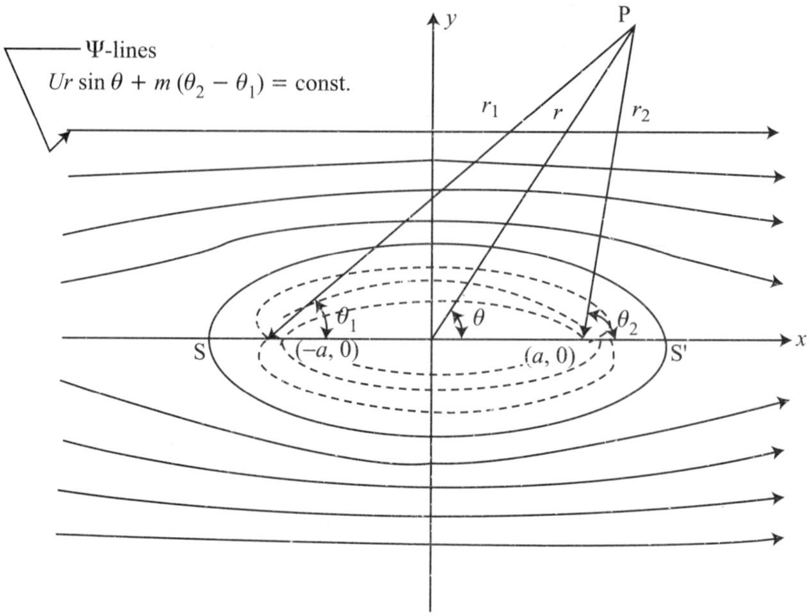

Fig. 5.9 Combination of a source, sink, and uniform flow

There are two stagnation points along the x-axis, and the equation to the stream line passing through the stagnation points is

$$Ur_b \sin\theta + m(\theta_2 - \theta_1) = 0$$

where r_b indicates as before the radius of the body. The oval-shaped body is called a Rankine body. Various shapes of bodies may be obtained by choosing different values for U, m and the distance $2a$ between the source and sink. It is interesting to note that the internal flow inside the shape of the body also satisfies the equation of continuity in that whatever discharge emanates from the source finds its way into the sink as they cannot flow across the stream line defining the shape of the body.

Uniform flow and doublet

A particular case of the above combination is when the source and sink merge together to form a doublet of strength μ. The velocity potential and stream function values will then become

$$\Phi = Ur\cos\theta + (\mu\cos\theta)/r$$

$$\Psi = Ur\sin\theta - (\mu\sin\theta)/r$$

The body shape represented by the stream line $\Psi = 0$ is given by $r^2 = \mu/U$. Since U and μ are constants, this shape represents a circle of radius $b = (\mu/U)^{1/2}$.

The flow about a circle or a cylinder placed with its axis normal to the uniform flow direction is given by the combination of a doublet and uniform flow (Fig. 5.10). The tangential velocity at the surface of the body is

$$(v_\theta)_b = \left(\frac{1}{r}\right)\frac{\partial\Phi}{\partial\theta} = -U\sin\theta - \mu\sin\theta/r_b^2 = -U\sin\theta - (\mu\sin\theta)/(\mu/U) = -2U\sin\theta$$

The pressure around the body is given by $p_0 - p = 1/2\rho U^2\sin^2\theta$, where p_0 is the stagnation pressure. Since the flow is symmetric, the resultant pressure on the body is zero in any direction. This is true only for ideal fluids and is known as the d'Alembert's paradox. In real fluids there will be a force exerted on the body by the flow which we shall discuss in later chapters.

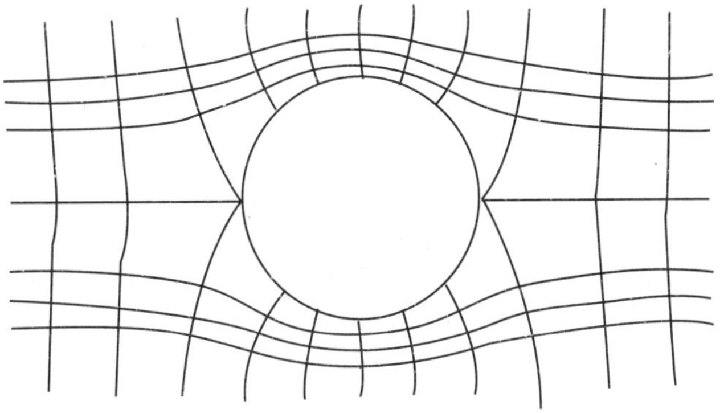

Fig. 5.10 Flow around a cylinder (doublet + uniform flow)

Doublet, uniform flow and vortex

The flow around a cylinder with circulation is obtained by the superposition of three elementary fields of flow, namely uniform flow, doublet and a vortex. For illustration, let the vortex have an anti-clockwise rotation.

$$\Phi = Ur\cos\theta + (\mu\cos\theta)/r + \left(\frac{\Gamma}{2\pi}\right)\theta$$

$$\Psi = Ur\sin\theta - (\mu\sin\theta)/r - \left(\frac{\Gamma}{2\pi}\right)\ln r$$

For a vortex with a clockwise rotation, the sign before the last term on the right-hand side in each of the expressions above is altered. From the above expressions, it is seen that when

$$r = \left(\frac{\mu}{U}\right)^{1/2}, \Psi = \left(\frac{\Gamma}{2\pi}\right)\ln\left(\frac{\mu}{U}\right)^{1/2} = \text{constant}$$

In other words, Ψ = constant along a circle of radius $b = (\mu/U)^{1/2}$. Note that Ψ is not equal to zero. The flow patterns are illustrated in Fig. 5.11. The effect of the vortex is to shift the stagnation points away from the axis of symmetry. The stagnation points and the pressure fields are best described by considering the velocity components in two orthogonal directions. For a clockwise circulation of the vortex, the velocity distributions are given by

$$u_r = \frac{\partial \Phi}{\partial r} = U \cos\theta - (\mu \cos\theta)/r^2$$

$$v_\theta = \left(\frac{1}{r}\right)\frac{\partial \Phi}{\partial \theta} = -U \sin\theta - \frac{(\mu \sin\theta)}{r^2} - \frac{\Gamma}{(2\pi r)}$$

At the stagnation point, $u_r = v_\theta = 0$ and the radial velocity $u_r = 0$ at $\theta = \pm\pi/2$ or along the circle of radius, $b = (\mu/U)^{1/2}$.

The tangential velocity is equal to zero when

$$v_\theta = 0 = \sin\theta\left(U + \mu/r^2\right) + \frac{\Gamma}{2\pi r}$$

If the stagnation points occur on the cylinder of radius $b = (\mu/U)^{1/2}$, then the positions of these points are defined by the angles

$$\sin\theta = -\Gamma/4\pi(\mu U)^{1/2} \quad \text{or} \quad \theta = \sin^{-1}[-\Gamma/4\pi(\mu U)^{1/2}]$$

There are two stagnation points on the cylinder itself when $\Gamma < 4\pi(\mu U)^{1/2}$ as shown in Fig. 5.11*a*. When $\Gamma = 4\pi(\mu v)^{\frac{1}{2}}, \theta = \frac{3\pi}{2}$, and there is only one single stagnation point at the lowest point of the cylinder as shown in Fig. 5.11*b*. However, when $\Gamma > 4\pi(\mu U)^{1/2}$, the

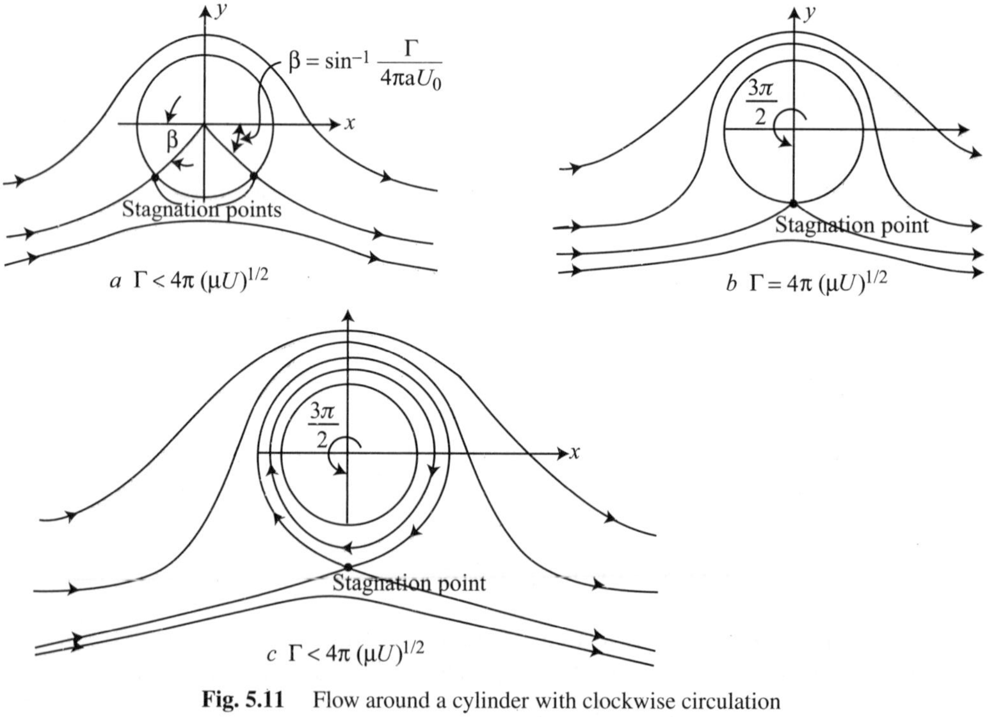

Fig. 5.11 Flow around a cylinder with clockwise circulation

stagnation can no longer remain on the surface of the cylinder. It moves away from it but always occurs at the same angle $\theta = 3\pi/2$ as shown in Fig. 5.11c. The radial distance of the point of stagnation is then obtained by substituting for $\theta = 3\pi/2$ in the expression $v_\theta = 0$, viz.,

$$U + \mu/r^2 - \Gamma/2\pi r = 0$$

Lift on the cylinder

The pressure distribution in the flow field is obtained by the expression

$$p_0 - p = \frac{1}{2}\rho(u_r^2 + v_\theta^2)$$

$$p_0 - p = \frac{1}{2}\rho\left[U^2 cos^2\theta + (\mu^2\cos^2\theta)/r^4 - (2U\mu\cos^2\theta)/r^2 + U^2\sin^2\theta\right.$$
$$+(\mu^2\sin^2\theta)/r^4 + \Gamma^2/4\pi^2r^2 + (2\mu U\sin^2\theta)/r^2 + (2\mu\Gamma\sin\theta)/2\pi r^3$$
$$\left.+(2U\Gamma\sin\theta)/2\pi r\right]$$

The pressure distribution, p_0 around the cylinder of radius $r = b = (\mu/U)^{1/2}$ is

$$(p_0 - p_b) = \frac{1}{2}\rho\,(2U\sin\theta + \Gamma/2\pi b)^2$$

The component of pressure force in a direction normal to the direction of uniform flow is

$$L = -\int_0^{2\pi} p_b\sin\theta b d\theta = -\int_0^{2\pi}\left[p_0 - \frac{1}{2}\rho\,(2U\sin\theta + \Gamma/2\pi b)^2\right]b\sin\theta d\theta$$
$$= -\int_0^{2\pi} p - \frac{1}{2}\rho(2U\sin\theta + \Gamma/2\pi b)^2 b\sin\theta\, d\theta$$

which after integration yields (since $\int_0^{2\pi}\sin\theta d\theta = \int_0^{2\pi}\sin^3\theta d\theta = 0$ and $\int_0^{2\pi}\sin^2\theta = \int_0^{2\pi}\frac{1}{2}(1-\cos 2\theta)d\theta$)

$$L = \rho U\Gamma \quad \text{per unit length of cylinder}$$

If the circulation is anti-clockwise, the lift force will be acting downwards. It is significant to note that the lift on the body is a product of the density of the fluid, the magnitude of the uniform velocity and the strength of the circulation around the body. The diameter of the body has not figured in the expression of the lift. The expression for lift though obtained for a cylinder, is independent of the shape or size of the body and is true for all two-dimensional bodies placed in an incompressible fluid. If the flow is symmetrical with respect to the y-axis, the force in the x-direction of flow is zero or in other words, the drag force in the direction of the flow is zero.

An important application of the lift induced due to circulation may be found in the flow about a lifting aerofoil or the wings of an aircraft. A table tennis ball may be observed to describe a curved path due to the spin imparted to it. Rotating cylinders were earlier used on small ships in place of sails.

Exercise

5.1. Show that the circulation about a doublet is zero.

5.2. A sink and a source of equal strength m are located at $(-a, 0)$ and $(a, 0)$ respectively.

(a) Show that the stream function, ψ passing through a point (x, y) is given by

$$\tan \frac{\psi}{m} = \frac{2ay}{x^2 + y^2 - a^2}$$

(b) If both are sources of equal strength m, find the values of ϕ and ψ.

5.3. Given the following values for the stream function, show that each of them satisfies Laplace's equation and obtain the velocity potential in each case. Sketch the stream lines and velocity potential lines.

(a) $\Psi = xy$
(b) $\Psi = x^2 - y^2$
(c) $\Psi = Ar^{\pi/\alpha} \sin(\pi\theta/\alpha)$

5.4. (a) Given $\Psi = x^3 - 3xy^2$, find ϕ.
(b) If $\Phi = x^2 - y^2$, find Ψ. Discuss the solution in the light of those obtained in Problems 5.3(*a*) and 5.3(*b*).

5.5. Two vortices having the same strength but opposite directions of rotation are placed at a distance b apart, shown in sketch (*a*). Show that the the induced velocity of one vortex acting upon the other vortex will cause the vortices to propagate through the fluid at a constant velocity. Find the magnitude of the velocity of propagation. What will happen if the vortices spin in the same direction (sketch (*b*))? If there are two pairs of vortices existing parallel to each other as shown in sketch (*c*), describe their behaviour.

5.6. Discuss the case of the combination of a source of strength $2m$ at $(-a, 0)$ and a sink at $(a, 0)$ of strength m placed in a uniform flow of velocity U. Find the stagnation point(s), for $U = 2$ m/s, $m = 2$ m^3/s/m, $a = 2$ m and the flow conditions downstream of the sink.

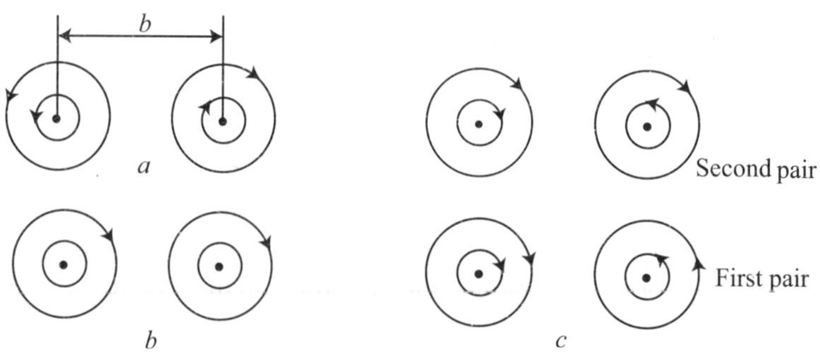

Figure for Problem 5.5

5.7. A source of strength m is located at $(-a, 0)$ from a wall situated along the y-axis. What is the total pressure force on the wall. The pressure at infinity and behind the wall is p_o. Hint: The wall may be replaced by an imaginary source located at $(a, 0)$.

5.8. What should be the strength of a doublet to simulate a flow of air at 15 m/s around a cylinder of diameter 3 m. If a clockwise vortex of strength 20 m^2/s is also superposed, find (a) the lift on the cylinder ($\rho = 1.236\,kg/m^3$). (b) Locate the stagnation points.

5.9. A circular cylinder of diameter 2 m rotates in an anti-clockwise direction at 300 rpm in air moving at a speed of 150 m/s. (a) What is the lift force on the cylinder per metre length. (b) Determine the location of the stagnation points. (c) At what speed should the cylinder rotate so that a single stagnation point occurs on the cylinder at $\theta = \pi/2$.

5.10. Find the stagnation points and lift on a cylinder of 50 cm radius placed in a uniform stream of air of 5 m/s and a clockwise circulation of 10 m^2/s. (a) What should be the circulation to have a stagnation point at a radius of 60 cm. (b) If the lift force is fixed at 10 kN, find the required circulation. ($\rho = 1.236\,kg/m^3$)

5.11. A clockwise vortex with circulation Γ is located above a flat plate at $(0, a)$ placed in a uniform stream of velocity U. The leading edge of the plate is at the origin. The flow represents the flow past an oval shaped body. Show that the pressure at any point at a distance x from the leading edge is

$$p_\infty - p = \frac{\rho\Gamma^2 a^2}{2\pi^2(x^2 + a^2)^2} - \frac{\rho U \Gamma a}{\pi(x^2 + a^2)}$$

and the upward force on the plate is $L = (\rho\Gamma^2/4\pi a) - \rho U\Gamma$ where p_∞ is the pressure at $x = \pm\infty$ and below the plate.
Hint: Place an imaginary vortex at $(0, -a)$ to replace the plate.

5.12. For a Rankine body, show that the points of stagnation occur at $x = \pm a(q/\pi a U + 1)^{1/2}$. Find an expression for pressure distribution along the surface of the Rankine body. Find also the length and maximum width of the body.

5.13. A vortex pair is located at $(0, +a)$ and $(0, -a)$ with circulation $-\Gamma$ and $+\Gamma$ respectively in a uniform stream of velocity U. Obtain the stream function for the combination and determine the stagnation points. The stream line $\psi = 0$ for $(\Gamma/2\pi)Ua = 1$ is known as the Kelvin oval.

6 Applications of Bernoulli's Theorem and Modifications

6.1 Introduction

It has been pointed out that the Bernoulli theorem of constant energy is true for irrotational motion; in fact it was a precondition to derive the Bernoulli theorem. In practical flow conditions, the motion is more often than not rotational, mainly on account of viscosity and turbulence present in the flow. However the Bernoulli equation finds many practical applications in situations such as

a) rapidly accelerated flows such as flow in convergent pipes or channels. In rapidly accelerated flows, the flow net is often used to predict fairly accurate distributions of velocity and pressure;
b) flow around streamlined bodies where little or no flow separation takes place;
c) where energy losses are negligible, for example, over short stretches of the flow regime;
d) or where the energy losses can be estimated to a fair degree of approximation.

The momentum equation based on Newton's second law of motion is used more extensively in rotational motion where viscous resistance and turbulence take predominance and there are considerable losses of energy which cannot be estimated *a priori*. The application of the momentum theorem is discussed in Chapter 7.

In this chapter, we will discuss applications of the Bernoulli theorem for simple cases of flow such as free jets, flow through orifices and mouthpieces, venturimeters, sluice gates, weirs etc.

6.2 Free Liquid Jet

Let us consider a jet issuing from a nozzle inclined to the horizontal as shown in Fig. 6.1. We will consider the datum as passing through the exit end of the nozzle.

Let the velocity of the jet issuing from the nozzle be V_0. Its horizontal and vertical components will be $V_0 \cos \theta_0$ and $V_0 \sin \theta_0$ respectively. The Bernoulli theorem applied to points, one at the nozzle exit end and at any point (x, y) on the trajectory of the jet is

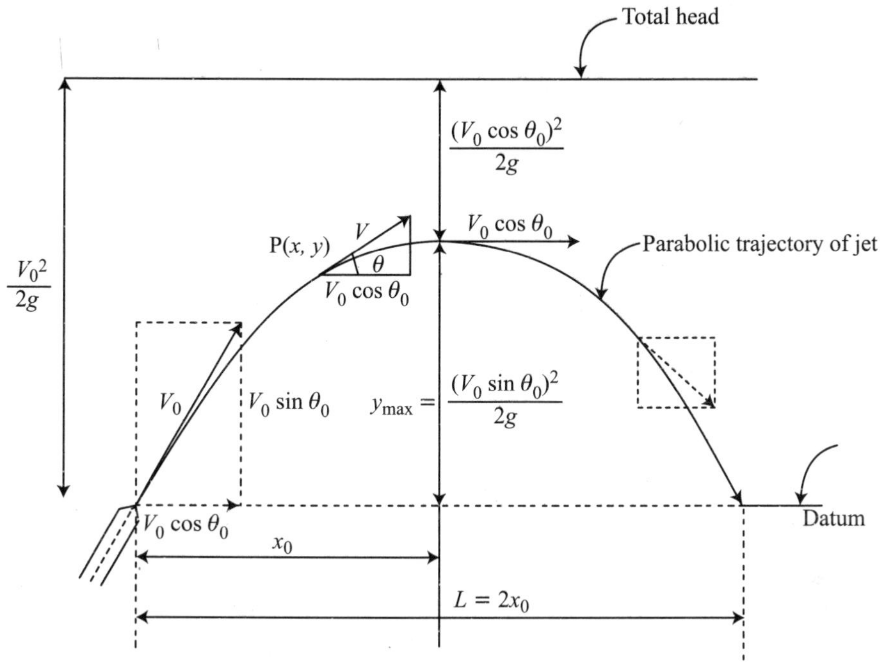

Fig. 6.1 Inclined jet

$$p_a/\gamma + 0 + V_0^2/2g = p_a/\gamma + y + (V_{x,y})^2$$

$$(V_{x,y})^2 = (V_{x,y} \cos \theta_{x,y})^2 + (V_{x,y} \sin \theta_{x,y})^2$$

But the horizontal component remains the same throughout the trajectory of the jet, provided the air resistance is zero, as there are no horizontal forces. Hence we have $V_{x,y} \cos \theta_{x,y} = V_0 \cos \theta_0$. But $V_{x,y} \sin \theta_{x,y}$ goes on decreasing as the jet rises higher due to gravitational effects. The Bernoulli equation can be rewritten as

$$\frac{(V_0 \sin \theta_0)^2}{2g} = y + (V_{x,y} \sin \theta_{x,y})^2 = \text{constant}$$

At the zenith of its trajectory, the vertical component is zero. Hence the maximum height to which the jet rises is given by

$$y_{max} = \frac{(V_0 \sin \theta_0)^2}{2g}$$

The horizontal distance travelled by the jet in time t is $x = (V_0 \cos \theta_0)t$ and the vertical distance travelled by the jet in the same time is

$$y = (V \sin \theta)t - \frac{1}{2}gt^2$$

Eliminating t from the two relationships, we get

$$y = x \tan\theta_0 - \frac{1}{2}g\left(\frac{x}{V_0 \cos\theta_0}\right)^2 \tag{6.1}$$

The trajectory is therefore a parabola. The distance x_0 travelled by the jet to reach its maximum height $y_{\max}$ is

$$\frac{(V_0 \sin\theta_0)^2}{2g} = x_0 \tan\theta_0 - \frac{1}{2}g\left(\frac{x_0}{V_0 \cos\theta_0}\right)^2$$

which on simplification yields

$$x_0 = \frac{(V_0^2 \sin 2\theta_0)}{2g} \tag{6.2a}$$

and the horizontal distance L where the downward trajectory passes the datum line is

$$L = 2x_0 = \frac{(V_0^2 \sin 2\theta_0)}{g} \tag{6.2b}$$

For a given initial velocity of the jet, the horizontal distance travelled by the jet is a maximum when $\theta = 45°$.

EXAMPLE 6.1 A nozzle of 10 cms diameter inclined at an angle of 30° above the horizontal discharges 0.05 cu.m of liquid. Determine the maximum height to which it raises and its horizontal travel at the nozzle level. If the jet was inclined at 45°, calculate the corresponding distances.

Velocity of the jet $= V_0 = 0.05 \times 100^3/(\pi\, 10^2/4) = 636.62$ cm/s
The maximum elevation to which the jet will rise is

$$y_{\max} = \frac{(V_0 \sin\theta_0)^2}{2g} = \frac{(636.62 \times 0.5)^2}{2 \times 981} = 51.64 \text{ cms}$$

The horizontal distance it will travel is $L = V_0^2 \sin 2\theta_0/g = (636.62^2 \times 0.866)/981 = 357.77$ cms

If the jet were inclined at 45°, $y_{\max} = (636.62 \times 0.707)^2/2 \times 981 = 103.25$ cms and the horizontal distance $L = (636.62^2 \times 1)/981 = 413.13$ cms

EXAMPLE 6.2 If the jet in the above example issues horizontally ($\theta_0 = 0$), and at some point in the path of the jet, the jet is inclined at 45° to the horizontal, find the area of cross-section of the jet at that point. Find the coordinates of this point.

The horizontal velocity remains unaltered, $V_0 = 636.62$ cms/s. Hence at the point where the jet velocity vector is inclined to the horizontal at 45°, the vertical velocity component is also $V_0 = 636.62$ cms/s and the resultant jet velocity is $\sqrt{2} \times 636.62 = 900.32$ cms/s.

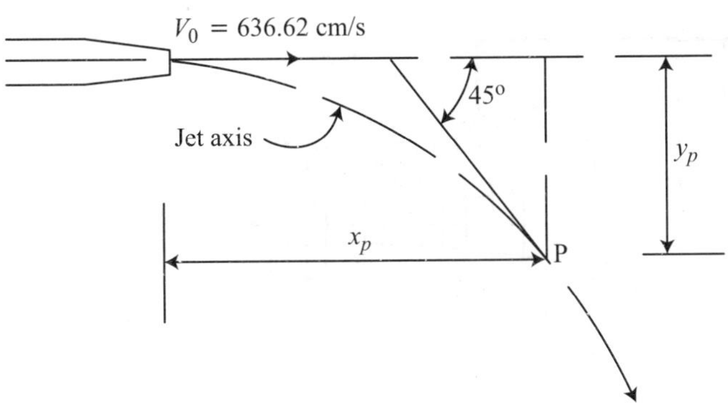

Sketch for Example 6.2

By equation of continuity, the diameter, d_p of the jet at P is

$$\left(\frac{\pi d_p^2}{4}\right) \times 900.32 = 0.05 \times 100^3 \quad \text{and} \quad d_p = 8.409\,\text{cms}$$

The trajectory of the jet is given by

$$y = \frac{1}{2}g\left(\frac{x}{V_0}\right)^2 \quad \text{and} \quad \frac{dy}{dx} = \frac{gx}{V_0^2}$$

At P the slope with $\theta = 45°$

$$\frac{dy}{dx} = 1 = \frac{gx}{V_0^2} \quad \text{and} \quad x = \frac{V_0^2}{g} = \frac{636.62^2}{981} = 413.13\,\text{cms}$$

$$y_p = \frac{1}{2}g\left(\frac{V_0}{g}\right)^2 = \frac{1}{2}\frac{V_0^2}{g} = 206.57\,\text{cms}$$

6.3 Efflux from an Opening at the End of a Pipe

If one draws the flow net for a jet issuing from a restricted opening or orifice at the end of a pipe, it will become evident that to satisfy the requirement of a flow net, the jet contracts after leaving the end and at a certain distance from the exit end, the stream lines become parallel. From a practical point of view also, it is seen that the boundary stream line PQR cannot physically turn through 90°, as it would mean creation of a singular point at R with zero radius of curvature or infinite curvature which would require infinite force. Nature avoids the anomaly (inertial effects play the role) by changing the direction gradually which leads to the contraction. In Fig. 6.2*a*, a typical flow net through a two-dimensional orifice at the end of a pipe is shown with a contraction from an initial width B to a contracted width b. The ratio of the area at the contracted section at which the stream lines become parallel to the area of the opening is known as the coefficient of contraction, C_c.

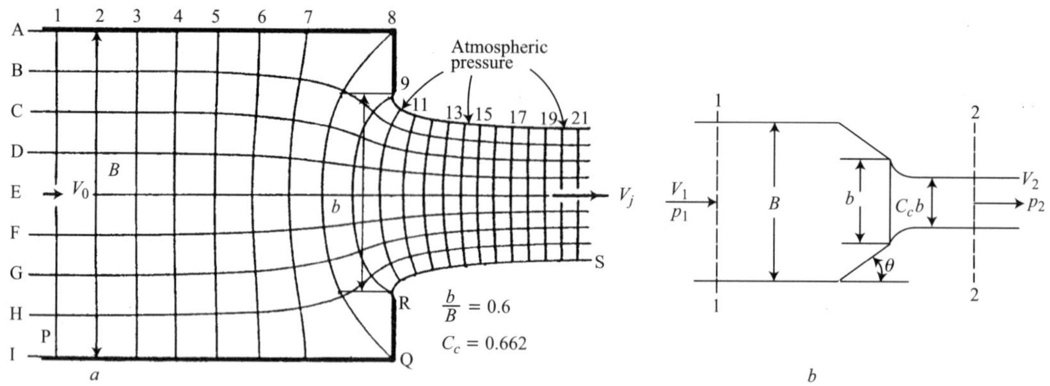

Fig. 6.2 (a) Flow net for an efflux from an pipe orifice; (b) pipe nozzle

Kirchoff showed that the coefficient of contraction for a two-dimensional irrotational flow from a very large container was theoretically equal to $\pi/(\pi+2) = 0.611$. Later von Mises found the values of C_c for the two-dimensional boundary form shown in Fig. 6.2*b* and reproduced in Table 6.1.

Table 6. 1 Values of C_c for different values of b/B and θ

b/B	$\theta = 45°$	$\theta = 90°$	$\theta = 135°$	$\theta = 180°$
0.0	0.746	0.611	0.537	0.500
0.1	0.747	0.612	0.546	0.513
0.2	0.747	0.616	0.555	0.528
0.3	0.748	0.622	0.566	0.544
0.4	0.749	0.631	0.580	0.564
0.5	0.752	0.644	0.599	0.586
0.6	0.758	0.662	0.620	0.613
0.7	0.768	0.687	0.652	0.646
0.8	0.789	0.722	0.698	0.691
0.9	0.829	0.781	0.761	0.760
1.0	1.000	1.000	1.000	1.000

Application of the energy theorem to two sections, Sections 1-1 and 2-2, as shown in Fig. 6.2*b* will yield

$$p_1 - p_2 = \frac{1}{2}\rho\left(V_2^2 - V_1^2\right)$$

and the equation of continuity gives $V_1 B = V_2 C_c b$

From these two equations we can eliminate V_1 and get the expression for

$$V_2 = \frac{1}{\sqrt{1 - C_c^2\left(\dfrac{b}{B}\right)^2}}\sqrt{\frac{2(p_1 - p_2)}{\rho}} \tag{6.3}$$

The discharge q can be obtained by multiplying the preceding equation by the area at the vena contracta, $C_c b$. We may rewrite the discharge equation as

$$q = C_d b \left(\frac{2\Delta p}{\rho}\right)^{1/2} \tag{6.4}$$

where C_d is the discharge coefficient replacing $1/[1 - C_c^2(b/B)^2]^{1/2}$

A similar equation can be obtained for a circular orifice of diameter d at the end of a circular pipe of diameter D. The equation of continuity for the pipe orifice is $(\pi D^2/4)V_1 = C_c[\pi d^2/4]V_2$. This is used to eliminate V_2 from the energy equation to obtain the expression for discharge.

$$Q = C_d \left(\frac{\pi d^2}{4}\right)\left(\frac{2\Delta p}{\rho}\right)^{1/2} \tag{6.5}$$

The discharge coefficient now becomes

$$C_d = \frac{C_c}{\left[1 - C_c^2\left(\frac{d}{D}\right)^4\right]^{1/2}}$$

6.4 Venturimeter

A venturimeter is used to measure the discharge in a pipe flowing full. It consists of a converging portion followed by a throat portion of constant diameter and then a diverging section. Both the converging and diverging transitions are gradual and it is assumed that the flow lines in the throat section are parallel and consequently $C_c = 1.0$. Pressure-tappings are made, one before the converging transition starts and the other at the throat as shown in Fig. 6.3.

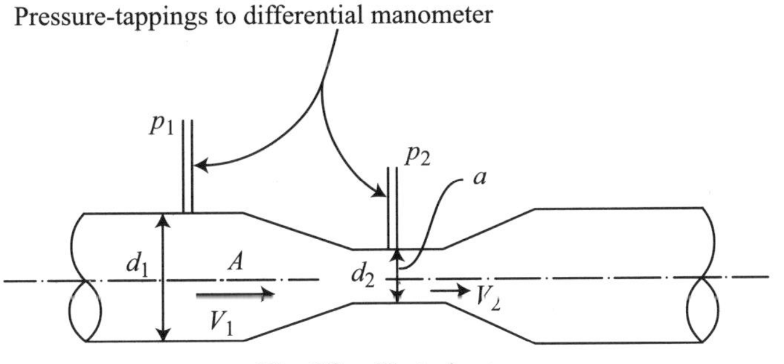

Fig. 6.3 Venturimeter

It is assumed that there is no energy loss between Sections 1-1 and 2-2. By applying the Bernoulli theorem for these two sections

$$p_1 - p_2 = \Delta p = \frac{1}{2}\rho(V_2^2 - V_1^2)$$

and the equation continuity yields $A \cdot V_1 = aV_2$, where A and a are the area of the venturimeter before the convergence and the throat respectively (Sections 1-1 and 2-2). Substituting for V_1, we get after dividing by $\gamma = \rho g$

$$2g\left(\frac{\Delta p}{\gamma}\right) = V_2^2(1 - a^2/A^2)$$

and the discharge is

$$Q = aV_2 = \frac{a}{\sqrt{1-\left(\frac{a}{A}\right)^2}}\sqrt{2gH} \qquad (6.6)$$

where $H = \Delta p/\gamma$ is the differential pressure head between the Sections 1-1 and 2-2. Certain losses are inevitable due to the viscosity of the fluid and a coefficient C_d generally in the range of 0.91 to 0.99 is introduced after due calibration of the venturimeter. The shape of the venturimeter is standardised by the Indian Standards Institute and manufacturers follow these guidelines and also supply calibration charts.

EXAMPLE 6.3 A venturimeter is installed in a 15 cms dia pipe, with an inclination as shown in the sketch. The throat diameter is 7.5 cms. The pressure tappings located at the upstream end and the throat are connected to a U-tube manometer with mercury as the manometer fluid. If the mercury deflection is measured as 40 cms, find the discharge.

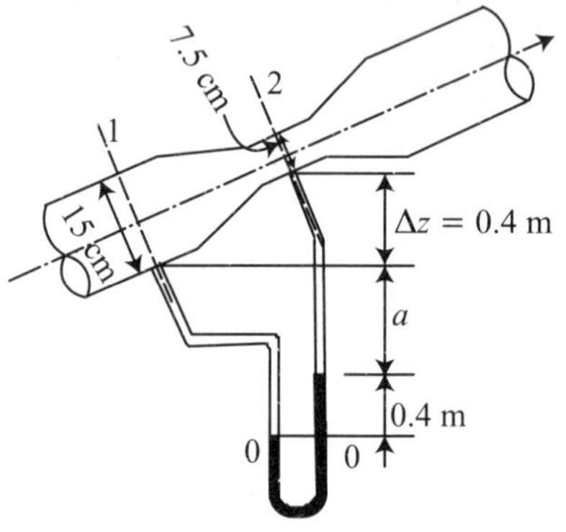

Figure for Example 6.3

The Bernoulli equation applied to the two sections 1-1 and 2-2 is

$$p_1/\gamma + V_1^2/2g = p_2/\gamma + V_2^2/2g + \Delta z$$

From the equation of continuity

$$V_2 = \left(\frac{D}{d}\right)^2 V_1 = 4V_1$$

The manometer equation gives (equating the pressures in both limits at 0-0)

$$p_1/\gamma + a + 0.4 = p_2/\gamma + \Delta z + a + 0.4 \times 13.6$$

Hence $$p_1/\gamma - p_2/\gamma - \Delta z = 5.44 - 0.4 = 5.04\,\text{m} = 15V_1^2/2g$$

$$V_1^2 = \frac{(2 \times 9.81 \times 5.04)}{15}$$

$$V_1 = 2.5676\,\text{m/s}$$

The discharge $Q = (\pi D^2/4)V_1 = [\pi(0.15)^2/4]2.5676 = 0.04537\,\text{m}^3/\text{s}$

$Q = 45.37$ litres/s

6.5 Orifice Meter

An orifice meter is also a discharge-measuring device. An orifice plate with a symmetrical circular opening is fitted between the flanges of a pipe line. A sharp-edged orifice plate is shown in Fig. 6.4*a* while Fig. 6.4*b* shows the commercially available version.

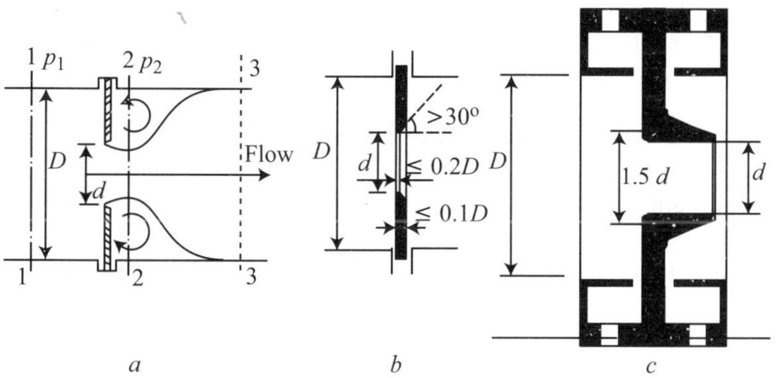

Fig. 6.4 Working of an orifice meter: (a) orifice plate; (b) orifice meter; (c) flow nozzle

The pressure-tapping upstream of the orifice plate is usually made at a distance D from the location of the plate but it may vary between $(D - d)$ to $2(D - d)$. D is the diameter of the main pipe and d is the diameter of the orifice opening. The general practice for the location of the downstream tapping is at a distance of $d/2$, though this may also vary in the range of $d/2$ to d. As shown in the figure, the flow separates from the sharp edge of the orifice and a vena contracta of area $C_c a (a = \pi d^2/4)$ is formed at a distance of approximately $0.5d$ downstream of the plate. Application of the Bernoulli equation between Sections 1-1 and 2-2 yields

$$V_1^2/2g + p_1/\gamma + z_1 = V_2^2/2g + p_2/\gamma + z_2$$

Equation of continuity: $AV_1 = C_c a V_2 = Q$

$$Q^2/2gC_c^2a^2 - Q^2/2gA^2 = \left(\frac{p_1}{\gamma} + z_1\right) - \left(\frac{p_2}{\gamma} + z_2\right) = H$$

where H is the piezometric head difference between the two points. Simplification of the above equation gives the discharge, Q

$$Q = \frac{C_c a}{\sqrt{1 - \left(\frac{C_c a}{A}\right)^2}} \sqrt{2gH} = C_d a (2gH)^{1/2} \tag{6.7}$$

The discharge coefficient, C_d is to be calibrated for each individual orifice plate but generally depends upon the ratio d/D and the Reynolds number, $\mathbf{R} = \rho V D/\mu$. There is flow separation downstream of the orifice and an eddy is formed around the issuing jet. There will be a head loss between Sections 1-1 and 3-3 which is a disadvantage of the orifice meter. The head loss cannot be predicted by the Bernoulli equation but can be estimated by the momentum theorem which is presented in the next section. The discharge coefficient C_d varies with the ratio of d/D (or a/A) and Reynolds number $\mathbf{R}$. However, at large values of $\mathbf{R} > 10^5$, the value of C_d depends only on d/D or a/A. For $a/A = 0.25, 0.375$ and 0.50, the values of C_d are 0.628, 0.662, and 0.720 respectively.

The flow nozzle shown in Fig. 6.4*c* has a coefficient of contraction of 1.0 but a correction coefficient for velocity C_v, varying between 0.94 and 1.18 depending on the Reynolds number and the area ratio a/A, is introduced and the discharge calculated.

6.6 Discharge from a Tank through Outlet Fittings

We will first consider a tank or reservoir in which the depth of water remains constant. The orifice is fitted in a side wall as shown in Fig. 6.5 and the head of water over the centre line of the orifice is constant. The outlet fittings can be of different shapes and sizes. The simplest is a sharp-edged

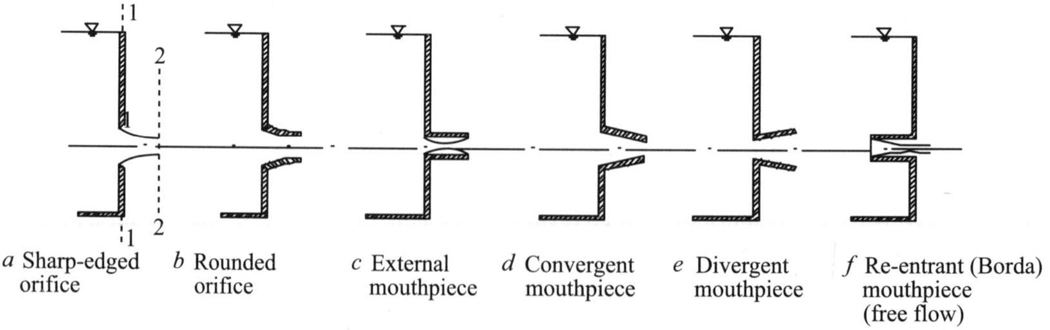

Fig. 6.5 Different types of reservoir outlets

orifice of area a. Other shapes commonly used are also shown in Fig. 6.5a–f. We will assume that these outlets are discharging freely into the atmosphere. For the sharp-edged orifice there is a vena contracta formed at Section 2-2 as shown in Fig. 6.5a. It is also assumed that the area of the tank or reservoir is so large compared to the area of the orifice that the velocity head at 1-1 is considered as zero. The Bernoulli equation applied to Sections 1 and 2 yields

$$H + 0 + 0 = 0 + V^2/2g + 0 \quad \text{or} \quad V = (2gH)^{1/2}$$

The velocity is however reduced due to friction and a coefficient of velocity C_v is introduced to modify the expression for velocity as $V = C_v(2gH)^{1/2}$

The area at the vena contracta is $C_c a$ and hence the discharge, Q is

$$Q = (C_c a)C_v(2gH)^{1/2} = C_c C_v a(2gH)^{1/2} = C_d a(2gH)^{1/2} \tag{6.8}$$

where $C_d = C_c C_v$ is known as the discharge coefficient.

Table 6.2 gives the values of the coefficients of contraction, and discharge for different types of orifices and mouthpieces. The values given are indicative as they vary slightly with the head of liquid over the outlet.

Table 6. 2 Values of C_c, C_v and C_d for different types of orifices

Type	C_c	C_v	C_d
Sharp-edged	0.62	0.98	0.61
Rounded	1.00	0.98	0.98
Straight	1.00	0.80	0.80
Borda flowing full	1.00	0.75	0.75
Borda flowing free	0.50	1.00	0.50

The values of C_c and C_v for a divergent mouthpiece depends on the angle of divergence and the tube length. For large angles of divergence, the flow separates from the boundary and may issue as a free jet or cling to one side of the mouthpiece. The values listed in the table are for free jets discharging into the atmosphere. For submerged (or drowned) orifices, the values of C_d will be slightly less and the head to be taken is the difference in levels of liquid on either side of the orifice or mouthpiece.

EXAMPLE 6.4 Show that the jets from two orifices in the side of a tank will intersect a plane through the base at the same distance if the head on the upper orifice is equal to the height of the lower orifice above the base. Assume $C_v = 1.0$ for both orifices.

The head on the lower orifice is $(a + b)$ and the head on the upper orifice is a. The horizontal velocity of the upper jet is $(2ga)^{1/2}$ and the horizontal velocity of the lower jet is $[2g(a + b)]^{1/2}$. The trajectory of the jet is given by $y = gx^2/V^2$.

For the lower jet $y = a$; therefore, $x = [2a(a + b)]^{1/2}$

For the upper jet, $y = a + b$ and the distance travelled is $x = [2a(a + b)]^{1/2}$

Hence the two jets meet at the same point.

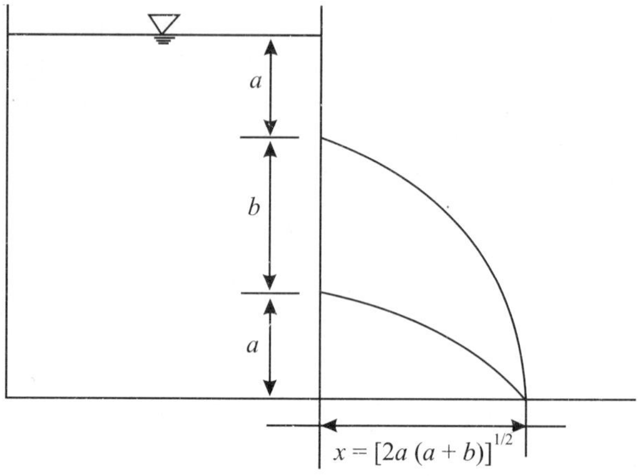

Figure for Example 6.4

6.7 Large Vertical Orifice

If the vertical dimension of the orifice is large compared to the head over the orifice, then we cannot assume that the velocity across the jet issuing from the orifice is uniform. The variation of head at different vertical sections of the orifice must be accounted for.

Figure 6.6 shows a large vertical rectangular orifice of width B in the side of a tank. Let the height of the liquid level be H_1 above the top of the orifice and let H_2 be the head above the lower edge of the orifice.

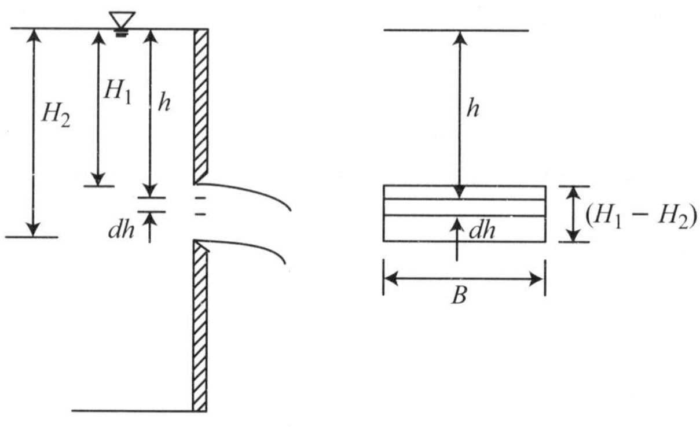

Fig. 6.6 Large orifice

The velocity through an elementary strip of depth dh (assumed small compared to the head above it and velocity is considered constant across this depth) located at a depth h below the free surface is $V = C_v(2gh)^{1/2}$. The elementary discharge through this strip is

$$dQ = B\, dh C_c C_v (2gh)^{1/2}$$

The total discharge is obtained by integrating the above between the limits H_1 and H_2.

$$Q = C_d B\sqrt{2g}\int_{H_1}^{H_2} h^{1/2}dh = 2/3C_d B\sqrt{2g}[h^{3/2}]_{H_1}^{H_2} = 2/3C_d B\sqrt{2g}[H_2^{3/2} - H_1^{3/2}] \quad (6.9)$$

The value of C_v may be taken as unity and $C_d = C_c$

If the velocity of approach, U_0 into the tank is taken into account, the heads over the top and bottom of the orifice are to be modified as

$$H_1' = (H_1 + U_0^2/2g)$$

and

$$H_2' = (H_2 + U_0^2/2g)$$

6.8 Time of Emptying a Tank

Suppose we have to determine the time required to lower the water level from H_1 to H_2, when there is no inflow into the tank through the discharge from an orifice or mouthpiece of area a. We will assume C_d to be constant over the varying heads. We shall first consider a tank of constant area A. Let the head above the orifice in the tank at time t be h. In time dt let the fall in level be dh. Then the volume discharged in time dt is $A\ dh$ (Fig. 6.7).

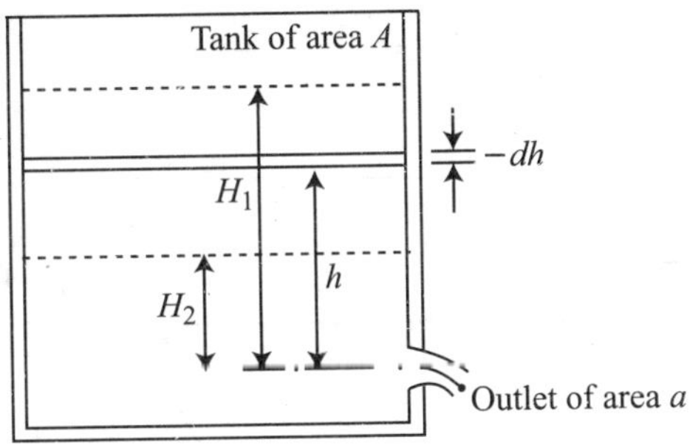

Fig. 6.7 Time taken to empty a tank

The rate of discharge through the orifice under a head h is $Q = C_d a(2gh)^{1/2}$. Hence the volume discharged through the orifice in time dt is

$$Q\ dt = C_d a(2gh)^{1/2} dt = -A\ dh$$

The minus sign appears since the water level is decreasing with time.

Transposing the terms, we get

$$dt = \frac{-A\ dh}{C_d a\sqrt{2gh}}$$

which when integrated over the heads H_1 and H_2 gives the required time T

$$\int_0^T dt = -\frac{A}{C_d a\sqrt{2g}}\int_{H_1}^{H_2} h^{-1/2}dh$$

$$T = \frac{2A}{C_d a\sqrt{2g}}\left[H_1^{1/2} - H_2^{1/2}\right] \tag{6.10}$$

If the tank is completely emptied, $H_2 = 0$, then

$$T = \frac{2A}{C_d a\sqrt{2g}}H_1^{1/2}$$

If A is not a constant but varies as a known function of h, then A is also to be included in the integration process. For example if A varies linearly with h, say $A = a + kh$

$$\int_0^T dt = T = -\frac{1}{C_d a\sqrt{2g}}\int_{H_1}^{H_2} (a + kh)h^{-1/2}dh$$

$$= \frac{1}{C_d a\sqrt{2g}}\left[2a(H_1^{1/2} - H_2^{1/2}) + \frac{2k}{3}(H_1^{3/2} - H_2^{3/2})\right] \tag{6.11}$$

EXAMPLE 6.5 Find the time taken to empty a hemispherical vessel of radius R through an orifice of area a located at the bottom.

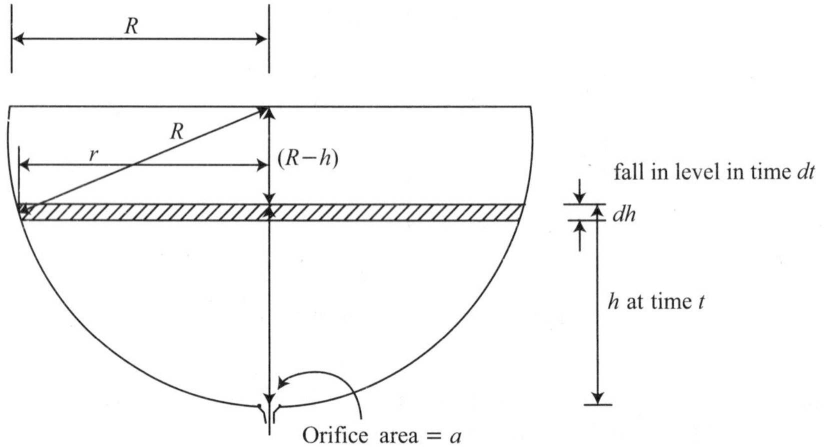

Figure for Example 6.5

As before, let the depth of water above the orifice level be h at time t. Since the instantaneous discharge is $Q = C_d a(2gh)^{1/2}$, the volume discharged in time dt is Qdt. Let the fall in level be dh in time dt. From the geometry of the figure, the radius r at height h above the base is

$$r^2 = R^2 - (R - h)^2 = 2Rh - h^2$$

and the volume of liquid drained in time dt is

$$\pi r^2 dh = \pi(2Rh - h^2)dh$$

Equating the two expressions for the volume discharged in time dt, we get

$$dt = -\frac{(2Rh - h^2)dh}{C_d a\sqrt{2gh}} = -\frac{(2Rh - h^2)h^{-1/2}dh}{C_d a\sqrt{2g}}$$

which on integration yields

$$T = \frac{\pi}{C_d a\sqrt{2g}}\int_{H_1}^{H_2}(2Rh^{1/2} - h^{3/2})dh = -\frac{\pi}{C_d a\sqrt{2g}}\left[\frac{4}{3}Rh^{3/2} - \frac{2}{5}h^{5/2}\right]_{H_1}^{H_2}$$

$$T = \frac{2\pi}{C_d a\sqrt{2g}}\left[\frac{2}{3}R(H_1^{3/2} - H_2^{3/2}) - \frac{1}{5}(H_1^{5/2} - H_2^{5/2})\right] \tag{6.12}$$

If the tank is initially full and completely emptied $H_1 = R$ and $H_2 = 0$

$$T = \frac{14\pi R^{5/2}}{15C_d a\sqrt{2g}} \tag{6.13}$$

6.9 Transfer Time from One Tank to Another

Liquid flows from one tank of area A_1 with a higher level to an adjacent tank of area A_2 with a lower level through an orifice of area a situated in the common wall of the two tanks as shown in Fig. 6.8. Let the difference in level initially be H_1 and after time T, the difference is H_2. It is required to find the value of T to lower the differential head from H_1 to H_2.

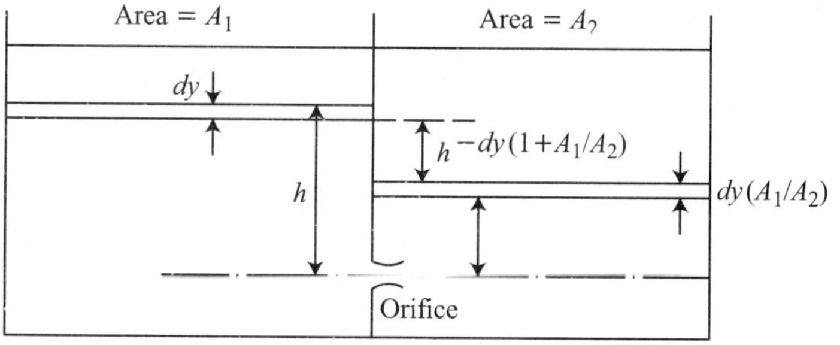

Fig. 6.8 Flow from one tank to another

The orifice is submerged and an appropriate value of C_d is chosen and is assumed constant over varying differential heads. Let the levels in the two tanks differ by h at any time t. If the level in tank 1 is lowered by dy in time dt, the volume depleted in the first tank is $A_1 dy$. Since this is transferred to the second tank, the corresponding rise in level in the second tank is $dy(A_1/A_2)$.

Hence the new difference in level is $h - dy - dy(A_1/A_2)$. The change in the head causing the flow is

$$dh = dy(1 + A_1/A_2)$$

or

$$dy = \frac{dh}{(1 + A_1/A_2)}$$

We now equate the volume of flow out of tank 1 in time dt to the discharge through the orifice in the same time.

$$-A_1 dy = -\frac{A_1 dh}{(1 + A_1/A_2)} = C_d a(2gh)^{1/2} dt$$

$$dt = -\frac{A_1 dh}{C_d a(1 + A_1/A_2)\sqrt{2gh}}$$

$$\int_0^T dt = T = -\frac{A_1}{C_d a(1 + A_1/A_2)\sqrt{2g}} \int_{H_1}^{H_2} h^{-1/2} dh = \frac{2A_1(H_1^{1/2} - H_2^{1/2})}{C_d a(1 + A_1/A_2)\sqrt{2g}} \tag{6.14}$$

Example 6.6 If the two tanks are of the same area $A_1 = A_2 = 5$ sq. m. and the area of the orifice is 50 sq.cms and $C_d = 0.62$, find the time required to bring an initial difference in level of 2 m between the two tanks to zero.

Since $A_1/A_2 = 1$ and $H_2 = 0$

$$T = (2 \times 5 \times 2^{1/2})/0.62 \times 0.005 \times 2 \times 19.62^{1/2}$$

$$= 514.96 \text{ seconds}$$

6.10 Inflow and Outflow

There is a constant inflow at the rate of Q m^3/s into a tank of area A sq.m. Liquid from the tank is also simultaneously being discharged at the rate of $q = C_d a(2gh)^{1/2}$ at a given instant t when the instaneous head over the orifice is h through an orifice of area a and discharge coefficient, C_d (Fig. 6.9).

The difference in discharges into and out of the tank accounts for the positive or negative storage in the tank. If $Q > q$, the level in the tank increases but if $Q < q$, then the tank level is depleted. If $Q = q$, then the tank level remains unaltered. Let the level in the tank increase by dh in time dt due to the difference in the values of Q and q. Then by the continuity equation

$$[Q - C_d a(2gh)^{1/2}]dt = A\, dh$$

If we write $C_d a(2g)^{1/2}$ = a constant k for convenience, then we have

$$dt = \frac{(A\,dh)}{(Q - kh^{1/2})}$$

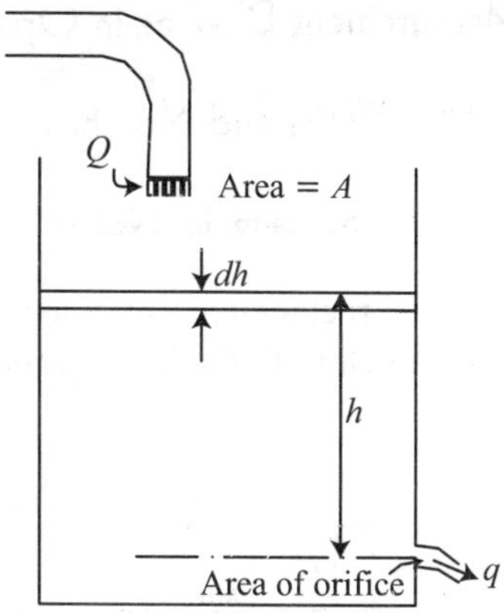

Fig. 6.9 Inflow and outflow

To facilitate the integration of the above equation we make another substitution $Q - kh^{1/2} = z$ or $h = (Q-z)^2/k^2$ which on differentiation with respect to z gives $dh = -[2(Q - z)dz]/k^2$. On substitution of this value for dh in the above equation, we get

$$dt = -\frac{[2A(Q - z)dz]}{k^2 z} = -\left(\frac{2A}{k^2}\right)\left[\left(\frac{Q}{z}\right) - 1\right]dz$$

Integration of this equation yields

$$\int_0^T dt = T = -\frac{2A}{k^2}(Q \ln z - z)$$

and we can substitute for z its original value.

$$T = -\frac{2A}{k^2}[Q\ \ln(Q - kh^{1/2}) - (Q - kh^{1/2})]_{H_1}^{H_2}$$

$$T = -\frac{2A}{k^2}\left[Q\ \ln\frac{(Q - k\sqrt{H_2})}{(Q - k\sqrt{H_1})} + k(\sqrt{H_2} - \sqrt{H_1})\right] \tag{6.15}$$

EXAMPLE 6.7 A tank of area 2.5 m^2 has an outlet 10 cm in dia. with a C_d value of 0.60. There is a constant inflow into the tank at the rate of 25 l/s. Determine (a) the depth above the outlet at which the level in the tank reaches an equilibrium condition and remains constant and (b) the time required to lower the level from 2 m to 1 m above the outlet.

A = area of tank = 2.5 m^2; Inflow Q = 25 l/s = 0.025 m^3/s; C_d = 0.60; $a = (\pi/4)$ 10^2 cm^2

or $a = 0.007854$ m^2; $k = 0.60 \times 0.007854 \times 19.62^{1/2} = 0.02087$

Equilibrium is established when the discharge from the orifice is equal to the inflow i.e., $Q = kh^{1/2}$ or $h = (Q/k)^2 = (0.025/0.02087)^2 = 0.7588$ m. To find the time required to lower the level from 2 m to 1.0 m

$$T = -2 \times 2.5/0.0287^2\left[0.025 \ln\left\{\frac{(0.025 - 0.02087 \times 1)}{(0.025 - 0.02087 \times 2^{1/2})}\right\} + 0.02087(1 - 2^{1/2})\right.$$

$$= -3035.12\left[\ln\left(\frac{0.0037}{0.0045}\right) + 0.02087 \times (-0.414)\right.$$

$$= -3035.12[(\ln 0.822) - 0.0118] = -3035.12(-0.1957 - 0.0118) = 629.8\ \text{s}$$

Measurement Devices in Open Channels

6.11 Weirs and Notches

6.11.1 Rectangular Notch

A typical rectangular notch located in the side wall of a tank is shown in Fig. 6.10. The breadth of the notch is B, the head of the bottom of the notch is H.

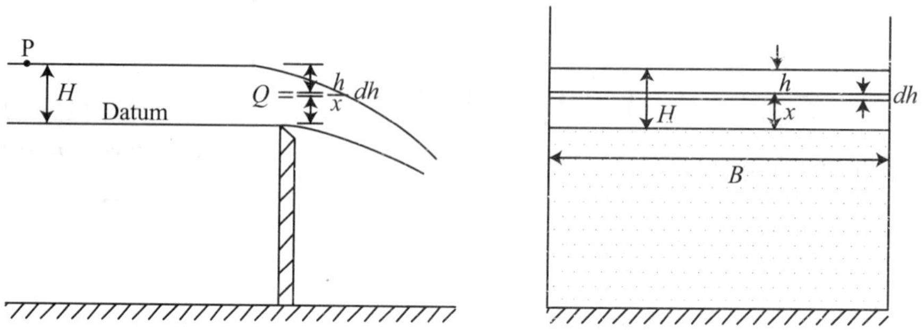

Fig. 6.10 Rectangular notch

Consider an elementary strip of width B and height dh located at a depth h below the free surface. Applying the Bernoulli theorem to the points P and Q with the bottom of the notch as the datum, we get (on the assumption that velocity at P is negligible)

$$0 + (x + h) + 0 = x + V^2/2g$$

or

$$V = (2gh)^{1/2}$$

and the discharge is $dQ = BV\,dh$

Hence the theoretical discharge is

$$Q = \int_0^H B(2gh)^{1/2} dh = \left(\frac{2}{3}\right) B(2g)^{1/2}(H)^{3/2}$$

However due to the contraction of the nappe and frictional effects, the discharge is less than the theoretical discharge and a discharge coefficient C_d is introduced to account for the losses. In the final form, the discharge is

$$Q = \frac{2}{3} C_d B\sqrt{2g} H^{3/2} \tag{6.16}$$

The coefficient of discharge is usually taken as 0.62

6.11.2 Sharp-crested Weir

In the previous example of a rectangular notch, it was implicitly assumed that the area of the tank was very large compared to the water area over the rectangular notch with the result that the velocity of water in the tank was assumed to be zero. In the case of a sharp-crested weir, we will consider that the height of the weir is of the same order as the depth in the channel in which the weir is constructed to measure the discharge. Hence we will have to consider the velocity of approach in the channel as shown in Fig. 6.11.

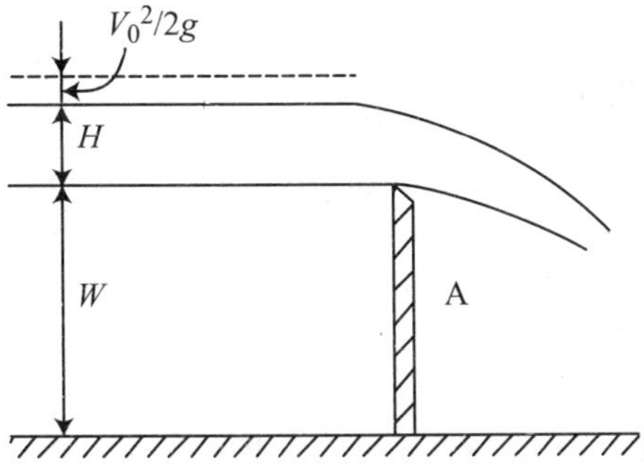

Fig. 6.11 Sharp-edged weir

The figure illustrates the shape of the falling nappe and it is assumed that the area below the nappe is fully ventilated so that the pressure on both sides of the nappe is atmospheric. Otherwise the falling nappe will remove the air in the enclosed area marked A under the nappe causing a negative pressure to develop and the trajectory of the falling jet to deviate and ultimately cling to the weir itself. In such a case, the head over the weir is augmented corresponding to the negative pressure created below the nappe. Though the discharge may increase on account of the increase in head, the safety of the weir structure is threatened due to the pulsating negative (cavitation) pressures imposed on it.

Let V_0 be the velocity in the approach channel. An additional head $V_0^2/2g$ now acts over the weir. The discharge through the elementary stream as before is $dQ = C_d B(2g)^{1/2}h^{1/2}dh$. We integrate this over the head over the weir to get the total discharge.

$$Q = C_d\sqrt{2g}B\int_{V_0^2}^{H+V_0^2} h^{1/2}dh = \frac{2}{3}C_d\sqrt{2g}B\left[\left(H+\frac{V_0^2}{2g}\right)^{3/2} - \left(\frac{V_0^2}{2g}\right)^{3/2}\right] \quad (6.17)$$

The value of C_d is a function of the relative ratio of head over the weir to the height of the weir and is given as $C_d = 0.611 + 0.075(H/W)$. The head over the weir is measured at a distance of at least $3H$ upstream of the location of the weir.

6.11.3 Allowance for Side Contraction

When water flows over a sharp-edged weir it was found that similar to the formation of the vena contracta, the width of the nappe is not equal to the width of the weir. Hence an allowance was made by Francis for the reduction of the width of the nappe. He estimated that the width is reduced by $0.1\,nH$, where n is the number of side contractions present in the weir. For a simple rectangular weir, there are only two side contractions ($n = 2$) one at each end of the weir. Supposing there is a bridge pier in the middle of the weir, then $n = 4$; if there are two bridge piers, $n = 6$. Hence we must replace B by $(B - 0.1nH)$ in the above equation for discharge over the sharp-crested weir.

$$Q = \frac{2}{3}C_d\sqrt{2g}(B - 0.1nH)\left[\left(H + \frac{V_0^2}{2g}\right)^{3/2} - \left(\frac{V_0^2}{2g}\right)^{3/2}\right] \tag{6.18}$$

6.11.4 Triangular Notch (V-notch)

The triangular notch (Fig. 6.12) is preferred sometimes because of its accuracy. The main advantage of the V-notch is that the wetted edge in this case is linearly proportional to the head over the vertex of the notch unlike in the rectangular notch where the bottom edge remains a constant addition to the wetted edge for all heads over the weir.

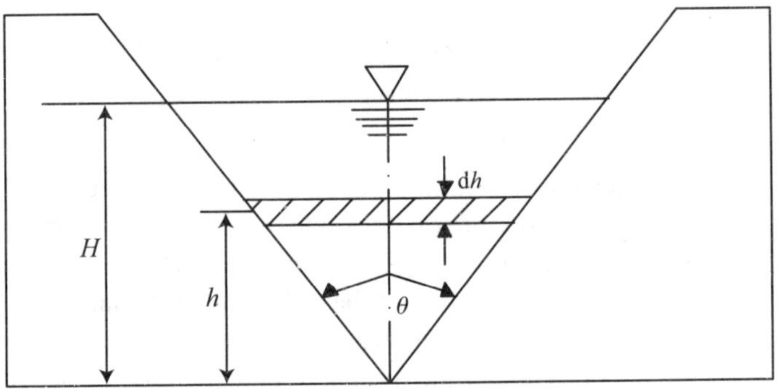

Fig. 6.12 Triangular notch

Let the central angle of the notch be θ and the head over the vertex be H. We will consider the discharge through an elementary strip of thickness dh located at a depth h below the free surface. The width of the strip is $2(H - h)\tan(\theta/2)$. And the discharge through the strip is

$$dQ = C_d 2(H - h)\frac{\tan\theta}{2}(2gh)^{1/2}dh$$

Integrating, we get

$$Q = 2C_d\sqrt{2g}\frac{\tan\theta}{2}\int_0^H (H - h)h^{1/2}dh = 2C_d\sqrt{2g}\frac{\tan\theta}{2}\left[\frac{2}{3}Hh^{3/2} - \frac{2}{5}h^{5/2}\right]_0^H$$

$$Q = \frac{8}{15}C_d\sqrt{2g}\left(\frac{\tan\theta}{2}\right)H^{5/2} \tag{6.19}$$

The value of C_d depends on the angle θ primarily but also depends on factors such as the height of the vertex, W above the channel bottom, width B of the channel, and the head H. For an angle of 90°, $C_d \sim 0.6$.

6.11.5 Cipoletti Weir

We have seen in Section 6.11.3 above that the end contractions cause a loss in discharge equal to $(2/3)C_d(2g)^{1/2}0.1nH.H^{3/2}$, where n stands for the number of end contractions. If there are only two end contractions, the loss in discharge is $(2/3)C_d(2g)^{1/2}0.2H^{5/2}$. This can be compensated by introducing sloping sides such that the resulting trapezoidal weir gives the discharge equal to the discharge over a rectangular weir without any loss due to side contractions. If the ends are inclined at $\theta/2$ to the vertical, the discharge in the triangular portions on either side taken together is $8/15C_d(2g)^{1/2}(\tan\theta/2)\ H^{5/2}$ and we can equate this to the loss in end contractions (C_d is taken as the same both for the rectangular and triangular weirs).

$$\left(\frac{2}{3}\right) C_d(2g)^{1/2}0.2H^{5/2} = \frac{8}{15}C_d(2g)^{1/2}\left(\frac{\tan\theta}{2}\right) H^{5/2}$$

or $\tan(\theta/2) = 1/4$ or the side slope is to be maintained at 1 horizontal to 4 vertical.

6.11.6 Proportional Weirs

We have seen that the discharge over a rectangular weir is proportional to $H^{3/2}$ and the discharge over a triangular weir is proportional to $H^{5/2}$. A parabolic weir gives the discharge proportional to H^2. Other shapes may be designed as shown in Fig. 6.13 to give the discharge as a desired function of H.

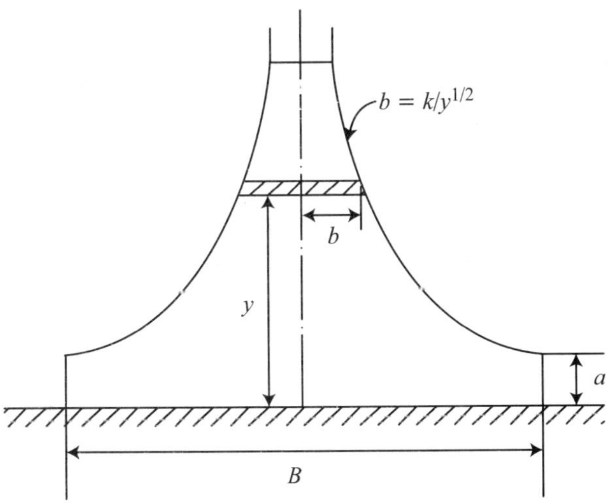

Fig. 6.13 Proportional weir

One of the earliest attempts was to get a profile such that the discharge was directly proportional to H. This may be designated as a linear proportional weir or Sutro weir named after the person who proposed it. We will first consider a weir profile described by the equation $b = k/(y)^{1/2}$. As shown in Fig. 6.13, let the discharge through an elementary strip of width b and depth dy be dQ. Then

$$dQ = C_d b[2g(H-y)]^{1/2}dy = C_d\left(\frac{k}{y^{1/2}}\right)[2g(H-y)]^{1/2}$$

which on integration yields $Q = C_d k(2g)^{1/2}(\pi/2)H$, thus making the discharge proportional to H. However, this is not a practical profile since the width at the bottom, $z = 0$ becomes infinite. Sutro proposed that a rectangular portion of width B and height a be provided at the bottom. He prescribed that the height be measured above a datum fixed at a distance $a/3$ above the base of the weir. The discharge equation then becomes

$$Q = C_d B(2ga)^{1/2}(H - a/3) \tag{6.20}$$

The profile of the Sutro weir can then be shown as

$$\frac{x}{B} = 1 - \left(\frac{2}{\pi}\right)\tan^{-1}\left(\frac{y'}{a}\right)^{1/2}$$

Similar devices with different weir profiles have been derived to get discharge as different functions of H. For example, a logarithmic weir makes the discharge proportional to log H. The aim in each case is to improve the accuracy of the discharge measurement despite small errors in measuring the head over the weir.

6.11.7 Broad-crested Weir

If the thickness of the weir is large, say $t > 0.66H$, then a free nappe cannot be formed and the flow adheres to the sill as shown in Fig. 6.14.

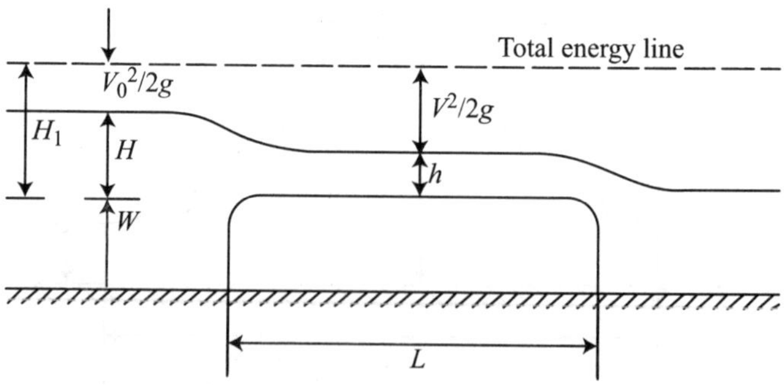

Fig. 6.14 Broad-crested weir

If the sill length, L is large, frictional resistance is also to be taken into account. If velocity of approach and other losses are neglected and the flow over the sill is assumed parallel to the crest, then

$$H = h + V^2/2g$$

$$V = [2g(H - h)]^{1/2}$$

where H is the head over the weir measured upstream of the weir and h is the depth over the weir. The discharge is

$$Q = C_d Bh[2g(H - h)]^{1/2}$$

The discharge depends on the value of h and it adjusts itself to give the maximum discharge. Hence differentiating the above equation

$$\frac{dQ}{dh} = C_d(2g)^{1/2} B \left(\frac{1}{2}\right)(Hh^2 - h^3)^{-1/2}(2Hh - 3h^2) = 0$$

$$\text{or} \quad 2Hh - 3h^2 = 0 \quad \text{or} \quad h = \left(\frac{2}{3}\right) H$$

Substituting this value in the equation for discharge, we get

$$Q = C_d B\sqrt{2g}\frac{2}{3}H\sqrt{\frac{1}{3}H} = 0.3849 C_d B\sqrt{2g}H^{3/2} \tag{6.21}$$

The discharge coefficient has absorbed all losses and corrections for the length of the sill, frictional losses, losses at the entrance, viscous and surface tension effects etc.

6.11.8 Submerged Weir

If the water depth downstream of the weir is above the top of the weir, the weir is said to be submerged (see Fig. 6.15).

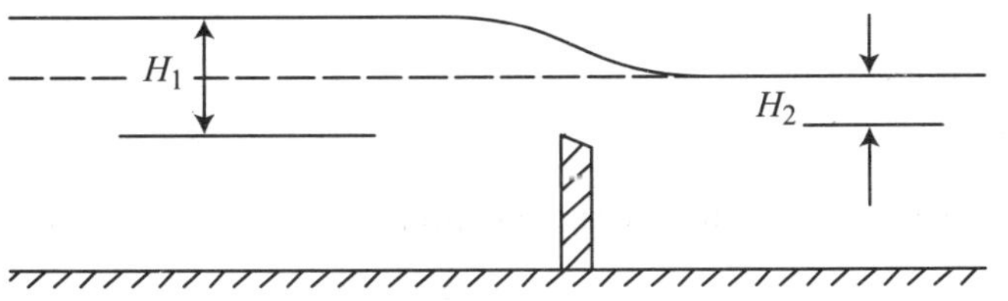

Fig. 6.15 Submerged weir

Then, it is usual to treat this as divided into two parts. The top portion between the upstream depth, H_1 and the downstream depth, H_2 as a free weir and the lower portion between the weir top and the downstream depth, H_2 as an orifice.

For the free weir

$$Q_1 = C_d B\sqrt{2g}\int_0^{H_1-H_2} h^{1/2}dh = \frac{2}{3}C_d B\sqrt{2g}(H_1 - H_2)^{3/2}$$

For the orifice portion

$$Q_2 = C_d B H_2[2g(H_1 - H_2)]^{1/2}$$

The total discharge is given as the sum of the two $Q = Q_1 + Q_2$.

Alternatively, we may use the equations developed by Villemonte, viz.,

$$\frac{Q_s}{Q_f} = \left[1 - \left(\frac{H_2}{H_1}\right)^{3/2}\right]^{0.385} \quad \text{for rectangular weirs and}$$

$$\frac{Q_s}{Q_f} = \left[1 - \left(\frac{H_2}{H_1}\right)^{5/2}\right]^{0.385} \quad \text{for triangular weirs}$$

where Q_s is the discharge under submerged conditions and Q_f is the discharge under free flow conditions under a head H_1 calculated as in Section 6.11.2.

6.11.9 Sluice Gate

Sluice gates are used as controls on spillways, canals and reservoir outlets. The flow takes place below the gate and the discharge can be controlled by raising or lowering the gate to the desired value of opening height a. Let the depth upstream of the gate be H_1 and the velocity of approach be V_1. If the flow below the gate is free (Fig. 6.16a), let the depth at the vena contracta be $C_c a$. Then the equation of continuity gives

$$V_1 H_1 = C_c a V_2$$

or

$$V_1 = \frac{V_2 C_c a}{H_1}$$

We substitute this relationship in the energy equation

$$H_1 + V_1^2/2g = C_c a + V_2^2/2g$$

$$H_1 + (V_2^2 C_c^2 a^2)/H_1^2 2g = C_c a + V_2^2/2g$$

or $\quad V_2^2/2g[C_c^2 a^2/H_1^2 - 1] = C_c a - H_1$

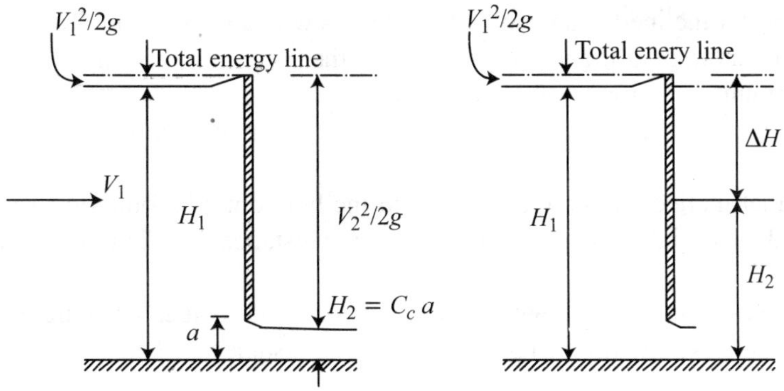

(a) Free flow (b) Submerged flow

Fig. 6.16 Flow below a sluice gate

Discharge per unit width

$$q = C_c a V_2 = C_c a \sqrt{\frac{C_c a - H_1}{C_c^2 a^2 - H_1^2}} \sqrt{2gH_1} \tag{6.22}$$

$q = C_d a(2ghH_1)^{1/2}$, where $C_d = \frac{C_c}{\sqrt{1+C_c a/H_1}}$, the value of C_v is taken as 1.0.

For submerged sluice gate flow, the head is taken as the difference between the upstream depth and the depth immediately downstream of the gate as shown in Fig. 6.16*b*, $\Delta H = H_1 - H_2$ and the discharge equation is

$$q = C_d a(2g\Delta H)^{1/2} \tag{6.23}$$

The value of C_d is a function of a/H_1 and varies in the range of about 0.6 for $a/H_1 = 0.2$ to about 0.69 for $a/H_1 = 0.8$. A value of 0.61 is often adopted.

Exercises

6.1 A fireman has two options to reach a window at a height of 36 m above the ground level with a fire hose nozzle of diameter 5 cm and having a jet velocity of 30 m/s. Determine the maximum distance from the building at which the fireman can stand to reach the window.

Hint to obtain the maximum distance:

$$x_0 = V_0 t \cos\theta; \; z = V_0 t \sin\theta - gt^2/2$$

Eliminate θ and obtain: $x = [V_0^2 t^2 - (z + gt^2/2)^2]^{1/2}$

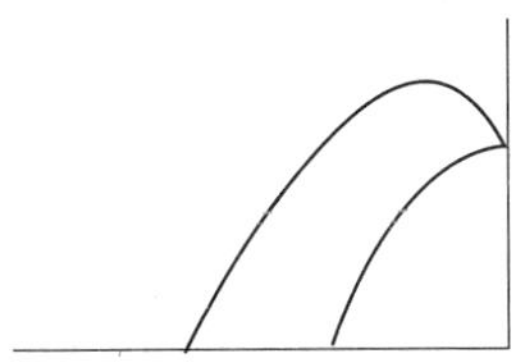

Differentiate and set $dx/dt = 0$ to obtain the value of t and then obtain $x_{\max}$ and the value of θ.

6.2 A jet of water inclined at an angle of 45° leaves with a velocity of 10 m/s. Determine the radius of curvature at the initial point and at the point of maximum elevation. Also find the maximum elevation to which the jet rises.
Hint: At the initial point $a_n = V_0^2/r = g \cos 45°$ and at maximum elevation $a_n = g = V_x^2/r$.

6.3 A liquid jet inclined at an angle of 60° to the horizontal is found to rise to a height of 5 m. Find the initial velocity, and the horizontal distance travelled to reach the maximum height.

6.4 Water leaves from a nozzle with an initial velocity of 10 m/s. What are the two possible nozzle inclinations to hit a target at a horizontal distance of 8 m.

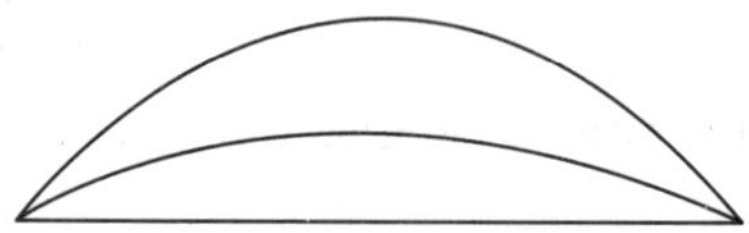

6.5 At what distance H below the free surface should an orifice be placed in the side of a tank so that it reaches the maximum distance in level with the base of the tank. The height of the water above the base of the tank is H_0.

6.6 Oil of specific gravity 0.90 and dynamic viscosity 8×10^{-1} kg/m s flows through a pipe 10 cm in diameter. The rate of flow is measured by a diaphragm orifice of diameter 7.5 cm. Determine the discharge if the differential manometer connected across the orifice showed a head equivalent to 5 cm of oil. $C_d = 0.73$

6.7 A pipe of diameter 20 cm ends in a nozzle with an exit diameter of 7.5 cm. The inclination of the sides of the nozzle is 45°. Assume a value of $C_c = 0.75$. If the discharge through the nozzle is 0.1 m^3/s, find the pressure in the pipe.

6.8 Water from a tank discharges through a divergent mouthpiece. If the outlet area is 4 times the minimum area, find the maximum head in the tank at which steady flow through the mouthpiece can take place without incipient cavitation which may be assumed to occur at an absolute pressure of 2.45 m of water. $C_d = 0.8$

6.9 A convergent mouthpiece of 10 cm exit diameter with $C_c = 1.0$ and $C_v = 0.98$ is operating under a head of 2.5 m. Calculate the discharge. In order to double the discharge, the original mouthpiece is replaced by a convergent–divergent mouthpiece (diverging after the 10 cm dia. section) with $C_v = 0.85$. Calculate the diameter of the mouthpiece at the outlet. What is the pressure at the plane of the original mouthpiece. What is the maximum theoretical discharge assuming that incipient cavitation takes place at 2 m of water absolute. Neglect losses.

6.10 A discharge of 200 l/s is delivered through a conical mouthpiece tapering from 25 cm to 10 cm in a length of 10 cm. Find the pressure at the entrance and at the mid-section of the mouthpiece. Neglect losses.

6.11 A venturimeter is fitted horizontally in a 15 cm dia. pipe and the pressure in the pipe corresponds to a head of 10 m. If the maximum flow through the meter is 0.15 m^3/s, find the smallest throat diameter so that the pressure does not fall below 2.45 m of water (absolute).

6.12 A venturimeter having a throat diameter of 7.5 cm is installed in a vertical pipe 15 cm in dia. carrying oil of specific gravity 0.80. If the flow is vertically upwards, what is the pressure difference between the entrance and throat if the flow rate is $0.10\,m^3/s$. The difference in elevations of pressure-tappings is 25 cm.

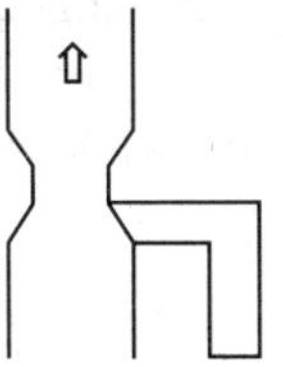

6.13 An air duct changes abruptly in diameter from 2 m to 1 m. If the pressure head in the 2 m dia. pipe is 5 cm of water when the flow is $10\,m^3/s$, what is the pressure intensity at the vena contracta in the smaller dia. pipe? $\rho_{air} = 1.2$

6.14 A rectangular large orifice measuring 15 cm high and 20 cm wide is located in the side of a tank with the water level in the tank 50 cm above the centre line of the orifice. Assuming a coefficient of discharge of 0.62, find the discharge through the orifice.

6.15 Two vertical sided basins each having a surface area of $200\,m^2$ are connected by a gate of area $0.30\,m^2$. The initial difference in level between the basins is 3 m. Find the time taken to reduce the difference to 1.5 m.

6.16 A tank is in the shape of a frustrum of a cone 4 m high. The diameter at the bottom is 1.5 m and the diameter at the top is 3 m. The tank is initially full of water. An orifice of area $0.10\,m^2$ is located in the bottom surface of the tank. Find the times required to lower the depth by 2 m as well as to empty the tank completely.

6.17 An overflow rectangular sharp-edged weir is used to spill excess inflow into a reservoir. The width of the weir is 25 m and the weir height is 3 m above the bed. The weir has two intermediate piers each 1 m in width. The approach channel is 40 m wide. Find the discharge when the head over the weir is 2.5 m.

6.18 A submerged weir built across a channel 10 m wide is 2 m high. The depths upstream and downstream of the weir are 3 m and 2.5 m respectively. Find the discharge. Assume $C_d = 0.60$

6.19 Determine the discharge through a sluice gate 2 m wide with a gate opening of 0.5 m under a head of 8 m (a) when the opening is not submerged and the discharge is free, $C_d = 0.60$ and (b) when the gate is submerged with a downstream depth of 3 m, $C_d = 0.52$.

6.20 A triangular weir of central angle 90° is used to measure the discharge in a channel 2 m wide having an approach depth of 2 m wide. The depth over the vertex of the weir is 1.2 m. Find the discharge.

6.21 A rectangular channel 2 m wide carries a maximum of $0.20\,m^3/s$ and a minimum of $0.1\,m^3/s$. At what level should the vertex of a 90° V-notch be placed to attain an upstream depth of 1.5 m. What then will be the minimum depth? Neglect velocity of approach and assume $C_d = 0.60$.

6.22 A sharp-crested opening in the form of an isosceles triangle with a horizontal base of 1.0 m and height of 0.75 m is located in the side of a tank. Calculate the rate of flow over the weir when the height of the water level above the base is 0.5 m. $C_d = 0.62$

6.23 Determine the discharge over a broad-crested weir 3 m in width if the total head upstream of the weir is 60 cm and the depth of flow over the weir is 50 cm. What is the maximum possible discharge under the given head of 60 cm and what is the velocity of flow over the crest? Assume $C_d = 0.85$.

6.24 A rectangular channel 5 m wide carries a discharge of 2.5 m^3/s of water at a depth of 0.8 m. How high should a broad-crested weir be constructed over the full width to double the depth. Assume $C_d = 0.85$.

7 Force–Momentum Relationship

7.1 Introduction

It has been emphasised earlier that the Bernoulli constant energy theorem is valid strictly for irrotational flows. However, it can be used in its modified form as was illustrated in Chapter 6 if the losses in the system can be evaluated and corrections introduced in the energy equation. We have seen that modifications may be made by introducing certain coefficients such as C_d which are assessed by carefully conducted experimental programs with the results presented in non-dimensional forms.

However, in a majority of fluid flow problems, it is not possible to evaluate losses in the system *a priori*. In some other problems, the forces acting on submerged bodies are to be determined. Again, in some other cases external forces acting on the system are prescribed and the resulting velocity fields are to be derived. In all these cases, we have to have recourse to the well-known relationship given by Newton's second law that the rate of change of momentum is proportional to the force and takes place in the direction of the force. To assess the force–momentum relationship, the control volume approach is very handy.

7.2 Control Volume

A control volume is defined as a volume fixed with respect to a coordinate system in a fluid flow field. We have already derived the equation of continuity for one-dimensional flow and for a differential element for two- and three-dimensional flows in Chapter 3.

7.2.1 Conservation of Mass

Consider a certain quantity of matter enclosed in space A at time t as shown in Fig. 7.1.

At time $t + dt$, it occupies the space B + A − C. If mass contained in spaces A, B, C are denoted by m_A, m_B, m_C, then we have the relation

$$m_A(t) = m_B(t + dt) + m_A(t + dt) - m_C(t + dt)$$

Rearranging and dividing by dt and taking limits as $\Delta t \to 0$, we get

$$\frac{\text{lt}}{\Delta\text{t} \to 0} \frac{m_A(t + dt) - m_A(t)}{\Delta t} = \frac{\text{lt}}{\Delta\text{t} \to 0} \left[\frac{m_C(t + dt) - m_B(t + dt)}{\Delta t} \right]$$

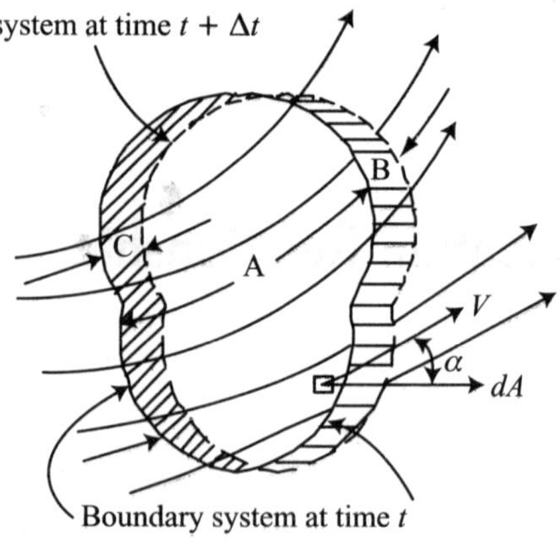

Fig. 7.1 Control volume (system moving through space)

$\partial(m)_{CV}/\partial t$ = rate of mass inflow – rate of mass outflow

$$\frac{\partial}{\partial t}\int_{CV}\rho d\forall = \int_{Ain}\rho V\cos\alpha dA - \int_{Aout}\rho V\cos\alpha d\mathrm{A} = -\int_{CS}\rho\mathbf{V}\cdot d\mathbf{A}$$

where $\mathbf{V}$ is the velocity vector and V is its magnitude.

$$\frac{\partial}{\partial t}\int_{CV}\rho d\forall + \int_{CS}\rho\mathbf{V}\cdot d\mathbf{A} = \rho\mathbf{V}\cdot d\mathbf{A} = 0 \quad \text{for steady flows.}$$

For steady, incompressible flows, we have

$$\int_{CS}\mathbf{V}\cdot d\mathbf{A} = \int_{CS}V_n dA = 0$$

V_n is the velocity normal to the elemental area dA.

7.2.2 Linear Momentum

The momentum equation for the control volume is developed in a similar manner. This equation is one of the most important analytical relationships in fluid mechanics. The equation will give forces acting on solid surfaces such as impact of jets on plates and curved vanes of turbines or pumps, pipe bends, lift and drag on submerged bodies including airplane wings, and forces on other fluids such as thrust of a jet engine etc.

By Newton's law, the net force, **F** acting on a system of particles of fixed mass is given as

$$\mathbf{F} = \frac{d\mathbf{M}}{dt}$$

where **M** is the linear momentum of the system. In Fig. 7.1 we replace velocity by momentum.

The change in momentum, $\Delta\mathbf{M}$ is given as

$$\Delta\mathbf{M} = \mathbf{M}_A(t+\Delta t) - \mathbf{M}_C(t+\Delta t) + \mathbf{M}_B(t+\Delta t) - \mathbf{M}_A(t)$$

Dividing by Δt and taking limits as $\Delta t \to 0$, we have

$$\frac{\text{lt}}{\Delta t \to 0}\frac{\Delta\mathbf{M}}{\Delta t} = \frac{\text{lt}}{\Delta t \to 0}\left[\frac{\mathbf{M}_A(t+\Delta t) - \mathbf{M}_A(t)}{\Delta t}\right] + \frac{\text{lt}}{\Delta t \to 0}\left[\frac{\mathbf{M}_B(t+\Delta t) - \mathbf{M}_C(t+\Delta t)}{\Delta t}\right]$$

$$= \frac{\partial}{\partial t}\mathbf{M}_{CV} + \frac{\Sigma_B[\Delta\mathbf{M}(t+\Delta t)]_B}{\Delta t} - \frac{\Sigma_C[\Delta\,\mathbf{M}(t+\Delta t)]_C}{\Delta t}$$

$$= \frac{\partial}{\partial t}\int_{CV}\rho\,\mathbf{V}\,d\forall + \sum_B \Delta\,\dot{\mathbf{M}} - \sum_C \Delta\,\dot{\mathbf{M}}$$

$$= \frac{\partial}{\partial t}\int_{CV}\rho\mathbf{V}d\forall + [\Sigma\Delta\dot{m}\mathbf{V}]_{out} - [\Sigma\Delta\dot{m}\mathbf{V}]_{in}$$

$$\mathbf{F} = \frac{\partial}{\partial t}\int_{CV}\rho\mathbf{V}d\forall + \int \mathbf{V}\rho V\cdot d\mathbf{A}$$

The total force is made up of the total surface contact forces, namely, the normal pressure force and the tangential shear force and a distant-acting body force such as gravitational force.

$$\mathbf{F}_p + \mathbf{F}_s + \mathbf{F}_B = \frac{\partial}{\partial t}\int_{CV}\rho\mathbf{V}d\forall + \int_{CS}\mathbf{V}\rho\mathbf{V}\cdot d\mathbf{A}$$

For steady flow, the first term on the RHS is zero.

The dot product of $\mathbf{V}\cdot d\mathbf{A}$ is $|\mathbf{V}||d\mathbf{A}|\cos\alpha$ and $V_n = V\cos\alpha$, we write for steady flow

$$\Sigma\mathbf{F} = \int_{CS}\rho\mathbf{V}V_n dA$$

If the density and velocity are uniform over the areas where the fluid is crossing the control surfaces 1 and 2, we have

$$\Sigma\mathbf{F} = \dot{m}(V_2 - V_1)$$

The forces may be resolved in the three coordinate directions to yield

$$\Sigma\mathbf{F}_x = \dot{m}(V_{x2} - V_{x1})$$

$$\Sigma\mathbf{F}_y = \dot{m}(V_{y2} - V_{y1})$$

$$\Sigma\mathbf{F}_z = \dot{m}(V_{z2} - V_{z1})$$

We may generalise the above derivations for mass conservation and linear momentum to extend to other quantities as well.

The rate of change of any quantity, X such as mass, momentum or energy from the system denoted by dX/dt is given by

$$\left.\frac{dX}{dt}\right|_{\text{system}} = \left.\frac{\partial X}{\partial t}\right|_{\text{control volume}} + \int_{\text{control surface}} (x\rho \mathbf{V}\cdot d\mathbf{A}) \tag{7.1}$$

where ρ is the density of the fluid and x is the flow quantity in question per unit mass.

For the law of conservation of mass,

$$X = M \text{ and } \frac{dM}{dt} = 0$$

For the law of conservation of momentum (Newton's second law),

$$X = (MV) \text{ and } F = \frac{dX}{dt} = \frac{d(MV)}{dt}$$

and for energy

$$X = E \text{ and } \frac{dE}{dt} = \frac{dQ}{dt} - \frac{dW}{dt}$$

where dQ/dt represents the rate of addition of heat energy and dW/dt represents the rate of work done by the system.

For conservation of mass, we must have $dM/dt = 0$. Since the total mass is

$$M = \int \rho d\forall \text{ and } X = \int x\rho d\forall$$

with $x = 1$. Hence we have

$$\left.\frac{\partial M}{\partial t}\right|_{\text{control volume}} + \int_{\text{control surface}} \rho V_n dA = 0$$

n is the normal to the surface dA.

For steady flows,

$$\int_{\text{control surface}} (\rho V_n dA) = 0$$

For conservation of momentum, $x = MV/M = V$

$$\mathbf{F} = \left.\frac{d(M\mathbf{V})}{dt}\right|_{\text{system}} = \left.\frac{\partial(M\mathbf{V})}{\partial t}\right|_{\text{control volume}} + \int_{\text{control surface}} \mathbf{V}(\rho \mathbf{V} dA) \tag{7.2}$$

Force and momentum are vector quantities possessing both direction and magnitude. We can express the above equation as three scalar momentum equations by resolving it in the three

Cartesian coordinates. In the x-, y- and z-directions, the force components are F_x, F_y, and F_z and the value of x in the corresponding directions are V_x, V_y, and V_z respectively. The equations are

$$\Sigma F_x = \left.\frac{d(M\mathbf{V})}{dt}\right|_{\text{system}} = \left.\frac{\partial(MV)_x}{\partial t}\right|_{\text{CV}} + \int_{\text{CS}} V_x \rho V_n dA$$

$$\Sigma F_y = \left.\frac{d(M\mathbf{V})}{dt}\right|_{\text{system}} = \left.\frac{\partial(MV)_y}{\partial t}\right|_{\text{CV}} + \int_{\text{CS}} V_y \rho V_n dA \tag{7.3}$$

$$\Sigma F_z = \left.\frac{d(M\mathbf{V})}{dt}\right|_{\text{system}} = \left.\frac{\partial(MV)_z}{\partial t}\right|_{\text{CV}} + \int_{\text{CS}} V_z \rho V_n dA$$

The force term on the left-hand side is the sum of all forces exerted on the fluid within the control volume and accounts for distant-acting forces such as gravity, magnetic, electrical, electrostatic forces and contact forces such as surface tension, pressure and shear etc. The first term on the right-hand side represents the rate at which linear momentum is stored in the control volume. The second term denotes the efflux of linear momentum from the control volume out of the surfaces.

For steady flow, the first term on the right-hand side namely the momentum stored in the volume is zero. If the flow is one-dimensional, say in the x-direction, then the linear momentum equation in the x-direction reduces to

$$\Sigma F_x = \int_{\text{CS}} V_x(\rho V_n dA) \tag{7.4}$$

With reference to Fig. 7.1, the term on the right-hand side can be written in a simpler form to facilitate easy application.

$\mathbf{V}$ is the velocity of the fluid, in any arbitrary direction, out of the area dA of the control surface. Its component normal to the area dA is V_n and if the angle between $\mathbf{V}$ and V_n is θ, then $V_n = V\cos\theta$. The rate of mass of the fluid flowing out of the area dA is $\rho V_n dA = \rho V \cos\theta dA$. Then the rate of momentum of this fluid in the x-direction is $(\rho V \cos\theta dA)V_x$, where V_x is the component of $\mathbf{V}$ in the x-direction as already stated. The momentum equation (7.4) in the x-direction becomes

$$\Sigma F_x = \int_{\text{CS}} \rho \mathbf{V} V_x \cos\theta dA \tag{7.5}$$

The application of the above equation is further simplified by a proper choice of the control surface both at entry and exit of the control volume. This will be illustrated by a number of examples in the following sections.

7.3 Impact of Jet on a Flat Plate

Let us consider a horizontal liquid (density ρ) jet of area A, moving with a velocity V_1 striking a stationary plate held normal to the jet as shown in Fig. 7.2.

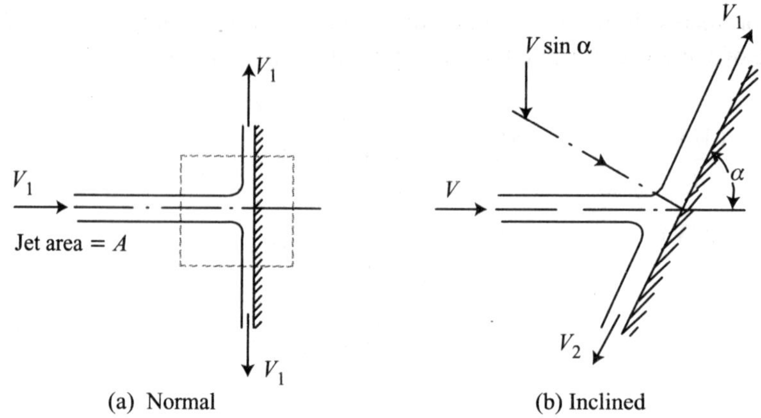

Fig. 7.2 Impact of horizontal jet on a plate

The jet after striking the plate is deflected sideways at right angles to the jet. The mass rate of flow is $\rho Q = \rho A V_1$ and its momentum is $\rho A V_1 V_1$. The velocity of the jet in the x-direction is zero after impact with the plate. Hence its momentum in the x-direction is zero. The entire system is subjected to atmospheric pressure and since there is no differential pressure, the force component due to pressure is zero. Further the gravitational component in the horizontal direction is also zero. Hence the total force on the plate is

$$F_x = \rho Q V_1 - 0 = \rho Q V_1$$

If the plate is advancing towards the jet with a uniform velocity V_2, then the relative velocity would be $V_r = (V_1 + V_2)$. With $Q = A V_r$, the force on the plate is $\rho A (V_1 + V_2)^2$ whereas in the case of the plate moving away from the jet with velocity V_2, the force would be $\rho A (V_1 - V_2)^2$. If the plate is moving away with the same velocity as the jet $(V_1 = V_2)$, the force on the plate will be zero.

For an inclined plate as shown in Fig. 7.2b, the force normal to the plate is $\rho Q V \sin\alpha$. The division of flow along the inclined plate is uneven unlike the symmetrical division of flow for the normal plate. The velocities of the two partitioned jets V_1 and V_2 are assumed equal in the absence of any loss. The momentum theorem applied parallel to the plate is

$$\rho Q V \cos\alpha = \rho Q_1 V_1 - \rho Q_2 V_2 \text{ and } V = V_1 = V_2; \text{ therefore } Q \cos\alpha = Q_1 - Q_2$$

By continuity

$$Q = Q_1 + Q_2$$

We obtain from these two relations

$$Q_1 = \frac{1}{2} Q(1 + \cos\alpha) \text{ and } Q_2 = \frac{1}{2} Q(1 - \cos\alpha)$$

7.4 Force on a Curved Vane

Let a jet of water of diameter d impinge on a stationary vane as shown in Fig. 7.3. The angle of entry at the entrance is θ_1 and at the exit θ_2 with respect to the x-axis.

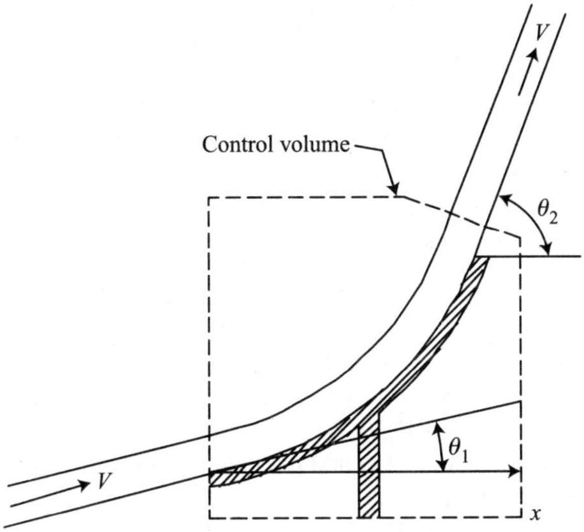

Fig. 7.3 Flow over a vane

The velocity is V throughout the flow over the vane assuming no friction losses. The pressure over the system is atmospheric. The mass rate of flow is $\rho(\pi d^2/4)V$ and its momentum in the x-direction is $\rho(\pi d^2/4)VV\cos\theta_1$ and the horizontal momentum in the x-direction at the exit is $\rho(\pi d^2/4)VV\cos\theta_2$. The force in the x-direction on the control volume or the reaction of the supporting structure equal to the rate of momentum change in the x-direction is therefore

$$F_x = \rho\left(\frac{\pi d^2}{4}\right)V^2(\cos\theta_2 - \cos\theta_1)$$

Similarly the force in the perpendicular z-direction (xz-plane is taken as the horizontal plane)

$$F_z = \rho\left(\frac{\pi d^2}{4}\right)V^2(\sin\theta_2 - \sin\theta_1)$$

The resultant force on the vane may be obtained as $F_R = (F_x^2 + F_z^2)^{1/2}$ making an angle $\theta = \tan^{-1}(F_z/F_x)$ with the horizontal.

EXAMPLE 7.1 A jet of diameter 2 cms with a velocity of 10 m/s enters a curved vane shown in Fig. 7.3 at an angle of 30° and leaves the vane at an angle of 60°. Find the force components in the horizontal and normal directions and the magnitude and direction of the resultant force. If the entry and exit angles are 0° and 180°, find the force components.

Let $d = 2$ cms, area of jet $= 0.00031416\,\text{m}^2$, $V = 10$ m/s, $\theta_1 = 30°$ and $\theta_2 = 60°$.
Mass rate of flow $= 1000 \times 0.00031416 \times 10 = 3.1416$ kg/s.

$$F_x = 3.1416 \times 10(\cos 60° - \cos 30°) = 31.416 \times -0.366 = -11.498\,\text{N}$$

$$F_z = 31.416 \times (\sin 60° - \sin 30°) = 11.498\,\text{N}$$

Resultant = 16.46 N acting at 45° to the horizontal.

If $\theta_1 = 0$ and $\theta_2 = 180°$, $F_x = 31.416 \times (-2) = -62.832$ N and $F_z = 0$ and this principle is used in the design of hemispherical cups in Pelton wheels to obtain the maximum forward thrust for a given jet's characteristics.

7.4.1 Force on a Moving Curved Vane

If the vane moves in the same direction as the jet with a velocity V_b, the jet is no longer deflected through an angle θ, since the exit velocity, V_2 is now a resultant of the velocity of the vane and of the velocity of the fluid. The velocity of the fluid is the relative velocity $V_R = (V_1 - V_b)$ and is assumed to be the same over the entire vane since friction is ignored. Hence the exit velocity V_2 is a vector sum of $(V_1 - V_b)$ and V_b.

The initial momentum is $\rho Q(V_1 - V_b)$ and the final momentum is $\rho Q(V_1 - V_b)\cos\alpha$

The change in the momentum is $\rho Q(V_1 - V_b)\cos\alpha - \rho Q(V_1 - V_b)$

From the vector diagram shown in Fig. 7.4, the force components with $[Q = (V_1 - V_b)A]$ are

$$-F_x = \rho Q(V_R\cos\alpha + V_b - V_1) = \rho Q[(V_1 - V_b)\cos\alpha - (V_1 - V_b)]$$

$$= \rho A(V_1 - V_b)[(V_1 - V_b)(\cos\alpha - 1)]$$

$$F_x = \rho A(V_1 - V_b)^2(1 - \cos\alpha)$$

$$F_y = \rho Q(V_1 - V_b)\sin\alpha = \rho A(V_1 - V_b)^2\sin\alpha$$

The absolute velocity V_2 has components in the x-and y-directions given by

$$V_{2x} = V_b + V_R\cos\alpha$$

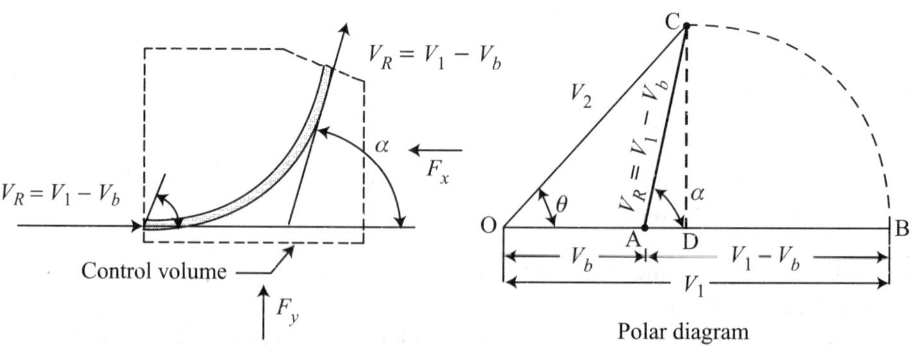

Fig. 7.4 Force on a moving curved vane

$$V_{2y} = V_R \sin\alpha$$

From these components, one can find

$$V_2^2 = (V_{2x})^2 + (V_{2y})^2$$

and

$$\theta = \tan^{-1}\frac{V_{2y}}{V_{2x}}$$

7.5 Pipe Bend

Figure 7.5 shows a reducing vertical bend. The control volume is also indicated. The discharge is Q. The weight of water in the bend is W. The area, velocity, and pressure at the entrance and exit ends of the bend are (A_1, V_1, p_1) and (A_2, V_2, p_2) respectively. p_1 is known. The normal to the entrance and exit surfaces make angles θ_1 and θ_2 respectively. The centre of the exit end is at a height of y above the centre of entrance.

$$V_1 = \frac{Q}{A_1} \text{ and } V_2 = \frac{Q}{A_2}$$

The pressure p_2 can be found by the Bernoulli theorem. Assuming no losses and that the datum passes through the centre of the entrance, we get

$$p_1/\gamma + V_1^2/2g + 0 = p_2/\gamma + V_2^2/2g + y$$

$$F_x = \rho Q(V_2\cos\theta_2 - V_1\cos\theta_1) + p_2A_2\cos\theta_2 - p_1A_1\cos\theta_1$$

$$F_y = \rho Q(V_2\sin\theta_2 - V_1\sin\theta_1) + p_2A_2\sin\theta_2 - p_1A_1\sin\theta_1 + W$$

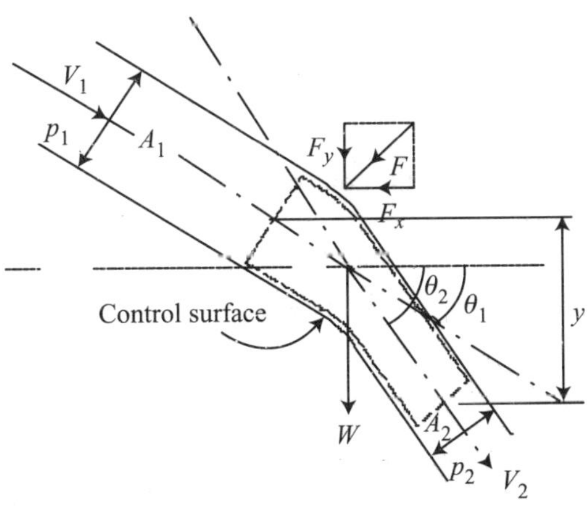

Fig. 7.5 Reducing bend

Example 7.2 A right-angled reducing bend has an entry diameter of 25 cms and reduces to a diameter of 15 cms at the exit. The velocity and pressure at the entry are 4 m/s and 200 kN/m^2 respectively. The weight of water in the length of the bend may be taken as 60 kN. Find the horizontal and vertical components as well as the magnitude and direction of the resultant force on the bend. The difference in height between the entrance and exit of the bend is 0.2 m.

Diameter of pipe at entrance = 25 cms; $A_1 = 0.0491\ \text{m}^2$

Exit diameter = 15 cms; $A_2 = 0.0177\ \text{m}^2$; $V_1 = 4$ m/s; $V_2 = 0.0491 \times 4/0.0177 = 11.11$ m/s; $Q = 0.1964\ \text{m}^3/\text{s}$; $p_1 = 200\ \text{kN/m}^2$; $y = 0.2$ m; $\theta_1 = 0$; $\theta_2 = 90°$

W = Weight of water in the bend = 60 kN

$$p_2 = 200 + \gamma[-0.2 + 4^2/2 \times 9.81 - 11.11^2/2 \times 9.81]$$

$$= 200 + 9.81(-5.676) = 144.32\ \text{kN/m}^2$$

$$F_x = 1000 \times 0.1964(-4 \times 1) - 200 \times 0.0491 \times 1 = -768.42\ \text{kN}$$

$$F_y = 1000 \times 0.1964(11.11) + 144.32 \times 0.0177 + 60 = 2244.56\ \text{kN}$$

$F = 2372.46$ kN acting at an angle $\theta = \tan^{-1} 71.1°$ to the horizontal.

7.6 Rocket

A rocket (Fig. 7.6) is moving vertically against the earth's gravity field. Let us assume that it is moving in a frictionless medium. We will also assume steady conditions as far as the exit pressure p_e, exit velocity, v_e and exit density, ρ_e of the discharge gas are concerned. Let V be the velocity of the rocket at any instant of time t and its acceleration be denoted by dV/dt. Let M_0 be the initial mass of the rocket including the initial mass of fuel. The fuel is burning at a constant rate of αt and the mass M at any instant t is $M_0 - \alpha t$. If the exit cross-sectional area is A_e, then $\alpha = \rho_e A_e v_e$.

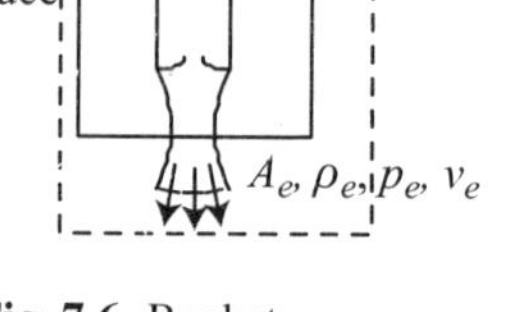

Fig. 7.6 Rocket

The momentum theorem applied to the control volume shown gives

$$M\left(\frac{dV}{dt}\right) = \rho_e A_e v_e^2 + p_e A_e - Mg$$

$$(M_0 - \alpha t)\left(\frac{dV}{dt}\right) = \alpha v_e + p_e A_e - (M_0 - \alpha t)g$$

$$dV = \left[\frac{\alpha v_e + p_e A_e}{M_o - \alpha t} - g\right] dt$$

Integrating we get,

$$V = -\left(\frac{\alpha v_e + p_e A_e}{\alpha}\right)[\ln(M_0 - \alpha t) - \ln M_0] - gt = \left(v_e + \frac{p_e A_e}{\alpha}\right)\ln\left(\frac{M_0}{M_0 - \alpha t}\right) - gt$$

7.7 Jet Propulsion

Rockets carry their own propelling fluid. In a turbojet engine, ambient air is taken in through the inlet and is compressed in an axial flow compressor. It then goes to the combustion chamber and the products of combustion expand first into the axial flow turbine (which drives the compressor) and then into the exhaust nozzle.

Let the engine be moving with a constant velocity V_1, as shown in Fig. 7.7.

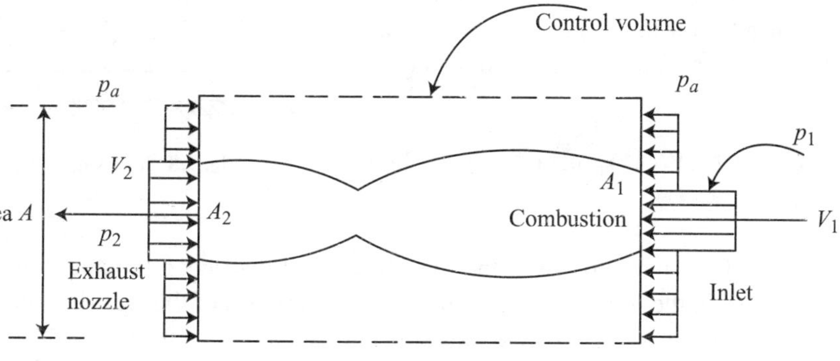

Fig. 7.7 Jet propulsion

Air enters the engine with a velocity V_1 in the opposite direction. Let V_2 be the velocity of the gases at the exit. The net pressure force acting on the system is

$$p_2 A_2 + p_a(A - A_2) - p_a(A - A_1) - p_1 A_1 = (p_2 - p_a)A_2 - (p_1 - p_a)A_1$$

$$\dot{m}_a(V_2 - V_1) + \dot{m}_f V_2 + (p_2 - p_a)A_2 - (p_1 - p_a)A_1 = \text{Thrust}$$

$$\dot{m}_a = \rho V_1 A_1$$

$$\dot{m}_f = \text{mass rate of fuel}$$

Let us assume that the fuel is mixed with air in the ratio 1 : N. The equation of continuity gives

$$\rho_1 V_1 A_1 + 1/N(\rho_1 V_1 A_1) = \rho_2 V_2 A_2$$

The thrust F is given by the momentum theorem

$$F + (p_1 - p_a)A_1 - (p_2 - p_a)A_2 = (\rho_2 V_2 A_2)V_2 - (\rho_1 V_1 A_1)V_1$$

$$F = \rho_1 V_1 A_1\left[\left(1 + \frac{1}{N}\right)V_2 - V_1\right] + (p_2 - p_a)A_2 - (p_1 - p_a)A_1$$

At subsonic speeds, the pressure at the inlet maybe taken as atmospheric pressure, $p_1 = p_a$ and the last term on the right-hand side disappears. Further, in most cases, the pressure at the exhaust is also close to p_a. The fuel percentage compared to the air mass is very low and hence N is large or $1/N$ is small compared to unity. With these simplifications, we get

$$F = \rho_1 V_1 A_1 (V_2 - V_1)$$

7.7.1 Force of a Jet

The propulsive force of a jet issuing from the side of a tank may be also assessed in a similar fashion (Fig. 7.8).

The velocity of the jet V issuing from a smooth bell mouth outlet ($C_c = 1.0$ and $C_v = 0$) can be got from Bernoulli's theorem, $V = (2gH)^{1/2}$, where H is the constant head in the tank. If the velocity in the tank is assumed as zero, the change in momentum is

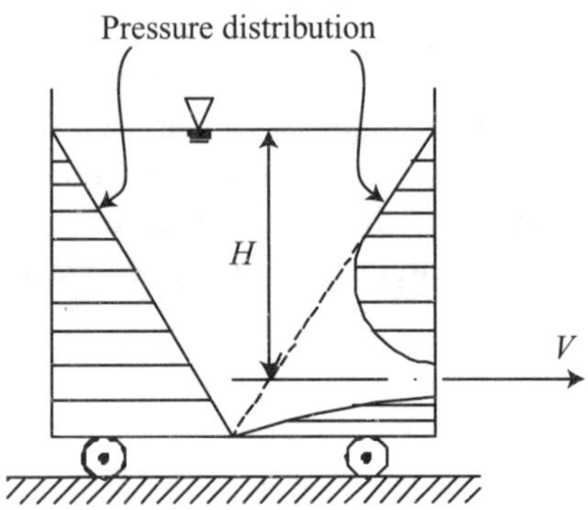

Fig. 7.8 Jet issuing from a tank

$$F = \rho Q V = \rho A V \cdot V = \rho A V^2 = \rho A \cdot 2gH$$
$$= 2\gamma A H$$

A being the area of the mouthpiece at the exit. If the tank had been mounted on friction-free wheels, it will move in the opposite direction due to the reaction of the jet. The force exerted on the tank will be equal to $F = 2\gamma A H$ and the velocity of the tank will be inversely proportional to its total mass. If the pressure distribution on the opposite wall of the tank is noted, it is observed that the pressure force on the opposite wall is larger than the pressure force on the wall in which the mouthpiece is located. The difference in force on the walls gives the thrust on the tank. Similarly, a boat can be propelled by a pump installed in the boat which draws water at the bow from the water body and pumps it out through a nozzle at the stern of the boat.

7.8 Loss of Energy in a Pipe Expansion

We had seen in Chapter 6 that the discharge coefficient in mouthpieces was less than 1.0 due to the formation of a vena contracta and subsequent expansion to fill the pipe. At that point a compensating coefficient C_d was introduced to take care of the loss since the energy theorem itself cannot predict the loss. The momentum theorem can be used with advantage to estimate the loss. Let us take the general case of a pipe expansion from area A_1 to A_2 as shown in Fig. 7.9.

The flow from the smaller pipe expands to the larger pipe and in doing so there is flow separation at the beginning of the expansion. There will be an eddy formed which is driven by the main flow through the free shear layer between the eddy and the main flow. The circulating pocket of water enclosed between the boundary and the main flow is called a standing eddy. The pressure in the eddy varies, being the lowest at the centre of the eddy. However, as an approximation the

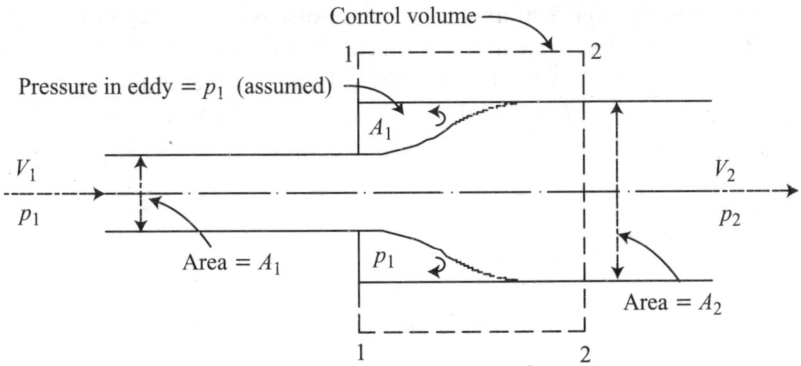

Fig. 7.9 Loss in pipe expansion

pressure in the standing eddy at Section 1-1 can be taken as the same pressure p_1 in the smaller pipe. Equating the pressure forces on the control surfaces 1-1-2-2-1 or bounded by 1-1 and 2-2 to the change in momentum between the two sections, we get

$$p_1A_1 + p_1(A_2 - A_1) - p_2A_2 = \rho A_2V_2(V_2 - V_1)$$

$$p_1A_2 - p_2A_2 = \rho A_2V_2(V_2 - V_1)$$

From continuity for incompressible flows, $A_1V_1 = A_2V_2$

$$p_1 - p_2 = \rho\frac{A_1}{A_2}V_1^2\left(\frac{A_1}{A_2} - 1\right)$$

We can now substitute this relation in the energy relationship between the two sections and find the loss in the expansion.

$$p_1/\gamma + V_1^2/2g = p_2/\gamma + V_2^2/2g + \text{Loss}$$

$$\text{Loss} = \frac{(p_1 - p_2)}{\rho} - \frac{1}{2g}(V_2^2 - V_1^2)$$

$$= \left(\frac{A_1}{A_2}\right)\frac{V_1^2}{g}\left[\left(\frac{A_1}{A_2}\right) - 1\right] - \frac{1}{2g}V_1^2\left[\left(\frac{A_1}{A_2}\right)^2 - 1\right]$$

$$\text{Loss} = \left(1 - \frac{A_1}{A_2}\right)^2\left(\frac{V_1^2}{2g}\right)$$

The loss may also be expressed in terms of V_2 from the equation of continuity. The pipe losses are usually expressed in terms of velocity head as loss $= KV_1^2$, where $K = (1 - A_1/A_2)^2$. The loss of head $h_L = KV_1^2/2g$. The loss of head in terms of the velocities at the two sections is

$$h_L = \frac{(V_1 - V_2)^2}{2g} \text{ since } \frac{A_1}{A_2} = \frac{V_2}{V_1}$$

Let us consider the flow through a straight mouthpiece discussed in Chapter 6. The flow expands immediately after the vena contracta and losses occur during this expansion similar to the loss in a pipe expansion. Here $A_1 = C_c A_2$ and $V_1 = V_2/C_c$. Hence the Bernoulli theorem applied to the two sections, one at the free surface and the other at the exit gives us (ignoring the loss between the entrance and the vena contracta since the flow is converging in that section),

$$H = V_2^2/2g + \text{Loss},\ h_L = V_2^2/2g + (V_1 - V_2)^2/2g$$

$$H = V_2^2/2g + (V_2/C_c - V_2)^2/2g = V_2^2/2g[1 + (1/C_c - 1)^2]$$

If $C_c = 0.62$, then $H = 1.3757V_2^2/2g$ or $V_2 = 0.8526(2gH)^{1/2}$ or

$Q = 0.8526A_2(2gH)^{1/2}$ and $C_d = 0.8526$ compared to the value of $C_d = 0.8$ normally adopted. The difference arises because of the assumptions made in the derivation of the expression for the loss in an expansion.

The pressure at the vena contracta, p_{VC} may be found now by applying the Bernoulli equation between the free surface and the vena contracta.

$$p_a/\gamma + H = p_{VC} + V_1^2/2g = p_{VC} + (V_2/C_c)^2/2g = p_{VC} + H/[C_c^2(1 + (1/C_c - 1)^2]$$

If $C_c = 0.62$, $p_{VC} = p_a/\gamma - 0.891H$ showing that the pressure at the vena contracta is less than atmospheric pressure.

7.9 Angular Momentum

It is sometimes necessary to find the moments exerted on a fluid volume such as in the case of rotating machinery like pumps and turbines. We will consider a two-dimensional flow. As illustrated in Fig. 7.10, the total momentum on the control volume is determined by integrating the moments on the differential elements comprising the total volume.

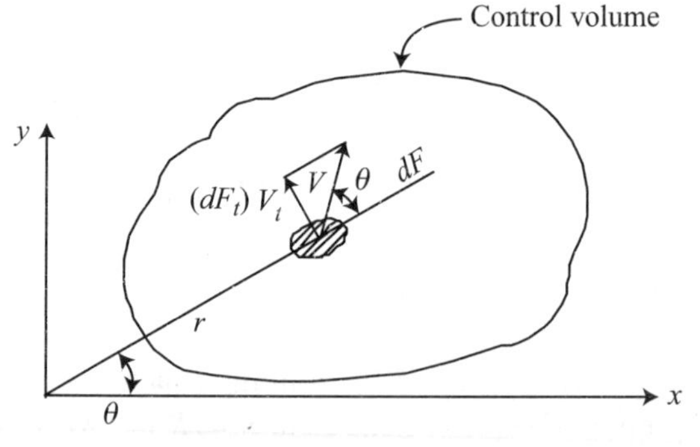

Fig. 7.10 Angular momentum

Let dT_0 be the moment exerted on the differential volume which is equal to the product of the tangential force dF_t normal to the radial distance, r between the force component and the differential volume and the moment is $dT_0 = dF_t r$. The differential force dF_t is obtained from the linear momentum equation applied in the tangential direction. ($dM = \rho d\forall$)

where M denotes the mass of volume $\forall$

$$dF_t = \partial(V_t dM)/\partial t + V_t(\rho V_n dA)$$

and the moment

$$dT_0 = r\partial(V_t dM)/\partial t + rV_t(\rho V_n dA)$$

The total torque is obtained by the integration of the above over the entire control volume.

$$T_0 = \frac{\partial}{\partial t}\int_{CV}(rV_t dM) + \int_{CS} rV_t(\rho V_n dA)$$

Since $V_t = V\sin\theta$, the above equation can be written as

$$T_0 = \frac{\partial}{\partial t}\int_{CV}(rV\sin\theta)dM + \int_{CS}(r.V\sin\theta)(\rho V_n dA)$$

The left-hand side stands for all the externally applied torques due to pressure, gravity, viscous or friction, magnetic forces and the first term on the right-hand side represents the rate at which angular momentum is stored in the control volume and the second term denotes the net rate of efflux of momentum out of the control volume. For steady flows, the first term is equal to zero and the torque becomes

$$T_0 = \int_{CS}(rV\sin\theta)(\rho V_n dA)$$

If the jet impinges on a vane of a turbine with a tangential velocity of V_{t1} and leaves with a tangential velocity of V_{t2} and the radial distances of entry and exit points from the central axis are r_1 and r_2 respectively and $Q = \int V_n dA$, we get the torque applied to the turbine as

$$T = \rho Q(r_2 V_{t2} - r_1 V_{t1})$$

7.9.1 Lawn Sprinkler

The sprinkler shown in Fig. 7.11 discharges at the rate of $Q/2$ through each nozzle (total discharge through the inlet at the centre $= Q$). The area of each nozzle located at a distance R from the centre is A. Find the relation between the torque due to friction and the speed of rotation.

Due to the reaction of the jet leaving the nozzle, the sprinkler will rotate in the clockwise direction as shown in the figure. The moment of momentum of the jets must be equal to the shaft friction which is the only externally applied torque.

$$T = \left(\frac{\rho Q}{2}\right)RV_t + \left(\frac{\rho Q}{2}\right)RV_t = \rho QRV_t$$

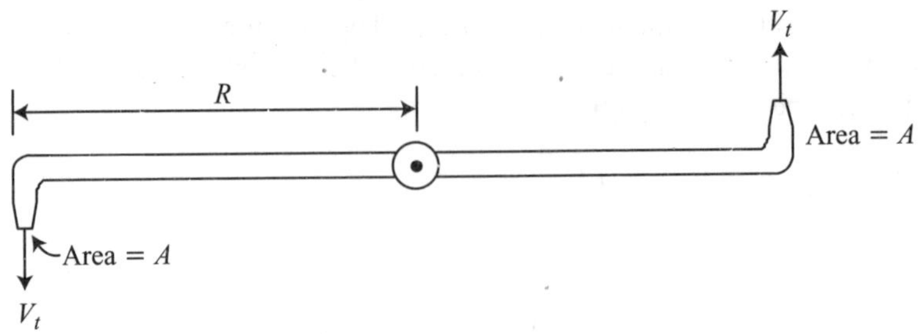

Fig. 7.11 Lawn sprinkler

V_t is the absolute velocity and $V_t = V - \omega R = Q/(2A) - \omega R$

$$T = \rho Q R[Q/(2A) - \omega R]$$

For a frictionless shaft $T = 0$, the maximum speed of rotation is

$$\omega = \frac{Q}{(2AR)}$$

and to find the value of T which prevents any rotation, we

put $\omega = 0$ and $T = \rho Q^2 R/(2A)$

If the jet is inclined at an angle, α to the radial direction, the expression for torque becomes

$$T = \rho Q R[Q/(2A)\sin\alpha - \omega R]$$

7.9.2 Pelton Wheel

Pelton wheel is an impulse type of turbine where a single jet of water impinges on a series of buckets mounted on the periphery of a wheel as shown in Fig. 7.12

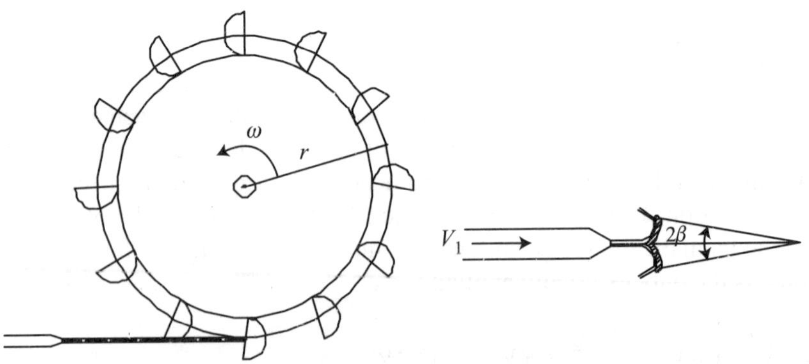

Fig. 7.12 Pelton wheel, bucket and vector diagram

The control volume chosen completely encloses the wheel. The jet velocity is V_j and enters the bucket at zero angle and it exits at an angle β to the line of the incoming jet as shown in figure. The mass of fluid entering the bucket is ρQ and its momentum is $\rho Q V_j$.

The exit velocity relative to the bucket is $V_j - \omega r$ but the exit velocity relative to the ground is found by the exit diagram shown in the figure

$$V_{\text{exit}} = -(V_j - \omega r)\cos\beta + \omega r$$

The fluid leaving the control volume has its momentum as $\rho Q[-(V_j - \omega r)\cos\beta + \omega r]$. The resultant force equal to the change of momentum in the axial direction is therefore

$$F = \rho Q V_j - \rho Q[-(V_j - \omega r)\cos\beta + \omega r] = \rho Q(V_j - \omega r)(1 + \cos\beta)$$

The torque on the water wheel is

$$T = \rho Q r(V_j - \omega r)(1 + \cos\beta)$$

Obviously the torque is a maximum for $\beta = 0^\circ$ or the jet turns through 180°.

The power developed is

$$P = T\omega = \rho Q\omega r(V_j - \omega r)(1 + \cos\beta)$$

The power becomes a maximum when $dP/d\omega = 0$ and this is found to occur when $V_j = 2\omega r$.

7 9.3 Reaction Turbines

Reaction turbines are of two types – mixed flow type (Francis turbine) and axial flow type (Kaplan or propeller turbines). In the Francis type turbine, the water first flows inwards and passes through fixed guide vanes (which serve to control the proper angle of entry of water to the runners) and then passes over rotating vanes called the runner. A torque is exerted on the runner in the direction of rotation. The basic difference between the Pelton wheel and the reaction turbine is that the fluid in the turbine enters the runner over the entire periphery at high pressure. Further, the velocity of the fluid is not constant over the runner vanes but actually increases along it. The runner in a reaction turbine is always full of water. In mixed flow turbines, the flow velocity leaving the rotor has both radial and axial components. In propellor type turbines, though the flow enters radially, the flow over the vanes is axial and leaves the rotor axially.

The basic equation for torque (developed first by Euler) in both types of turbine is the same. For steady flow, the equation for torque is

$$T = \int_{\text{CS}} rV_t(\rho V_n dA) = \rho Q[r_2 V_{t2} - r_1 V_{t1}]$$

We have to assess in each case the values of V_{t1} and V_{t2} from the vector diagram of velocities at inlet and exit of the runner vane. The diagram shown in Fig. 7.13 shows the passage formed by two vanes of a radial flow turbine which rotates at a constant speed about the vertical axis

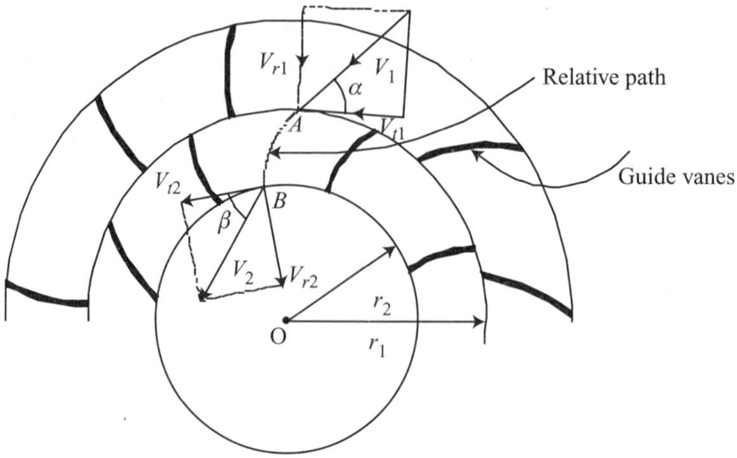

Fig. 7.13 Reaction turbine

through O. The absolute velocities are V_1 and V_2 and the tangential velocities are V_{t1} and V_{t2} at the entry and exit respectively. The difference in the moment of fluid momentum entering at A and that leaving at B is equal to the moment of all forces acting on the fluid in the passage. Since gravity does not contribute to this moment, it is the force of the vanes exerted on the fluid and hence equal to the torque exerted by the fluid on the turbine shaft.

$$V_{t1} = V_1 \cos\alpha \text{ and } V_{t2} = V_2 \cos\beta$$

The torque developed by the fluid is opposite to the torque of the shaft and hence

$$T = \rho Q[r_1 V_1 \cos\alpha - r_2 V_2 \cos\beta]$$

Multiplication of each term by angular velocity ω then yields the power, P delivered to the runner

$$P = T\omega = \rho Q\omega(r_1 V_1 \cos\alpha - r_2 V_2 \cos\beta)$$

Further applications of the force–momentum relationships may be found in the succeeding chapters.

EXAMPLE 7.3 In a turbine (Fig. 7.13), the absolute velocities at entry and exit are 15 m/s and 5 m/s respectively. The outer and inner radii of the blades are 50 cm and 25 cm respectively with $\alpha = 25^\circ$ and $\beta = 75^\circ$. The discharge is 5 m^3/s and the turbine is rotating at 250 rpm. Find the torque exerted by the discharge and the power delivered to the runner.

$$\text{Torque} \quad T = 998 \times 5[15 \times \cos 25^\circ \times 0.50 - 5 \times \cos 75^\circ \times 0.25]$$

$$= 4990[6.797 - 0.113] = 33353 \text{ Nm}$$

$$\text{Power} \quad P = T\omega = 33353 \times \frac{2\pi \times 250}{60} = 1173119\,\text{W} = 1173.19\,\text{kW}$$

EXAMPLE 7.4 Design a Pelton wheel to develop a maximum power of 1000 kW under a head of 150 m operating at 300 rpm. Assume that the coefficient of velocity of the jet $C_v = 0.98$. Assume that the diameter of the jet is equal to one-tenth of the diameter of the wheel.

Torque is maximum for $\beta = 0$; the jet is turned through 180°

$$V_j = C_v\sqrt{2gH}$$
$$= 0.98\sqrt{2 \times 9.81 \times 150}$$
$$= 53.16\,\text{m/s}$$

For maximum power, peripheral velocity of buckets $= \omega r = \dfrac{V_j}{2} = \dfrac{53.16}{2} = 26.58\,\text{m/s}$

$$\text{But} \quad \omega = \frac{2\pi N}{60} = \frac{2\pi \times 300}{60} = 31.42\,\text{radians}$$

$$r = \frac{26.58}{31.42} = 0.846\,\text{m}; \quad D = 1.692\,\text{m}$$

$$d = \text{Diameter of jet} = \frac{1.692}{10} = 0.1692\,\text{m}$$

$$Q = V_j A_j = 53.16 \times \frac{\pi}{4} \times (0.1692)^2 = 61.195\,\text{m}^3/\text{s}$$

Number of buckets required is empirically given as $(D/2d) + 15 = 20$
Similarly the width and depth of the buckets are empirically specified as

Width of buckets $= 5 \times d = 0.846\,\text{m}$

Depth of bucket $= 1.2d = 1.2 \times 0.1692 = 0.203\,\text{m}$

EXERCISES

7.1. A jet of oil of specific gravity 0.80 strikes a stationary plate held normal to the jet direction (Fig. 7.2*a*).

(a) If the velocity of the jet is 10 m/s, find the force on the stationary plate.
(b) What is the force if the plate is moving away from the jet at a uniform velocity of 2 m/s?
(c) Determine the force if the plate is moving towards the jet at a velocity of 2 m/s.

7.2. A jet of water with a velocity of 15 m/s strikes a plate inclined at an angle of 60° to the axis of the jet (Fig. 7.2*b*). Find the components of the discharge in the two directions parallel to the plate. Find also the force component normal to the plate. If the inclined plate is moving towards the jet at a uniform velocity of 5 m/s, find the power required to move the plate.

7.3. Determine the force components on a single moving vane shown in the figure due to a jet impinging on the vane at a velocity of 40 m/s and the vane is moving at a velocity of 10 m/s. The diameter of the jet is 2 cms and the deflection angle of the vane is 150°.

(a) What are the components of the absolute velocity V_2 leaving the vane.

(b) Find the magnitude of V_2 and the kinetic energy leaving the vane and compare it with the kinetic energy of the jet entering the vane.

(c) At what vane angle does the horizontal component of absolute velocity become zero.

(d) If the absolute velocity is inclined at an angle of 120°, find the guide vane angle.

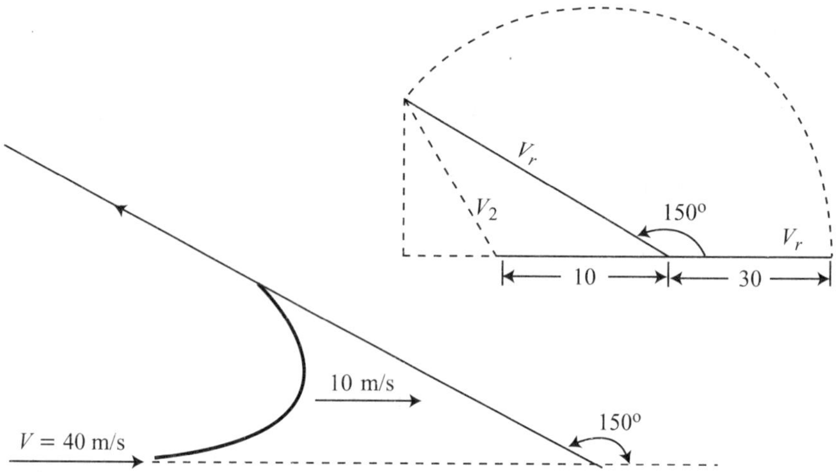

Figure for Problem 7.3

7.4. Show that the power developed by a moving vane when the vane speed V_b is one-third of the jet speed, $V_1 (V_b = V_1/3)$

7.5. Water flows up a vertical reducing section shown in the figure. The pressure at the larger section is equal to 10 m of water. Neglecting the losses and weight of water in the pipe, determine the force on the contraction.

7.6. A horizontal 90° bend located in a pipe of 20 cm diameter carries a discharge of 0.5 m^3/s. The pressure in the pipe is 1.2×10^4 Pa.

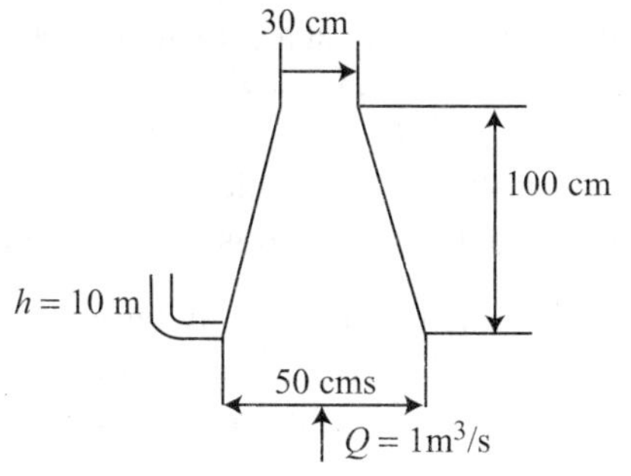

Figure for Problem 7.4

(a) Find the force components and the resultant force.

(b) If the bend at the end section 2-2 is open to the atmosphere, find the force components.

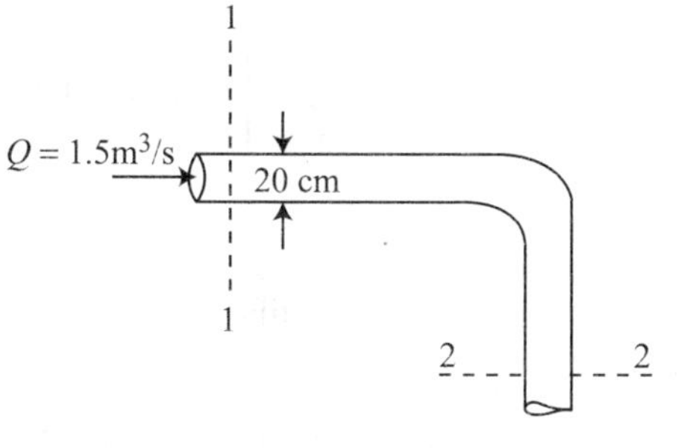

Figure for Problem 7.5

7.7. A pipe with an initial diameter of 20 cm is bent on the horizontal plane through an angle of 60° and reduces to an exit diameter of 10 cm. If the initial velocity is 5 m/s and the initial pressure 200 kN/m^2, find the force components and the resultant force on the bend. Ignore the losses and weight of water in the bend. If the bend is in the vertical plane and the difference in elevation between the entry and exit ends of the bend is 1 m, find the force components and resultant force.

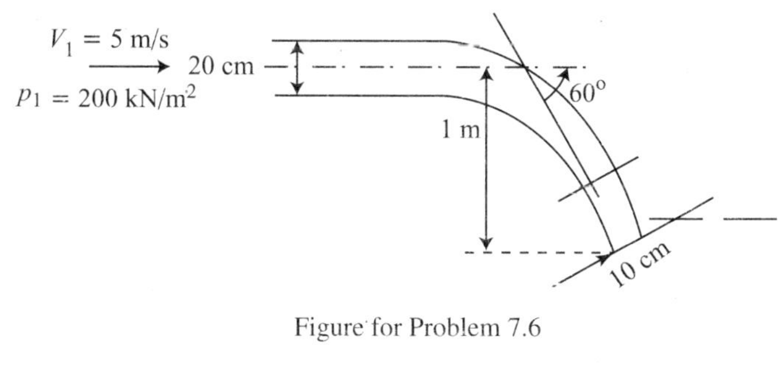

Figure for Problem 7.6

7.8. A stationary rocket discharges gases at a velocity of 800 m/s. The mass flow rate of gases is 6 kg/s. The exit area is 0.10 m^2. The nozzle exit pressure is 200 kN/m^2 and the outside ambient pressure is 100 kN/m^2. Determine the thrust on the rocket.

7.9. A rocket weighs $W_0 = 120000$ kN of which 75% is fuel weight. The fuel is consumed at a constant rate and the exit velocity, v_e is 3000 m/s. Find

(a) the time taken for the fuel to be fully burned out [Hint: Initial thrust should be greater than the weight. Assuming the thrust required is equal to the weight, $(p_e A_2 v_e) v_e = M_0$. Hence, find $\alpha_e 300 = 120000$ i.e., $\alpha_e = 40$ kN/s; $T = 0.75 W_0/g\alpha_e$],

(b) speed of the rocket at the instant the fuel is exhausted

(c) the height of the rocket at this stage (assume $p_e = 0$) and

(d) the maximum cruising height that the rocket will reach.

$$\text{Hint:} \quad h = \int_0^T v dt = -\frac{gt^2}{2}\bigg|_0^T - \int_0^T \ln\left(1 - \frac{\alpha_e}{M_0}t\right) dt$$

$$= -\frac{gt^2}{2}\bigg|_0^T - \left\{\left[-t - \ln\left(1 - \frac{\alpha_e}{M_0}t\right)\right]\left[\frac{M_0}{\alpha_e} - t\right]\right\}_0^T$$

7.10. A lawn sprinkler with a radial arm of 15 cm has nozzles at the end of the rotating arm. The nozzle diameter is 1.5 cm and is inclined at 60° as shown in the figure. The velocity at the exit is 10 m/s. Find

(a) the torque required to hold the rotating arm stationary;

(b) if the arm is free to rotate without any friction, find the speed of rotation of the arm;

(c) if there are 2 arms at right angles as shown in Fig. (*b*), will the arm rotate faster, if the total discharge remains the same?;

(d) if the nozzles are directed as shown in Fig. (*c*), will there be any rotation?;

(e) if the two nozzles are located at different radii as shown in (*d*), will there be any rotation? What is its direction?

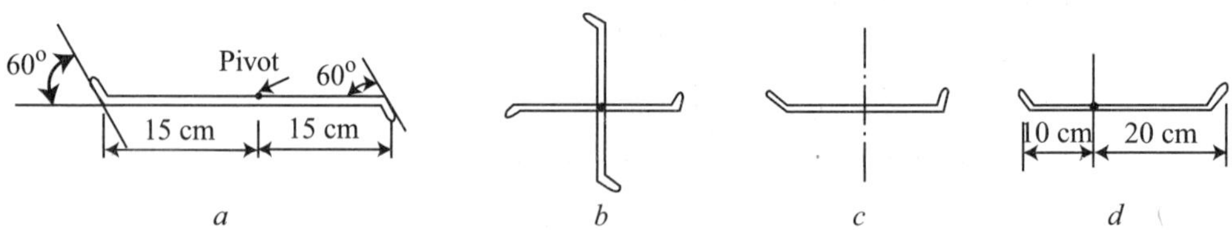

Figure for Problem 7.9

7.11. A pipe 5 cm in dia. has an expansion followed by a contraction as shown in the figure. Find the losses at the expansion and contraction if the velocity in the 5 cm pipe is 5 m/s. Assume $C_c = 0.6$.

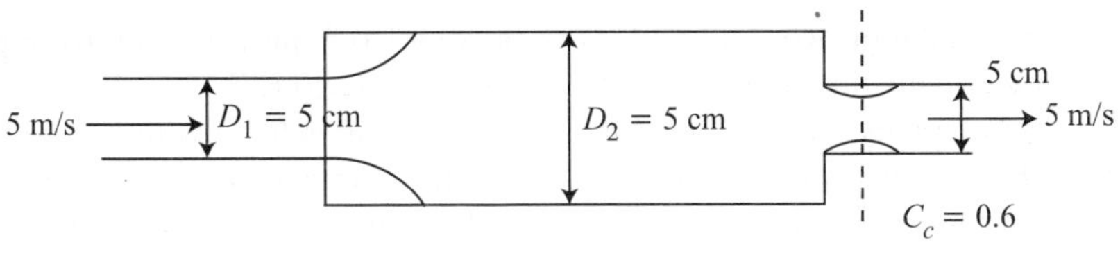

Figure for Problem 7.10

7.12. The absolute velocities of water entering and leaving a reaction turbine are 30 m/s and 10 m/s respectively. The inner and outer radii of the blades are 40 cm and 60 cms respectively. The discharge is 4.5 m^3/s. The speed of rotation is 300 rpm. The angles at entry and exist are 30° and 75° respectively. Find the torque exerted by the discharge and the power delivered to the runner.

7.13. Show that the maximum power from a Pelton wheel impulse turbine is obtained when the peripheral speed is half of the jet speed.

7.14. A Pelton wheel has to develop 5000 kW under an effective head of 250 m. Find the wheel diameter for maximum power delivery with an efficiency of 90%. The runner speed is 300 rpm. Estimate the discharge.

7.15. A water nozzle is directed vertically downwards against a flat fixed plate as shown in the sketch. Find the force exerted on the plate by the water jet. Find V_j and V.

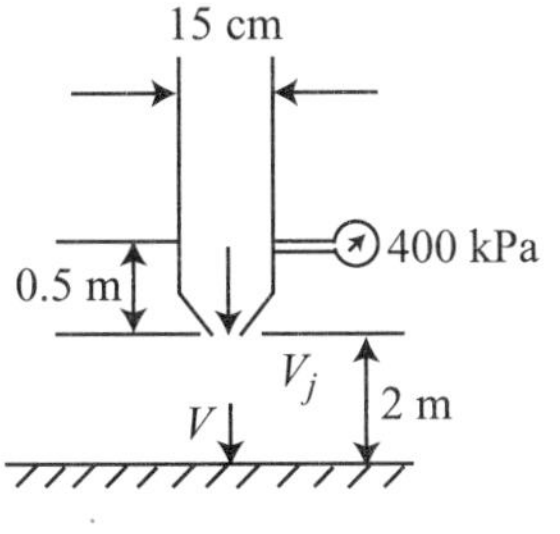

Figure for Problem 7.14

7.16. A vertical jet of diameter 5 cms strikes a wooden float of diameter 1 m and height 1.5 m and of specific gravity 0.8. The float has a top curved surface deflecting the jet by 30°. Determine the velocity of the jet so that the top of the wooden float is just level with water surface.

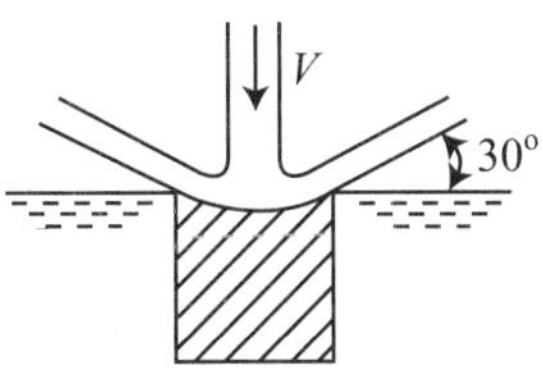

Figure for Problem 7.15

7.17. A 10 cm circular orifice is located at the end of a 15 cm pipe ($C_c = 0.6$). What force will be exerted on the orifice if the pressure in the pipe is 200 kPa.

7.18. A rectangular tank measuring 1.2 m × 2 m × 2.75 m is filled with water to the required height H, and the level is maintained constant. The tank is mounted on wheels whose rolling friction is given as 0.05. An orifice of diameter 10 cm is located in the side of the tank at a height 0.5 m above the bottom. Find the minimum head in the tank to set the tank in motion. Neglect the weight of the tank itself.

8 Laminar Flow

Part I Laminar Motion

8.1 Introduction

In laminar flows, fluid particles travel along distinct layers and a particle from one layer does not cross over to another layer. This occurs essentially at low velocity in confined spaces and in fluids having high viscosity. In such cases, any small disturbance, either naturally present in the system or externally induced, will be suppressed by the viscous shear and will not be allowed to grow and disturb the layered flow. Any disturbing force will tend to cause acceleration of flow around the disturbance, overcome the inertia of the particles and may move them in random directions, thus transporting particles from one layer to another. If the viscous forces are strong enough to overcome the disturbing forces, then the flow will remain laminar. On the other hand, if the inertial forces are stronger than the resisting viscous forces, then the flow will be turbulent and particles from different layers mingle with each other. The relative values of inertial and viscous forces will therefore determine the nature of flow. The ratio of inertial forces to viscous forces is known as the Reynolds number.

$$\mathbf{R} = \frac{\text{Inertial forces}}{\text{Viscous forces}}$$

We have seen in the derivation of Euler's equation, the inertial force terms are represented by terms such as $\rho u\ \partial u/\partial x$ or more generally $\rho U(U/L)$, where ρ is the mass density of the fluid, U is a characteristic velocity and L is a characteristic length of the flow. The Newton's law of viscosity is

$$\tau = \mu \frac{\partial u}{\partial y}$$

The shear force is $\partial\tau/\partial y$ $(\mu\partial^2 u/\partial y^2)$ or again in general terms $\mu(U/L^2)$. The ratio of the inertial and viscous forces then becomes

$$\frac{\rho(U^2/L)}{(\mu U/L^2)} = \frac{\rho U L}{\mu}$$

The Reynolds number is thus defined as

$$\mathbf{R} = \frac{\rho U L}{\mu} \tag{8.1}$$

8.2 Shear Stresses

In laminar flows, viscous shear forces play a dominant role. In Chapter 1, the property of viscosity of fluids had been discussed. Viscosity is the property of a fluid that enables it to develop resistance to deformation under the action of shear forces. Viscosity may also be seen as a momentum transfer coefficient. The change in momentum of a unit volume of fluid across a distance dy is (Fig. 8.1)

$$(\rho + \partial\rho)(u + \partial u) - \rho u = \partial(\rho u) \tag{8.2}$$

and the momentum gradient or potential across the depth dy is $\partial(\rho u/\partial y)$.

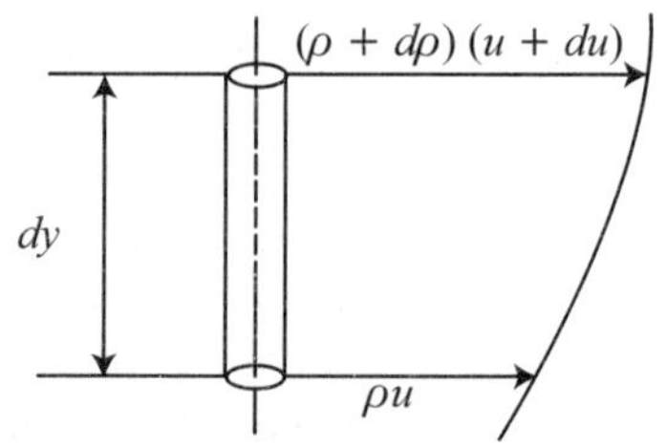

Fig. 8.1 Velocity gradient and shear stress

This can be caused by a shear stress $\tau_{yx} = \nu\partial(\rho u)/\partial y$, where ν, the kinematic coefficient of viscosity may be interpreted as a momentum transfer coefficient. The dynamic coefficient of viscosity, μ has also been defined as $\tau/(\partial u/\partial y)$ and $\nu = \mu/\rho$. Newton's law of viscosity has also been presented in Chapters 1 and 3 and the relation between shear stress, τ and angular deformation have been derived. In one-dimensional motion, the relationship is

$$\tau_{yx} = \mu\left(\frac{du}{dy}\right) \tag{8.3}$$

and in three-dimensional motion

$$\tau_{yx} = \mu\left(\frac{\partial u}{\partial y} + \frac{\partial v}{\partial x}\right);\ \tau_{zy} = \mu\left(\frac{\partial v}{\partial z} + \frac{\partial w}{\partial y}\right);\ \tau_{xz} = \mu\left(\frac{\partial w}{\partial x} + \frac{\partial u}{\partial z}\right) \tag{8.4}$$

As defined earlier, the first subscript indicates the direction of the normal to the plane on which the shear stress is acting and the second subscript refers to the direction of action of the shear stress. In all there are six shear stresses on the six faces of the three-dimensional element but there are only three independent stresses since

$$\tau_{yx} = \tau_{xy};\ \tau_{zy} = \tau_{yz};\ \tau_{xz} = \tau_{zx} \tag{8.5}$$

To prove equation (8.5) consider a slice of element in the xy-plane with sides dx and dy. The stresses acting on the element are shown in Fig. 8.2. We take moments of all the forces acting on the element about the centroid. The normal forces and the body forces, if any, do not have moments as they all pass through the centroid. The moments are only due to the shear forces acting on the four faces as shown. The moments are (anti-clockwise direction is positive)

$$-\left[\tau_{yx} + \left(\frac{\partial \tau_{yx}}{\partial y}\right)\left(\frac{dy}{2}\right)\right] dx\, dz \left(\frac{dy}{2}\right) - \left[\tau_{yx} - \left(\frac{\partial \tau_{yx}}{\partial y}\right)\left(\frac{dy}{2}\right)\right] dx\, dz \left(\frac{dy}{2}\right)$$

$$+\left[\tau_{xy} + \left(\frac{\partial \tau_{xy}}{\partial x}\right)\left(\frac{dx}{2}\right)\right] dy\, dz \left(\frac{dx}{2}\right) + \left[\tau_{xy} - \left(\frac{\partial \tau_{xy}}{\partial x}\right)\left(\frac{dx}{2}\right)\right] dy\, dz \left(\frac{dx}{2}\right)$$

$$= \frac{d(\text{angular momentum})}{dt}$$

On simplification, we get

$(\tau_{xy} - \tau_{yx}) dx\, dy\, dz = d/dt$ (angular momentum) $= d/dt$ (moment of inertia $\times$ angular speed)

As the size of the element is made smaller and smaller, the right-hand side containing the moment of inertia having an order of $(\delta x)^4$ disappears faster than the left-hand side which is of the order $(\delta x)^3$. In the final limit as $(\delta x) \to 0$, we have $\tau_{xy} = \tau_{yx}$. By similar arguments and taking slices of element in the other two planes, it can be shown that $\tau_{yz} = \tau_{zy}$ and $\tau_{xz} = \tau_{zx}$.

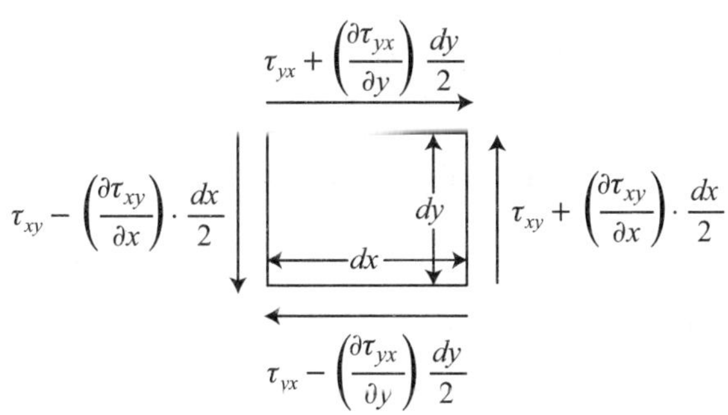

Fig. 8.2 Complementary shear stresses

8.3 Navier–Stokes Equations

We now apply Newton's second law to the element shown in Fig. 8.2 including the shear forces, normal forces, and body forces in the x-direction (mass $\times$ acceleration $=$ force).

$$\rho\,(dx\;dy\;dz)\left(\frac{du}{dt}\right) = dx\;dy\;dz\left[X + \frac{\partial\sigma_{xx}}{\partial x} + \frac{\partial\tau_{yx}}{\partial y} + \frac{\partial\tau_{zx}}{\partial z}\right] \tag{8.6}$$

where X is component of the body force in the x-direction. In a purely gravitational field, $X = 0$, $Z = 0$ and $Y = -g$

The acceleration term du/dt can be expanded as

$$\frac{du}{dt} = \frac{\partial u}{\partial t} + u\cdot\frac{\partial u}{\partial x} + v\cdot\frac{\partial u}{\partial y} + w\cdot\frac{\partial u}{\partial z}$$

$$\tau_{yx} = \mu\left(\frac{\partial v}{\partial x} + \frac{\partial u}{\partial y}\right) \text{ and } \tau_{zx} = \mu\left(\frac{\partial u}{\partial z} + \frac{\partial w}{\partial x}\right)$$

It can be shown that for incompressible flows (this proof is quite involved and it is not presented here)

$$\sigma_{xx} = -p + 2\mu\left(\frac{\partial u}{\partial x}\right); \sigma_{yy} = -p + 2\mu\left(\frac{\partial v}{\partial y}\right); \text{ and } \sigma_{zz} = -p + 2\mu\left(\frac{\partial w}{\partial z}\right)$$

where $-p = (\sigma_{xx} + \sigma_{yy} + \sigma_{zz})/3$

Substituting these values in the Newton's equation (8.6), we obtain

$$\frac{\partial u}{\partial t} + u\frac{\partial u}{\partial x} + v\frac{\partial u}{\partial y} + w\frac{\partial u}{\partial z}$$

$$= X - \left(\frac{1}{\rho}\right)\frac{\partial p}{\partial x} + 2\nu\frac{\partial^2 u}{\partial x^2} + \nu\left(\frac{\partial^2 u}{\partial y^2} + \frac{\partial^2 v}{\partial x\partial y}\right) + \nu\left(\frac{\partial^2 u}{\partial z^2} + \frac{\partial^2 w}{\partial x\partial z}\right)$$

$$\text{R.H.S} = X - \left(\frac{1}{\rho}\right)\frac{\partial p}{\partial x} + \nu\left(\frac{\partial^2 u}{\partial x^2} + \frac{\partial^2 u}{\partial y^2} + \frac{\partial^2 u}{\partial z^2}\right) + \nu\frac{\partial}{\partial x}\left(\frac{\partial u}{\partial x} + \frac{\partial v}{\partial y} + \frac{\partial w}{\partial z}\right)$$

The last term on the right-hand side vanishes because of the equation of continuity. The final form of the Navier–Stokes equation (8.7) for incompressible flows in the x-direction is

$$\frac{\partial u}{\partial t} + u\frac{\partial u}{\partial x} + v\frac{\partial u}{\partial y} + w\frac{\partial u}{\partial z} = X - \left(\frac{1}{\rho}\right)\frac{\partial p}{\partial x} + \nu\left(\frac{\partial^2 u}{\partial x^2} + \frac{\partial^2 u}{\partial y^2} + \frac{\partial^2 u}{\partial z^2}\right) \tag{8.7}$$

Similar equations are developed in the y- and z-directions. Viz.,

$$\frac{dv}{dt} = Y - \left(\frac{1}{\rho}\right)\frac{\partial p}{\partial y} + \nu\left(\frac{\partial^2 v}{\partial x^2} + \frac{\partial^2 v}{\partial y^2} + \frac{\partial^2 v}{\partial z^2}\right) \tag{8.8}$$

$$\frac{dw}{dt} = Z - \left(\frac{1}{\rho}\right)\frac{\partial p}{\partial z} + \nu\left(\frac{\partial^2 w}{\partial x^2} + \frac{\partial^2 w}{\partial y^2} + \frac{\partial^2 w}{\partial z^2}\right) \tag{8.9}$$

The terms $\nu\nabla^2 u$, $\nu\nabla^2 v$, $\nu\nabla^2 w$ are the viscous terms known as the Navier–Stokes terms and they define the viscous forces per unit mass. These were not present in Euler's equations for ideal fluids since $\mu = 0$ for ideal fluids. However, the viscous terms also vanish for irrotational flows of real fluids as shown in the example below.

EXAMPLE 8.1 Show that for irrotational flows, the expressions in the viscous terms $\nabla^2 u = \nabla^2 v = \nabla^2 w = 0$.

For irrotational flows, we have

$$\frac{\partial u}{\partial y} = \frac{\partial v}{\partial x}; \text{ therefore } \frac{\partial^2 u}{\partial y^2} = \frac{\partial^2 v}{\partial x \partial y}$$

$$\frac{\partial u}{\partial z} = \frac{\partial w}{\partial x}; \text{ therefore } \frac{\partial^2 u}{\partial z^2} = \frac{\partial^2 w}{\partial x \partial z}$$

$$\frac{\partial^2 u}{\partial x^2} + \frac{\partial^2 u}{\partial y^2} + \frac{\partial^2 u}{\partial z^2} = \frac{\partial}{\partial x}\left(\frac{\partial u}{\partial x} + \frac{\partial v}{\partial y} + \frac{\partial w}{\partial z}\right) = 0$$

since the term within the bracket is zero by virtue of the equation of continuity.

The Navier–Stokes equations are a set of non-linear equations for which closed form solutions for complex boundary conditions are difficult to find. However for particular cases of flow with simple boundary conditions, exact solutions are available which will be presented in the succeeding sections.

8.4 Navier–Stokes Equations in Cylindrical Coordinates

We can derive these equations in cylindrical or spherical coordinates either from first principles or transformation from Cartesian coordinates to the desired coordinate system. The derivation is not presented in this text but the final form in the cylindrical coordinate system is presented here.

Radial:

$$\frac{\partial u}{\partial t} + u\left(\frac{\partial u}{\partial r}\right) + \left(\frac{v}{r}\right)\left(\frac{\partial u}{\partial \theta}\right) + w\left(\frac{\partial u}{\partial z}\right) - \frac{v}{r^2} = R - \left(\frac{1}{\rho}\right)\left(\frac{\partial p}{\partial r}\right)$$
$$+ \nu\left[\frac{\partial^2 u}{\partial r^2} + \left(\frac{1}{r}\right)\left(\frac{\partial u}{\partial r}\right) + \left(\frac{1}{r^2}\right)\left(\frac{\partial^2 u}{\partial \theta^2}\right) + \frac{\partial^2 u}{\partial z^2} - \frac{u}{r^2} - \left(\frac{2}{r^2}\right)\left(\frac{\partial v}{\partial \theta}\right)\right]$$

Tangential:

$$\frac{\partial v}{\partial t} + u\left(\frac{\partial v}{\partial r}\right) + \left(\frac{v}{r}\right)\left(\frac{\partial v}{\partial \theta}\right) + w\left(\frac{\partial v}{\partial z}\right) + \frac{uv}{r} = \theta - \left(\frac{1}{\rho r}\right)\frac{\partial p}{\partial \theta}$$
$$+ \nu\left[\left(\frac{\partial^2 v}{\partial r^2}\right) + \left(\frac{1}{r}\right)\left(\frac{\partial v}{\partial r}\right) + \left(\frac{1}{r^2}\right)\left(\frac{\partial^2 v}{\partial \theta^2}\right) + \frac{\partial^2 v}{\partial z^2} + \left(\frac{2}{r^2}\right)\left(\frac{\partial v}{\partial \theta}\right) - \frac{v}{r^2}\right]$$

Axial:

$$\frac{\partial w}{\partial t}+u\left(\frac{\partial w}{\partial r}\right)+\left(\frac{v}{r}\right)\left(\frac{\partial w}{\partial \theta}\right)+w\left(\frac{\partial w}{\partial z}\right)=Z-\left(\frac{1}{\rho}\right)\left(\frac{\partial p}{\partial z}\right)$$

$$+\nu\left[\frac{\partial^2 w}{\partial r^2}+\left(\frac{1}{r}\right)\left(\frac{\partial w}{\partial r}\right)+\left(\frac{1}{r^2}\right)\left(\frac{\partial^2 w}{\partial \theta^2}\right)+\frac{\partial^2 w}{\partial z^2}\right] \tag{8.10}$$

8.5 Exact Solutions of Navier–Stokes Equations

There are two types of exact solutions available in literature as of date. The first class belongs to the type where non-linear terms on the left-hand side of the Navier–Stokes equation (the inertial or acceleration terms) disappear due to the simple boundary conditions present and only linear terms are left which can be easily solved. The other class belongs to the type where non-linear terms are retained but closed form solutions are available which satisfy the boundary conditions. There is a third type which may be classified as those which have approximate solutions omitting some terms in the derivation. We shall discuss solutions to linearised equations in detail.

8.6 Flow between Horizontal Parallel Plates (Hagen–Poiseuille Flow)

Hagen, a German engineer, and Poiseuille, a physician working on the resistance to flow through tubes were the first to investigate laminar flow through small tubes. The latter was interested in the flow of blood through arteries and veins. We shall start our study of laminar flows by observing a steady uniform flow between parallel horizontal plates under the action of a pressure gradient. The spacing between the plates is $2a$ as shown in Fig. 8.3.

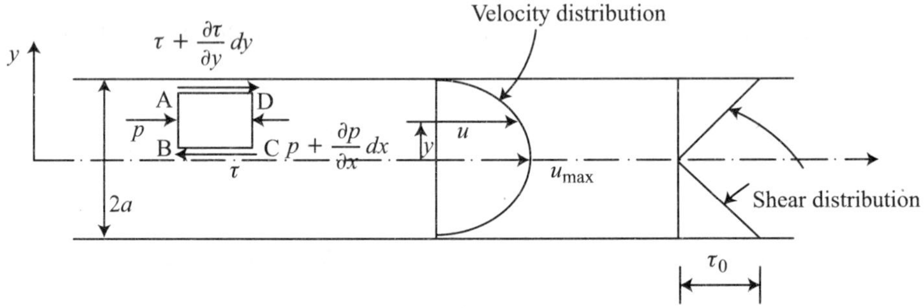

Fig. 8.3 Flow between parallel plates

We shall derive the expression for velocity distribution from first principles and later demonstrate the use of the Navier–Stokes equations for the same problem.

Consider an element, ABCD of sides dx and dy and unit depth as shown in the figure. The flow is steady and uniform. The discharge between the plates is constant. There is no acceleration of the fluid in any direction. At the boundaries the velocity is zero, as in real fluids there cannot be any slip between the fluid in contact with the boundary and the stationary boundary (slip is possible in an ideal fluid flow). We also realise that along the centre line, the fluid particle being farthest removed from the boundaries, the velocity must be maximum. The velocity distribution must also be symmetric across the width between the plates. The only forces acting on the element are the pressure gradient and viscous shear forces and these two forces must be in equilibrium since there is no acceleration.

$$p \cdot dy - \left[p + \left(\frac{\partial p}{\partial x}\right) dx\right] dy - \tau \cdot dx + \left[\tau + \left(\frac{\partial \tau}{\partial y}\right) dy\right] dx = 0$$

which on simplification yields $\partial p/\partial x = \partial \tau/\partial y$. Since p is a function of x only and τ is a function of y only, we can write the result as

$$\left(\frac{dp}{dx}\right) = \frac{d\tau}{dy} \text{ and by Newton's law } \tau = \mu \frac{du}{dy}$$

$$\frac{d(\mu du/dy)}{dy} = \left(\frac{dp}{dx}\right) \text{ or } \frac{d^2u}{dy^2} = \left(\frac{1}{\mu}\right)\left(\frac{dp}{dx}\right) \tag{8.11}$$

Integrating once,

$$\frac{du}{dy} = \left(\frac{1}{\mu}\right)\left(\frac{dp}{dx}\right) y + C_1$$

The constant $C_1 = 0$, since $du/dy = 0$ at $y = 0$.
Integrating again we get

$$u = \left(\frac{1}{\mu}\right)\left(\frac{dp}{dx}\right)\frac{y^2}{2} + C_2$$

The constant C_2 is evaluated from the condition $u = 0$ at $y = \pm a$. The final expression for the velocity distribution is

$$u = -\left(\frac{1}{2\mu}\right)\left(\frac{dp}{dx}\right)(a^2 - y^2) \tag{8.12}$$

The distribution is parabolic. The maximum velocity at the centre is

$$u_{\max} = -\left(\frac{1}{2\mu}\right)\left(\frac{dp}{dx}\right) a^2$$

The negative sign occurs since the pressure decreases in the direction of flow. The shear stress $\tau = (dp/dx)y$ and therefore varies linearly from a zero value at the centre $y = 0$ and a maximum value at the boundary $\tau_0 = \pm(dp/dx)a$. The discharge can be found by integrating udy across the spacing.

$$q = \int u\,dy = \int_{-a}^{+a} -\left(\frac{1}{2\mu}\right)\left(\frac{dp}{dx}\right)(a^2 - y^2)\,dy = \left(\frac{2}{3\mu}\right)\left(-\frac{dp}{dx}\right)a^3$$

The average velocity is

$$V = \frac{q}{2a} = \left(\frac{1}{3\mu}\right)\left(-\frac{dp}{dx}\right)a^2 \text{ and } V = \left(\frac{2}{3}\right)u_{\max}$$

The force on one plate of length L due to the shear stress is

$$\int_0^L \tau_0 dx = \left(\frac{\Delta p}{L}\right)aL = (\Delta p \cdot a)$$

where Δp is the pressure loss over a length L. For both plates, the combined force is $(2\Delta p \cdot a)$. The power, P required is

$$\text{Force} \times \text{average velocity} = 2\Delta\ p \cdot a\left[\left(\frac{1}{3\mu}\right)\left(-\frac{dp}{dx}\right)a^2\right] = \left(\frac{2}{3\mu}\right)\left(\frac{\Delta p}{L}\right)^2 a^3 L$$

This is also equal to the power required $P = \gamma q h_f$ to overcome the head loss $h_f = \Delta p/\gamma$.

8.6.1 Hagen–Poiseuille flow with x-axis along the Lower Plate

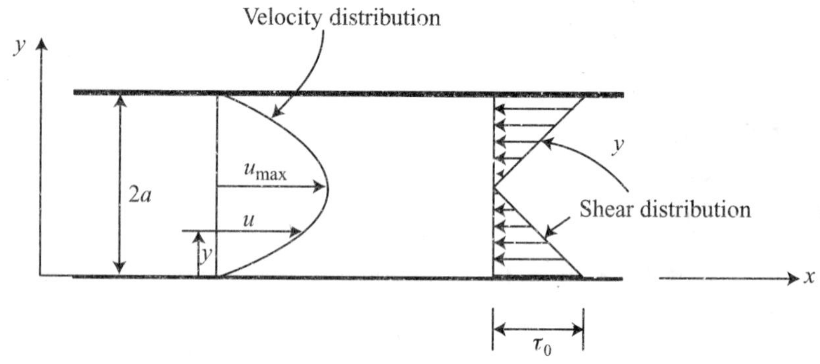

Fig. 8.4 Flow between parallel plates with the x-axis along the lower plate

If the x-axis is made to coincide with the lower plate (Fig. 8.4), then we have $u = 0$ at $y = 0$ and $y = 2a$, then $du/dy = 0$ at $y = a$.

$$\frac{du}{dy} = \left(\frac{1}{\mu}\right)\left(\frac{dp}{dx}\right)y + C_1 \text{ and } C_1 = \left(\frac{1}{\mu}\right)\left(-\frac{dp}{dx}\right)a$$

$$u = \left(\frac{1}{\mu}\right)\left(\frac{dp}{dx}\right)\left(\frac{y^2}{2}\right) - \left(\frac{1}{\mu}\right)\left(\frac{dp}{dx}\right)ay + C_2$$

But $C_2 = 0$ since $u = 0$ at $y = 0$. Hence the expression for the velocity distribution with the x-axis lying on the lower plate is

$$u = \left(\frac{1}{\mu}\right)\left(\frac{dp}{dx}\right)\left(\frac{y^2}{2} - ay\right) \tag{8.13}$$

8.6.2 Derivation from Navier–Stokes Equations

One can readily obtain the above solution for flow between parallel plates by the direct use of the Navier–Stokes equation in the x-direction.

$$\frac{\partial u}{\partial t} + u\frac{\partial u}{\partial x} + v\frac{\partial u}{\partial y} + w\frac{\partial u}{\partial z} = X - \left(\frac{1}{\rho}\right)\left(\frac{\partial p}{\partial x}\right) + \nu\left(\frac{\partial^2 u}{\partial x^2} + \frac{\partial^2 u}{\partial y^2} + \frac{\partial^2 u}{\partial z^2}\right)$$

We can eliminate some of the terms for the specific case of flow between parallel plates.

The flow is steady: $\partial u/\partial t = 0$; flow is uniform in the x-direction: $\partial u/\partial x = 0$; $\partial^2 u/\partial x^2 = 0$. There are no changes in the z-direction: $\partial u/\partial z = 0, \partial^2 u/\partial z^2 = 0$ and $\partial w/\partial z = 0$; by the equation of continuity, if $\partial u/\partial x = \partial w/\partial z = 0$, then $\partial v/\partial y = 0$ or integrating $v =$ constant. But $v = 0$ at the boundaries and hence $v = 0$ everywhere. By virtue of these arguments, all the terms on the left-hand side disappear. Thus the equation is now a linear equation. Further the body force, X in the x-direction is zero. After introducing all these simplifications, the equation becomes

$$0 = -\left(\frac{1}{\rho}\right)\frac{\partial p}{\partial x} + \mu\frac{\partial^2 u}{\partial y^2} \text{ or } \frac{\partial^2 u}{\partial y^2} = \left(\frac{1}{\mu}\right)\frac{\partial p}{\partial x} \text{ or } \frac{d^2 u}{dy^2} = \left(\frac{1}{\mu}\right)\frac{dp}{dx}$$

which is the same relation as (8.11). Further derivations are identical to the ones outlined above

8.7 Inclined Parallel Plates

Let the planes be inclined at an angle θ to the horizontal, and the x-axis is chosen to lie along the bottom plate, as shown in Fig. 8.5. The additional force acting on the fluid is to the body force, namely the gravitational component along the inclined direction which is equal to $g \sin\theta$ per unit mass.

The equation of equilibrium of forces is now written as

$$p \cdot dy - (p + dp)dy - \tau \cdot dx + \left(\tau + \frac{\partial \tau}{\partial y}\right)dx + \gamma\, dx\, dy \sin\theta = 0$$

Also we have $dx \sin\theta = dh$. On substitution and simplification, we get

$$d\tau = \frac{d(p + \gamma h)}{dx} \text{ or } \frac{d^2 u}{dy^2} = \left(\frac{1}{\mu}\right)\left[\frac{d(p + \gamma h)}{dx}\right]$$

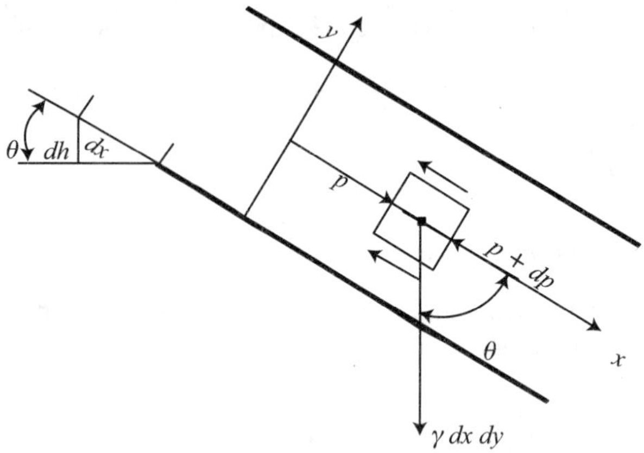

Fig. 8.5 Inclined plates

and the solution is

$$u = \left(\frac{1}{\mu}\right)\left[\frac{d(p+\gamma h)}{dx}\right]\left(\frac{y^2}{2} - ay\right) \tag{8.14}$$

The preceding equation is similar to equation (8.13) except that the pressure p is replaced by the piezometric pressure $(p + \gamma h)$.

8.8 Couette Flow

Couette studied the flow caused by the movement of one of the parallel plates relative to the other at a uniform velocity, U (as was the case in Newton's experiments to establish the law of viscosity). Let us consider the case of the lower plate being stationary and the upper plate moving at a uniform velocity, U (Fig. 8.6). A force is applied to the top plate to move it and the flow is caused by transmission of shear force from the moving plate to the interior of the fluid.

In this case the pressure gradient is zero. The only forces acting on the element are the two shear forces acting on the opposite faces.

$$\left(\tau + \frac{\partial \tau}{\partial y}\right) - \tau = 0; \text{ or } \frac{d\tau}{dy} = \mu\frac{d^2u}{dy^2} = 0$$

which on integration yields

$du/dy = C_1$ and the second integration gives $u = C_1 y + C_2$;

Since $u = 0$ at the stationary plate or $y = 0$, $C_2 = 0$; and $u = U$ at $y = 2a$, hence $C_1 = U/2a$ and the velocity distribution is linear

$$u = \left(\frac{U}{2a}\right)y \tag{8.15}$$

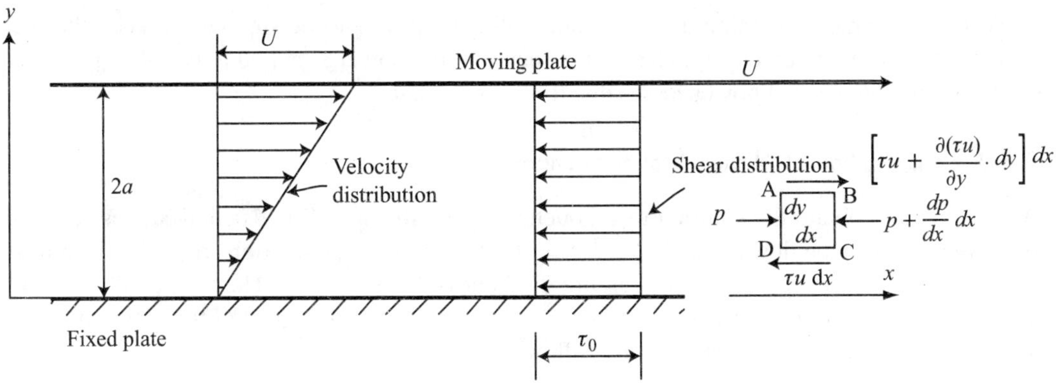

Fig. 8.6 Couette flow

and the shear stress is constant across the spacing

$$\tau = \tau_0 = \frac{\mu U}{2a}$$

The average velocity is $U/2$ and the discharge is $(U/2)2a = U.a$

The force required to move the top plate of unit width and length, L is

$$F = \tau_0 L = \left(\frac{\mu U}{2a}\right) L$$

and the power required is $F \cdot U = (\mu U/2a)L \cdot U = \mu U^2 L/2a$

The power input can also be calculated by considering the powers supplied to the sides of the element ABCD (Fig. 8.6).

Power P = stress × area × velocity

$$dP = -\tau u dx + \tau u dx + \frac{d(\tau u)}{dy} dy\, dx$$

The net power input in the absence of pressure gradient $(dp/dx = 0)$ is $[d(\tau u)/dy]dx\, dy$ or per unit length, it is $(d(\tau u)/dy) \cdot dy$. Substituting the expressions for $\tau = \mu U/2a$ and $u = Uy/2a, dP = [d(\tau u)/dy]dy$

$$dP = \left[\frac{d(\mu U^2 y/4a^2)}{dy}\right] dy = \left(\frac{\mu U^2}{4a^2}\right) \cdot dy$$

which on integration across the spacing from $y = 0$ to $y = 2a$, yields

$$P = \int_0^{2a} \left(\frac{\mu U^2}{4a^2}\right) dy = \left\{\frac{\mu U^2}{4a^2}\right] 2a = \frac{\mu U^2}{2a} \text{ per unit length of plate.}$$

This is the same expression as that obtained for the power to be supplied to keep the plate moving at a constant velocity U. Hence it is seen that the power supplied to the plate is entirely consumed by the internal power losses due to viscous shear.

8.8.1 Application of Navier–Stokes Equation

We may readily obtain the solution to a Couette flow by the use of the Navier–Stokes equation. Here we know that the flow is steady ($\partial u/\partial t = 0$), flow is in the x-direction only and it is uniform ($\partial u/\partial x = 0$, $v = w = 0$) and the pressure gradient ($\partial p/\partial x$) is zero. Hence the only term left in the equation in the x-direction is $\partial^2 u/\partial y^2 = d^2u/dy^2 = 0$ and the rest of the solution for this linear second-order differential equation follows as explained earlier.

8.9 Combined Hagen–Poiseuille and Couette Flows

Here, we will consider that the lower plate is stationary, the top plate is moving with a constant velocity, U and there is a pressure gradient dp/dx in the direction of flow. Since the two flows are solutions of linear differential equations, we may superpose the two linear solutions and obtain the solution for the velocity distribution (Fig. 8.7).

$$u = \left(\frac{1}{\mu}\right)\left(\frac{dp}{dx}\right)\left(\frac{y^2}{2} - ay\right) + \frac{Uy}{2a} \tag{8.16}$$

$$q = \int_0^{2a} = u \cdot dy = Ua - \frac{\left[(dp/dx)(2a)^3\right]}{12\mu} \tag{8.17}$$

The velocity distribution can be expressed in a non-dimensional form

$$\frac{u}{U} = \frac{y}{2a} + K\left(\frac{y}{2a}\right)\left(1 - \frac{y}{2a}\right)$$

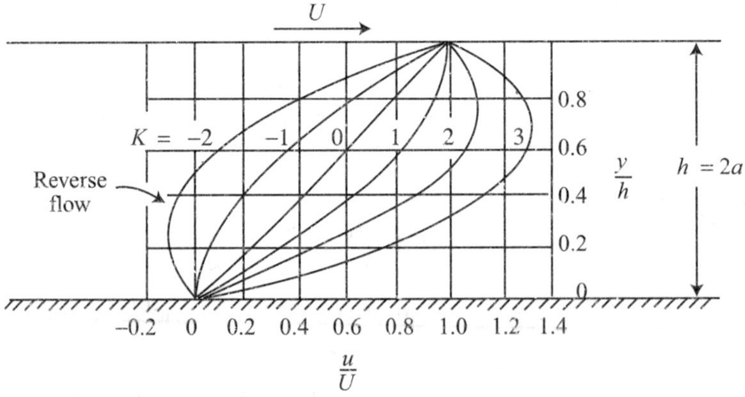

Fig. 8.7 Combined Hagen–Poiseuille and Couette flow

where

$$K = \frac{(2a)^2}{(2\mu U)}\left(-\frac{dp}{dx}\right)$$

We may have dp/dx as positive in which case the flow due to pressure gradient opposes the Couette flow. The velocity patterns for different values of K are shown in the figure. The average velocity, V is

$$V = \frac{q}{2a} = \left(\frac{1}{2a}\right)\int_0^{2a} u \cdot dy = \left(\frac{1}{2} + \frac{K}{6}\right) U \text{ and the discharge } q = 2aV$$

In the special case when $(1/2 + K/6) = 0$, the average velocity and the discharge will become zero since there is no net flow or the flow in the positive direction due to the movement of the plate is balanced by the flow in the negative direction due to the positive pressure gradient or

$$K = \frac{(2a)^2}{(2\mu U)}\left(-\frac{dp}{dx}\right) = -3$$

$$\frac{dp}{dx} = \frac{6\mu U}{(2a)^2}$$

$$\frac{du}{dy} = \left(\frac{1}{\mu}\right)\left(\frac{dp}{dx}\right)(y - a) + \frac{U}{2a}$$

and u is maximum when $du/dy = 0$ or
when $y = a - \mu U/[2a(dp/dx)] = a\,(1 + 1/K)$ and the shear stress $\tau = \mu du/dy$.

8.9.1 Slider Bearing

The combined Hagen–Poiseuille and Couette flow is of interest in slider bearings. Here, the former type of flow is caused by the slight inclination between the two plates as shown in Fig. 8.8 and the latter by the movement of the top plate at a uniform velocity, U.

Since the spacing between the plates changes very slightly, there will be acceleration of the flow and consequent pressure gradient. However, an assumption is made that the acceleration term is negligible and the derivations made in Section 8.9 are valid. The length of the bearing is L, and the spacing between the plates at the two ends are $2a_1$ and $2a_2$ respectively and the angle between the plates is $\alpha = (2a_1 - 2a_2)/L$.

At any distance x from the left end, the spacing is $2a = 2a_1 - \alpha x$.

The discharge from (8.17),

$$q = \frac{U(2a)}{2} - \frac{[(dp/dx)(2a)^3]}{12\mu}$$

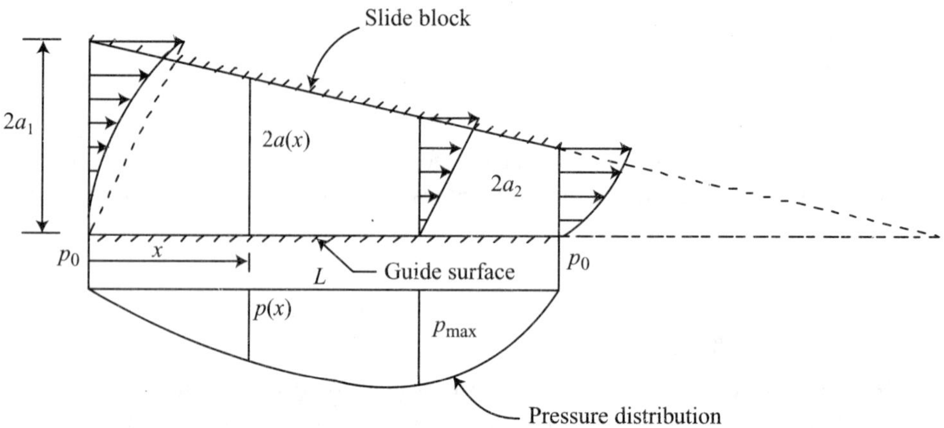

Fig. 8.8 Slider bearing

and solving for

$$\frac{dp}{dx} = \frac{6\mu U}{(2a_1 - \alpha x)^2} - \frac{12\mu q}{(2a_1 - \alpha x)^3}$$

Further integration and evaluation of constants of integration using the conditions that $p = p_0$ at $x = 0$ and $x = L$, yields

$$p = p_0 + \frac{6\mu U x\,(L - x)\,(2a_1 - 2a_2)}{L\,(2a)^2\,(2a_1 + 2a_2)} \tag{8.18}$$

The discharge is $q = 2Ua_1a_2/(a_1 + a_2)$

Integration of the pressure intensity over the length L yields the normal force per unit width of bearing.

$$P = \frac{6\mu U L^2}{(2a_1 - 2a_2)^2}\left[\ln\frac{a_1}{a_2} - 2\frac{a_1 - a_2}{a_1 + a_2}\right] \tag{8.19}$$

Finally, after elimination of the pressure gradient we get the velocity distribution

$$\frac{u}{U} = \frac{y}{2a}\left[1 - \frac{3\,(2a_1 - 2a_2)\,(2a_1 L - 2x(2a_1) + \alpha x L)}{L(2a_1 + 2a_2)}\left(1 - \frac{y}{2a}\right)\right] \tag{8.20}$$

8.10 Flow along an Inclined Channel with a Free Surface

The laminar flow in a channel (Fig. 8.9) can be considered as one-half of the Hagen–Poiseuille flow between parallel plates, the free surface coinciding with the centre-line or the x-axis.

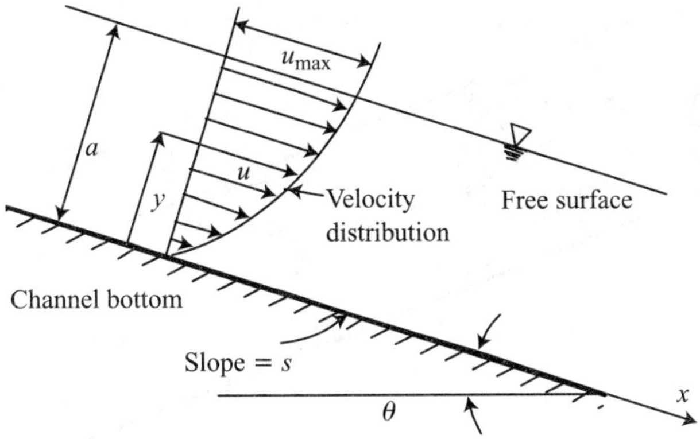

Fig. 8.9 Laminar flow along an inclined channel

The velocity distribution is parabolic given by the expression

$$u = -\left(\frac{1}{2\mu}\right)\left[\frac{d(p+\gamma h)}{dx}\right](a^2 - y^2)$$

Since the pressure is atmospheric, we write

$$\left[\frac{d(p+\gamma h)}{dx}\right] = -\gamma dh/dx = \gamma \sin\theta = \gamma(\text{slope } S) = \gamma S u = \left(\frac{\gamma S}{2\mu}\right)(a^2 - y^2) \qquad (8.21)$$

The maximum velocity occurs at the free surface $u_{max} = (\gamma S/2\mu)a^2$

$$q = \int_0^a u\, dy = \left(\frac{\gamma S}{2\mu}\right)(a^2 y - y^3/3)\Big|_0^a = \frac{\gamma a^3 S}{3\mu}$$

The average velocity is

$$V = \frac{q}{a} = \frac{\gamma a^2 S}{3\mu} \qquad (8.22)$$

8.11 Hagen–Poiseuille Flow in Cylindrical Tubes

The flow is steady and uniform. We follow the usual convention in the cylindrical coordinate system (r, θ, z) that the z-axis is perpendicular to the $(r\theta)$-plane and w is the axial velocity. The radial velocity u and the tangential velocity v are both zero. On the elementary cylinder of radius a, as shown in Fig. 8.10, the element is in dynamic equilibrium under pressure, gravity and viscous shear forces.

$$p \cdot \pi r^2 - (p + dp)\pi r^2 + 2\pi r \tau dz + \gamma \pi r^2 dz \sin\theta = 0$$

But $dz \sin\theta = -dh$ and on simplification we get $[d(p+\gamma h)/dz]r/2 = \tau$

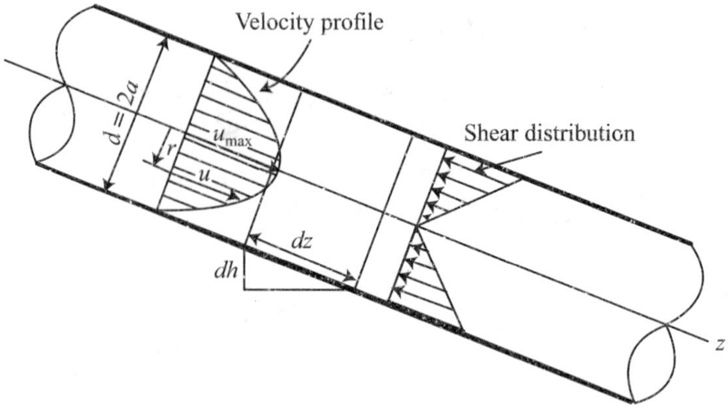

Fig. 8.10 Laminar flow in cylindrical tubes

$$\text{Since } \tau = \mu \frac{dw}{dr} = \left[\frac{d(p + \gamma h)}{dz}\right]\frac{r}{2} \tag{8.23}$$

which shows that the shear stress distribution is linear with radius and the maximum shear stress occurs at the boundary, $r = a$. Integration of (8.23) yields

$$w = \left(\frac{1}{4\mu}\right)\left[\frac{d(p + \gamma h)}{dz}\right]r^2 + C$$

and the integration constant C is evaluated from the condition $u = 0$ at $r = +a$; $C = -(1/4\mu)[d(p + \gamma h)/dz]a^2$

$$w = \left(\frac{1}{4\mu}\right)\left[\frac{-d(p + \gamma h)}{dz}\right](a^2 - r^2) \tag{8.24}$$

The discharge,

$$Q = \int_0^a w \cdot 2\pi r dr = \left(\frac{\pi a^4}{8\mu}\right)\left[\frac{-d(p + \gamma h)}{dz}\right]$$

The average velocity,

$$V = \frac{Q}{\pi a^2} = \left(\frac{a^2}{8\mu}\right)\left[\frac{-d(p + \gamma h)}{dz}\right]$$

The head loss in piezometric head over a finite length $L(dz = L)$ of the pipe is

$$h_f = \frac{\Delta(p + \gamma h)}{\gamma} = \left(\frac{8\mu V}{\gamma a^2}\right)L$$

Multiplying and dividing the right-hand side by a factor $2V$ and substituting $a = d/2$, and $\gamma = \rho g$, we get after rearranging the terms

$$h_f = \frac{64V^2L}{(\rho Vd/\mu)d2g}$$

The factor $\rho Vd/\mu$ is known as the Reynolds number, **R**.

$$h_f = \left(\frac{64}{\mathbf{R}}\right)\left(\frac{V^2L}{2g}\right) \tag{8.25}$$

This relationship may be compared to the Darcy–Weisbach expression for the loss of head in a length of pipe L in terms of a friction or pipe resistance factor f, the average velocity V and diameter d of the pipe $h_f = fV^2L/d \cdot 2g$ and comparing with (8.25), we conclude that the friction factor f for laminar flow is given as

$$f = \frac{64}{\mathbf{R}} \tag{8.26}$$

8.11.1 Navier–Stokes Equations in Cylindrical Coordinates

The above derivation for flow through cylindrical tubes could have been got directly from the N–S equations in cylindrical coordinates (8.10). As explained earlier, the radial and tangential components are zero and all their derivatives are also zero with the result that the equations of motion in the radial and tangential directions need not be considered for flow in the axial directions. In the axial directions also, the flow is uniform and steady and all the terms on the left-hand side of the equation of motion in the axial direction disappear. On the right-hand side of the equation in the axial direction, the axial velocity does not change with θ. Hence we have

$$\frac{\partial^2 w}{\partial r^2} + \left(\frac{1}{r}\right)\left(\frac{\partial w}{\partial r}\right) = \left(\frac{1}{\mu}\right)\left[\frac{\partial(p+\gamma h)}{\partial z}\right]$$

or since w changes only with r and the piezometric pressure changes only with z, we write

$$\frac{d^2 w}{dr^2} + \left(\frac{1}{r}\right)\left(\frac{dw}{dr}\right) = \left(\frac{1}{r}\right)\frac{d[r(dw/dr)]}{dr} = \left(\frac{1}{\mu}\right)\left[\frac{d(p+\gamma h)}{dz}\right] \tag{8.27}$$

Successive integrations and substitutions of the boundary conditions

$$w = 0 \text{ at } r = a \text{ and } \frac{dw}{dr} = 0 \text{ at } r = 0 \text{ yields}$$

$$w = \left(\frac{1}{4\mu}\right)(r^2 - a^2)\left[\frac{d(p+\gamma h)}{dz}\right]$$

which is the same as (8.24) and the rest of the derivations is similar to those outlined in the previous section.

8.12 Laminar Flow in an Annulus

Laminar flow in the annular space between two concentric pipes of radii a and b may be found by integrating (8.27) with the boundary conditions that $w = 0$ at $r = a$ (radius of the external pipe) and $r = b$ (radius of the internal cylinder).

First integration yields

$$r\left(\frac{dw}{dr}\right) = \left(\frac{1}{\mu}\right)\left[\frac{d(p+\gamma h)}{dz}\right]\left(\frac{r^2}{2}\right) + C_1$$

A second integration gives

$$w = -\left(\frac{1}{4\mu}\right)\left[\frac{d(p+\gamma h)}{dz}\right](r^2) + C_1 \ln r + C_2$$

The constants may be evaluated from the given boundary conditions to give

$$w = -\left(\frac{1}{4\mu}\right)\left[\frac{d(p+\gamma h)}{dz}\right]\left\{a^2 - r^2 + (a^2 - b^2)\frac{\ln(a/r)}{\ln(b/a)} \ln a/r^3\right\} \tag{8.28}$$

It is seen that when $b = 0$, the above expression reverts to that of (8.24) for flow through pipe of radius a.

Example 8.2 For the same discharge, compare the pressure drop for ratios of $b/a = 0$ and $b/a = 1/1000$. Similarly, calculate the discharges for the same pressure drop. The radius a of the outer pipe is held constant.

The discharge through the annulus is

$$Q = \int_b^a w \cdot 2\pi r dr = -\frac{\pi a^4}{8\mu}\frac{d}{dz}(p+\gamma h)\left[1 - \frac{b^4}{a^4} - \left(\frac{1 - b^2/a^2}{\ln a/b}\right)\right] \tag{8.29}$$

The ratio of discharges for $b/a = 0$ and $b/a = 1/1000$ is

$$\frac{(Q)_{b/a=1/1000}}{(Q)_{b/a=0}} = 1 - \left(\frac{1}{1000}\right)^4 - \left[\frac{1 - 1/1000^2}{\ln 1000}\right] = 0.855$$

For the same discharge, the ratio of the pressure drops is

$$\frac{\Delta(p+\gamma h)_{b/a=1/1000}}{\Delta(p+\gamma h)_{b/a=0}} = 1.17$$

The above example illustrates that the introduction of even a thin wire causes appreciable decrease in discharge for the same pressure gradient or an increase in the pressure gradient for similar discharge.

8.13 Co-Axial Couette Flow

Let us consider another type of axi-symmetric Couette flow in which a cylinder of radius b is dragged at a uniform velocity, U through a co-axial pipe of radius a, the annular space being filled with a fluid density ρ and dynamic viscosity μ.

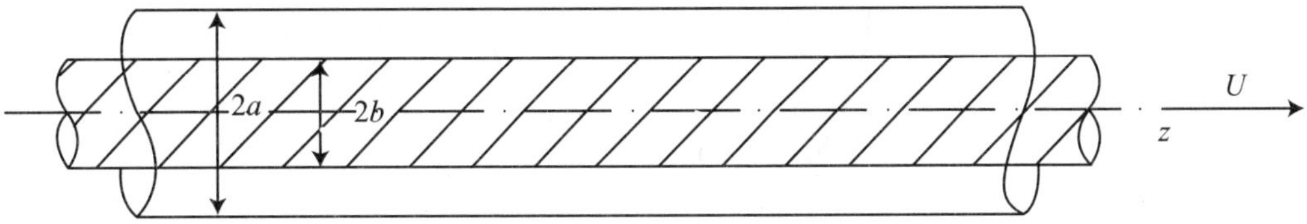

Fig. 8.11 Co-axial couette flow

The common axis of the two pipes is chosen as the z-axis and the equation of motion in the direction of flow becomes

$$\frac{d^2w}{dr^2} + \left(\frac{1}{r}\right)\left(\frac{dw}{dr}\right) = 0$$

as there is no pressure gradient and the flow is caused by the movement of the inner cylinder

$$\left(\frac{1}{r}\right)\frac{d[r(dw/dr)]}{dr} = 0$$

and integrating successively we get

$$w = C_1 \ln r + C_2$$

The boundary conditions are $w = 0$ at $r = a$ and $w = U$ at $r = b$ and after evaluations of the constants, the velocity distribution is given as

$$w = \frac{U}{\ln a/b} \ln a/r \tag{8.30}$$

8.14 Rotating Cylinders

Let us now consider the flow caused in the annular space of two concentric vertical cylinders. Let the outer cylinder of radius a rotate at an angular speed of Ω_1 and the inner cylinder of radius b rotate at an angular speed of Ω_2. The directions of rotation need not be the same.

The velocity distribution in the annulus is readily obtained by the equation of motion in the tangential direction from the N–S equation in the cylindrical coordinate system. The flow is caused by the rotation of the cylinders and hence the pressure gradient in the tangential direction

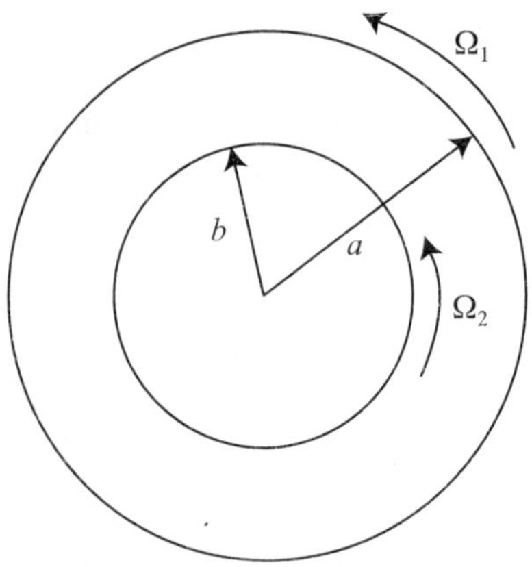

Fig. 8.12 Rotating cylinders

is zero. The radial and axial velocity components are absent and only the tangential component, v is present. The equation of motion in the radial direction is

$$\frac{\rho v^2}{r} = \frac{dp}{dr}$$

The pressure gradient in the radial direction is dependent upon the centrifugal force. The equation of motion in the tangential direction is

$$\frac{d^2 v}{dr^2} + \frac{d(v/r)}{dr} = 0$$

First integration yields

$$\frac{dv}{dr} + \frac{v}{r} = \left(\frac{1}{r}\right)\frac{d(vr)}{dr} = C_1$$

Second integration leads us to $v = (C_1/2)r + C_2/r$. The constants of integration can be found by the conditions that $v = \Omega_1 a$ at $r = a$ and $v = \Omega_2 b$ at $r = b$. After simplification, the tangential velocity becomes

$$v = \frac{1}{\left(a^2 - b^2\right)}\left[\left(a^2\Omega_1 - b^2\Omega_2\right) r - \frac{a^2 b^2 \left(\Omega_1 - \Omega_2\right)}{r}\right] \tag{8.31}$$

An interesting case occurs when the rotational speeds of the outer and inner cylinders are controlled such that $\Omega_1 = C/a^2$ or $v = C/a$ and $\Omega_2 = C/b^2$ or $v = C/b$. The tangential velocity will then reduce to $v = C/r$, the velocity distribution for a free vortex which is an example of irrotational flow.

The angular deformation in the $r\theta$-plane is

$$h = \left(\frac{1}{r}\right)\frac{\partial u}{\partial \theta} + \frac{r\partial(v/r)}{\partial r}$$

and if $v = C/r$ and $u = 0$, then $h = -2C/r^2$ and hence the shear stress is $\tau = \mu h = -2\mu C/r^2$ and the torque resulting from this shear stress is

$$T = (2\pi r\tau)r = -4\pi\mu C \tag{8.32}$$

The power, P required to overcome this torque is $T\Omega$ and $\Omega = C/r^2$

$$P = -4\pi\mu C^2 r^2$$

and the difference in power required to drive the inner and outer cylinders is thus

$$P = 4\pi\mu C^2(1/b^2 - 1/a^2)$$

The rate of energy dissipation in the system must be exactly equal to this if the flow is to remain irrotational and the energy in the system to be constant.

8.15 Non-Linear Exact Solutions

The inertial terms on the left-hand side of the Navier–Stokes equations being non-linear, it is difficult to obtain closed form solutions to all flow problems where these terms cannot be ignored for complex boundary conditions. In literature, one may find complete solutions for the following situations.

1. Flow between convergent and divergent plates. The flow between convergent plates being continuously accelerated does not separate from the boundary walls whereas flow deceleration between divergent plates causes flow separation at a certain distance from the origin in the presence of adverse pressure gradient.
2. Flow against a normal infinite wall and a corresponding axi-symmetric problem.
3. Laminar jet.
4. Flow around a rotating disk and flow between two rotating disks.

However, the solutions are indeed very complicated and will not be attempted here.

8.16 Approximate Solution for Slow Motion Around a Sphere

There are several categories of approximations possible but we will now consider only one of them belonging to the class of very slow motion in which the inertial terms in the Navier–Stokes equation of motion may be ignored in comparison with the pressure and viscous terms. Evidently this implies that the Reynolds number is very low, of the order of $\mathbf{R} = 0.1$ or less. In Cartesian coordinates, the equations reduce, in the absence of body forces and steady flow conditions, to

$$\frac{\partial p}{\partial x} = \mu\left(\frac{\partial^2 u}{\partial x^2} + \frac{\partial^2 u}{\partial y^2} + \frac{\partial^2 u}{\partial z^2}\right) = \mu\nabla^2 u$$

$$\frac{\partial p}{\partial y} = \mu \nabla^2 v; \quad \frac{\partial p}{\partial z} = \mu \nabla^2 w$$

Differentiating the first equation with respect to x, the second with respect to y and the third with respect to z and adding, we get

$$\frac{\partial^2 p}{\partial x^2} + \frac{\partial^2 p}{\partial y^2} + \frac{\partial^2 p}{\partial z^2} = \mu \nabla^2 \left(\frac{\partial u}{\partial x} + \frac{\partial v}{\partial y} + \frac{\partial w}{\partial z} \right) = 0$$

$$\nabla^2 p = 0 \tag{8.33}$$

by virtue of the equation of continuity. The pressure satisfies the Laplacian equation. Let us consider a sphere of radius $a(d = 2a)$ placed in a uniform flow of velocity U (the motion is very slow) or $\rho Ud/\mu < 0.1$.

It has been found that the pressure distribution is given by

$$p = -\frac{(3\mu a U \cos\theta)}{r^2} \tag{8.34}$$

The pressure force on the sphere is obtained by integrating the pressure component in the direction of flow over the surface of the sphere

$$F_p = \int [(p)_{r=a}] \cos\theta (2\pi a \sin\theta) d\theta = 2\pi a \mu U$$

It may also be shown that the integral of the shear stresses (τ_{zx} and τ_{yx}) on the surface area of the sphere is $F_\tau = 4\pi a \mu U$.

Hence the total drag force on the sphere is

$$F = 2\pi a \mu U + 4\pi a \mu U = 6\pi a \mu U$$

$$F = 3\pi \mu U d \tag{8.35}$$

which is the Stokes expression for the total drag force and is known as the Stokes law.

8.16.1 Settling Velocity or Fall Velocity

If a sphere of diameter d and specific weight γ_s is falling freely at a uniform velocity U in a stagnant fluid of specific weight γ, the drag experienced by the falling sphere is the same as that of the sphere placed in a uniform flow by the principle of superposition. Since there is no acceleration of the sphere, the two forces namely the buoyant weight of the sphere acting downwards must be equal to the drag force opposing the motion. Hence we have

$$\frac{4}{3}(\gamma_s - \gamma)\pi a^3 = 6\pi a \mu U$$

or the terminal velocity

$$U = \frac{2a^2(\gamma_s - \gamma)}{9\mu} \tag{8.36}$$

Part II Laminar Stability

8.17 Introduction

The solutions presented in Part I of this chapter have assumed the existence of laminar flow conditions. We shall now examine the conditions for the persistence of such a state of flow or the transformation to an entirely different state of flow known as turbulent flow.

In practical problems, small disturbances are invariably and inherently present in the system itself. These disturbances may be present as initial disturbances at the source of the flow itself; or they may arise out of boundary irregularities; or they may be caused by the nature of the pressure gradient; or by external mechanisms. When a small disturbance occurring in the flow is damped by the action of viscosity, then the flow is stable. On the other hand, if the disturbance grows in magnitude and extent spreading throughout the flow, then the flow is inherently unstable. In the former case, the flow returns to the laminar state. But in the latter case, subsequent flow degenerates into turbulent flow. Whether the superposition of the accelerating or the inertial effects of the disturbance on the primary or the mean flow causes the transformation from laminar to turbulent flow depends on the properties of the fluid (density and viscosity), pattern of primary flow (velocity, pressure gradient, boundary dimensions) and the nature of disturbances (frequency and amplitude).

Stability of laminar flow depends on whether the rate of viscous dissipation of energy of the disturbance is greater than the rate at which transfer of energy between the disturbance and the primary flow takes place.

8.18 Reynolds Number

Physically a disturbance may be looked upon as a superposition of the accelerative effects of the disturbance on the primary flow. If the fluid viscosity is high and the density and velocity of the flow is low, then the effect of viscous resistance to changes imposed by the disturbance may be greater than the influence of the disturbing force to cause acceleration normal to the laminar layers. Then the disturbance is damped and laminar flow continues to prevail. Whereas if the flow velocity and fluid density are high and viscosity is low, mass acceleration caused by the disturbance may outweigh the viscous resistance. The disturbance may get amplified and create turbulence in the flow and the flow will no longer remain in laminar layers.

A dimensional quantity expressing the relative magnitude of inertial to viscous forces

$$\frac{(\rho V^2/L)}{(\mu V/L^2)} = \frac{\rho V L}{\mu}$$

known as the Reynolds number is an essential parameter to describe the stability of laminar flow.

Reynolds in his classical experiments on the diffusion of a dye in a cylindrical tube (Fig. 8.13) had established that there exists a critical Reynolds number at which the flow may change over from laminar to turbulent flow. The value of the lower critical Reynolds number is around 2300 (various values varying from 2000 to 2400 are quoted).

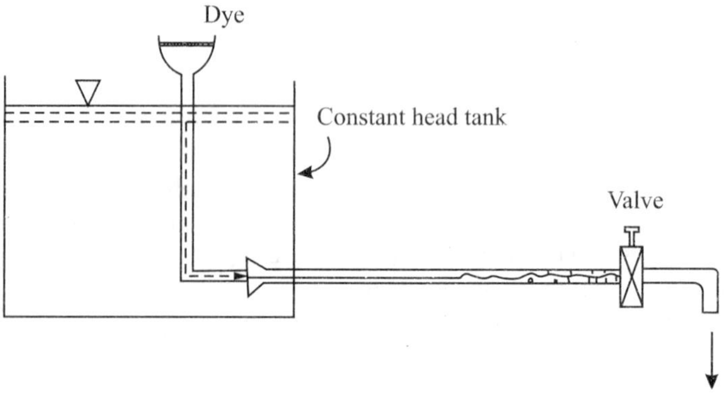

Fig. 8.13 Reynolds apparatus

Below this critical value of the Reynolds number, the dye carefully injected at the bell mouth entrance remained along a straight line. Any further increase in the discharge caused the Reynolds number to increase and the dye line to break up into a series of eddies leading to diffusion of the dye over the entire cross-section of the tube. It is possible, however, under controlled laboratory conditions to maintain laminar flow even at higher Reynolds number of the order of 40,000 by eliminating initial disturbances, careful design of boundary geometry, control of temperature and vibration-free mounting of the apparatus. This is however of theoretical interest and only the lower value of the Reynolds number, below which laminar flow is stable, is of practical importance.

8.19 Stability Parameter

The role of the inertial property of the disturbance in causing instability may be brought out significantly by the Helmholtz model of two superposed ideal fluids illustrated schematically in Fig. 8.14.

If a small disturbance occurs at the interface, it may result in the deviation of the interface and disrupt the parallel nature of the stream lines. In the first stage a wavy interface as shown in Fig. 8.14*b* is formed. The stream lines are seen to converge at the crests and diverge at the troughs. Since an increase in velocity in the crest region is accompanied by a lowering of pressure and a decrease in velocity in the trough region causes an increase in pressure, there is a pressure gradient established across the interface. The direction of the pressure gradient changes sign alternately as shown in the figure. This will serve to aggravate the waviness of the interfacial profile as shown in Fig. 8.14*c* and will eventually lead to the formation of vortices or eddies along the interface.

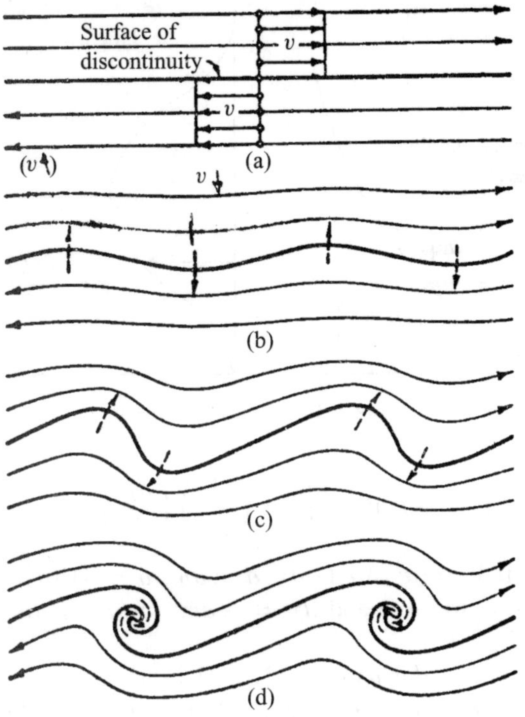

Fig. 8.14 Interfacial instability

If the two superposed fluids are now considered as viscous, the velocity gradients created across the interface is accompanied by viscous shear leading to dissipation of energy of the disturbance. If the disturbance occurs close to a solid boundary, the presence of the boundary inhibits the growth of the disturbance (velocity at the boundary is zero). And as such, proximity to the boundary is a stabilising factor. Combination of the density ρ, viscosity μ, velocity gradient du/dy, and the distance from the boundary, y yields the non-dimensional stability number

$$\chi = \frac{(\rho y^2 du/dy)}{\mu} \tag{8.37}$$

known as the Rouse number. As an example applied to the Hagen–Poiseuille flow between parallel plates or a pipe, it may be seen that $\chi = 0$ at the boundaries and at the axis of symmetry and it reaches maximum at some intermediate point between the wall and the axis of symmetry as shown in Fig. 8.15. It has been found that if $\chi_{max} < 500$ in the flow regime, the flow is stable. For values of $\chi > 500$, the flow may break into turbulence if disturbances of sufficient magnitude are present in the flow.

For flow in circular pipes, we can find the location or the distance from the boundary at which the value of χ becomes a maximum.

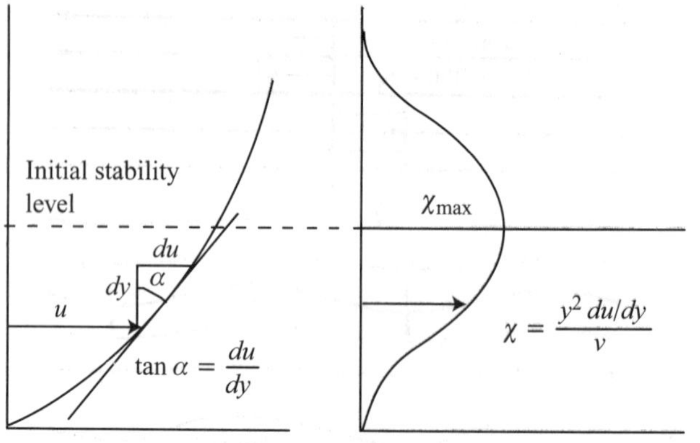

Fig. 8.15 Rouse parameter

$$w = \left(\frac{1}{4\mu}\right)\left(-\frac{dp}{dx}\right)(a^2 - r^2); \quad \frac{dw}{dr} = \left(\frac{1}{2\mu}\right)\left(\frac{dp}{dx}\right)r$$

The average velocity is $V = (a^2/8\mu)(-dp/dx)$; $dw/dr = 4V(a^2/r)$

We can substitute $y = (a - r)$ and $dy = -dr$; $dw/dr = (dw/dy)(dy/dr)$; $dw/dy = -4V[(a - y)/a^2]$

Therefore we have $\chi = (\rho y^2 dw/dy)/\mu = [\rho y^2 4V(a^2)]/\mu$

The maximum value of χ occurs when $d\chi/dy = 0 = (4V/\nu a^2)(2ay - 3y^2)$

χ_{max} occurs at $y = 2a/3$.

In terms of Reynolds number, we have $\chi_{max} = (4V/\nu a^2)(8a^3/27) = (16/27)(Va/\nu) = (8/27)VD/\nu = (8/27)\mathbf{R}$.

If the onset of turbulence occurs at $\chi = 500$, this corresponds to a Reynolds number of $\mathbf{R} = 1687.5$.

8.20 Stability Curves

A state of flow is unstable only if there is a transfer of energy from the primary motion to the disturbance. Since all disturbances are subject to viscous dissipation, the stability of the primary flow depends essentially on whether the rate of dissipation of the disturbance more than offsets the rate at which energy is received from the primary flow.

As pointed out earlier, disturbances are characterised by their magnitude and frequency or wave length. The hydrodynamic instability of real flows has been investigated analytically and experimentally by various investigators but are too involved for presentation in this book. The investigators have established that there is a neutral stability curve linking the disturbance wave length ($\lambda = 2\pi/\alpha$) and the characteristic Reynolds number, $\mathbf{R} = \rho U\delta/\mu$ of the flow, where U is the velocity of the uniform flow outside the boundary layer of thickness δ. Such a stability curve for a laminar boundary layer is shown in Fig. 8.16.

The region interior of the curve represents the class of disturbances for which the mean flow is unstable while the region exterior to the neutral curve indicates the stable form of disturbances. The

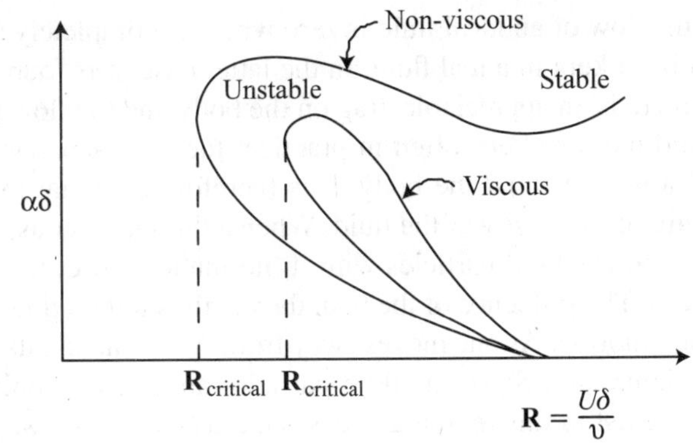

Fig. 8.16 Typical curves of neutral stability for (a) non-viscous stability and (b) viscous stability

lowest value of the curve on the **R**-axis corresponds to the critical Reynolds number which signifies that the flow is stable for small disturbances of any frequency below the critical Reynolds number. In the boundary layer, the value of **R** (critical) has been found to be about 420 corresponding to a value of $\alpha\delta = 0.16$. It is seen that inviscid solutions possess lower critical Reynolds number than the viscous solution, and in a similar fashion, the presence of an adverse pressure gradient or points of inflexion in the velocity distributions will lead to an earlier instability of laminar flows than those without these features.

Schubauer and Skramstad were able to confirm the theoretical predictions of the transition mechanism from laminar to turbulent flow. Their experimental investigations were conducted in a specially designed wind tunnel and they studied the stability characteristics of the laminar boundary layer along a flat plate at zero angle of incidence. The turbulence level in the initial stream was as low as 0.02% of the uniform velocity outside the boundary layer. The disturbances were superposed on the main stream by electromagnetically vibrating a thin metal strip at different frequencies. The strip 30 cm long, 2.5 mm wide, and 0.05 mm thick was placed at 0.15 mm above the boundary. It was found that for every Reynolds number (other than the critical value) there were two values of frequencies in the range of which the flow was unstable as predicted by theoretical investigations.

Part III Laminar Boundary Layer

8.21 Introduction

The concept of flow of a real fluid close to a boundary was first given by Prandtl in 1904 though several other investigators like Froude had earlier noted the distinctiveness of flow characteristics near a boundary. It has been shown in the chapter on examples of potential flow that the drag on a body submerged in the flow of an ideal fluid is zero which is completely at variance with what is experienced by an object kept in a real fluid. In the latter case, it is found that, however small the viscosity may be, there is an appreciable drag on the body and the flow pattern deviates from that found in ideal fluid motion. Very often in practice, the flow separates from the boundary with the formation of a wake behind the body. It is therefore essential to distinguish between zero viscosity and negligible viscosity of the fluid. When a fluid is viscous, there cannot be a slip between the boundary and the fluid particles with it, no matter however small. The velocity at the boundary is thus zero. The influence of the boundary is thus felt on fluid particles close to the boundary, retarding their motion. As one moves away from the boundary, its influence on the fluid motion becomes less significant. Since the fluid particles are retarded only near the boundary, steep velocity gradients exist in this region and eases out as we move away from the boundary. Viscous forces are thus confined to a thin layer close to the boundary where these forces are comparable in their order of magnitude to the inertial forces. This layer close to a solid boundary in which maximum changes in velocity take place and viscous forces are no less significant than the inertial forces is known as the boundary layer. The flow in this region is essentially rotational. Certain essential characteristics of the boundary layer developing along a flat plate kept at zero angle of incidence to the flow are illustrated in Fig. 8.17.

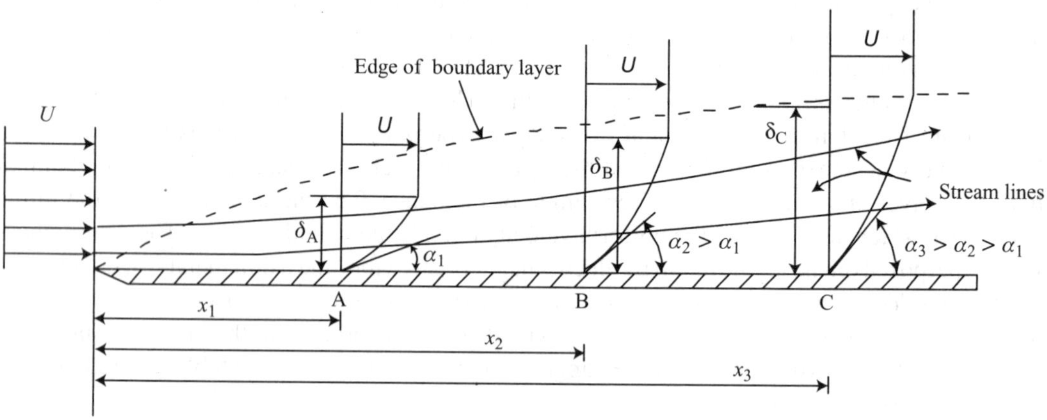

Fig. 8.17 Boundary layer characteristics

The flow before reaching the boundary is essentially uniform and parallel. In nature the singularity at the leading edge is avoided by slight modifications in the velocity distribution of the approach flow ahead of the leading edge. As we proceed downstream, more and more of the solid surface is brought in contact with the fluid and there is a continuous increase in the total shear force along the length of the plate. Increase in shear force with distance causes further retardation of the fluid mass and the range of influence spreads laterally. This is referred to as the growth of the boundary-layer thickness. At the same time, it must be realised that the velocity gradients become less steep with distance from the leading edge. The changes in slope of the velocity profile at successive sections are shown in the figure. Retardation of the fluid particles in the thickening boundary layer causes divergence of the stream lines as shown. The boundary layer characteristics may be summarised as follows.

(a) The velocity of the real fluid particles in contact with the boundary must have the same velocity as the boundary and there cannot be any slip between the fluid and the boundary. Velocities in the boundary layer are less than those corresponding to ideal fluid motion. The presence of such velocity defects is a feature of the boundary layer.
(b) The flow is rotational or vorticity is present in the boundary layer and is gradually transferred to the undisturbed regions

8.22 Boundary-Layer Thickness

The velocity defect in the boundary layer (normal to the boundary) disappears asymptotically as one moves away from the boundary. Hence it is difficult to define a practical limit to the boundary-layer thickness corresponding to the zero-velocity defect. The boundary-layer thickness is thus conveniently (though arbitrarily) defined as the distance δ from the boundary at which the velocity is within 1% of the local mainstream velocity. This value is found to agree fairly well with an alternative definition that it is the thickness at the extremity of which the total energy content $(p/\gamma + V^2/2g + z)$ is within 2% of the energy content of the local free stream.

8.23 Displacement, Momentum and Energy Thickness

8.23.1 Displacement Thickness

It is seen that the velocity defect in the boundary layer causes a reduction in the discharge within a given normal distance from the boundary compared to the discharge in the same thickness in an ideal fluid flow. The discharge may be thought of as due to an imaginary displacement of the boundary through a distance δ_1. The displacement thickness δ_1 of the boundary layer may be found by equating the discharge in the boundary layer to the discharge of an ideal fluid in a thickness $(\delta - \delta_1)$ as shown in Fig. 8.18.

If the free stream velocity is U and velocity at a distance y from the boundary is u and δ is the boundary-layer thickness, then we have

$$\int_0^\delta u dy = U(\delta - \delta_1) \quad \text{or} \quad \delta_1 = \int_0^\delta (1 - u/U) dy \tag{8.38}$$

The displacement thickness δ_1 may be found by the geometric method of finding a line a–a such that the hatched areas below and above the line are equal in area. The physical significance of the displacement thickness is that the flow conditions around a submerged body can be found by enhancing the dimensions of the body laterally by an amount equal to local displacement thickness. The flow around the new body may be analysed as an irrotational flow and flow conditions external to the boundary layer may be determined.

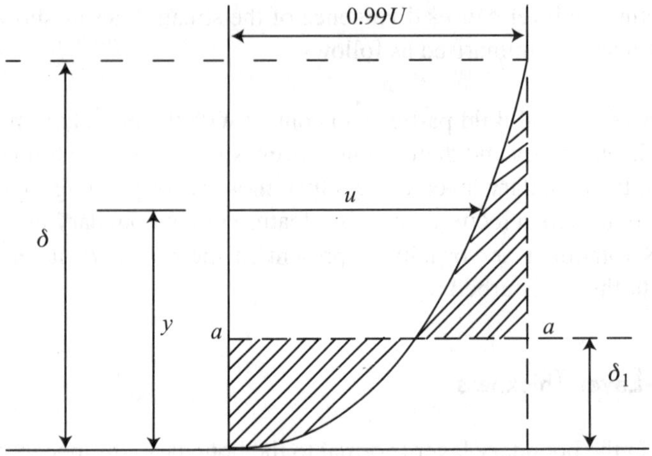

Fig. 8.18 Displacement thickness

8.23.2 Momentum Thickness

The boundary shear forces cause a loss of momentum of the flow in the boundary layer. The rate of mass flow at level y above the boundary is $\rho \cdot u \cdot dy$ and its momentum is $\rho \cdot u^2 \cdot dy$. The same mass had an original momentum of $(\rho \cdot u \cdot dy)U$. Hence the loss of momentum over the length x as shown in Fig. 8.19 is $\rho \cdot udy(U - u)$.

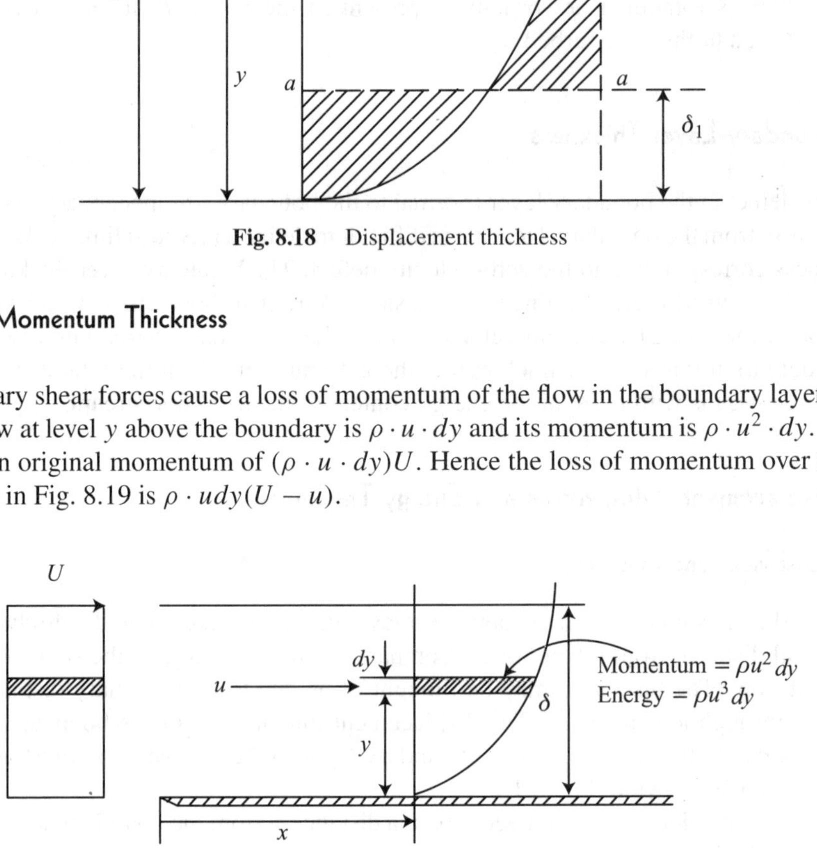

Fig. 8.19 Momentum and energy thickness

We can now define a momentum thickness δ_2 at any section in which the rate of flow of momentum of the fluid moving with an uniform velocity is equal to the loss of momentum in the boundary layer. Thus we have

$$\rho U^2 \delta_2 = \int_0^{\delta} \rho udy(U - u)dy \text{ or } \delta_2 = \int_0^{\delta} \frac{u}{U}\left(1 - \frac{u}{U}\right) dy \tag{8.39}$$

8.23.3 Energy Thickness

A similar definition for the energy thickness δ_3 in which the rate of flow of energy of a fluid moving at velocity U is equal to the energy lost by the fluid in the boundary layer (as shown in Fig. 8.19) gives

$$\delta_3 = \int_0^{\delta} \frac{u}{U}\left(1 - \frac{u^2}{U^2}\right) dy \tag{8.40}$$

8.24 Boundary-Layer Equations

In accordance with the boundary-layer characteristics stated earlier, the following assumptions may be made to simplify the equations of motion as applied to the boundary layer.

(a) The boundary-layer thickness is considered small ($\delta << L$) in comparison with a characteristic linear dimension L of the boundary (for example, the length of the plate); (b) the divergences of the stream lines in the boundary layer are sufficiently small to enable us to assume that the normal velocity component is also of the order of δ; (c) the longitudinal velocity component u is of the same order as the external flow velocity so that we may define the order of magnitude of u/U as unity; (d) the Reynolds number of the flow $\rho UL/\mu$ is of the order $1/\delta^2$ or $1/\mathbf{R}$ is of the order δ^2

The equation of continuity in the non-dimensional form is

$$\frac{\partial(u/U)}{\partial(x/L)} + \frac{\partial(v/U)}{\partial(y/L)} = 0 \tag{8.41}$$

It is seen from the equation that if u/U and x/L are of order of unity, O(1) and y/L is of the order of δ, O(δ), then v/U should also be of the order of δ, O(δ) as was stated earlier. In the non-dimensional form, the equations of motion for two-dimensional incompressible steady flow may be written as

$$\left(\frac{u}{U}\right)\left[\frac{\partial(u/U)}{\partial(x/L)}\right] + \left(\frac{v}{U}\right)\left[\frac{\partial(u/U)}{\partial(y/L)}\right] = -\left[\frac{\partial(p/\rho U^2)}{\partial(x/L)}\right] + \frac{\mu}{\rho UL}\left[\frac{\partial^2(u/U)}{\partial(x/L)^2} + \frac{\partial^2(u/U)}{\partial(y/L)^2}\right] \tag{8.42}$$

$$\mathrm{O}(1) \qquad \mathrm{O}(1) \qquad \mathrm{O}(\delta) \quad \frac{\mathrm{O}(1)}{\mathrm{O}(\delta)} \qquad\qquad \mathrm{O}(\delta^2)\left[\frac{\mathrm{O}(1)}{\mathrm{O}(1)} \qquad \frac{\mathrm{O}(1)}{\mathrm{O}(\delta^2)}\right]$$

y-direction

$$\left(\frac{u}{U}\right)\left[\frac{\partial(v/U)}{\partial(x/L)}\right] + \left(\frac{v}{U}\right)\left[\frac{\partial(v/U)}{\partial(v/L)}\right] = -\frac{\partial(p/\rho U^2)}{\partial(y/L)} + \frac{\mu}{\rho UL}\left[\frac{\partial^2(v/U)}{\partial(x/L)^2} + \frac{\partial^2(v/U)}{\partial(y/L)^2}\right] \tag{8.43}$$

$$\mathrm{O}(1) \quad \frac{\mathrm{O}(\delta)}{\mathrm{O}(1)} \qquad \mathrm{O}(\delta) \quad \frac{\mathrm{O}(\delta)}{\mathrm{O}(\delta/L)} \qquad\qquad \mathrm{O}(\delta^2) \quad \frac{\mathrm{O}(\delta)}{\mathrm{O}(1)} \quad \frac{\mathrm{O}(\delta)}{\mathrm{O}(\delta^2)}$$

The order of magnitude of each term is indicated below it. In the first equation, $\partial^2(u/U)/\partial(y/L)^2$ is very large when compared to the term $\partial^2(u/U)/\partial(x/L)^2$ with the result that one may drop the latter term in relation to the former. In the second equation also $\partial^2(v/U)/\partial(x/L)^2$ may be neglected in comparison with $\partial^2(v/U)/\partial(y/L)^2$. More importantly, it is observed that all terms in the second equation of motion in the y-direction are of the order of $O(\delta)$. This leads to significant deductions: (a) The gradient of pressure in the y-direction is also of the order of $O(\delta)$ and as such may be neglected. This assumption would imply that the pressure across the boundary layer at any section is constant and is equal to the pressure outside the boundary layer or that prevailing in the free stream. In other words, it may be said that the pressure outside the boundary layer is impressed on the boundary layer. The pressure, therefore, may be considered to vary only with x. Further, since all terms in the equation of motion in the y-direction are small, the equality or inequality of this equation is unimportant or satisfying this equation is irrelevant and this equation can be completely ignored as an approximation. Hence only one equation in the x-direction is of relevance and all terms in this equation is of order $O(1)$. With these arguments stemming from the consideration of orders of magnitude of different terms, we get a single equation in the x-direction

$$u\left(\frac{\partial u}{\partial x}\right) + v\left(\frac{\partial u}{\partial y}\right) = -\left(\frac{1}{\rho}\right)\left(\frac{dp}{dx}\right) + \nu\left(\frac{\partial^2 u}{\partial y^2}\right) \tag{8.44}$$

which along with the equation of continuity form the essential boundary-layer equations. The equation of motion in the y-direction, each term of which is of order $O(\delta)$ is deleted from further consideration. Since the flow outside the boundary layer is considered as potential flow, we write the Bernoulli equation for two-dimensional flow

$$p + \frac{1}{2}\rho U^2 = \text{constant}$$

which on differentiation with respect to x yields

$$-\left(\frac{1}{\rho}\right)\left(\frac{dp}{dx}\right) = U\left(\frac{dU}{dx}\right) \tag{8.45}$$

Substitution of the preceding relation in the boundary-layer equation gives the final form of the boundary-layer equation for steady incompressible flow.

$$u\frac{\partial u}{\partial x} + v\frac{\partial u}{\partial y} = U\left(\frac{dU}{dx}\right) + \nu\frac{\partial^2 u}{\partial y^2} \tag{8.46}$$

$$\frac{\partial u}{\partial x} + \frac{\partial v}{\partial y} = 0$$

Boundary-layer assumptions are not valid in some cases, for example: (a) in creeping motion similar to the one considered for the slow motion past a sphere where distortion of the flow field is not restricted to a thin layer but extends widely, and (b) near the leading edge where δ and x may be of the same order of magnitude.

8.25 Boundary Layer along a Flat Plate with Zero Pressure Gradient

An exact solution of the above equation was given by Blasius for the particular case of the development of the laminar boundary layer along a flat plate at zero angle of incidence. The flow field outside the boundary layer is uniform ($dU/dx = 0$) and hence according to (8.45), the pressure gradient dp/dx is also zero. Equation 8.44 reduces to

$$u\left(\frac{\partial u}{\partial x}\right) + v\left(\frac{\partial u}{\partial y}\right) = \nu\frac{\partial^2 u}{\partial y^2} \tag{8.47}$$

The discussion on the orders of magnitude has shown that the Reynolds number is of the order of $1/\delta^2$. Hence the boundary-layer thickness δ is proportional to $1/\mathbf{R}_x^{1/2}$. We now define a thickness δ_0 such that it is proportional to δ and $\delta_0 = 1/\mathbf{R}_x^{1/2} = (\mu x/\rho U)^{1/2}$. We shall assume that the velocity profiles are similar at all successive sections; the velocity distribution is thus specified at any section by a single functional relationship between u/U and y/δ_0.

$$\frac{u}{U} = F\left(\frac{y}{\delta_0}\right) = F(\eta), \text{ where } \eta = \frac{y}{\delta_0}$$

$$\Psi = U\int \delta_0 F(\eta)d\eta = U\delta_0 f(\eta)$$

Further we define $f(\eta) = \int F(\eta)d\eta$ such that the velocity components and their derivatives are obtained as follows

$$u = Uf'(\eta), \frac{\partial u}{\partial x} = -Uf''(\eta)\left(\frac{\eta}{\delta_0}\right)\left(\frac{d\delta_0}{dx}\right), \quad \frac{\partial u}{\partial y} = \left(\frac{U}{\delta_0}\right)f''(\eta)$$

$$\frac{\partial^2 u}{\partial y^2} = \left(\frac{U}{\delta_0^2}\right)f'''(\eta), \quad v = -\frac{\partial \Psi}{\partial x} = \frac{d\delta_0}{dx}[\eta f'(\eta) - f(\eta)]$$

Substituting the foregoing relationships in (8.47) and simplifying, we get

$$f'''(\eta) + \left(\frac{\rho U\delta_0}{\mu}\right)\frac{d\delta_0}{dx}f(\eta)f''(\eta) = 0$$

Since $\delta_0^2 = \mu x/\rho U$, we get $\delta_0(d\delta_0/dx) = \mu/(2\rho U)$ and $(\rho U\delta_0/\mu)d\delta_0/dx = 1/2$
The boundary-layer equation becomes

$$2f''' + ff'' = 0 \tag{8.48}$$

The solution of the above equation was given by Blasius by assuming a series of the type

$$f(\eta) = a_0 + a_1\eta + \frac{a_2\eta^2}{2!} + \frac{a_3\eta^3}{3!} + \ldots + \frac{a_n\eta^n}{\eta!}$$

Application of the boundary conditions $u = v = 0$ at the boundary $\eta = 0$ and evaluating the constants, he found that $a_0 = a_1 = a_3 = a_4 = a_6 = a_7 = a_9 = a_{10} = \ldots\ldots = 0$ and all the remaining constants can be expressed as powers of a_2.

$$f(\eta) = \frac{a_2\eta^2}{2} - \frac{a_2^2\eta^5}{2.5!} + \frac{11a_2^3\eta^8}{4.8!} - \frac{375a_2^4\eta^{11}}{8.11!} + \frac{27897a_2^5\eta^{14}}{16.14!} - \frac{3,817,137a_2^6\eta^{17}}{32.17!} + \text{etc}$$

The value of the constant a_2 was evaluated from the third boundary conditions that $f'(\eta) = 1$ at $\eta =$ infinity. Though the actual process of evaluation of a_2 (which was done by numerical integration) will not be discussed here, it was found that $a_2 = 0.332$.

The shear stress at the boundary is

$$\tau_0 = \mu\left(\frac{\partial u}{\partial y}\right)_0 = \mu\left(\frac{U}{\delta_0}\right) f''(0) = \frac{0.332U\mathbf{R}_x^{1/2}}{x} \quad \text{or}$$

$$\frac{\tau_0}{\left(\frac{1}{2}\rho U^2\right)} = \frac{0.664}{\mathbf{R}_x^{1/2}} \tag{8.49}$$

The drag F on the plate per unit width on one side of the plate of length x is

$$F = \int_0^x \tau_0 dx = \frac{0.332}{\rho U^2}\int \frac{1}{\mathbf{R}_x^{1/2}}dx = \left(\frac{1.328}{\mathbf{R}_x^{1/2}}\right) x\frac{1}{2}\rho U^2 \tag{8.50}$$

From the velocity profile shown in Fig. 8.20a, we can accurately predict the boundary-layer thickness, δ at which the u/U is 0.99

$$\frac{\delta}{x} = \frac{5}{\mathbf{R}_x^{1/2}} \tag{8.51}$$

The values of f, f' and f'' are also given in Fig. 8.20b.

8.26 Boundary-Layer Integral Momentum Equation

The differential form of the boundary-layer equation (8.46) may now be integrated to yield further interesting results leading to the solution of the rate of growth of the boundary layer. We may add and subtract the term $u(\partial u/\partial x)$ to the boundary-layer equation without altering its value, thus

$$2u\left(\frac{\partial u}{\partial x}\right) - u\left(\frac{\partial u}{\partial x}\right) + v\left(\frac{\partial u}{\partial y}\right) = U\left(\frac{dU}{dx}\right) + \nu\left(\frac{\partial^2 u}{\partial y^2}\right)$$

Since by the equation of continuity, $\partial u/\partial x = -\partial v/\partial y$, the above equation may be written as

$$2u\left(\frac{\partial u}{\partial x}\right) + \frac{\partial(uv)}{\partial y} = U\left(\frac{dU}{dx}\right) + \nu\left(\frac{\partial^2 u}{\partial y^2}\right)$$

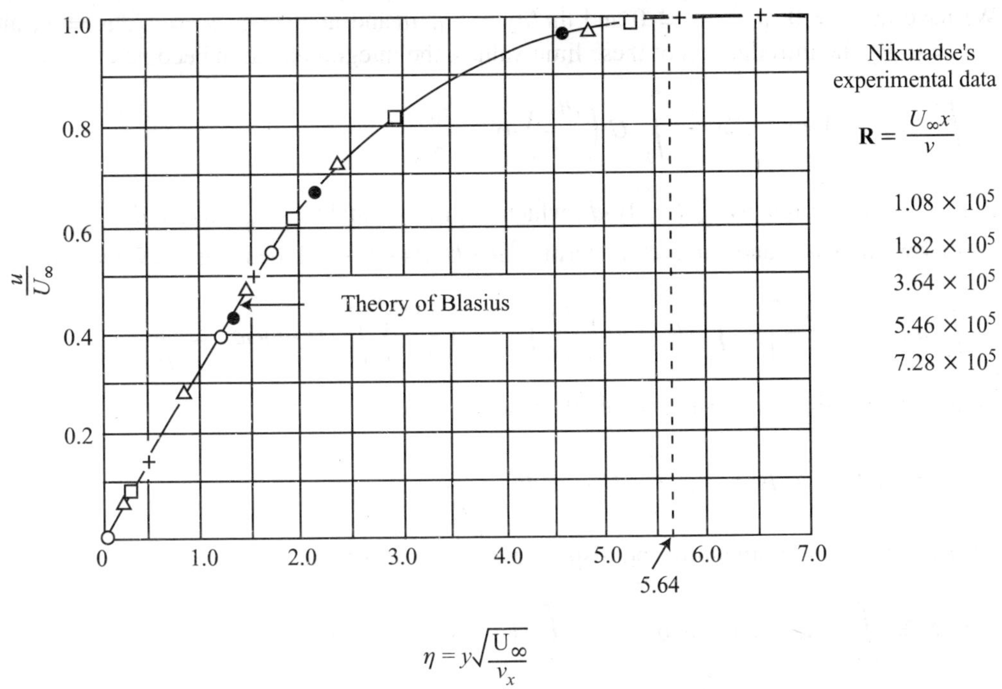

a. Velocity profile

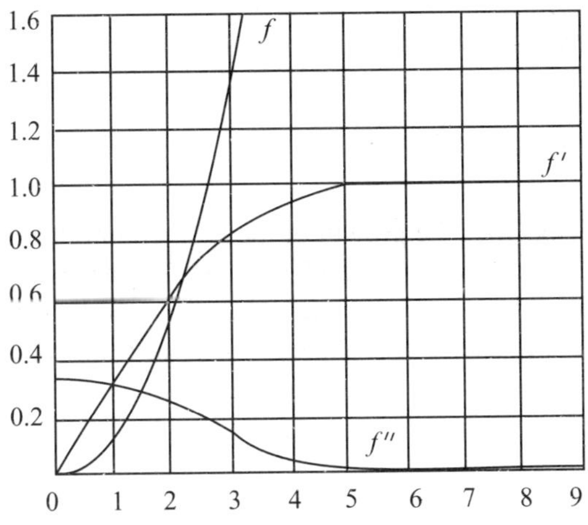

b Values of *f*, *f'* and *f''*

Fig. 8.20 Results of Blasius exact theory

Integration with respect to y between limits $y = 0$ and $y = \delta$ yields

$$\int_0^\delta 2u\left(\frac{\partial u}{\partial x}\right)dy + uv|_0^\delta = \int_0^\delta U\left(\frac{dU}{dx}\right)dy + \nu\left(\frac{\partial u}{\partial y}\right)|_0^\delta$$

We have at $y = 0$: $u = v = 0$ and $\partial u/\partial y = \tau_0/\mu$ and at $y = \delta$: $u = U$, $v = v_\delta$ and $\partial u/\partial y = 0$. By the introduction of these limit values, the integral equation becomes

$$\int_0^\delta 2u\left(\frac{\partial u}{\partial x}\right)dy + Uv_\delta = \int_0^\delta U\left(\frac{dU}{dx}\right)dy - \frac{\tau_0}{\rho}$$

But v_δ may be written as $\int_0^\delta(\partial v/\partial y)dy$ which in turn is equal to $-\int_0^\delta(\partial u/\partial x)dy$ Further we may add and subtract the term $\int_0^\delta u(\partial U/\partial x)$ to yield the integral expression

$$\int_0^\delta u\left[\frac{\partial(U-u)}{\partial x}\right] + \int_0^\delta (U-u)\left(\frac{\partial u}{\partial x}\right)dy + \left(\frac{dU}{dx}\right)\int_0^\delta (U-u)dy = \frac{\tau_0}{\rho}$$

Combining the first two terms on the left-hand side, we get

$$\frac{d}{dx}\int_0^\delta u(U-u)dy + \frac{dU}{dx}\int_0^\delta (U-u)dy = \frac{\tau_o}{\rho}$$

The definitions for momentum and displacement thickness yield

$$U^2\delta_2 = \int_0^\delta u(U-u)dy \text{ and } U\delta_1 = \int_0^\delta (U-u)dy$$

Hence we can write the momentum integral equation as

$$\frac{d}{dx}\left(U^2\delta_2\right) + \frac{dU}{dx}\left(U\delta_1\right) = \frac{\tau_o}{\rho}$$

$$\frac{d\delta_2}{dx} + \frac{1}{U}\frac{dU}{dx}(2\delta_2 + \delta_1) = +\frac{\tau_o}{\rho U^2} \tag{8.52}$$

The preceding equation specifies the growth of the momentum thickness with distance along the boundary. Since δ, δ_1. δ_2 are all related to each other, (8.52) may be considered as the equation for the rate of growth of boundary-layer thickness. For the specific case of the development of the boundary layer along a flat plate at zero angle of incidence ($dU/dx = 0$), we have

$$\frac{d\delta_2}{dx} = \frac{\tau_0}{\rho U^2} \tag{8.53}$$

On integration,

$$\frac{\delta_2}{x} = \frac{\tau_0}{\rho U^2} = \frac{0.664}{\mathbf{R}_x^{1/2}} \tag{8.54}$$

and the displacement thickness,

$$\frac{\delta_1}{x} = \frac{1.72}{\mathbf{R}_x^{1/2}} \tag{8.55}$$

EXAMPLE 8.3 Assume that the velocity profile is given by the fourth-degree polynomial $u/U = A\eta + B\eta^2 + C\eta^3 + D\eta^4 + E$, where $\eta = y/\delta$. Find the coefficients A, B, C and D, the displacement, the moment thickness and the growth of the boundary layer.

All velocity profiles must satisfy the essential boundary conditions: $u = 0$ at $\eta = 0$; $u = U$, $\partial u/\partial y = 0$ at $\eta = 1$ and $\partial^2 u/\partial y^2 = 0$ at $\eta = 1$ and at $\eta = 0$.

$u = 0$ at $y = 0$ gives $E = 0$; $u = U$ at $\eta = 1$ gives $A + B + C + D = 1$

$\partial u/\partial y = 0$ at $\eta = 1$ gives $A + 2B + 3C + 4D = 0$; $\partial^2 u/\partial y^2 = 0$ at $\eta = 0$ gives $B = 0$ and $\partial^2 u/\partial y^2 = 0$ at $\eta = 1$ gives $C + 2D = 0$

Hence $A = 2$, $B = 0$, $C = -2$, $D = 1$

$$\frac{u}{U} = 2\eta - 2\eta^3 + \eta^4$$

Displacement thickness $= \delta_1/\delta = \int_0^1 (1 - u/U)d\eta = [\eta - 2\eta^2/2 + 2\eta^4/4 - \eta^5/5]_0^1 = 3/10$

Momentum thickness: $\delta_2/\delta = \int_0^1 u/U(1 - u/U)d\eta = 37/315$

The shear stress at the boundary: $\tau_0 = \mu(du/dy)_{y=0} = 2\mu U/\delta$

From (8.53) we have $d\delta_2/dx = \tau_0/\rho U^2 = 2\mu U/\rho U^2\delta = 2\mu/\rho U\delta$

Since $\delta_2 = 37\delta/315$, we write $(37/315)(\delta d\delta/dx) = 2\mu/\rho U$ and on integration, we get $\delta = (1260/37)^{1/2}(\mu x/\rho U)^{1/2}$

$$\frac{\delta}{x} = \frac{5.84}{\mathbf{R}_x^{1/2}}$$

$$\frac{\tau_0}{\rho U^2} = \frac{2\mu}{\rho U\delta} = \frac{2\mu \mathbf{R}_x^{1/2}}{5.84\rho U x} = \frac{0.343}{\mathbf{R}_x^{1/2}}$$

The following table gives the values of displacement, thickness, δ, and $\tau_0/\rho U^2$ for various assumptions of the velocity profile in the laminar boundary layer though they may not satisfy all the boundary conditions.

Compare all these values with the Blasius derivation

$$\delta_1 = 1.7208\left(\frac{\nu x}{U}\right)^{1/2}, \delta_2 = 0.664\left(\frac{\nu x}{U}\right)^{1/2}, \frac{\tau_0}{\rho U^2} = \frac{0.332}{\mathbf{R}_x^{1/2}}$$

u/U	δ_1/δ	δ_2/δ	δ_1/x	$\tau_0/\rho U^2$
y/δ	1/2	1/6	$3.464/\mathbf{R}_x^{1/2}$	$0.289/\mathbf{R}_x^{1/2}$
$2y/\delta - y^2/\delta^2$	1/3	2/15	$5.48/\mathbf{R}_x^{1/2}$	$0.365/\mathbf{R}_x^{1/2}$
$(3/2)y/\delta - (1/2)(y/\delta)^3$	3/8	39/280	$4.64/\mathbf{R}_x^{1/2}$	$0.323/\mathbf{R}_x^{1/2}$
$\sin \pi y/2\delta$	$(\pi - 2)/\pi$	$(4 - \pi)/\pi$	$2.91/\mathbf{R}_x^{1/2}$	$0.54/\mathbf{R}_x^{1/2}$

8.27 Transition to Turbulence

As the boundary-layer thickness increases, the flow in the boundary layer may become unstable and break into turbulence. The actual point of transition to turbulence is dependent on several factors such as plate roughness, initial value of turbulence and heat transfer to and from the plate. Hansen found that the transition occurred at $\mathbf{R}_x = 3 \times 10^5$ corresponding to $\mathbf{R}_\delta = 2740$ but more recent experiments have shown that the laminar boundary layer along a flat plate may exist up to $\mathbf{R}_x = 3 \times 10^6$ for a very small value of initial turbulence $(u'^2/U < 0.001)$ in the main stream. The turbulent boundary layer and its characteristics will be discussed in the next chapter.

EXAMPLE 8.4 A thin plate of dimensions $a = 0.2\,\text{m}$ and $b = 0.5\,\text{m}$ is held in water ($\nu = 10^{-6}\ \text{m}^2/\text{s}$ and $\rho = 1000\,\text{kg/m}^3$ flowing parallel to it at a velocity of $U = 1\,\text{m/s}$. Find the drag force on the plate. Which edge of the plate should be aligned in the direction of flow to obtain a lower drag force? Assume that the laminar boundary layer exists throughout the length of the plate.

Area of the plate on both sides: $2ab = 2 \times 0.2 \times 0.5 = 0.2\,\text{m}^2$

Force: $F = \text{Area} \times (1.328/\mathbf{R}_L^{1/2}) \times (\rho U^2/2) = [0.2 \times 1.328 \times 1000 \times 1^2/2]/\mathbf{R}_L^{1/2} = 132.8/\mathbf{R}_L^{1/2}$ and if the plate is held with $a = 0.2\,\text{m}$ as the length, then $UL/\nu = 1 \times 0.2/10^{-6} = 2 \times 10^5$ and $F = 132.8/(2 \times 10^5)^{1/2} = 0.297\,\text{N}$

If the plate is held with $b = 0.5\,\text{m}$ as the length, then $\mathbf{R}_L = 1 \times 0.5/10^{-6} = 5 \times 10^5$ and $F = 132.8/(5 \times 10^5)^{1/2} = 0.188\,\text{N}$

Hence the alignment with length $= 0.5\,\text{m}$ gives lower resistance.

EXAMPLE 8.5 A flat plate 4 m wide and 1 m long in the direction of flow is kept in keresone ($\nu = 2.29 \times 10^{-6}\,\text{m}^2/\text{s}$, and $\rho = 800\,\text{kg/m}^3$) flowing with a uniform velocity of 0.5 m/s. Find the thickness of the boundary layer and the wall shear stress τ_0 at the centre of the plate as well as at the trailing edge. Find also the total drag on the plate.

$$\frac{\delta}{x} = \frac{5}{\mathbf{R}_x^{1/2}}$$

At the centre of the plate, $\mathbf{R}_x = (UL/2)/\nu = 0.5 \times 0.5/2.29 \times 10^{-6} = 109200$ and at the trailing edge $\mathbf{R} = 218000$

At the centre of the plate, $\delta = 5 \times 0.5/(109200)^{1/2} = 0.756\,\text{cms}$ and at the trailing edge, $\delta = 5 \times 1/(218000)^{1/2} = 1.07\,\text{cms}$

The shear stress at the centre, $\tau_0 = (0.664/\mathbf{R}_x^{1/2})(\rho U^2/2) = [0.664/(109200)^{1/2}](800 \times 0.5^2/2) = 0.20\,\text{N/m}^2$

τ_0 at trailing edge $= [0.664/(218000)^{1/2}](800 \times 0.5^2/2) = 0.142\,\text{N/m}^2$

The total drag on the plate, $F = [1.328/(218000)^{1/2}](4 \times 1)(800 \times 0.5^2/2) = 1.14\,\text{N}$ on one side of the plate.

EXAMPLE 8.6 Air ($\rho = 1.167\,\text{kg/m}^3$, $\nu = 1.5 \times 10^{-5}\text{m}^2/\text{s}$ flows over a thin plate with a free stream velocity of 6 m/s. At a point 15 cms from the leading edge, determine the value of y at which $u/U = 0.456$. Also calculate v and $\partial u/\partial y$ at this point.

The solution depends on the type of velocity profile assumed. If Blasius solution is used, then referring to Fig. 8.20, we find that for the given value of $u/U = 0.456$, $y(U/\nu x)^{1/2} = 1.4$ and hence

$$y = 1.4\left(1.5 \times 10^{-5} \times \frac{0.15}{6}\right)^{1/2} = 0.857\,\text{mm}$$

Other values are $f(\eta) = 0.4$, $f'(\eta) = 0.5$, $f''(\eta) = 0.31$. With these values we get

$$v = 0.5\left(\frac{\nu U}{x}\right)^{1/2} [\eta f'(\eta) - f(\eta)]$$

$$= 0.5\left(1.5 \times 10^{-5} \times \frac{6}{0.15}\right)^{1/2} \times (1.4 \times 0.5 - 0.4) = 0.392\,\text{cm/s}$$

and

$$\frac{\partial u}{\partial y} = U\left(\frac{U}{\nu x}\right)^{1/2} f''(\eta) = 6\left(\frac{6}{1.5} \times 10^{-5} \times 0.15\right)^{1/2} \times 0.31 = 3038\,\text{s}^{-1}$$

EXAMPLE 8.7 Liquid of density 1200 kg/m^3 and $\mu = 0.90$ kg/m s flows in a circular pipe of diameter 5 cm. The pipe is inclined at 30° to the horizontal. Pressures measured at two points differing in vertical elevation of 1 m are 200 kN/m^2 and 180 kN/m^2 respectively. Determine the velocity distribution, maximum velocity, discharge in the pipe, and shear stress as the boundary.

The velocity distribution is given by

$$w = \left(\frac{1}{4\mu}\right)\left[\frac{-d(p + \gamma h)}{dx}\right](a^2 - r^2)$$

The piezometric pressures at the two points are

$$p_1 + \gamma h = 200000 + 1200 \times 9.81 \times 1 = 211772 \text{ and } p_2 = 180000\,\text{N/m}^2$$

With $dx = 2$ m, we have $-d(p + \gamma h)/dx = -31772/2 = -15886$ N/m

$$w = (1/4 \times 0.90)[15886(0.025^2 - r^2)] = 2.75 - 4412r^2$$

Maximum velocity occurs at the centre, where $r = 0$ and $w_{\text{max}} = 2.75$ m/s. Average velocity, $V = (a^2/8\mu)[-d(p + \gamma h)/dx] = (0.025^2/8 \times 0.98)15886 = 1.379$ m/s

The discharge $Q = (\pi d^2/4)V = (\pi \times 0.05^2/4) \times 1.379$.

The shear stress at radius r is $\tau = \mu dw/dr = \mu(2.75 - 8824r)$.

The boundary shear stress, $\tau_0 = 0.98(2.75 - 8824 \times 0.025) = -216.22\,\text{N/m}^2$.

EXAMPLE 8.8 Two parallel plates are inclined at an angle θ to the horizontal and the distance between the plates is $2a$. A liquid of density ρ and dynamic viscosity μ occupies the space between the plates. The lower plate is held stationary and the upper plates moves with a uniform velocity, U parallel to the lower plate. Find an expression for the velocity distribution across the plates in times of U and piezometric pressures.

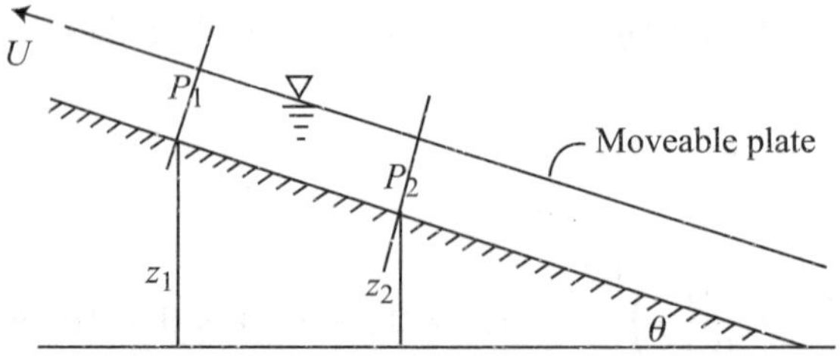

Figure for Example 8.8

This is an example of a combined Hagen–Poiseuille and Couette flow. The results obtained in Section 8.9 can be used here with the modification that the pressure gradient is replaced by the piezometric pressure gradient.

$$\frac{u}{U} = \frac{y}{2a} + K\left(\frac{y}{2a}\right)\left(1 - \frac{y}{2a}\right)$$

where K now stands for

$$K = \left[\frac{(2a)^2}{2\mu U}\right]\left[\frac{-d(p+\gamma h)}{dx}\right]$$

$$\frac{d(p+\gamma h)}{dx} = \frac{[(p_1+\gamma z_1)-(p_2+\gamma z_2)]}{L}$$

EXAMPLE 8.9 A conveyor belt with the lower end submerged in a chemical tank is used to transport a liquid from the lower elevation to a higher elevation as shown in the figure. Derive an expression for velocity distribution and discharge.

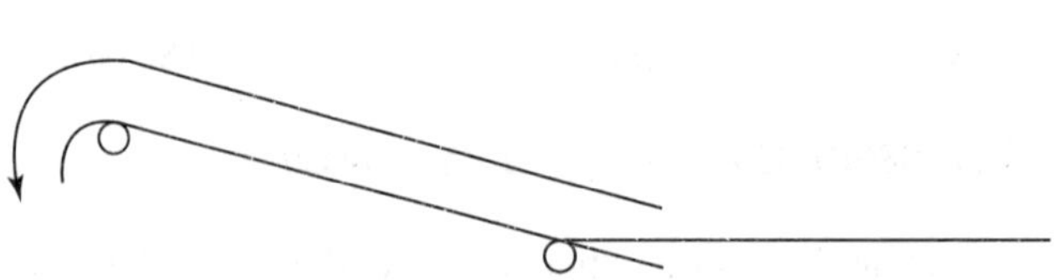

Figure for Example 8.9

The flow is uniform and steady and the depth of flow is uniform. The same hydrostatic pressure exists at all sections. The equation of motion reduces to

$$\frac{d\tau}{dy} = \frac{\mu d^2\mu}{dy^2} = \gamma \sin\theta$$

Successive integration yields

$$\frac{du}{dy} = \left(\frac{\gamma}{\mu}\right) \sin\theta y + C_1$$

$$u = \left(\frac{\gamma}{\mu}\right) \sin\theta \left(\frac{y^2}{2}\right) + C_1 y + C_2$$

The boundary conditions are $u = U$ at $y = 0$ which gives $C_2 = U$.
At $Y = h$, the velocity is a maximum and $du/dy = 0$; $C_1 = -(\gamma/\mu)\sin\theta h$

$$u = \left(\frac{\gamma}{\mu}\right) \sin\theta \left[\frac{y^2}{2} - hy\right] + U$$

The discharge per unit width: $q = \int_0^h u dy = Uh - g \sin\theta h^3/3$.

The first term stands for the discharge dragged up by the conveyor belt and the second term accounts for the flow downwards due to gravitational action. Students may verify that the efficiency of the system is 1/3.

EXAMPLE 8.10 A rotating cylinder viscosimeter has two concentric cylinders of radius R_1 and R_2 with a very narrow annular space. The inner cylinder has a height of h and the bottom clearance is a as shown in the figure. The outer cylinder rotates at a constant known revolutions per minute. The inner cylinder, which is freely suspended by a wire, experiences a torque to cause an angular deflection of the suspended wire for which the calibration of torque vs angular deflection is available. Derive the relation between torque experienced by the suspension wire and the viscosity of the liquid.

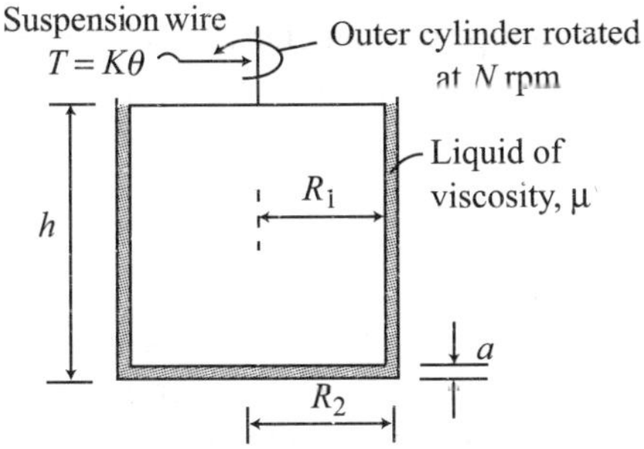

Figure for Example 8.10

Let the outer cylinder rotate at N rpm. Then $\omega_2 = 2\pi N/60$ and the tangential velocity is $v_2 = 2\pi N R_2/60$.

Since the annular space $(R_2 - R_1)$ is small, we assume that the tangential velocity varies linearly in the annular space. Therefore we write the velocity gradient as

$dv/dr = 2\pi NR_2/60(R_2 - R_1)$ and the shear stress is $\tau = \mu dv/dr$

$$\tau = 2\pi\mu NR_2/60(R_2 - R_1)$$

Torque, T_1 on the inner cylinder of radius R_1 of contact peripheral area of $2\pi R_1 h$ is $(2\pi R_1 h)\tau R_1$. Substituting for τ we get

$$T_1 = 4\pi^2\mu R_1^2 R_2 hN/60(R_2 - R_1)$$

The bottom of the inner cylinder is also subject to shear stresses. Let the tangential velocity at any radius r from the centre be $v = \omega r$.

The shear stress $\tau = \mu v/a = \mu\omega r/a = 2\pi\mu Nr/60a$

Force on elementary area $2\pi r dr$, $dF = \tau 2\pi r dr = 4\pi^2\mu Nr^2/60a$

The elementary torque, dT_2 is $dF \cdot r = 4\pi^2\mu Nr^3/60a$

The total torque T_2 from the bottom is obtained by integrating the preceding expression between the limits 0 to R_1.

$$T_2 = \int_0^{R_1} \frac{4\pi^2\mu Nr^3}{60a} = \frac{\pi^2\mu NR_1^4}{60a}$$

The total torque on the inner cylinder is $T = T_1 + T_2$

$$T = \left[\frac{\pi^2\mu NR_1^2}{60}\right]\left[\frac{4R_2 h}{(R_2 - R_1)} + \frac{R_1^2}{a}\right]$$

Generally T_2 is small compared to T_1. Since we know the values of N, R_1, R_2, h, a, and $T = K\theta$ is measured, we can obtain the value of μ.

EXERCISES

8.1. In a laminar flow of oil in a pipe of diameter 2.5 cm, the velocity at the centre is 2.5 m/s. Calculate the average velocity, discharge and the pressure gradient ($\mu = 4 \times 10^{-1}$ Pa s; $\rho = 850\,\text{kg/m}^3$).

8.2. At what radius of the pipe is the velocity equal to the average velocity?

8.3. The shear stress in a pipe of 10 cm dia. is 2 kN/m^2 at a radius of 2.5 cm. Calculate the discharge ($\mu = 4 \times 10^{-1}$Pa s).

8.4. An oil of specific gravity 0.90 flows through a 10 cm dia. pipe. If the velocity at 2.5 cm from the wall is 0.5 m/s, what head loss occurs in a pipe length of 50 m. Calculate the wall shear stress and the discharge in the pipe ($\mu = 7.5$ poises).

8.5. A continuous moving belt passes through a liquid bath (density ρ and viscosity μ) vertically upwards at a velocity U and picks up a thin film of liquid of thickness t. Find an expression for the rate at which the liquid flows up the belt.

8.6. A 10 cm dia. shaft runs at 900 rpm in a sleeve with a radial clearance of 1 mm. If the sleeve length is 15 cm long and the space is filled with an oil of dynamic viscosity $\mu = 0.10\,\text{kg/ms}$, determine the torque resistance.

8.7. A slender outlet tube is located at the bottom of a constant head tank. The diameter and length of the tube are 2 mm and 30 cm. If the head in the tank is 0.5 m and the flow collected in 30 seconds is 0.01 litres, find the viscosity of the liquid. (Sp. gr. of the liquid = 0.8)

8.8. Crude oil is pumped through a 10 cm dia. inclined pipe (inclined downwards at an angle of 30° to the horizontal) at the rate of 0.015 m^3/s.. What will be the pressure head difference between two points separated by a distance of 300 m. What pumping head is required if the pipe is inclined upwards over the same length and discharge (Sp.gr. = 0.90 and $\mu = 80 \times 10^{-3}$ Pa s).

8.9. Two plane parallel boundaries are 0.5 cm apart and filled with a liquid of viscosity 80×10^{-3} Pa s. A 0.30 cm thick square plate of sides 0.3 m is placed symmetrically in the gap and moved at a velocity of 15 cm/s. Determine the force required to move the plate.

8.10. A G I pipe 10 cm in diameter carries a discharge of 8 litres of water per minute. Is the flow laminar? What is the value of the friction factor, f. Find the head lost due to friction in a length of 200 m.

8.11. Determine the fall velocity of mist droplets of dia. 0.025 mm. ($\mu_{air} = 1.85 \times 10^{-5}$ kg/m s, $\rho = 1.2$ kg/m^3)

8.12. In Section 8.13, the inner cylinder was moved while the outer cylinder was stationary. Derive an expression for axial velocity distribution in the annulus if the inner cylinder is stationary and the outer cylinder is moved at an uniform velocity, U in the axial direction.

8.13. Water flows through an annular section 10 m long. The inner diameter is 10 cm and the outer diameter is 10.5 cm. If the average velocity in the annulus is 10 cm/s, is the flow laminar or turbulent? Determine the friction factor and the loss of head drop per metre. ($\nu = 1 \times 10^{-6}\ m^2/s$). The hydraulic radius of the pipe is defined as the ratio of area, A to wetted perimeter, $P (R = A/P)$. For a circular pipe $R = D/4$ and hence $D = 4R$. For non-circular sections, we define the Reynolds number as $\mathbf{R} = VD/\nu$ and hence for non-circular sections $\mathbf{R} = V(4R)/\nu$

8.14. The annular space between two rotating cylinders having inner and outer diameters of 10 cm and 12 cm respectively is filled with a liquid of dynamic viscosity 2×10^{-2} Pa s. If the outer cylinder is rotating at a speed of 60 rpm, what should the rotational speed of the inner cylinder be so that the flow in the annulus is irrotational in the annular space? Find the power required to rotate the cylinders.

8.15. The space between two parallel plates is filled with two immiscible fluids of equal thickness a but which have different values of density and viscosity as shown in the figure. If the top plate is moved with a velocity, U and the lower is held fixed, find the velocity distribution in the two liquids.

Hint: With the lower plate as the x-axis, the boundary conditions are $u_1 = 0$ at $y = 0$, $u_2 = U$ at $y = 2a$, $u_1 = u_2$ at the interface $y = a$. The shear stress should also match at the interface $y = a$; $\tau_1 = \tau_2$.

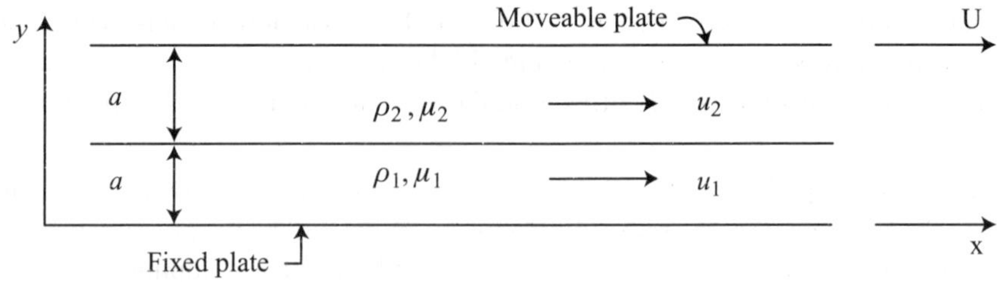

Figure for Problem 8.15

8.16. In Problem 8.15, both plates are stationary but the flow is caused by a pressure gradient, $K = -dp/dx$. Find the velocity distribution in the two liquids.

8.17. For each of the values of u/U given in the table in Section 8.26, derive the relationships for displacement thickness, momentum thickness, boundary layer thickness and $\tau_0/\rho U^2$.

8.18. A flat plate 0.75 m long and 2.0 m wide is kept in an air stream with a velocity of 5 m/s. Determine the boundary layer thickness at a distance of 0.375 m from the leading edge of the plate and at the end of the plate. Find also the drag on both sides of the plate and the shear stress at the mid-point. $\nu = 1.5 \times 10^{-5} \text{m}^2/\text{s}$ and $\rho = 1.2\,\text{kg/m}^3$. Use Blasius theory.

8.19. A rectangular plate 2.0 m long and 1.5 m wide is towed on water edgewise at a velocity of 0.5 m/s. Determine the distance up to which the laminar boundary layer will persist. $\nu = 1 \times 10^{-6}\,\text{m}^2/\text{s}$. Find the total drag on the plate up to this distance and the boundary layer thickness as well as the shear stress at the end.

8.20. Find the ratio of boundary layer resistance experienced by the first one-third of the plate to the latter one-third of the plate assuming that the boundary layer remains laminar throughout the length of the plate.

9 Turbulent Flow

Part I Reynolds Equations

9.1 Introduction

In the preceding chapter, an elementary discussion on the stability of laminar flows was presented. It has been observed that beyond a certain Reynolds number (the magnitude of which depends on the frequency of the disturbance, the initial value of the turbulence, the type of pressure gradient, the shape and roughness of the boundary etc.) the laminar flow becomes unstable. Once instability sets in, turbulence in the form of eddies spreads rapidly covering the entire flow region. By turbulence, it is implied that the flow parameters such as velocity and pressure at any point in the flow medium are continuously fluctuating in a random manner. The fluctuations should be in fact random both with respect to time at a given time and space over a given time interval. Periodic flows though their characteristics change with time are not random in nature and hence are not classified as turbulent flow. In steady flows, the velocity variation with respect to time at a given point may be illustrated as shown in Fig. 9.1.

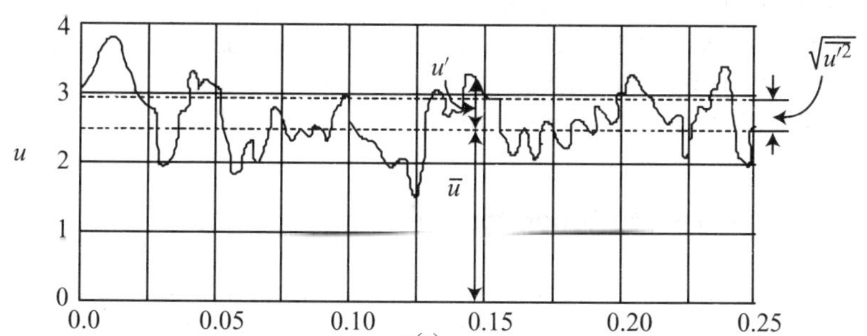

Fig. 9.1 Fluctuations in velocity

If u is the instantaneous value of the velocity component in the x-direction, the time average of the velocity component known as its mean value may be obtained as

$$\overline{u} = \frac{1}{T}\int_0^T u dt \tag{9.1}$$

where the bar over u indicates the mean value. The true mean is obtained when T is sufficiently large. The instantaneous value may be conceived as consisting of two parts, namely the mean value and the fluctuating part u'

$$u = \overline{u} + u' \tag{9.2}$$

A measure of turbulence may be found in the statistical description of u'. Since the average of fluctuations $\overline{u'} = 0$, we may obtain the average of the absolute values of the fluctuating quantity $|\overline{u'}|$ which may be obtained graphically by rectifying the velocity diagram about the mean value (all negative values are taken as positive). Alternatively another measure of the intensity of turbulence is obtained by performing the averaging process after the fluctuating quantities are first squared to remove the negative signs. Then, $\overline{u'^2}$ is the mean square of the fluctuations and the square root of the mean square known as the root mean square has the same dimension as velocity.

$$\sqrt{\overline{u'^2}} = \left[\operatorname*{lt}_{T\to\infty} \frac{1}{T}\int_0^T u'^2 dt\right]^{1/2} \tag{9.3}$$

The root mean square (rms for short) is an eminently suitable statistical measure of the intensity of turbulence. The foregoing statements are true for any fluctuating quantity. If the turbulence can be envisaged as a superposition of a large number of eddies of various shapes, sizes and strengths on the mean motion, then the intensity of turbulence indicated by the rms value is a characteristic velocity of the eddy motion. The eddy motion in fluid turbulence may be compared to molecular motion, except for the fact that the scale of motion differ widely—the former being macroscopic. When any two points are chosen close to each other in the flow field, then a positive fluctuation at one point is also matched by a fluctuation of the same sign at the other point. As the two points chosen are farther and farther apart, there will a limiting distance in the statistical sense beyond which the signs of the fluctuations at these points are no longer related. This radius of influence within which the fluctuations are dependent on each other may be termed as the scale of turbulence or the eddy scale. The scales may vary at different locations in the flow field. Before any discussion can be attempted on the scales of turbulence or wave lengths of the disturbances, it is necessary to introduce the equations of motion for turbulent flow.

9.2 Reynolds Equations of Motion for Turbulent Flow

In this and succeeding sections, we shall restrict the discussion to steady two-dimensional incompressible flows. The equation of continuity and the Navier–Stokes equations of motion are true at any instant of time. Hence the variables u, v and p that appear in these equations are taken as instantaneous values. These instantaneous values may be written as sums of the mean and fluctuating components as in equation (9.2). Hence we write $u = \overline{u}+u'$, $v = \overline{v}+v'$, $w = \overline{w}+w'$. The definition of a two-dimensional flow is that the derivatives of mean quantities with respect to z are zero. A special case is of course if $\overline{w} = 0$ but even in this case w' is always present. The instantaneous equation of continuity takes the form ($\overline{w} = 0$ and all variations of the mean and fluctuating quantities with respect to z is zero).

$$\frac{\partial \overline{u}}{\partial x} + \frac{\partial u'}{\partial x} + \frac{\partial \overline{v}}{\partial y} + \frac{\partial v'}{\partial y} + \frac{\partial w'}{\partial z} = 0$$

We can take the average of the preceding instantaneous equation of continuity. In doing so we observe the rules: (a) the average of a differential is the differential of the average; (b) the average of the average is the average itself; (c) the average of a fluctuating quantity is zero; (d) the average of the product of two fluctuating quantities is not equal to zero.

The equation of continuity for the mean two-dimensional flow is

$$\frac{\partial \overline{u}}{\partial x} + \frac{\partial \overline{v}}{\partial y} = 0 \tag{9.4}$$

The instantaneous equation of motion may be likewise written after substituting $u = \overline{u} + u'$, $v = \overline{v} + v'$, $w = w'$, $p = \overline{p} + p'$ in the Navier–Stokes equation and then average the entire equation to obtain the equation for the mean flow. After some manipulation and rearranging of the terms, we get the equation for the mean flow (see Appendix to this chapter for detailed derivation).

$$\overline{u}\frac{\partial \overline{u}}{\partial x} + \overline{v}\frac{\partial \overline{u}}{\partial y} = X - \frac{1}{\rho}\frac{\partial \overline{p}}{\partial x} + \frac{\mu}{\rho}\nabla^2\overline{u} - \frac{\partial \overline{u'^2}}{\partial x} - \frac{\partial \overline{u'v'}}{\partial y} \tag{9.5}$$

$$\overline{u}\frac{\partial \overline{v}}{\partial x} + \overline{v}\frac{\partial \overline{v}}{\partial y} = Y - \frac{1}{\rho}\frac{\partial \overline{p}}{\partial y} + \frac{\mu}{\rho}\nabla^2\overline{v} - \frac{\partial \overline{u'v'}}{\partial x} - \frac{\partial \overline{v'^2}}{\partial y}$$

$$0 = -\frac{\partial \overline{u'v'}}{\partial x} - \frac{\partial \overline{v'w'}}{\partial y}$$

The above equations are known as the Reynolds equations for a mean two-dimensional flow. The terms $\rho\overline{u'^2}$, $\rho\overline{u'v'}$, $\rho\overline{v'^2}$ are referred to as Reynolds stress terms, but it should be noted that the terms originated from the accelerative terms on the left-hand side of the N–S equations but for convenience have been shifted to the right-hand side and combined with shear stresses. It is more appropriate to call them as apparent shear stresses. In three-dimensional flows these stresses are nine in number but three are dependent. The independent Reynolds stresses are $\rho\overline{u'^2}$, $\rho\overline{u'v'}$, $\rho\overline{u'w'}$, $\rho\overline{v'^2}$, $\rho\overline{v'w'}$, $\rho\overline{w'^2}$. The terms $\rho\overline{u'^2}$, $\rho\overline{v'^2}$, $\rho\overline{w'^2}$ are called normal turbulent stresses and the remaining ones are turbulent shear stresses. The physical significance of the normal turbulent stresses($-\rho\overline{u'^2}$ in the x-direction etc.) may be interpreted as the additional momentum leaving in the respective directions per unit surface in unit time due to the respective fluctuating component in respective directions (e.g., u' in the x-direction etc.). This is equivalent to an additional dynamic pressure in the respective direction. The shearing stress $\rho\overline{u'v'}$ etc. can be explained with the aid of a simple example shown in Fig. 9.2.

Two open freight cars loaded with material are moving in the same direction parallel to each other at velocities u_A and u_B. Some material of mass m per unit time is thrown continuously from car B with a velocity $v_1 (m = \rho v_1)$ to car A. If the difference in the velocities of the two cars is

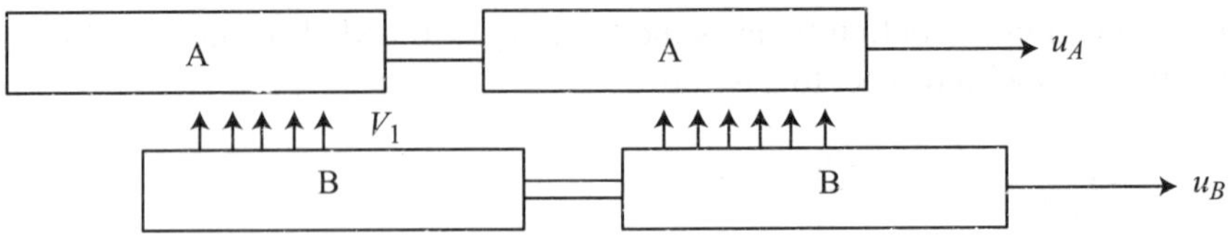

Fig. 9.2 Exchange of momentum

$u_1 = u_B - u_A$, then the momentum change at car A in the direction of motion is mu_1 which is equal to the tangential force acting on car A in the direction of motion. Thus

$$\text{Shearing stress} = \frac{\text{Tangential force}}{\text{Surface area}} = \rho v_1 u_1$$

In a similar manner, we can extend the analogy to explain that the shearing stress is the result of the rate of momentum transport of fluid masses due to random motion. In Fig. 9.3, the velocity of fluid particles at level A is u.

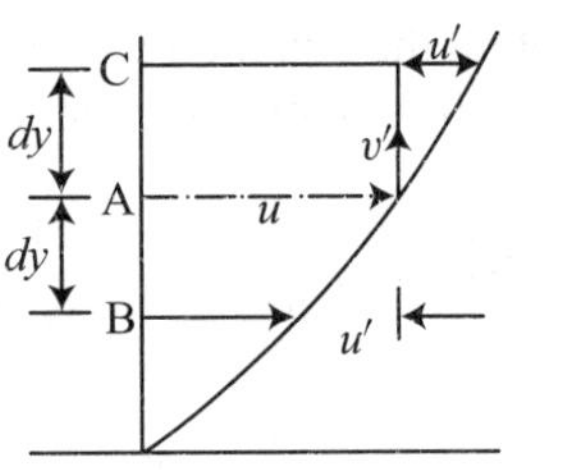

Fig. 9.3 Shear stress due to exchange of momentum

The velocity at level B is $u - u'$ and at level C it is $u + u'$. The fluctuating velocity component v' brings about a mean rate of momentum transfer of $-\rho\overline{v'u'}$ from level C to A and from level B to A. The mean rate of momentum transfer per unit area is equated to the turbulent shear stress

$$\tau_{yx} = -\rho\overline{u'v'} \tag{9.6}$$

Equations (9.5) reveal certain interesting features: (a) the equation for mean flow contains turbulent fluctuation terms indicating that the mean motion is influenced by turbulence; (b) instantaneous viscous terms are not present; (c) the coefficient of viscosity, μ is associated with terms $\nabla^2 u$, and $\nabla^2 v$ only and not with turbulent terms. Hence dissipation of both mean and turbulent flow can take place through the action of viscosity in the presence of mean velocity gradients.

EXAMPLE 9.1 Apply the Reynolds equations of motion for turbulent flow between parallel plates where the flow is caused by the movement of one of the plates (Couette flow), the other being stationary. The pressure gradient is zero in the direction of the flow.

Let the top plate move with a constant velocity U and let the x-axis be along the stationary bottom plate. The distance between the plates is a. The flow is steady, uniform and two-dimensional. Hence $\partial\overline{u}/\partial x = 0$; $\overline{v} = \overline{w} = 0$; $\partial\overline{p}/\partial x = 0$.

The Reynolds equation reduces to

$$
\begin{aligned}
0 &= \frac{\partial}{\partial y}\left(\mu\frac{\partial \overline{u}}{\partial y} - \rho\overline{u'v'}\right)\\
0 &= -\frac{\partial}{\partial y}\left(\overline{p} + \rho\overline{v'^2}\right)\\
0 &= -\frac{\partial}{\partial y}\left(\rho\overline{v'w'}\right)
\end{aligned}
\tag{9.7}
$$

Integration of the second equation yields $\overline{p} + \rho\overline{v'^2} = f(x, z)$. Since p does not vary with either x or z, $f(x, z) = \text{constant}$, say p_0.

$$\overline{p} = p_0 - \rho\overline{v'^2}$$

Since $\rho\overline{v'^2}$ is always positive, the maximum value of pressure p occurs at the wall where turbulence is zero. Integrating the first equation and noting that u is a function of y only and $\mu = \text{constant}$

$$\frac{d\overline{u}}{dy} = C + \rho\overline{u'v'}$$

where C is a constant. However the turbulence shear term vanishes at the walls and hence by Rolles's theorem, there must be a point of inflexion in the mean velocity distribution curve. The velocity distribution may be obtained by a second integration

$$\overline{u} = Cy + \int_0^y \rho\overline{u'v'}dy$$

If experimental measurement of the mean velocity is available, it is then possible to estimate the value of the turbulent shear. Finally, the integration of the last equation gives $\overline{v'w'} = \text{constant}$, but its value at the walls is zero and hence the value of the constant must be zero; $\overline{v'w'} = 0$ everywhere.

9.3 Characteristics of Turbulence

In the preceding sections, the equations of motion and continuity have been derived for the mean flow. It is also evident that introduction of the turbulence components have increased the number of unknowns while the number of equations have remained the same. The problem as such becomes indeterminate and to gain additional insight into the character of turbulence, assumed random, statistical or phenomenological approaches are necessary.

9.3.1 Eddy Scales

In a turbulent flow field, observation over a given interval of time will show that eddies of different sizes, strengths, and orientation will pass a given point of observation. Each of these characteristics of eddies should be related to, in a statistical sense, to the boundary geometry in which the flow is taking place, velocity of the mean flow, fluid characteristics, distance from the source of turbulence and the point of observation. At sufficiently high Reynolds numbers, the average size of the eddy, at any point in the wake of a body for example, should be dependent on the dimensions of the body itself. Likewise the average size of the eddies behind a wind tunnel grid is dependent on the mesh size of the grid. The sizes of the eddies vary over a wide range, and it is necessary to introduce a statistical parameter known as scale of turbulence which is a measure of the size of the eddy.

9.3.2 Correlations

In turbulence, even though fluctuations in velocity or pressure may be random, there does exist a certain correlation between two random quantities at the same point (since they are interdependent) or between neighbouring points. The correlation between any two randomly varying quantities is defined as the ratio that the mean of the product of these variables bears to the product of the rms value of these variables. For example, the correlation between u' and v' is written as

$$E = \frac{\overline{u'v'}}{\sqrt{\overline{u'^2}}\sqrt{\overline{v'^2}}} \tag{9.8}$$

The correlation just discussed among the variables at a given point is known as auto-correlation. If the variables involved are measured at different points but at the same instant of time, then we obtain the spatial correlation of the Eulerian type

$$E_{u',x} = \frac{\overline{u'_x u'_{x+dx}}}{\sqrt{\overline{(u'^2)_x}}\sqrt{\overline{(u'^2)_{x+dx}}}} \tag{9.9}$$

Following the Lagrangian approach, we may relate the velocity fluctuations of a given fluid particle at two different instants of time to obtain the Lagrangian correlation

$$L_{u',t} = \frac{\overline{u'_t u'_{t+dt}}}{\overline{u'^2}} \tag{9.10}$$

The averaging in this case has been done for a large number of particles. Finally there is the Eulerian time correlation where a relationship is sought between fluctuations at the same point but at two different instants of time separated by a time interval dt

$$E_{u',t} = \frac{\overline{u'_t u'_{t+dt}}}{\overline{u'^2}} \tag{9.11}$$

A typical spatial correlation is shown in Fig. 9.4. It may be seen that when the distance between the points is zero ($dx = 0$), the correlation coefficient is unity and as the distance between the two points increases, the correlation decreases and approaches zero asymptotically or in some cases after a certain distance, negative correlation may occur. The distance at which the correlation becomes zero is indicative of the size of the eddy and suggests that correlations exist within an individual eddy and that if the two points lie in different eddies, the correlation is poor. Since it may be difficult sometimes to locate the point at which the correlation becomes zero, the area under the curve is a good measure of the size of the eddy. Thus the size of a large eddy, l may be expressed as

$$l_{u',x} = \int_0^{\infty} E_{u',x} dx \tag{9.12}$$

These scales are known as macro-scales or integrated scales of turbulence. The shape of the correlation curve at $x = 0$ may be used to find the size of small-scale eddies (since the correlation is close to unity in small-scale eddies). The radius of curvature, $1/\lambda^2$ at the vertex of the correlation curve may be used as a measure of the micro-scales of turbulence and it is proportional to the size of the small-scale eddy. Since the correlation curves are approximately parabolic near the vertex ($x = 0$), the intercept of the parabola (drawn with it coinciding with the vertex of the correlation curve) on the x-axis gives a measure of the micro-scale as shown in Fig. 9.4.

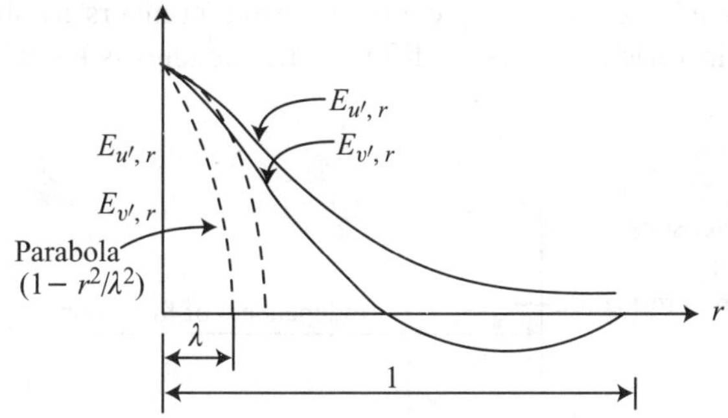

Fig. 9.4 Correlation curves

The shape of the correlation curve gives us information about the mix of different sizes of eddies—small, average and large. For example, if the curve is very sharp, then small-scale eddies are predominant. On the other hand, if the curve is flat, then large-scale eddies are predominant.

9.4 Role of Eddies in Fluid Motion

The formation of large-scale eddies may be assumed to occur from chance fluctuations of relatively small-scale turbulent motion amplified in the presence of mean velocity gradients. These large-scale eddies are produced due to the instability of the mean velocity profiles to a certain type of disturbance present in the small-scale turbulent motion. Large-scale eddies absorb energy from the mean flow. They have low frequencies and low energy content; it should not be assumed, however, that their energy content is negligible as it may be as high as 20% of the total turbulent energy. These large-scale eddies have for their primary role, the process of diffusion or mixing. They are primarily responsible for transporting transferable quantities such as turbulence, momentum, vorticity, heat, matter, and discrete particles. Large-scale eddies convect these transferable properties as they move through space and spread the region of turbulence into the ambient or surrounding regions. The nominal boundary separating the turbulent fluid region from the non-turbulent one is neither sharp nor regular. Turbulent fluid is bounded by contorted surfaces which are moved by the convective action of large eddies. These contorted surfaces provide greater surface area for entrainment of the surrounding fluid into the body of the turbulent fluid. Large eddies are not permanent in character. Due primarily to the inertial interaction of neighbouring eddies, in the presence of mean velocity gradients, these large-scale eddies are stretched and elongated with their cross-sectional area decreasing. The stretching and contorting process also contributes to the spreading of the mixing zone. As they stretch and twist, narrow necks are formed and they break up into smaller eddies. These smaller size eddies contain higher vorticity strengths and turbulent energy. Thus there is a transfer of energy from large-scale eddies to smaller scales. It may be assumed that as the eddy size breaks down progressively into smaller and smaller sizes, these small-scale eddies so generated become independent of the external forces that initiated the formation of large-scale eddies. The average size eddies have higher frequencies and are richer in energy than the very small-scale eddies. Energy content is found to be maximum at certain frequencies or wave numbers as shown in Fig. 9.5. The distribution of energy in various eddies of different frequencies is known as turbulence energy spectrum.

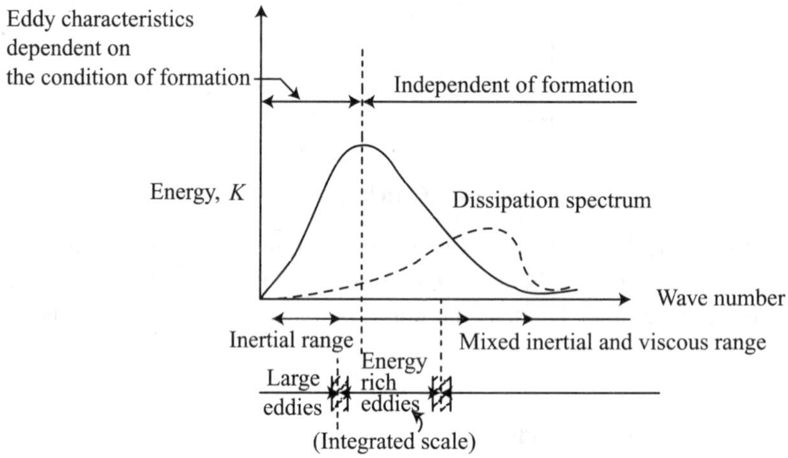

Fig. 9.5 Turbulent energy spectrum

The integral scales referred to earlier indicate the average size of energy-rich eddies. As the eddies become progressively smaller, the velocity gradient across the eddy becomes steep leading to dissipation of turbulent energy into heat by viscous action. While some energy dissipation takes place in large and average eddies also, it is mainly in the small-scale eddies above a certain wave number that maximum dissipation takes place by viscous action. There is thus a cascading transfer of energy from the mean flow to the large size eddies, on to the average size eddies and finally to the small-scale eddies where it is dissipated.

9.5 Classification of Turbulence

Since the study of general types of turbulence is extremely complex, special types of turbulence fields may be examined to provide additional insight into the structure of turbulence. A turbulence in which the mean flow is zero is said to be homogeneous, if the mean of the product of two functions of fluctuating quantities at two different points A and B is a function of the distance r between them and does not depend on the position or location of the points. For example, in a homogeneous field, $\overline{u'_A v'_B}$ is constant for any two points separated by the same distance r. As a special case we find in such fields that at the same point $\overline{u'_A v'_B}$, $\overline{u'_A{}^2}$ etc are constants over the entire field and their derivatives are zero. Thus we may say that in a homogeneous turbulence, the intensity and scales of turbulence are independent of the coordinate position.

Homologus turbulence occurs when shear stresses are constant as in the case of a Couette flow.

The turbulence is said to be isotropic if the turbulence characteristics are independent of direction also. Thus the rotation of the coordinate axes through 90° does not alter the value of the mean product: $\overline{u'_A v'_A} = -\overline{u'_A v'_A}$. This is possible only when $\overline{u'v'}$ at all points in the isotropic field is zero. Further in an isotropic turbulence field, we have $\overline{u'^2} = \overline{v'^2} = \overline{w'^2}$. Nearly isotropic turbulence is found to occur (a) when a uniform flow passes through a regular grid in a wind tunnel, (b) in the central region of a conduit flow and (c) in wide open channels.

Some of the results obtained by analytical study of isotropic turbulence are that

(a) the lateral integral scale of the eddy is half that of the longitudinal scale: $l_n = 1/2\, l_s$,
(b) the rate of dissipation of energy per unit mass is $\varepsilon = 15\mu\overline{u'^2}/\rho\lambda^2$,
(c) the rate of decay of turbulent energy is $(\partial\overline{u'^2}/\partial x) = -10(\mu\overline{u'^2}/\rho\lambda^2\overline{u})$ and
(d) if ε is the rate of energy dissipation of turbulent energy per unit mass, most of the viscous dissipation occurs in eddies whose wave numbers lie approximately in the range $0.5(\rho\varepsilon/\mu)^{1/4} > k > 0.2(\rho\varepsilon/\mu)^{1/4}$.

The limiting size of the small-scale eddy has been shown to be of the order of $(\mu^3/\rho^3\varepsilon)^{1/4}$.

Though statistical theories have proved useful in understanding the structure of turbulence, they do not offer solutions to mean flow characteristics. Solutions to mean flow equations are therefore sought by certain empirical theories or approximate physical models. Phenomenological theories are described briefly in the next section.

9.6 Phenomenological Theories

9.6.1 Prandtl's Mixing Length Theory

The kinetic theory of gases outlined in Chapter 1, while considering the molecular exchange of momentum defines the property of viscosity of a fluid as a product of the mean free path of the molecules and a characteristic velocity of the molecules. An analogous model was proposed by Prandtl for the turbulent exchange of momentum. Prandtl's mixing length theory will be presented here for two-dimensional flows.

Let us consider a two-dimensional flow in which the flow is essentially in the x-direction and a mean velocity gradient transverse to the flow direction exists as shown in Fig. 9.6.

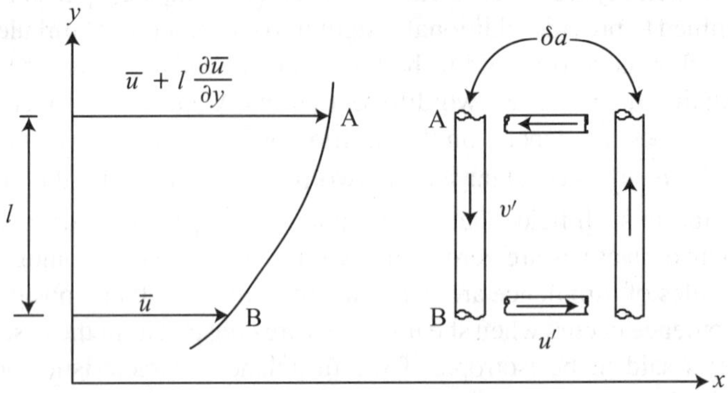

Fig. 9.6 Turbulent exchange of momentum

In a turbulent flow, we shall consider the transfer of momentum by lumps or agglomeration of fluid particles. If a lump of fluid on arriving at A receives a turbulent fluctuation v' in the direction AB (transverse to the flow), a mass of fluid $\rho(\delta a)v'$ travels along the column AB of area δa. This mass had a horizontal momentum of

$$\rho(\delta a)v'\left(u + l\frac{\partial u}{\partial y}\right)$$

at level A. Prandtl made the assumption that this mass of fluid will preserve its identity in regard to the transferable property of momentum until it reaches the level B. The distance AB across which the momentum is assumed to be conserved is known as the mixing length, l. On arriving at B, the fluid particles are assumed to take on the identity of the environments at level B. In other words, the horizontal momentum of the vertically moving mass will now correspond to the local mean velocity at level B, viz., $\rho(\delta a)v'u$. Hence the rate of change of momentum is given as the difference in momentum of the mass $\rho(\delta a)v'$ between levels A and B

$$\rho(\delta a)v'\left(u + l\frac{\partial u}{\partial y}\right) - \rho(\delta a)v'u$$

viz., $\rho(\delta a)v'l(\partial u/\partial y)$. The force per unit area causing the change of momentum defines the turbulent shear stress responsible for this transport.

$$\tau_{\text{turb}} = -\rho v' l\frac{\partial \overline{u}}{\partial y}$$

Since we are interested in the average shear stress, and not merely in the instantaneous shear stress, we get on averaging

$$\overline{\tau_{\text{turb}}} = -\rho\overline{v' l}\frac{\partial \overline{u}}{\partial y} \tag{9.13}$$

Since the equilibrium mean velocity profile is prescribed and continuity is satisfied (there can be no accumulation of fluid particles at level B due to transfer from level A), the downward fluctuation at level A must be accompanied by a return fluctuation at level B. Also a pair of longitudinal fluctuations as shown in Fig. 9.6 are required to complete the cycle in the formation of the eddy. Prandtl assumed that the fluctuation u' is proportional to the mean velocity difference between A and B

$$u' \sim l\frac{\partial \overline{u}}{\partial y}$$

The velocity fluctuations u' and v' are dependent quantities ($u' \sim v'$) and hence we have

$$\overline{\tau_{\text{turb}}} = -\rho\overline{u'v'} = \rho\overline{l^2}\left(\frac{\partial \overline{u}}{\partial y}\right)^2 \tag{9.14}$$

The proportionality constants are assumed absorbed in length l.

The mixing length was assumed constant and proportional to the width of turbulent mixing zone in free shear turbulent flows. In boundary layer problems, l is obviously zero at the boundary and in regions close to the boundary, the mixing length was assumed to be linearly proportional to the distance from the boundary ($l = \kappa y$, where κ is known as Karman's universal constant). With these assumptions we may write

$$\begin{aligned}\overline{\tau_{\text{turb}}} &= \rho\overline{l^2}\left|\left(\frac{\partial \overline{u}}{\partial y}\right)\right|\frac{\partial \overline{u}}{\partial y} \\ &= \rho\kappa^2 y^2\left|\frac{\partial \overline{u}}{\partial y}\right|\frac{\partial \overline{u}}{\partial y}\end{aligned} \tag{9.15}$$

the sign of $(\partial\overline{u}/\partial y)$ determines the direction of the shear stress.

9.6.2 Taylor's Vorticity Transfer Theory

GI Taylor proposed that the transferable property that is conserved over the mixing length is vorticity and not momentum. If $\overline{\varsigma}$ is the mean vorticity at level B and $\overline{\varsigma} + l\frac{\partial\overline{\varsigma}}{\partial y}$ is the mean vorticity at level A, a fluctuation v' causes the rate of exchange of vorticity to be $\rho v' l\frac{\partial\overline{\varsigma}}{\partial y}$ between the two levels. Following the method adopted earlier, we have the proportionalities

$$v' \sim l\frac{\partial \overline{u}}{\partial y}; \quad \varsigma' \sim l\frac{\partial \overline{\varsigma}}{\partial y}; \quad \overline{\varsigma} = \frac{\partial \overline{u}}{\partial y}$$

Considering the mean flow to occur in the x-direction only and that the mean velocity u varies only with y, the equation of motion for steady flow may be written as

$$-\frac{\partial}{\partial x}\left\{\frac{\overline{p}}{\rho} + \frac{(\overline{u}+u')^2 + v'^2}{2}\right\} = -v'\left(\varsigma' - \frac{\partial \overline{u}}{\partial y}\right)$$

where ζ' is the turbulent vorticity. Taking the mean value of the above equation, we get since $(\overline{u'^2} + \overline{v'^2})$does not change with x

$$\frac{1}{\rho}\frac{\partial \overline{p}}{\partial x} = \overline{v'\zeta'}$$

$$\frac{\partial \overline{p}}{\partial x} = \frac{\partial \overline{\tau}}{\partial y} = \rho l^2 \frac{\partial \overline{u}}{\partial y}\frac{\partial \overline{\varsigma}}{\partial y} = \rho l^2 \frac{\partial \overline{u}}{\partial y}\frac{\partial^2 \overline{u}}{\partial y^2} \tag{9.16}$$

If l is independent of y, the above expression for shear is similar to that obtained by Prandtl's mixing length hypothesis.

9.6.3 Karman's Similarity Hypothesis

Karman argued that the turbulent transport process is dependent on local conditions and the mixing length is dependent on the local characteristics of flow. If two points are chosen, the fluctuation of velocities must differ only in time and length scales. Expanding the velocity distribution around the chosen points in Taylor series he showed that the velocity derivatives at the two points should be proportional. From these arguments he derived the relationships

$$\frac{d\overline{u}/dy}{d^2\overline{u}/dy^2} = \frac{1}{\kappa}l$$

where κ is a universal constant whose value is usually taken as 0.4. The expression for shear stress becomes

$$\tau = \rho l^2 \left(\frac{d\overline{u}}{dy}\right)^2 = \rho\kappa^2 \frac{(d\overline{u}/dy)^4}{(d^2\overline{u}/dy^2)^2} \tag{9.17}$$

Part II Turbulent Boundary Layer

9.7 Turbulent Boundary Layer along a Flat Plate

It was stated in the concluding part of Chapter 8 that as the laminar boundary layer thickness progressively increases, there occurs a certain critical boundary layer thickness at which the flow is unstable for small disturbances present in the flow. There will however be a transition region between the laminar boundary layer and a fully turbulent one. Even in the fully turbulent layer, the flow is likely to be laminar in a region very close to the boundary due to the preponderance of the viscous effects. This inner layer next to the boundary is called the laminar sub-layer as shown in Fig. 9.7.

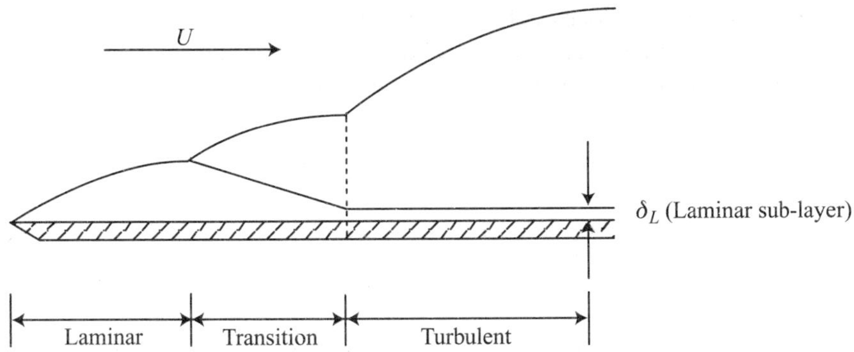

Fig. 9.7 Development of boundary layer

In the laminar sub-layer the mean velocity, u at any height y above boundary u is dependent on the boundary shear stress, τ_0 and the fluid properties. A combination of these parameters yield by dimensional reasoning

$$\frac{\overline{u}}{\sqrt{\tau_0/\rho}} = f\left(\frac{\rho y\sqrt{\tau_0/\rho}}{\mu}\right) \tag{9.18}$$

Close to the wall, the shear stress is assumed equal to the boundary shear stress

$$\overline{\tau} = \tau_0 = \frac{\mu du}{dy} \sim \frac{\mu\overline{u}}{y}$$

Hence the law of the wall is written in the form

$$\frac{\overline{u}}{u^*} = \frac{\rho y u^*}{\mu} \tag{9.19}$$

where $u^* = (\tau_0/\rho)^{1/2}$ is known as the shear–stress velocity (Fig. 9.8*a*). The parameter on the right-hand side is a form of Reynolds number and the law of the wall is valid up to $\rho y u^*/\mu = 4$. In the transverse transition range, between the laminar boundary layer and the fully turbulent layer, the viscous and turbulent stresses are of the same order, and an approximate formula for velocity distribution is obtained and is applicable in the region $4 < \rho y u^*/\mu < 30$

$$\frac{\overline{u}}{u^*} = 0.414\left\{\frac{1-\sqrt{1+z^4}}{z^3} + 2\int_0^z \frac{dz}{\sqrt{1+z^4}}\right\} \tag{9.20}$$

where $z = 0.161\rho y u^*/\mu$ (Fig. 9.8*a*).

At distances far from the boundary, where the Reynolds stresses are large compared to the viscous stresses, we may assume that similarity in velocity profiles exist with

$$\frac{U-\overline{u}}{u^*} = f\left(\frac{y}{\delta}\right) \tag{9.21}$$

This is also known as the velocity defect law (Fig. 9.8*b*). There is a range of distance in which both the law of the wall and the velocity defect law must both be applicable. It can be shown that the velocity profile in the overlapping region is logarithmic

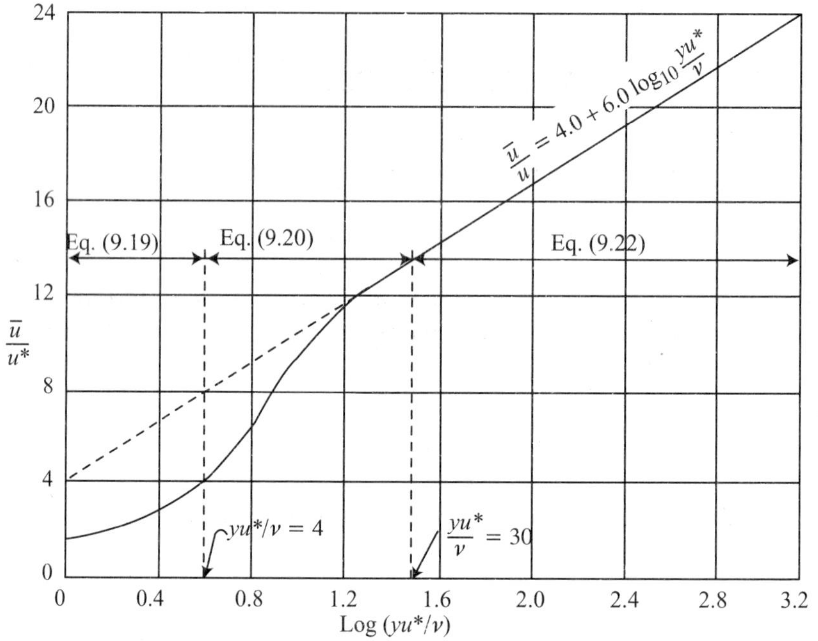

a Law of the wall and overlapping region

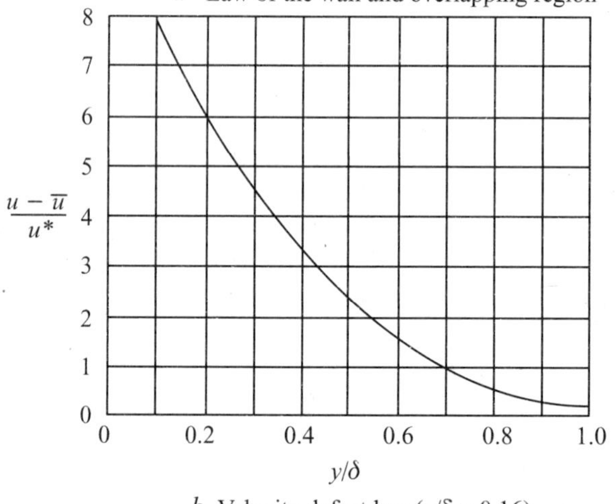

b Velocity defect law ($y/\delta > 0.16$)

Fig. 9.8 Experimental mean velocity distribution in a turbulent boundary layer along a flat plate

$$\bar{u} = 4.0 + 6.0 \log_{10} \frac{\rho y u^*}{\mu}$$
$$\frac{U - \bar{u}}{u^*} = 2.0 - 6.0 \log_{10} \frac{y}{\delta} \tag{9.22}$$

valid in the range $\rho y u^*/\mu > 30$ and $y/\delta < 0.16$ (Fig. 9.8*a*).

The velocity distribution in the turbulent layer along a flat plate is illustrated in Fig. 9.8*a* and *b*

9.8 Momentum Equation

Karman's momentum integral equation applied to a turbulent boundary layer in the same manner as obtained for a laminar boundary layer but including the turbulence terms yields the following equation for the growth of the turbulent boundarv laver along flat plates at zero angle of incidence (U = constant)

$$\frac{d\delta_2}{dx} = \frac{\tau_0}{\rho U^2} + \frac{1}{U^2}\frac{d}{dx}\int_0^\delta (\overline{u'^2} - \overline{v'^2})dy \tag{9.23}$$

The form of the preceding equation differs from that obtained for the laminar boundary layer by the presence of the last integral. Should this term be negligible in comparison to other terms, the equations for the laminar and turbulent boundary layers become identical. However it should be noted that the boundary shear stress τ_0 differs considerably in magnitude in the two types of boundary layers. Blasius obtained an empirical formula for boundary shear stress for smooth plates

$$\tau_0 = 0.0225 \rho U^2 \left(\frac{\mu}{\rho U \delta}\right)^{1/4} \tag{9.24}$$

Further if we use Prandtl's power law distribution of velocity in the turbulent boundary layer

$$\frac{u}{U} = \left(\frac{y}{\delta}\right)^n$$

we find the relationships for displacement, momentum and energy thicknesses:

$$\frac{\delta_1}{\delta} = \frac{n}{(n+1)}; \quad \frac{\delta_2}{\delta} = \frac{n}{(n+1)(2n+1)}; \quad \frac{\delta_3}{\delta} = \frac{2n}{(n+1)(3n+1)}$$

The value of n varies with the Reynolds number.

For a particular case of $n = 1/7$, the solution of the integral moment equation in conjunction with the empirical relation of Blasius for the boundary shear stress yields

$$\frac{\delta}{x} = 0.376 \left(\frac{\rho U x}{\mu}\right)^{-1/5} = 0.376(\mathbf{R}_x)^{-1/5} \tag{9.25}$$

and

$$\tau_0 = 0.0286(\mathbf{R}_x)^{-1/5} \rho U^2 \tag{9.26}$$

$$F \text{ per unit area} = 0.036(\mathbf{R}_x)^{-1/5} \rho U^2$$

9.9 Turbulent Flow in Pipes

9.9.1 Velocity Distribution from Prandtl's Hypothesis

Smooth pipes

The velocity distribution in circular pipes may be found by employing Prandtl's mixing length hypothesis

$$\overline{\tau} = \rho l^2 \left(\frac{\partial \overline{u}}{\partial y}\right)^2$$

The distribution of shear stress is not known before hand and as such further assumptions are necessary to obtain the velocity distribution. Prandtl assumed that the shear stress may be regarded as nearly constant over the entire cross-section of a fully established turbulent flow in a pipe. If the shear stress at any point across the boundary layer is equal to the local boundary shear stress $\tau_0/\rho = (u^*)^2 = l^2(\partial u/\partial y)^2$ and the mixing length is defined by

Karman's similarity hypothesis, viz., $l = \kappa \dfrac{\partial \overline{u}/\partial y}{\partial^2 u/\partial y^2}$

Close to the boundary, we may approximate $l = \kappa(u/y)/(u/y^2) = \kappa y$ and hence $l = \kappa y$ as postulated earlier. Hence the expression for velocity gradient becomes

$$\frac{d\overline{u}}{dy} = \frac{u^*}{\kappa y}$$

which on integration yields

$$\overline{u} = \frac{u^*}{\kappa} \ln y + C_1$$

The above expression may be rendered dimensionless by writing the constant $C_1 = u^* + (u^*/\kappa)\ln(\rho u^*/\mu)$. The velocity distribution is then given by the logarithmic law

$$\frac{\overline{u}}{u^*} = \frac{1}{\kappa} \ln \frac{\rho u^* y}{\mu} + C$$

In the neighbourhood of the boundary, the value of the universal constant is very nearly equal to 0.4. The value of integration constant C was found experimentally to be 5.5. The velocity distribution across the turbulent region of the flow in smooth pipes is therefore

$$\begin{aligned}\frac{\overline{u}}{u^*} &= 2.5 \ln \frac{\rho y u^*}{\mu} + 5.5 \\ \frac{\overline{u}}{u^*} &= 5.75 \log_{10} \frac{\rho y u^*}{\mu} + 5.5\end{aligned} \tag{9.27}$$

The velocity distribution given by the above equation apparently has two anamolies. At the boundary $y = 0$, the velocity goes to negative infinity; this arises due to the fact that the logarithmic distribution is true only for turbulent flow, and hence is valid only in the region above the laminar sub-layer. In the laminar sub-layer, the law of the wall prevails.

$$\frac{\bar{u}}{u^*} = \frac{\rho y u^*}{\mu} \tag{9.28}$$

The velocity distribution in the intermediate transition zone between the laminar sub-layer and the fully developed turbulent region is empirically given by von Karman as

$$\frac{\bar{u}}{u^*} = 11.5 \log_{10} \frac{u^* y}{\nu} - 3.05$$

Valid in the region $70 > u^* y/\nu > 5.0$

The velocity distribution in a pipe is shown in Fig. 9.9 consisting of the law of the wall (9.19) and the logarithmic law (9.27) in the turbulent region and a gradual transition between the two laws.

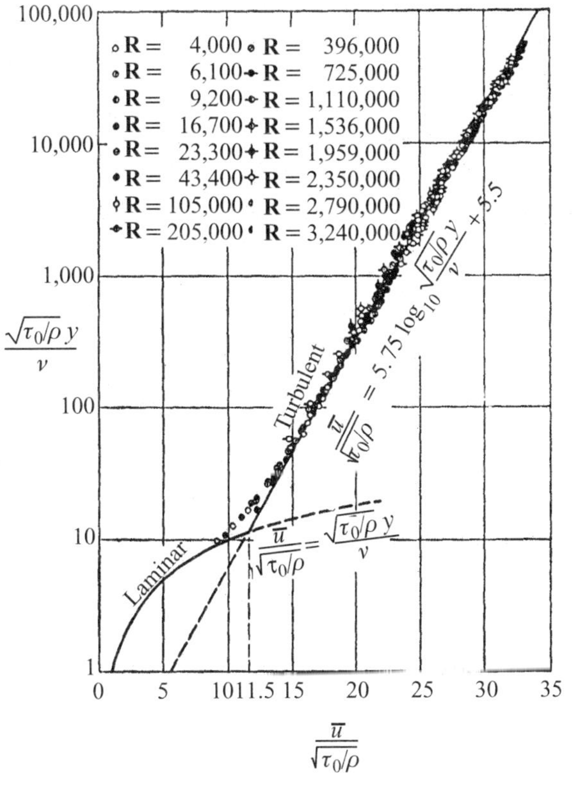

Fig. 9.9 Velocity distribution in a pipe

Further the velocity gradient at the centre of the pipe given by (9.27) is not zero due to the fact that the mixing length l has been assumed to be linearly proportional to the distance from the boundary over the entire cross-section of the pipe but is applicable only close to the boundary. At large distances from the boundary, the mixing length tends become constant, and it is more appropriately given by the equation

$$l = a[0.14 - 0.08(1 - y/a)^2 - 0.06(1 - y/a)^4] \quad (9.29)$$

However, despite all the simplifying assumptions used in the derivation of the logarithmic distribution, (9.27) agrees very well with experimental results.

An estimate of the laminar sub-layer thickness, δ_l may be obtained by finding the intersection of the two profiles describing the law of the wall (9.28) and the logarithmic law (9.27). With $y = \delta_l$, the common point on the two curves is obtained thus

$$\frac{\overline{u}}{u^*} = \frac{\rho \delta_l u^*}{\mu} = 5.75 \log_{10} \frac{\rho \delta_l u^*}{\mu} + 5.5$$

or

$$\frac{\rho \delta_l u^*}{\mu} = 11.6 \quad (9.30)$$

Rough pipes

The above equations are valid for smooth pipes only. When the roughness of the boundary is sufficiently large enough to protrude into the laminar sub-layer, the eddies formed in the wake of these roughness elements lead to the breaking up of the laminar sub-layer as shown in Fig. 9.10.

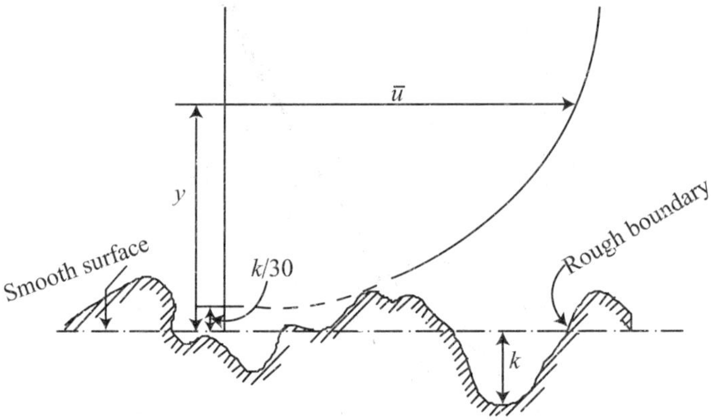

Fig. 9.10 Rough boundary

In such a case, it is estimated that the velocity reaches a zero value at a distance equal to $k/30$ from the boundary distance where k is the average height of the roughness. Introducing this boundary condition in the equation

$$\frac{d\overline{u}}{dy} = \frac{2.5u^*}{y}$$

we obtain the rough pipe law

$$\frac{\overline{u}}{u^*} = 5.75 \log_{10} \frac{y}{k} + 8.5 \tag{9.31}$$

The boundary is said to be hydraulically smooth if $k/\delta_l < 0.25$ and is considered rough if $k/\delta_l > 6(\delta_l = 11.6\nu/u^*)$. There is a transition between these limits.

9.9.2 Velocity Distribution from Karman's Similarity Hypothesis

We shall first assume that the variation in shear stress is linear with y.

$$\tau = \tau_0(1 - y/a)$$

where a is the radius of the pipe. From Karman's hypothesis, we get

$$\tau_0\left(1 - \frac{y}{a}\right) = \frac{\rho\kappa^2(d\overline{u}/dy)^4}{(d^2\overline{u}/dy^2)^2}$$

$$\frac{(d^2\overline{u}/dy^2)}{(d\overline{u}/dy)^2} = -\frac{\kappa}{u^*}\sqrt{\frac{1}{1 - y/a}}$$

The negative sign occurs due to the fact that the second derivative is negative throughout the range of integration. On integration we get

$$\frac{1}{d\overline{u}/dy} = -2\kappa\frac{a}{u^*}\sqrt{1 - y/a} + \text{Constant}$$

In the neighbourhood of the boundary du/dy is extremely large with the result that the left-hand side tends to zero for $y = 0$. Hence the value of the integration constant is $2\kappa a/u^*$.

$$\frac{d\overline{u}}{dy} = \frac{u^*}{2\kappa a}\frac{1}{1 - \sqrt{1 - y/a}}$$

A second integration yields

$$\overline{u} = \frac{u^*}{\kappa}\left\{\sqrt{1 - y/a} + \ln(1 - \sqrt{1 - y/a})\right\} + C$$

The integration may be evaluated in terms of the maximum velocity u_m at the centre of the pipe at $y = a$. The final expression for velocity distribution is

$$\frac{\overline{u_m} - \overline{u}}{u^*} = \frac{1}{\kappa}\left\{\ln(1 - \sqrt{1 - y/a}) + \sqrt{1 - y/a}\right\} \tag{9.32}$$

The preceding formula has the same drawbacks as that obtained by Prandtl's mixing length hypothesis while the latter is not only simpler but also in closer agreement with experimental results than the former.

9.10 Mean Velocity and Maximum Velocity

The discharge in the pipe is

$$Q = \int_0^a 2\pi r dr \bar{u}$$

where the mean velocity is given by (9.27) for smooth pipes and (9.31) for rough pipes. The average velocity $V = Q/\pi a^2$ is

$$\frac{V}{u^*} = 5.75 \log_{10}\left(\frac{\rho a u^*}{\mu}\right) + 1.75 \text{ for smooth pipes} \tag{9.33}$$

$$\frac{V}{u^*} = 5.75 \log_{10}\left(\frac{a}{k}\right) + 4.75 \text{ for rough pipes} \tag{9.34}$$

The velocity distribution for both smooth and rough pipe flow becomes

$$\frac{\bar{u} - V}{u^*} = 5.75 \log_{10} \frac{y}{a} + 3.75 \tag{9.35}$$

The location at which the mean velocity is equal to the average velocity is obtained by setting the left-hand side equal to zero. We get

$$\log_{10} \frac{y}{a} = -\left(\frac{3.75}{5.75}\right) = -0.652 \text{ and } \frac{y}{a} = 0.223$$

The maximum velocity u_m occurs at the centre where $y = a$ and $\log_{10}(y/a) = 0$. Hence we have

$$\frac{(u_{\max} - V)}{u^*} = 3.75 \tag{9.36}$$

$$\text{But} \quad u^* = V\sqrt{\frac{f}{8}}$$

Hence we can write

$$u_{\max} = 3.75u^* + V = 3.75V\sqrt{\frac{f}{8}} + V = V(1.326\sqrt{f} + 1) \tag{9.37}$$

The velocity at any distance from the boundary can be expressed in terms of the maximum velocity. Substituting the expression for V from equation (9.35) in (9.36), we get

$$\frac{\bar{u}}{u^*} - \frac{u_{\max}}{u^*} = 5.75 \log_{10} \frac{y}{a}$$

$$\frac{u_{\max} - \bar{u}}{u^*} = 5.75 \log_{10} \frac{a}{y} = 5.75 \log_{10} \frac{a}{a - r}$$

9.11 Pipe Resistance Factor

Darcy found the head losses in pipes in which the uniform flow occurs at an average velocity V. Considering equilibrium of a control volume as shown in Fig. 9.11, the application of the momentum theorem to the control volume yields

$$(p + dp)\pi a^2 - p \cdot \pi a^2 + \tau_0 2\pi a dl - \gamma \pi a^2 \cdot dl \sin\theta = 0$$

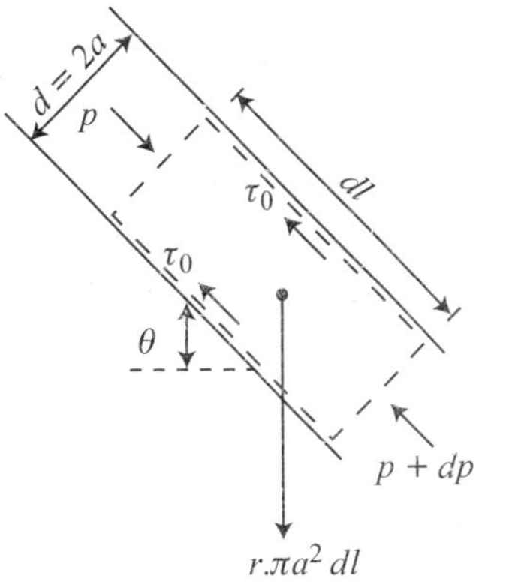

Fig. 9.11 Force diagram for pipe flow

which on simplification with $dl \cdot \sin\theta = dz$

$$d(p/\gamma + z) = dh = \frac{2\tau_0 \cdot dl}{(a\gamma)} = \frac{(4\tau_0 \cdot dl)}{(d \cdot \gamma)}$$

where h is the piezometric head.

The boundary shear stress may be expressed in terms of the Newtonian drag equation: $\tau_0 = c_f \cdot \rho V^2/2$, where c_f is the local resistance coefficient.

The piezometric head loss over a length l is

$$h = \frac{4c_f(\rho V^2/2)l}{d \cdot 2g} = \frac{4c_f l V^2}{d \cdot 2g} = \frac{f l V^2}{d \cdot 2g} \tag{9.38}$$

where $f = 4c_f$ is known as the pipe resistance factor.

The shear stress may now be expressed as

$$\tau_0 = \left(\frac{f}{4}\right)\left(\frac{\rho V^2}{2}\right) \tag{9.39}$$

and the shear velocity

$$u^* = \sqrt{\frac{\tau_0}{\rho}} = V\sqrt{\frac{f}{8}} \tag{9.40}$$

Substitution of this value for u^* in the expressions for average velocity in smooth and rough pipes, we get

$$\frac{1}{\sqrt{f}} = 2.03 \log_{10}\left(\frac{\rho V d}{\mu}\sqrt{f}\right) - 0.91 \tag{9.41}$$

$$\frac{1}{\sqrt{f}} = 2.03 \log_{10}\left(\frac{a}{k}\right) + 1.68 \tag{9.42}$$

The constants have been adjusted to suit the experimental data to give us the following expressions for smooth and rough pipes respectively.

$$\frac{1}{\sqrt{f}} = 2\log_{10}(\mathbf{R}\sqrt{f}) - 0.8 \qquad \text{for a smooth pipe} \tag{9.43}$$

$$\frac{1}{\sqrt{f}} = 2\log_{10}\left(\frac{a}{k}\right) + 1.74 \qquad \text{for a rough pipe} \tag{9.44}$$

It is to be noted that the pipe resistance factor is dependent on the value of the Reynolds number for smooth pipes. The laminar sub-layer thickness δ_l by (9.30) is: $\delta_l = 11.6\nu/u^*$ and the value of u^* is given by (9.40). Hence we have

$$\delta_l = \frac{11.6}{V\sqrt{f/8}} \tag{9.45}$$

and the relative thickness of the laminar sub-layer with respect to the radius of the pipe, a is

$$\frac{\delta_l}{a} = \frac{11.6\nu}{V\sqrt{(f/8)}(d/2)} = \frac{65.6}{\mathbf{R}\sqrt{f}} \tag{9.46}$$

The pipe resistance factor is therefore dependent on the relative thickness of the laminar sub-layer to the radius of the pipe. Blasius obtained an empirical relation for the resistance factor for smooth pipes

$$f = \frac{0.316}{\mathbf{R}^{1/4}} \tag{9.47}$$

which is valid between $2000 < \mathbf{R} < 10^5$. For Reynolds numbers less than 2000, i.e., in the laminar flow regime, the pipe resistance factor is $f = 64/\mathbf{R}$ as outlined in Chapter 8.

Again it is to be noted that for rough pipes f is independent of the Reynolds number and is dependent upon only the relative value of roughness, k/a. The plot of pipe resistance factor f against Reynolds number for various values of k/a ratio is shown in Fig. 9.12.

Colebrook and White established the empirical law in the transition region from smooth to rough pipes.

$$\frac{1}{\sqrt{f}} = 1.74 - 2\log\left(\frac{k}{a} + \frac{18.7}{\mathbf{R}\sqrt{f}}\right) \tag{9.48}$$

The transition begins when the roughness height is approximately one-fourth of the laminar sub-layer thickness ($\delta_l/4$) and ends when $k = 6\delta_l$. A hydraulically smooth pipe may therefore be defined as one whose surface roughness heights are fully submerged in the laminar sub-layer and $k < \delta_l/4$.

9.12 Explicit Formula for Pipe Resistance Factor

Nikuradse, Moody and other experimenters have established the variation of f with the Reynolds number, relative roughness height k/a (Fig. 9.12) from which solutions to a large number of pipe

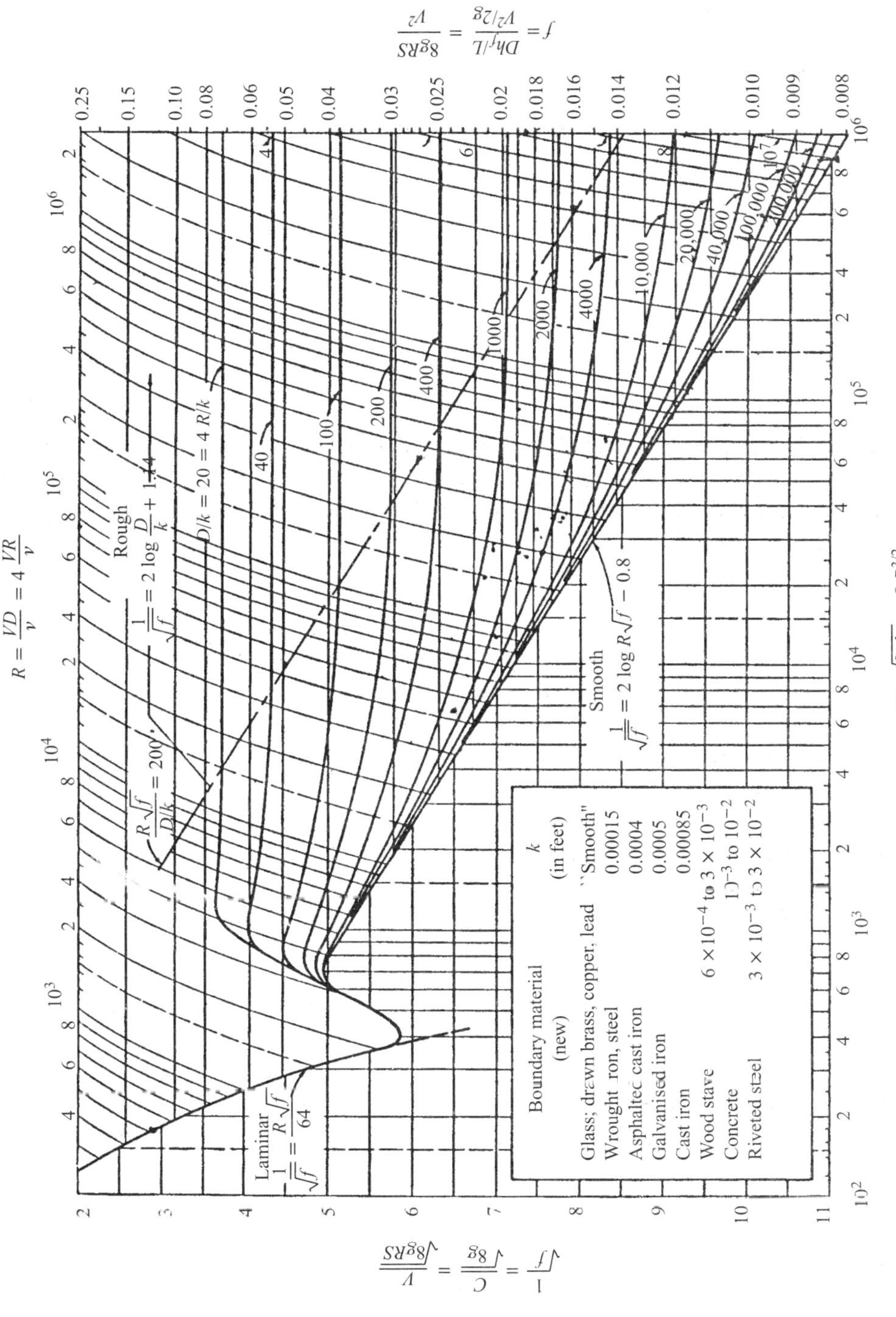

Fig. 9.12 Relationship between f and **R** for various values of k/a

flow problems may be obtained by trial and error methods. An explicit relationship connecting f, $\mathbf{R}$, and k/a was given by Wood in the form

$$f = m + n\mathbf{R}^{-p}$$

where the coefficients m, n and p are functions of k/a

$$m = 0.094\left(\frac{k}{a}\right)^{0.225} + 0.53\left(\frac{k}{a}\right)$$

$$n = 88\left(\frac{k}{a}\right)^{0.44}$$

$$p = 1.62\left(\frac{k}{a}\right)^{0.134}$$

EXAMPLE 9.2 Water flows over a flat plate 30 m long and 15 m wide with a free stream velocity of 2 m/s. Verify that the Reynolds number is greater than the critical Reynolds number for transition to turbulence. Find the total drag on one side of the plate if the boundary layer is turbulent right from the beginning.

Assume

$$\mathbf{R}_{cr} = 5 \times 10^5$$

$$\mathbf{R}_L = UL/\nu = 2 \times 30/10^{-6} = 60 \times 10^6 > \mathbf{R}_{cr}$$

$$F = (\text{area})\, 0.036\, \mathbf{R}_L^{-1/5}\, \rho U^2 = (30 \times 15) \times 0.036 \times (60 \times 10^6)^{-1/5} \times 1000 \times 2^2$$

$$= 1802.8\ \text{kN}$$

The boundary layer thickness at the trailing edge

$$\delta = 0.376\, L\, \mathbf{R}_L^{-1/5} = 0.376 \times 30 \times (60 \times 10^6)^{-1/5} = 0.3138\ \text{m}$$

If the plate is held with the 15 m edge along the flow direction, $\mathbf{R}_L = 2 \times 15 \times 10^6$.
$F = (30 \times 15) \times 0.036 \times (30 \times 10^6)^{-1/5} \times 1000 \times 2^2 = 2070.9\ \text{kN}$ and $\delta = 0.1569$ m

EXAMPLE 9.3 Air ($\rho = 1.2\ \text{kg/m}^3$ and $\nu = 1.5 \times 10^{-5}\ \text{m}^2/\text{s}$) flows at a velocity of 2 m/s over a thin plate 7.5 m long and 2 m wide. Find the length over which the laminar boundary layer exists and the total drag over the plate due to combined effects of laminar and turbulent boundary layers.

The general formula for determining the drag due to such a part laminar and part turbulent boundary layer is

Total drag = (Turbulent drag from $x = 0$ to $x = L$) − (Turbulent drag from $x = 0$ to $x = x_{cr}$)+ (Laminar drag from $x = 0$ to $x = x_{cr}$)

If $\mathbf{R}_{cr} = 5 \times 10^5$, then we have

$$\frac{C_D(BL)\rho U^2}{2} = (BL)0.036\,\mathbf{R}^{-1/5}\,\rho U^2 - 0.036 \times (5 \times 10^5)^{-1/5} \frac{\rho U^2(BL)x_{cr}}{L} + \left[\frac{1.328}{(5 \times 10^5)^{1/2}}\right] \frac{(\rho U^2/2)(BL)x_{cr}}{L}$$

Since $x_{cr}/L = \mathbf{R}_{cr}/\mathbf{R}_L$, the relationship can be simplified to yield

$$C_D = C_{D,\text{turbulent}} - \frac{1700}{\mathbf{R}_L}$$

where $C_{D,\text{turbulent}} = 0.072/\mathbf{R}_L^{1/5}$ in the equation $F = (\text{area})C_{D,\text{turbulent}}(\rho U^2/2)$.

The above derivation is approximate since the characteristics of the turbulent layer developing after the transition will be different from that developing from the beginning at $x = 0$. This approximation is valid up to 1×10^7.

Assuming $\mathbf{R}_{cr} = 5 \times 10^5 = x_{cr}U/\nu$; $x_{cr} = (5 \times 10^5 \times 1.5 \times 10^{-5})/2 = 3.75$ m

Hence the laminar boundary layer exists over half the length of the plate. The drag due to the laminar boundary layer is

$$F = (\text{area})\left[\frac{1.328}{(\mathbf{R}_{cr})^{1/2}}\right]\left(\frac{\rho U^2}{2}\right) = 3.75 \times 2 \times \left(\frac{1.328}{500000^{1/2}}\right) \times 1.2 \times \frac{2^2}{2} = 0.0338\text{ N}$$

If the boundary layer had been turbulent throughout, then the drag on the entire plate is

$$F = (7.5 \times 2) \times 0.036 \times \left[\frac{2 \times 7.5}{(1.5 \times 10^{-5})}\right]^{-1/5} \times 1.2 \times 2^2 = 0.1635\text{ N}$$

Subtract the drag on the first half of the plate where the laminar boundary layer exists:

$$F - (3.75 \times 2) \times 0.036 \times \left(\frac{2 \times 3.75}{1.5 \times 10^{-5}}\right)^{-1/5} \times 1.2 \times 2^2 = 0.094\text{ N}$$

The turbulent drag on the rear half of the plate is $0.1635 - 0.094 = 0.0695$ N

The total drag = laminar drag + turbulent drag = $0.0338 + 0.0695 = 0.1033$ N

If we use

$$C_D = (C_D)_{\text{turbulent}} - \frac{1700}{\mathbf{R}_L} \quad \text{with} \quad \mathbf{R}_L = \frac{UL}{\nu} = \frac{2 \times 7.5}{1.5 \times 10^{-5}} = 10^6$$

$$= \frac{0.072}{\mathbf{R}_L^{1/5}} - \frac{1700}{\mathbf{R}_L} = 0.00454 - 0.0017 = 0.00284$$

$$F = 0.00284 \times 2 \times 7.5 \times \frac{1}{2} \times 1.2 \times 2^2 = 0.1022\text{ N}$$

EXAMPLE 9.4 Find the maximum size of roughness element in a 50 cm pipe so that the pipe may still be classified as smooth at a Reynolds number of 50000.

For a hydraulically smooth pipe $k < \delta_l$ and ku_*/ν should be less than or equal to $11.6/4 \sim 3$. (Some researchers recommend a value of 5)

For $\mathbf{R} = 50,000$, we find from Fig. 9.11, $f = 0.021$ for smooth pipes.

$$u^* = \left(\frac{\tau_0}{\rho}\right)^{1/2} = V\left(\frac{f}{8}\right)^{1/2}; \quad \frac{ku^*}{\nu} = \left(\frac{k}{D}\right)\left(\frac{VD}{\nu}\right)\left(\frac{f}{8}\right)^{1/2} = 3$$

Hence with $f = 0.021$, we get $k_{max} = (50 \times 3)/[50000 \times (0.021/8)^{1/2}] = 0.0586\,\text{cm} \sim 0.6\,\text{mm}$

EXAMPLE 9.5 Water at 20°C($\nu = 10^{-6}\,\text{m}^2/\text{s}$) flows at the rate of $0.5\,\text{m}^3/\text{s}$ in a 2 cm asphalt coated CI pipe. What is the head loss per km length of pipe.

$V = 0.5/0.7854(0.20)^2 = 1.59\,\text{m/s}$ and $\mathbf{R} = 1.59 \times 0.2/10^{-6} = 3.18 \times 10^{-5}$

k/D for asphalted pipe $= 0.0006$ and from Fig. 9.11 we read off $f = 0.019$

The head loss in 1000 m length of pipe $h_f = flV^2/2gD = (0.019 \times 1000 \times 1.59^2)/(2 \times 9.81 \times 0.2) = 12.2\,\text{m}$

Note: If h_f is given along with D, L, ν we can calculate $\mathbf{R}f^{1/2}$ from the relation

$$\mathbf{R}f^{1/2} = \left(\frac{D^{3/2}}{\nu}\right)\left(\frac{2gh_f}{L}\right)^{1/2}$$

and enter the pipe resistance diagram (Fig. 9.11) to read off f. Then we can calculate $V = (2gDh_f/fL)^{1/2}$ and determine the discharge. This is one of the advantages of Fig. 9.11 in comparison with the traditional Moody diagram.

EXAMPLE 9.6 Water ($\rho = 100\,\text{kg/m}^3$ and $\nu = 10^{-6}\,\text{m}^2/\text{s}$) flows in a smooth pipe of diameter 15 cm at a velocity of 5 m/s. Find (a) the thickness of the laminar sub-layer, (b) the location where the turbulent core begins, (c) velocity distribution in the pipe, (d) velocity at the edge of the laminar sub-layer, (e) velocity at the beginning of the turbulent core and (f) velocity at the centre line.

$$\mathbf{R} = \frac{VD}{\nu} = \frac{5 \times 0.15}{10^{-6}} = 0.75 \times 10^6;$$

From Fig. 9.11, we get

$$f = 0.0122$$

The shear stress velocity is

$$u^* = V\left(\frac{f}{8}\right)^{1/2} = 5\left(\frac{0.0122}{8}\right)^{1/2} = 0.1952\,\text{m/s}$$

The laminar sub-layer

$$\delta_l \frac{u^*}{\nu} = \left(\frac{\delta_l}{D}\right)\left(\frac{VD}{\nu}\right)\left(\frac{f}{8}\right)^{1/2}$$

$$= \left(\frac{\delta_l}{D}\right)\mathbf{R}\left(\frac{f}{8}\right)^{1/2}$$

$$\left(\frac{\delta_l}{0.15}\right)(0.75 \times 10^6)\left(\frac{0.0122}{8}\right)^{1/2} = 0.1952 \times 10^6 \delta_l$$

We will assume that the laminar sub-layer ends at $\delta_l(u^*/\nu) = 5$

$$\delta_l = \frac{5 \times 10^{-6}}{0.1952} = 2.56 \times 10^{-5} \text{ m}$$

The velocity distribution in the laminar sub-layer is given by $u/u^* = yu^*/\nu$

$$u = V^2\left(\frac{f}{8}\right)\frac{y}{\nu} = 5^2\left(\frac{0.0122}{8}\right) \times 10^6 y = 38125y \text{ m/s}$$

Velocity at the edge of the laminar sub-layer

$$u = 38125 \times 2.56 \times 10^{-5} = 0.976 \text{ m/s}$$

We will assume that the turbulent core begins at $yu^*/\nu = 30$ or

$$y = 30 \times 10^{-6}/0.192 = 0.156 \times 10^{-3} \text{ m} = 0.156 \text{ mm}$$

For the buffer zone, the velocity distribution given by Karman is

$$u = 5u^* \ln\left(\frac{yu^*}{\nu}\right) - 3.05u^* = 5 \times 0.1952 \ln(195200y) - 3.05 \times 0.1952$$

$$u = 0.976 \ln(195200y) - 0.5954$$

Velocity at the beginning of the turbulent core i.e., at $y = 0.156 \times 10^{-3}$

$$u = 3.334 - 0.5954 = 2.739 \text{ m/s}$$

Velocity distribution in the turbulent core is

$$u = 2.5u^* \ln\left(\frac{yu^*}{\nu}\right) + 5.5u^* = 2.5 \times 0.1952 \ln(0.1952 \times 10^6 y) + 5.5 \times 0.1952$$

At the beginning of the turbulent core i.e., at $y = 0.156 \times 10^{-3}$

$$u = 1.659 + 1.0736 = 2.733 \text{ m/s}$$

which is in agreement with the velocity calculated in the buffer zone.

At the centre of the pipe, $y = 0.075$ m

$$u = 4.6807 + 1.0736 = 5.7543 \text{ m/s}$$

EXERCISES

9.1. An aircraft is flying at a velocity of 120 m/s. Assume that the wings can be treated as flat plates with the boundary layer being turbulent from the leading edge. The chord length of the wing is 2.5 m and the width is 20 m. Find the values of the thickness of the boundary layer and the displacement thickness at the trailing edge [given $u/U = (y/\delta)^{1/7}$]. Find the total skin friction drag. If the boundary layer is initially laminar up to the Reynolds number of 5×10^5, find the total drag. $\nu = 1.40 \times 10^{-5}\,\text{m}^2/\text{s}$, $\rho = 1.2\,\text{kg/m}^3$

9.2. Water flows over a flat plate 0.5 m wide and 0.5 m long at a velocity of 5 m/s. Find the drag on both sides on the portion of the plate on which the boundary layer is laminar. What is the total drag on both sides of the entire length of the plate? $\rho = 998\,\text{kg/m}^3$ and $\nu = 1 \times 10^{-6}\,\text{m}^2/\text{s}$.

9.3. Find the surface resistance on the top and sides of a trailer 3 m wide, 2.5 m high and 15 m long travelling at 100 km/hour. (For air, $\nu = 0.12\,\text{cm}^2/\text{s}$ and $\rho = 1.20\,\text{kg/m}^3$). Find the thickness of the boundary layer and shear stress at the trailing edge assuming that the boundary layer is turbulent from the leading edges.

9.4. The boundary layer thickness along a flat plate in air was found to be 20 cm at point A, the free stream velocity being 100 m/s. If the boundary layer thickness is 30 cm at point B downstream of A, find the distance between A and B and also the shear stress exerted on the length AB. $\nu = 1.4 \times 10^{-5}\text{m}^2/\text{s}$

9.5. The length of establishment, L_e for laminar flow at pipe entrance is given as $L_e/D = 0.0575\,\mathbf{R}$, where D is the diameter of the pipe and $\mathbf{R}$ is the Reynolds number based on the average velocity, V in the pipe ($\mathbf{R} = VD/\nu$). For turbulent flow, the length of establishment at pipe entry is $L_e/D = 4.4\,\mathbf{R}^{0.167}$

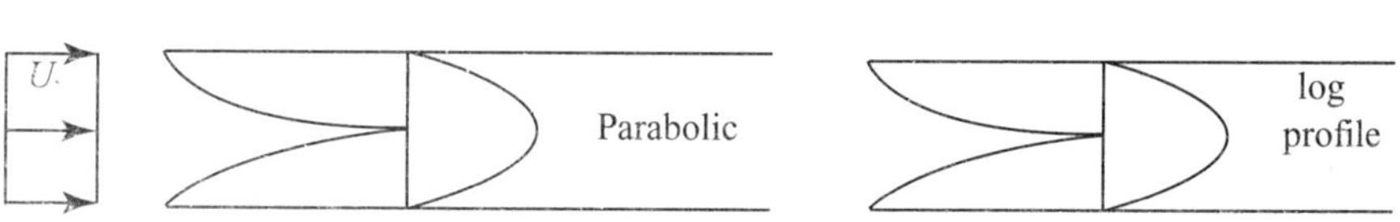

Figure for Exercise 9.5

If we base our calculation on the growth of boundary layer thickness in laminar flow as $\delta/x = 5/\mathbf{R}_x^{1/2}$ (though it is applicable for flat plates only and not for pipes), find the length required for the boundary layer thickness to reach a value of $D/2$.

Similarly, find the establishment length for turbulent flow ($\delta/x = 0.37\mathbf{R}_x^{-1/5}$)

9.6. The pressure drop in a pipe of 20 cm diameter and a length of 1000 m is 300 kN/m². The roughness height is 0.015 mm. Is the pipe smooth or rough. Calculate the maximum and average velocity and the discharge in the pipe.

9.7. Sand grains of 0.75 mm are cemented on the surface of a 15 cm diameter pipe so as to produce the maximum effect of roughness. At what velocity of water will (i) the sand grains first begin to disturb the laminar sub-layer and (ii) completely render the boundary layer turbulent.

9.8. If the average velocity in a smooth 50 cm diameter pipe is 4.0 m/s, find the value of f, the maximum velocity, and velocities at 5 cm, 10 cm and 20 cm from the wall ($\nu = 1 \times 10^{-6}$ m^2/s).

9.9. A 50 cm smooth pipe carries a discharge of 0.4 m^3/s, find the average velocity and the friction factor, head loss in a one-kilometre length of pipe and the power required to maintain the supply. You can use (9.43) or $f = 0.0032 + 0.220/\mathbf{R}^{0.237}$.

9.10. In a 25 cm diameter pipe, the Reynolds number is found to be $\mathbf{R} = 5 \times 10^4$. Determine the values of average velocity, f, τ_0, u_*, head loss per kilo metre length of pipe and the thickness of the laminar sub-layer ($\nu = 10^{-6}$ m^2/s).

9.11. In a turbulent flow in a pipe, at what distance from the boundary is the velocity equal to the average velocity?

9.12. In a rough pipe of diameter 10 cm and roughness height $k = 0.05$ cm the discharge is 0.015 m^3/s. Determine the values of f and the maximum velocity at the centre of the pipe ($\nu = 10^{-6}$ m^2/s).

Appendix

Derivation of equation for mean turbulent flow

The equation of motion for steady ($\partial/\partial t = 0$) two-dimensional ($w = 0$; $\partial/\partial z = 0$) turbulent flow is derived as follows:

The general equation for steady two-dimensional flow is

$$u\frac{\partial u}{\partial x} + v\frac{\partial u}{\partial y} = X - \frac{1}{\rho}\frac{\partial p}{\partial x} + \nu\left(\frac{\partial^2 u}{\partial x^2} + \frac{\partial^2 u}{\partial y^2}\right)$$

Substituting $u = \overline{u} + u'$, $\quad v = \overline{v} + v'$, $\quad p = \overline{p} + p'$, we get

$$(u + u')\left[\frac{\partial(\overline{u} + u')}{\partial x}\right] + (\overline{v} + v')\left[\frac{\partial(\overline{u} + u')}{\partial y}\right] = X - \frac{1}{\rho}\frac{\partial(\overline{p} + p')}{\partial x}$$

$$+ \nu\left[\frac{\partial^2(\overline{u} + u')}{\partial x^2} + \frac{\partial^2(\overline{u} + u')}{\partial y^2}\right]$$

Expanding, we obtain

$$\overline{u}\frac{\partial \overline{u}}{\partial x} + \overline{u}\frac{\partial u'}{\partial x} + u'\frac{\partial \overline{u}}{\partial x} + u'\frac{\partial u'}{\partial x} + \overline{v}\frac{\partial \overline{u}}{\partial y} + \overline{v}\frac{\partial u'}{\partial y} + v'\frac{\partial \overline{u}}{\partial y} + v'\frac{\partial u'}{\partial y} = X - \frac{1}{\rho}\frac{\partial \overline{p}}{\partial x} - \frac{1}{\rho}\frac{\partial p'}{\partial x}$$

$$+ \nu\frac{\partial^2 \overline{u}}{\partial x^2} + \nu\frac{\partial^2 u'}{\partial x^2} + \nu\frac{\partial^2 \overline{u}}{\partial y^2} + \nu\frac{\partial^2 u'}{\partial y^2}$$

The above equation is averaged throughout to get the equation for mean turbulent flow. In doing so, we apply the laws of averages, viz.,

(a) $\overline{u'} = \overline{v'} = \overline{w'} = \overline{p'} = 0$; note that the average of products of turbulent quantities are not zero:

$$\overline{u'v'} \neq 0;\ \overline{u'\frac{\partial u'}{\partial x}} \neq 0; \quad \overline{v'\frac{\partial u'}{\partial y}} \neq 0$$

(b) Average of the average is the average itself:

$$\overline{\overline{u}} = \overline{u}; \quad \overline{\overline{u}\frac{\partial \overline{u}}{\partial x}} = \overline{u}\frac{\partial \overline{u}}{\partial x}; \quad \overline{\overline{v}\frac{\partial \overline{u}}{\partial x}} = \overline{v}\frac{\partial \overline{u}}{\partial x}$$

(c) Average of the differential is the differential of the average:

$$\overline{\frac{\partial u'}{\partial x}} = \frac{\partial \overline{u'}}{\partial x} = 0; \quad \overline{\frac{\partial v'}{\partial x}} = \frac{\partial \overline{v'}}{\partial x} = 0$$

(d) While averaging a product of a mean quantity and a turbulent one, the mean quantity can be taken out of the averaging process;

$$\overline{\overline{u}\frac{\partial u'}{\partial x}} = \overline{u}\overline{\frac{\partial u'}{\partial x}} = \overline{u}\frac{\partial \overline{u'}}{\partial x} = 0; \quad \overline{\overline{v}\frac{\partial u'}{\partial y}} = 0$$

After application of the above averaging laws and simplifying we get

$$\overline{u}\frac{\partial \overline{u}}{\partial x} + \overline{v}\frac{\partial \overline{u}}{\partial y} + \overline{u'\frac{\partial u'}{\partial x}} + \overline{v'\frac{\partial u'}{\partial y}} = X - \frac{1}{\rho}\frac{\partial \overline{p}}{\partial x} + \nu\left(\frac{\partial^2 \overline{u}}{\partial x^2} + \frac{\partial^2 \overline{u}}{\partial y^2}\right)$$

We add the terms $\overline{u'(\partial u'/\partial x) + u'(\partial v'/\partial y)} = \overline{u'(\partial u'/\partial x + \partial v'/\partial y)}$ to the left-hand side of the preceding equation without altering its value since by virtue of the equation of continuity, the value of the additional terms is zero. Adding and grouping the terms appropriately we get

$$\overline{u}\frac{\partial \overline{u}}{\partial x} + \overline{v}\frac{\partial \overline{u}}{\partial y} + \overline{u'\frac{\partial u'}{\partial x} + u'\frac{\partial u'}{\partial x}} + \overline{u'\frac{\partial v'}{\partial y} + v'\frac{\partial u'}{\partial y}} = X - \frac{1}{\rho}\frac{\partial \overline{p}}{\partial x} + \nu\left(\frac{\partial^2 \overline{u}}{\partial x^2} + \frac{\partial^2 \overline{u}}{\partial y^2}\right)$$

$$\overline{u}\frac{\partial \overline{u}}{\partial x} + \overline{v}\frac{\partial \overline{u}}{\partial y} + \frac{\partial \overline{u}'^2}{\partial x} + \frac{\partial(\overline{u'v'})}{\partial y} = X - \frac{1}{\rho}\frac{\partial \overline{p}}{\partial x} + \nu\left(\frac{\partial^2 \overline{u}}{\partial x^2} + \frac{\partial^2 \overline{u}}{\partial y^2}\right)$$

We transfer the third and fourth terms on the left-hand side to the right and couple them with the viscous shear stress terms.

$$\overline{u}\frac{\partial \overline{u}}{\partial x} + \overline{v}\frac{\partial \overline{u}}{\partial y} = X - \frac{1}{\rho}\frac{\partial \overline{p}}{\partial x} - \frac{\partial \overline{u'^2}}{\partial x} - \frac{\partial \overline{u'v'}}{\partial y} + \nu\left(\frac{\partial^2 \overline{u}}{\partial x^2} + \frac{\partial^2 \overline{u}}{\partial y^2}\right)$$

The terms $\rho\overline{u}'^2$, $\rho\overline{u'v'}$ are designated as turbulent shear stresses or apparent shear stresses or Reynolds shear stresses.

The derivation of equations of motion in the y and z directions may be carried out in a similar manner and is left as an exercise to the student.

One may derive the equations of motion for unsteady, three-dimensional turbulent flow in a similar manner, giving six Reynolds stresses $\rho\overline{u'^2}$, $\rho\overline{u'v'}$, $\rho\overline{u'w'}$, $\rho\overline{v'^2}$, $\rho\overline{v'w'}$, $\rho\overline{w'^2}$.

10 Flow in Pipes

Part I Steady Flow

10.1 Losses in Pipes

In the previous chapters, expressions for loss of head in pipes flowing full due to boundary resistance have been derived, viz.,

$$h_f = \frac{flV^2}{D2g}$$

where the pipe resistance factor f for laminar flow is $f = 64/\mathbf{R}$, and for turbulent flow, f is a function of the Reynolds number and the roughness height k/a. In steady flow problems in pipes, there are several variables such as fluid properties (ρ, μ or ν), discharge Q, diameter d, (or velocity $\bar{V}$), roughness of pipe k and losses in pipe fixtures and boundary resistance. We shall first assess the minor losses in pipe fixtures.

10.1.1 Loss in Pipe Expansion

We have derived the loss in pipe expansion in Chapter 7. With reference to Fig. 10.1a, we have $h_e = (V_1 - V_2)^2/2g$ and $V_1A_1 = V_2A_2$

$$h_e = \left[1 - \left(\frac{A_1}{A_2}\right)\right]^2 \left(\frac{V_1^2}{2g}\right) = \left[1 - \left(\frac{d_1}{d_2}\right)^2\right]^2 \left(\frac{V_1^2}{2g}\right) = K\left(\frac{V_1^2}{2g}\right) \tag{10.1}$$

where $K = [1 - (d_1/d_2)^2]^2$

10.1.2 Exit Loss

If d_2 is very large as in the case of a pipe discharging into a very large reservoir (Fig. 10.1d) then the exit loss is $V^2/2g$.

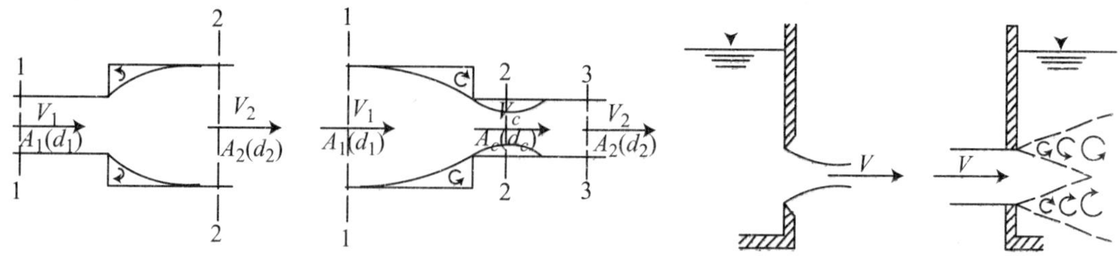

Fig. 10.1 Losses in pipe expansion and contraction

10.1.3 Loss at Contraction

The head loss due to a sudden contraction is also derived from the same analysis applied to sudden expansion. The loss between Sections 1–1 and 2–2 is small compared to that between Sections 2–2 and 3–3 (Fig. 10.1*b*) and hence generally ignored.

Immediately downstream of the contraction, a vena contracta is formed with the area of the jet at the vena contracta being given as $C_c A_2$ (Fig. 10.1*b*) and the velocity at that section is by continuity $V_c = V_2 A_2 / C_c A_2$. The loss due to the expansion from the Section 2–2 at vena contracta to Section 3–3 is given by (10.1).

$$h_e = \frac{(V_c - V_2)^2}{2g} = \left[\left(\frac{1}{C_c}\right) - 1\right]^2 \left(\frac{V_2^2}{2g}\right) \tag{10.2}$$

The value of C_c is given by Table 10.1 for different values of the contraction ratio

Table 10. 1

A_2/A_1	0.1	0.2	0.3	0.4	0.5	0.6	0.7	0.8	0.9	1.0
C_c	0.624	0.632	0.643	0.659	0.681	0.712	0.755	0.813	0.892	1.0

10.1.4 Loss at Entrance

The head loss at entrance (Fig. 10.1*c*) is usually taken as $0.5V^2/2g$ for a sharp-edged entrance. For a well-rounded entrance, the loss is as low as $0.01\,V^2/2g$ and may be neglected. For the Borda type re-entrant pipe entry, the loss is usually taken as $1.0V^2/2g$.

10.1.5 Loss in a Nozzle

At the exit of the nozzle (Fig. 10.2), the velocity is V_j and the pressure is assumed atmospheric. In the approach pipe, the velocity is V_0 and the pressure in it is p_0. If h_e is the loss in the nozzle, we have

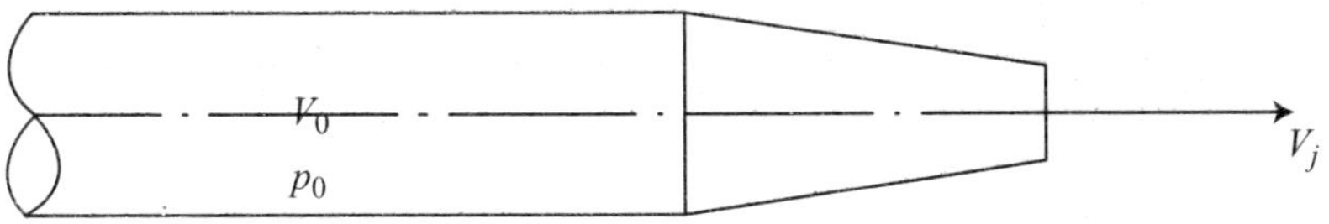

Fig. 10.2 Nozzle at the end of a pipe

$$\frac{p_0}{\gamma} + \frac{V_0^2}{2g} = \frac{V_j^2}{2g} + h_e$$

We can also express the jet velocity by the orifice equation ($C_v =$ coefficient of velocity)

$$V_j = C_v \left[2g\left(\frac{p_0}{\gamma} + \frac{V_0^2}{2g}\right)\right]^{1/2}$$

Eliminating $(p_0/\gamma + V_0^2/2g)$, we get

$$h_e = \frac{V_j^2}{2g}\left[\left(\frac{1}{C_v}\right)^2 - 1\right] \tag{10.3}$$

Most modern nozzles have their coefficients of velocity C_v between 0.95 and 1.0.

10.2 Losses in Pipe Fittings

Other pipe fixtures in bends, mitred joints, valves and other pipe fittings cause head or energy losses for which no rigorous analytical solutions are available but have been estimated by experimental methods. In all cases, the head loss is expressed in terms of the kinetic energy head through a dimensionless hydraulic coefficient, K. These values are relatively insensitive to the Reynolds number. A brief table of pipe losses in commonly used pipe fittings are given in Table 10.2, quoted from books by Babbit et al. and S. Whitaker.

10.3 Equivalent Length

It is some times advantageous in the analysis of pipe systems, especially for unsteady flows in pipes, to express these losses in pipe fixtures as occurring as friction loss in a straight pipe of certain length L_e. In other words, the head loss due to friction in length L_e is equal to the loss in the pipe fixtures for the same discharge and pipe diameter and resistance factor f.

$$fL_eV^2/(2gd) = KV^2/2g$$

$$\text{or } L_e = \frac{(Kd)}{f} \tag{10.4}$$

For example, if all the minor losses in a pipe system add up to $K = 15$ and if the pipe diameter is 15 cms and the value of the resistance factor $f = 0.015$, then the equivalent length which

Table 10. 2

Description	Babbit	Whittaker
Gate valve		
Fully open	0.19	0.20
$\frac{3}{4}$ open	1.1	0.90
$\frac{1}{2}$ open	5.6	4.5
$\frac{1}{4}$ open	24	24
Globe valve		
Open	10	6.4
$\frac{1}{2}$ open		9.5
Angle valve, open	5.0	
Switch valve, fully open	2.4	
90° elbow, short radius	0.9	
medium radius	0.75	0.75
long radius	0.6	0.45
Return bend 180°	2.2	
45° elbow	0.42	0.35
22.5° elbow	0.13	
Tee joint (branch flow)	1.25	1.5
Along run, line flow		0.4
Reducer (valve at small end)	0.25	
Increaser	$0.25(V_1^2 - V_2^2)/2g$	
Bellmouth reducer	0.10	
Open orifice	1.80	
Coupling or union		0.04

causes the same head loss as the pipe fixtures is $L_e = K \cdot d/f = 15 \times 0.15/0.015 = 150$ m. We can analyse the pipe system now by enhancing the length of the pipe by 150 m and not considering the existence of pipe fixtures.

Similarly if the pipe system consists of two pipes of different diameters d_1 and d_2 having pipe resistance factors f_1 and f_2 and lengths L_1 and L_2 respectively. If the second pipe is to be replaced by an equivalent length to suit the properties of the first pipe, then we have

$$f_1 V_1^2 L_e/(d_1 2g) = f_2 V_2^2 L_2/(d_2 2g)$$

Since the discharge is the same in both pipes, $V_2 = (d_1/d_2)^2 V_1$, the equivalent length for the second pipe is

$$L_e = \left(\frac{f_2}{f_1}\right)\left(\frac{d_1}{d_2}\right)^5 L_2 \tag{10.5}$$

The system is now reduced to a single pipe of diameter d_1, resistance factor f_1 and length $(L_1 + L_e)$. The arguments may be extended to more than two pipes.

10.4 Hydraulic and Energy Grade Lines

At each point along a pipe system, if the pressure head, p/γ is plotted above the centre of the pipe line, the locus of these end points is known as the hydraulic grade line. This line gives the piezometric head, $(p/\gamma + z)$ above a chosen datum and z is the elevation of the centre of the pipe above the datum as shown in Fig. 10.3. This actually represents the potential energy per unit mass of the fluid.

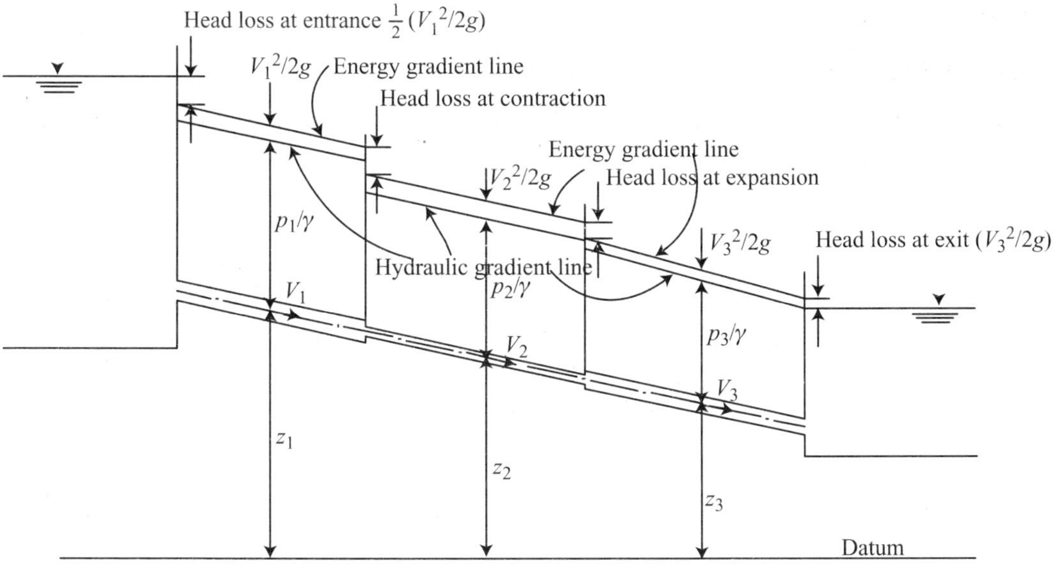

Fig. 10.3 Hydraulic and energy grade lines

The hydraulic grade line is also the locus of the heights to which the liquid would rise in vertical glass tubes connected to piezometric openings in the pipe. It may be realised that if the pressure is negative, the hydraulic grade line will be below the pipe line.

If we plot the local velocity head, $V^2/2g$ at each point vertically above the hydraulic grade line, we get the energy grade line. The locus of the top of the energy grade line measured vertically from the datum line gives the total energy per unit mass of the fluid, $(p/\gamma + z + V^2/2g)$. The slope of the energy grade line is always negative as the total energy decreases in the direction of the flow (energy is required to overcome the resistance to flow by way of pipe resistance and losses caused by pipe fixtures).

10.5 Problems in Pipe Flows (Pipes in Series)

Generally speaking, there are three types of problems that arise in the analysis and solutions to flow in pipes. These are

1. given Q, L, d, k, ν to find h_f;
2. given h_f, L, d, k, ν to find Q;
3. given Q, L, h_f, k, ν to find d.

10.5.1 Given Q, L, d, k, ν to find h_f

The first is the simplest of all preceding problems. Since Q and d are given, the Reynolds number is readily computed: $\mathbf{R} = 4Q/\pi\nu d$. With k/d also known, we can enter the pipe resistance diagram and find the pipe resistance factor, f. We can thus determine the loss due to pipe resistance, $h_f = 16f \cdot L \cdot Q^2/\pi^2 d^5 \cdot 2g$ and all other losses to pipe fixtures are also found by $[K(4Q/\pi d^2)^2/2g]$

EXAMPLE 10.1 Head loss to be determined given Q, L, d, k, ν

A pipe alignment is shown in the sketch. The pressure head at the beginning is given as 20 m and the discharge is 0.015 m^3/s. Assume $\nu = 1 \times 10^{-6}\ m^2/s$ and $k = 0.25 \times 10^{-3}$ m for all pipes (cast iron). Find the pressure head at section 2–2 if it is 10 m above the first point.

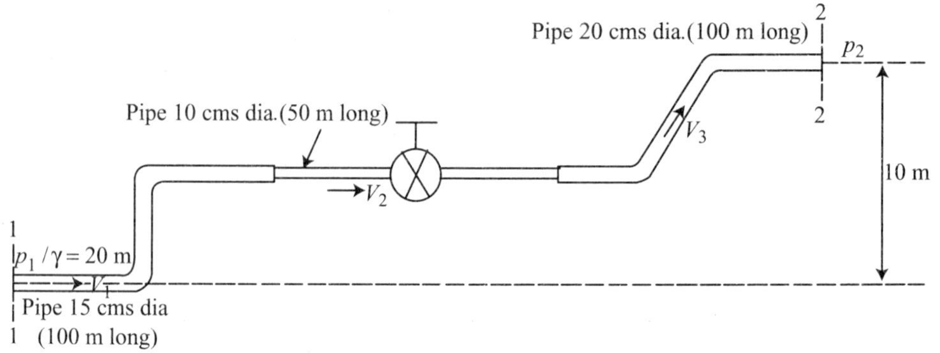

Sketch for Example 10.1

$$\text{Velocity in the 15 cm pipe } V_1 = \frac{0.015}{0.7854(0.15)^2} = 0.8488\ \text{m/s}$$

$$\text{Velocity in the 10 cm pipe } V_2 = \frac{0.015}{0.7854(0.10)^2} = 1.9099\ \text{m/s}$$

$$\text{Velocity in the 20 cm pipe } V_3 = \frac{0.015}{0.7854(0.20)^2} = 0.4775\ \text{m/s}$$

Pipe 1: $k/d_1 = 0.00025/0.15 = 0.001667$, $\mathbf{R}_1 = 0.8488 \times 0.015/10^{-6} = 127320$
From the pipe resistance chart, $f_1 = 0.024$,

$$h_{f1} = \frac{0.024 \times 100 \times (0.8488)^2}{0.15 \times 2 \times 9.81} = 0.5875\,\text{m}$$

Pipe 2: $k/d_2 = 0.00025/0.10 = 0.0025$, $\mathbf{R}_2 = 1.9099 \times 0.10/10^{-6} = 190990$
From the chart, $f_2 = 0.029$,

$$h_{f2} = \frac{0.029 \times 50 \times 1.9099^2}{0.10 \times 19.62} = 2.6958\,\text{m}$$

Pipe 3: $k/d_3 = 0.00025/0.20 = 0.00125$, $\mathbf{R}_3 = 0.4775 \times 0.20/10^{-6} = 95500$
From the chart, $f_3 = 0.020$,

$$h_{f3} = \frac{0.020 \times 100 \times 0.4775^2}{0.20 \times 19.62} = 0.1162\,\text{m}$$

Loss in two 90° elbows in the 15 cm pipe $= 2 \times 0.75 \times 0.8488^2/19.62 = 0.0551$

Loss in two 45° elbows in the 20 cm pipe $= 2 \times 0.35 \times 0.4755^2/19.62 = 0.0080$

Loss in fully opened gate valve of a 10 cm pipe $= 0.19 \times 1.9099^2/19.62 = 0.0353\,\text{m}$

Loss at contraction from 15 cm to 10 cm: Area ratio $(0.10/0.15)^2 = 4/9$; $C_c = 0.667$;

Loss $= (1/0.667 - 1)^2 \times (1,9099^2/19.62) = 0.0465\,\text{m}$

Loss at expansion from 10 cm to 20 cm $= (1 - A_2/A_3)^2 V_2^2/2g$
$= (9/16)\,(1.9099^2/19.62) = 0.1046\,\text{m}$

Exit velocity head $= 0.4775^2/2 \times 9.81 = 0.0116\,\text{m}$

Energy equation between the two points 1 and 2 (datum is passing through point 1)

$$20 + \frac{(0.8488)^2}{2 \times 9.81} = 20.0367 = \frac{p_2}{\gamma} + 0.5875 + 2.6958 + 0.1162 + 0.0551 + 0.0080$$
$$+ 0.0353 + 0.0465 + 0.1046 + 0.0116 + 10.0 = \frac{p_2}{\gamma} + 13.661$$

The pressure head at point 2 is $p_2/\gamma = 20.0367 - 13.661 = 6.3757\,\text{m}$

10.5.2 Given h_f, L, d, k, ν to find Q

EXAMPLE 10.2 Discharge to be found given head loss, L, d, k, ν

The two reservoirs as shown in the sketch are connected by pipes. The difference in the level of reservoirs is 10 m. We are required to find the discharge in the pipes.

We will express all losses in the pipe system in terms of one velocity head say $V_1^2/2g$. By continuity, we know $V_2 = V_1(d_1/d_2)^2 = 25V_1/9$

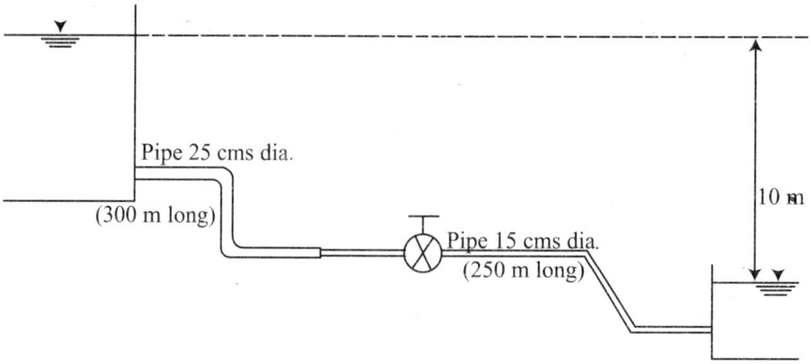

Sketch for Example 10.2

Losses are

$$\frac{V_1^2}{2g}\left[\underset{\text{Entrance loss}}{0.5} + \underset{\text{2 Elbows}}{2 \times 0.75} + \underset{\text{Contraction loss}}{0.534\left(\frac{25}{9}\right)^2} + \underset{\text{Globe valve}}{6.4\left(\frac{25}{9}\right)^2} + \underset{\text{2 45° Elbows}}{2 \times 0.35\left(\frac{25}{9}\right)^2} + \underset{\text{Exit}}{\left(\frac{25}{9}\right)^2} + \underset{\text{Loss in pipe 1}}{f_1 \times \frac{300}{0.25}} + \underset{\text{Loss in pipe 2}}{f_2 \times \frac{250}{0.15}}\right]$$

$$= \frac{V_1^2}{2g}[0.5 + 1.5 + 4.12 + 49.38 + 5.40 + 7.72 + f_1(1200) + f_2(1666.67)]$$

The total losses must be equal to the difference in the reservoir level viz., 10 m

$$10 = \frac{V_1^2}{2g}[68.62 + 1200 f_1 + 1666.67 f_2]$$

We have to solve the preceding equation by trial and error. Assume $V_1 = 1.25$ m/s, then $\mathbf{R}_1 = 1.25 \times 0.25/10^{-6}$ and $k_1/d_1 = 0.00025/0.25 = 0.001$ and $f_1 = 0.02$.

$\mathbf{R}_2 = (25/9) \times 1.25 \times 0.15/10^{-6} = 520833$; $k_2/d_2 = 0.00025/0.15 = 0.00167$ and $f_2 = 0.0225$

$$\frac{130.12 V_1^2}{2g} = 10 \text{ and } V_1 = \left(\frac{10 \times 19.62}{130.12}\right)^{1/2} = 1.228 \text{ m/s}$$

This is close enough to the assumed value of 1.25 m/s. A second assumption of $V_1 = 1.23$ m/s will satisfy the requirements. The discharge is $Q = 0.7854(0.25)^2(1.23) = 0.0604$ m^3/s

10.5.3 Given Q, h_f, L, k, ν to find diameter d

Example 10.3 Diameter to be determined given Q, h_f, L, k and ν

Find the minimum diameter required to deliver a discharge of 150 litres/s of water through a height of 25 m(h_f), the cast iron pipe ($k = 0.00025$ m) length being 2000 m. $\nu = 1 \times 10^{-6}$ m^2/s.

The head loss, $h_f = 8fLQ^2/g\pi^2d^5$ or $d^5 = 8fLQ^2/g\pi^2h_f$

$$d^5 = \frac{8 \times 2000 \times (0.15)^2 f}{\pi^2 \times 9.81 \times 25} = 0.1487f$$

The Reynolds number is expressed as

$$\mathbf{R} = \left(\frac{4Q}{\pi\nu}\right)\left(\frac{1}{d}\right) = \frac{4 \times 0.15}{\pi \times 10^{-6}d} = \frac{190986}{d}$$

Assume for the first trial $f = 0.02$, $d = (0.1487 \times 0.02)^{1/5} = 0.312$ m
$\mathbf{R} = 190986/0.312 = 612135$ and $k/d = 0.00025/0.312 = 0.000803$ and $f = 0.019$ from the chart.

Revised diameter $d = (0.1487 \times 0.019)^{1/5} = 0.309$ m. Further trials are not necessary and standard pipes close to this value (0.309 m) may be used.

10.6 Parallel Pipes

The main pipe may branch out at point A to two or more pipes and rejoin again at point B as shown in Fig. 10.4.

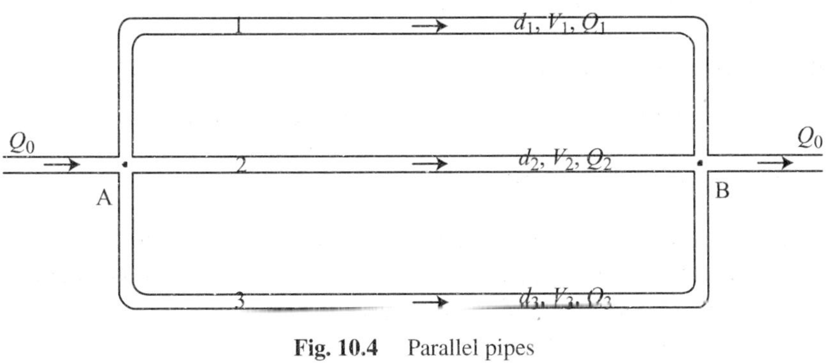

Fig. 10.4 Parallel pipes

As the points A and B are common to all branch pipes, the head loss is the same in each pipe: $h_f = h_{f1} = h_{f2} = h_{f3}$. The total discharge is the sum of all discharges in each branch pipe $Q_0 = Q_1 + Q_2 + Q_3 = \Sigma Q_i$. If the head loss is known, then the discharge in each individual pipe is $Q_i = (\pi^2 d_i^5 g h_f / 8 f_i L_i)^{1/2}$.

Calculate $\mathbf{R}(f)^{1/2} = (D^{3/2}/\nu)(2gh_f/L)^{1/2}$ and find $(1/f^{1/2})$ from Fig. 9.11. Once the value of f is known, we can find the discharge in the pipe.

If the total discharge Q_0 is known, one has to find individual discharges in each pipe Q_i and h_f. Assume a certain discharge in the first pipe Q_1' and find f_1' and determine $h_f' = 8f_1'L_1(Q_1')^2/\pi^2 d_1^5 g$. Since the head loss is the same for all pipes, using this value of h_{f1}, find the discharges in the other pipes. Then check for continuity $Q_0' = \Sigma Q_i'$. This may not match the actual discharge Q_0. Then distribute the error proportionately thus

$$Q_1 = \left(\frac{Q'_1}{Q'_0}\right) Q_0;\ Q_2 = \left(\frac{Q'_2}{Q'_0}\right) Q_0 = \ldots = \left(\frac{Q'_i}{Q'_0}\right) Q_0$$

After correcting for the discharges, calculate h_{fi} for each pipe and check whether the same is obtained for all pipes. Otherwise further iteration is necessary.

EXAMPLE 10.4 The total flow in the three-parallel pipe system shown in Fig. 10.4 is $5.0\,\text{m}^3/\text{s}$. The pipe lengths are $L_1 = 1500\,\text{m}$, $L_2 = 1200\,\text{m}$, $L_3 = 1000\,\text{m}$ and the corresponding diameters are $d_1 = 0.75\,\text{m}$, $d_2 = 0.50\,\text{m}$, $d_3 = 1.0\,\text{m}$. Find the discharges in the three pipes [$\nu = 1 \times 10^{-6}\,\text{m}^2/\text{s}$, $k = 0.00025\,\text{m}$].

Assume the discharge $Q'_1 = 1.5\,\text{m}^3/\text{s}$, then $\mathbf{R}'_1 = (4 \times 1.5)/(\pi \times 10^{-6} \times 0.75) = 2546479$, and $k/d = 0.00025/0.75 = 0.00033$, $f'_1 = 0.015$

$$h'_{f1} = (8 \times 0.015 \times 1500 \times 1.5^2)/(\pi^2 \times (0.75)^5 \times 9.81) = 17.627\,\text{m}$$

With this value of h'_f, we find the discharges in the other two pipes. Assume $f'_2 = 0.017$

$$Q'_2 = [(\pi^2 \times (0.5)^5 \times 9.81 \times 17.627)/(8 \times 1200 \times 0.017)]^{1/2} = 0.572\,\text{m}^3/\text{s}$$

$$Q'_3 = [(\pi^2 \times 1 \times 9.81 \times 17.627)/(8 \times 1000 \times 0.016)]^{1/2} = 3.651\,\text{m}^3/\text{s}$$

$$\Sigma Q'_i = 1.5 + 0.572 + 3.651 = 5.723\,\text{m}^3/\text{s}$$

Adjust the discharges proportionately:

$$Q_1 = 1.5 \times 5.0/5.72 = 1.311;\ Q_2 = 0.572 \times 5.0/5.72 = 0.5;$$

$$Q_3 = 3.651 \times 5.0/5.72 = 3.191\,\text{m}^3/\text{s}$$

The total adjusted discharge is $1.311 + 0.5 + 3.191 = 5.002\,\text{m}^3/\text{s}$ which is close to the given discharge. With these discharges in each pipe, we can calculate the Reynolds number and verify whether the resistance factors assumed for each pipe is reasonably accurate or not. Otherwise finer adjustments may have to be made.

10.7 Branch Pipes

Often we have pipes interconnecting three or more reservoirs as shown in Fig. 10.5. At the junction, we use the principle of continuity that the inflow into the junction is equal to the outflow from the junction. The hydraulic grade line is assumed to coincide with the reservoir surface (as the velocity in the reservoir is negligible). The value of the piezometric head, $h_j = (p_j/\gamma + z_j)$ at the junction J is unique.

The loss in head between any reservoir and the junction is $h_j - z_r = h_f$, where z_r is the elevation of the reservoir surface above the datum.

If the elevations of the reservoirs A and B, and Q_1 are given, then the problem is easily solved by calculating the head loss between reservoir A and the junction J, $h_{f1} = 8fL_1Q_1^2/\pi^2 d_1^5 g$

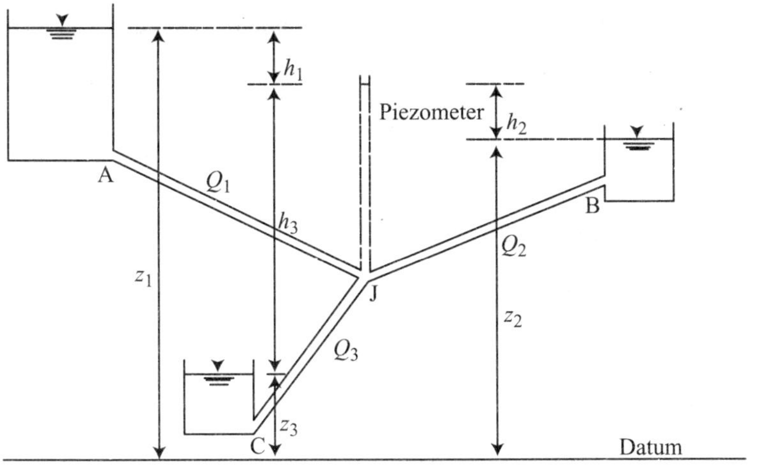

Fig. 10.5 Branch pipes

and the head at junction, J is determined, $h_j = z_1 - h_{f1}$. The discharge in pipe 2 is easily calculated by the head difference between J and the surface of the second reservoir. Continuity at junction J establishes the discharge in the third pipe and hence the level of the surface of the third reservoir.

The more common problem is to find the discharges in each of the branch pipes, given the reservoir levels above a certain datum. We initially make a trial value for the piezometric head at junction J and find the discharges in all the branch pipes. Then we should satisfy the equation of continuity at the junction say for three pipes as shown in the figure either as $Q_1 = Q_2 + Q_3$ or $Q_1 + Q_2 = Q_3$. If the flow into the junction is considered as positive and the discharge going away from the junction as negative, then $Q_1 + Q_2 + Q_3 = 0$. If the discharge into the junction is higher than the discharge going away, then we assume a higher trial value of h_j at the junction and recalculate the discharges till continuity at J is satisfied. The method is illustrated in the following example.

Example 10.5 Pipe AJ: $d_1 = 1.0$ m, $L_1 = 1500$ m, $k/d_1 = 0.00025$, assume $f_1 = 0.015$
Pipe BJ: $d_2 = 0.75$ m, $L_2 = 1200$ m, $k/d_2 = 0.00025/0.75$, let $f_2 = 0.016$
Pipe JC: $d_3 = 1.2$ m, $L_3 = 1800$ m, $k/d_3 = 0.00025/1.20$, let $f_3 = 0.014$
$z_1 = 40$ m, $z_2 = 25$ m, $z_3 = 15$ m
Find the discharges in the three pipes.

Initially let us assume piezometric head at J, $h_j = (p_j/\gamma + z_j) = 20$ m.

$$h_{f1} = 20 = (8 \times 0.015 \times 1500 \times Q_1^2)/(\pi^2 \times 1 \times 9.81) = 1.859 Q_1^2; \quad Q_1 = 3.28 \text{ m}^3/\text{s}$$

$$h_{f2} = 5 = (8 \times 0.016 \times 1200 \times Q_2^2)/(\pi^2 \times (0.75)^5 \times 9.81) = 6.685 Q_2^2; \quad Q_2 = 0.865 \text{ m}^3/\text{s}$$

$$h_{f3} = 5 = (8 \times 0.014 \times 1800 \times Q_3^2)/(\pi^2 \times (1.2)^5 \times 9.81) = 0.837 Q_3^2; \quad Q_3 = 2.44 \text{ m}^3/\text{s}$$

Since $Q_1 + Q_2 > Q_3$, we have to increase the piezometric head at J. Let us try $h_j = 24$ m.

$$h_{f1} = 16 = 1.859Q_1^2; \quad Q_1 = 2.93\ \text{m}^3/\text{s}$$

$$h_{f2} = 1 = 6.685Q_2^2; \quad Q_2 = 0.387\ \text{m}^3/\text{s}$$

$$h_{f3} = 9 = 0.837Q_3^2; \quad Q_3 = 3.279\ \text{m}^3/\text{s}$$

Since $Q_1 + Q_2$ is marginally greater than Q_3, a third trial of $h_j = 24.1$ m gives $Q_1 = 2.924\ \text{m}^3/\text{s}$ and $Q_2 = 0.367\ \text{m}^3/\text{s}$ and $Q_3 = 3.297\ \text{m}^3/\text{s}$.

Further refinements may be made by correctly assessing the values of resistance factors f_i corresponding to the Reynolds numbers and pipe roughness factors.

10.8 Pipe Networks

In water supply schemes, we invariably have interconnected pipes serving different outlets as shown in Fig. 10.6. Solutions to problems are often complicated requiring trial solutions ideally solved by computers for which software is available.

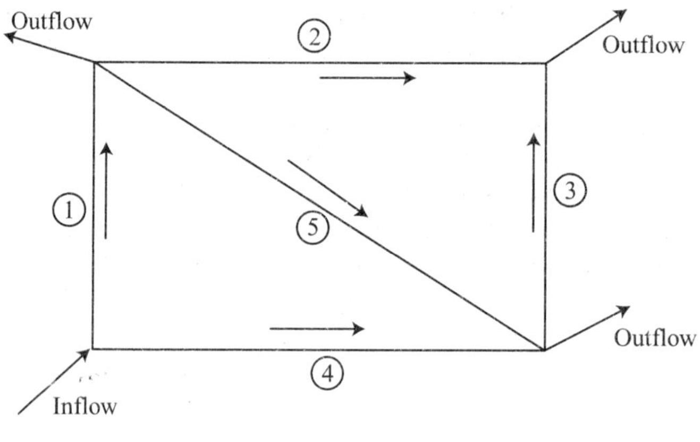

Fig. 10.6 A typical example of pipe network

Solutions to pipe network problems are based on the following principles.

1. Algebraic sum of pressure drops around each circuit must be zero.
2. At each junction, continuity must be satisfied: $Q_{\text{inflow}} = Q_{\text{outflow}}$
3. For each pipe, the head loss is given by the Darcy–Weisbach equation

$$h_f = fLV^2/d \cdot 2g = 8fLQ^2/\pi^2 gd^5$$

f is a function of the Reynolds number and k/d; it is often expressed as $f = aQ^b$ within a small range of Reynolds numbers. If we incorporate this approximation in the expression for head loss, we can write $h_f = KQ^n$, where K stands for $(8aL/\pi^2 d^5 g)$ and the exponent, n stands for $(2 + b)$. The value of K and the exponent n have to be assessed individually for each pipe.

Since analytical solutions are not possible we resort to successive trial solutions known as the Hardy–Cross method. The steps used in this method are

1. Assume the distribution of flow in each pipe which intuitively seems to suit the network. At each junction, continuity must be satisfied.
2. Compute $h_f = KQ^n$ for each pipe segment after assessing the values for K and n for that segment.
3. Calculate the net head loss for each circuit around each elementary circuit. It should be zero.
4. If the correct discharge in each pipe is Q and the assumed discharge is Q_0, the error in discharge is ΔQ; $Q = Q_0 + \Delta Q$. If we substitute this value in the expression for head loss $h_f = K(Q + \Delta Q)^n$, binomial expansion gives $h_f = K(Q_0^n + nQ_0^{n-1}\Delta Q +$ higher order terms of ΔQ which may be neglected as the error is assumed small.) In moving round a loop, the total head loss is zero $\Sigma h_f = 0$. Hence we have $\Sigma - KQ_0^n = \Sigma KnQ_0^{n-1}\Delta Q$. Since the discharge correction factor ΔQ is the same for all pipes in the circuit, it can be removed from the summation and

$$\Delta Q = -\frac{\Sigma KQ_0^n}{\Sigma nKQ_0^{n-1}} = -\frac{\Sigma h_f}{\Sigma\left(nh_f/Q_0\right)}$$

5. Compute for each circuit $\Sigma|nKQ^{n-1}|$ i.e., all terms are considered as positive.
6. A correction discharge factor $\Delta Q = -\Sigma KQ_0^n/\Sigma|nKQ^{n-1}|$ is introduced for each circuit.
7. It is convenient to work in a fixed direction either clockwise or anti-clockwise. The head loss is positive if the flow is in the chosen fixed direction and negative if the flow is against the fixed direction of the traverse.

The application of the Hardy–Cross method is illustrated by the following example.

EXAMPLE 10.6 The network is shown in the sketch and the data for each leg of the pipe network is given below. It is assumed that $h_f = KQ^2$

Pipe	Diameter (m)	Length (m)	f	$K = 8fL/\pi^2 d^5 g = 0.0826 fL/d^5$
AB	1.0	4035	0.015	5.0
BE	1.0	1614	0.015	2.0
EF	1.0	3027	0.015	4.0
FA	1.0	1614	0.015	2.0
BC	0.75	1795	0.016	10.0
CD	0.75	3532	0.016	20.0
DE	0.75	1795	0.016	10.0

The inflow at A is 12 m^3/s and the outflows at B, C, D, E, and F are 2, 1, 1, 6, 2 m^3/s respectively. We have to determine the discharges in each segment of the pipe in the network.

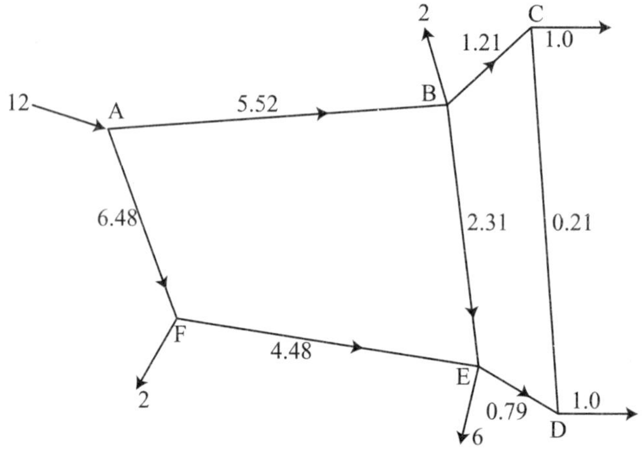

Sketch of network for Example 10.6

Pipe	K	Q	KQ^2	KQ^2/Q_0	New Q_0
AB	5	5.5	151.25	27.5	5.55
BE	2	2.0	8	4	$2.0 + 0.05$ $+0.25 = 2.30$
EF	4	−4.5	−81	18	−4.45
FA	2	−6.5	−84.5	13	−6.45
Sum			−6.25	62.5	

$\Delta Q = -(-6.25/2 \times 62.5) = +0.05$

Pipe	K	Q	KQ^2	KQ^2/Q_0	New Q_0
BC	10	1.5	22.5	15	1.25
CD	20	0.5	5	10	0.25
DE	10	−0.5	−2.5	5	0.75
EB	2	−2	−8	4	$-2.0 - 0.25 - 0.05$ $= -2.30$
Sum			+17.0	34	

$\Delta Q = -17/2 \times 34 = -0.25$

Note the transfer of corrections from one loop to another to the common member BE. Using the new values of discharges in the last column for each loop and recalculating, we get the corrected discharges in m^3/s.

AB: 5.52, BE: 2.31, EF: 4.48, FA: 6.48, BC: 1.21, CD: 0.21, DE: 0.79. The directions of flow are shown in the sketch.

Part II Unsteady Flow in Pipes

10.9 Establishment of Flow

In an unsteady flow, the mean velocity and pressure at any section will vary in magnitude with time. The local acceleration term $\partial v_s/\partial t$ along a stream line must be taken into account. In one-dimensional flows, the unsteady flow equation becomes

$$\left(\frac{1}{g}\right)\left(\frac{\partial V}{\partial t}\right) = -\frac{\partial H}{\partial s}$$

The equation implies that if the rate of flow varies with time, the mean total head will vary with distance along the filament. In a steady flow, on the other hand, we have seen that the total head remains constant along a stream line.

10.9.1 Establishment of Flow after Gate Opening

We will apply the above unsteady equation to the case of establishment of flow in a pipe leading from a reservoir after the valve at the end of the pipe is suddenly opened as shown in Fig. 10.7. We assume that the pipe is rigid and the flow pertains to ideal fluids.

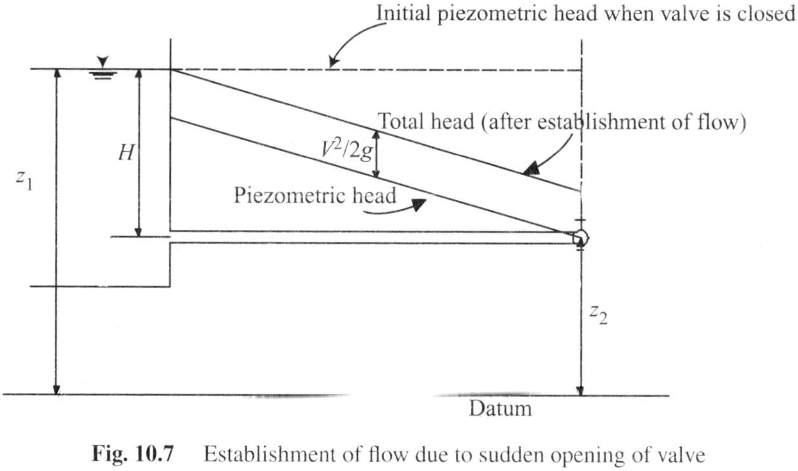

Fig. 10.7 Establishment of flow due to sudden opening of valve

We will consider the changes taking place at the end of the pipe. In such a case, ∂s is replaced by the length of the pipe, L and the difference in head, ∂H by the difference in the heads at the reservoir and the pipe end, viz., $H - V^2/2g$, remembering that V is a function of time. The unsteady flow equation becomes (since the velocity at the end of the pipe is a function of time only, $\partial V/\partial t$ may be replaced by dV/dt)

$$\left(\frac{L}{g}\right)\left(\frac{dV}{dt}\right) = H - \frac{V^2}{2g}$$

$$dt = 2L\left[\frac{dV}{(2gH - V^2)}\right]$$

Integration of the above equation yields

$$t = \frac{L}{\sqrt{2gH}} \ln \frac{\sqrt{2gH} + V}{\sqrt{2gH} - V}$$

The constant of integration is zero since $V = 0$ at $t = 0$. The final velocity is $V_{max} = (2gH)^{1/2}$ and hence we can write the preceding equation in the non-dimensional form

$$t\frac{\sqrt{2gH}}{L} = \ln \frac{1 + V/V_{max}}{1 - V/V_{max}}$$

The maximum velocity is reached asymptotically but V will reach $0.99V_{max}$ and $t = 5.293L/(2gH)^{1/2}$

10.9.2 Establishment of Flow after Gate Closure

Consider the case where there is a steady of flow of fluid with velocity V_0. We will now investigate the variation in velocity and pressure head when the gate is closed. Let the gate be closed in such a manner that

$$V = V_0\left(1 - \frac{t}{t_0}\right)$$

where t_0 is the time required to completely close the gate.Then we have

$$\frac{dV}{dt} = -\frac{V_0}{t_0}$$

$$\left(\frac{L}{g}\right)\frac{dV}{dt} = H - H_{gate}$$

But H_{gate} = pressure head + velocity head $= z + V^2/2g$ as shown in Fig. 10.8.

$$\left(\frac{L}{g}\right)\frac{dV}{dt} = H - z - \frac{V^2}{2g}$$

$$\left(\frac{L}{g}\right)\left(-\frac{V_0}{t_0}\right) = H - z - \left(\frac{V_0^2}{2g}\right)\left(1 - \frac{t}{t_0}\right)^2$$

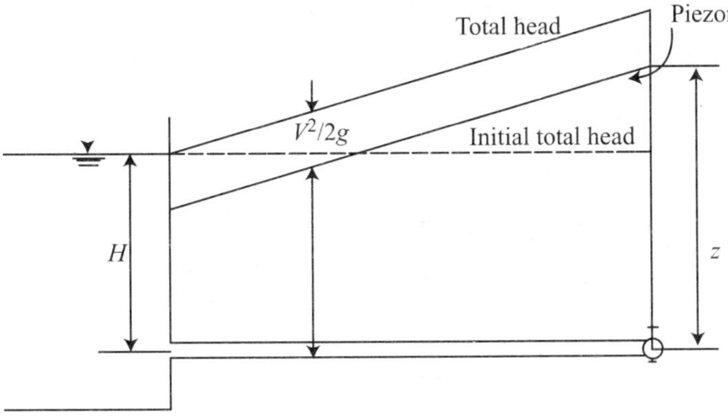

Fig. 10.8 Establishment of flow after gate closure

$$z = H + \frac{LV_0}{gt_0} - \left(\frac{V_0^2}{2g}\right)\left(1 - \frac{t}{t_0}\right)^2$$

It is seen that at the time when the gate is completely closed, the piezometric head is given by $z = H + LV_0/gt_0$. The simplified treatment shows that if the pipe length is long, the velocity high and the time of closure rapid, then the pressure rise in the pipe is very high. This qualitative statement is true even when we consider the compressibility of liquid and elasticity of pipe and pipe resistance. A more rigorous approach is presented in the next section.

10.10 Water Hammer

The term, water hammer, is commonly used to cover all pressure transients in hydraulic pipe lines that demonstrate both inertial and elastic effects. The Russian physicist Joukowsky and the Italian engineer Allievi, working independently were the first to demonstrate that the water hammer phenomenon was indeed an acoustic wave action.

A force applied at a point in a fluid medium leads to an immediate compression and propagates in the form of a compression wave in directions it is free to travel. A stone thrown into a still pond creates circular waves travelling radially outwards from the source of disturbance. In the confines of a pipe, the wave propagates in either direction along the length of the pipe. Thus the effect of a disturbance introduced at a point is felt some distance away after some time depending on the distance between the points and the speed of the propagating wave. If the disturbing force is applied in a period of time which is long compared to the travel time of the elastic wave, the fluid will react as if it were incompressible. Conversely, a relative rapid application of the force leads to a transient state of motion in which pressure will vary from point to point in the fluid. Since the terms gradual and rapid are relative, it is significant to determine the rate at which a compressive stress is transmitted in a given medium; this is also known as the celerity of an elastic wave.

If a solitary wave of compression or expansion is travelling in a fluid medium which is at rest, the course of the wave may be traced by the local change in pressure intensity as shown in Fig. 10.9a.

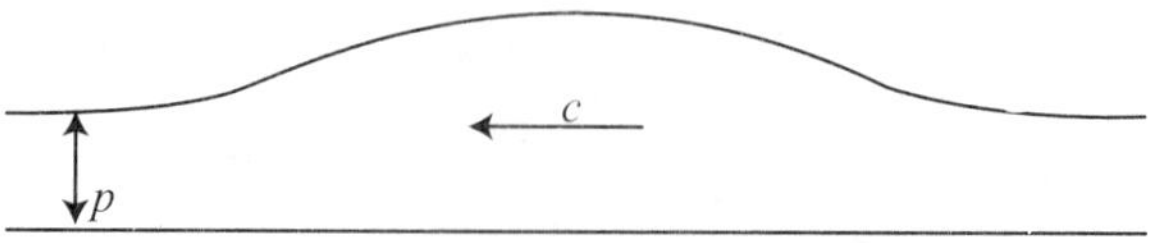

(a) Moving elastic wave

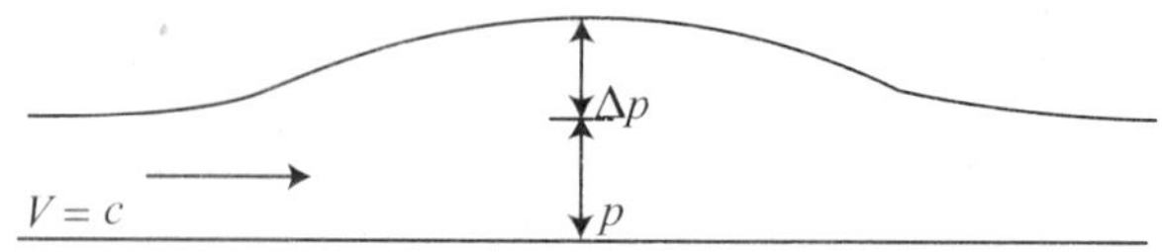

(b) Stationary wave

Fig. 10.9 Movement of an elastic wave

An elastic wave must be accompanied by a local displacement of fluid particles which would imply a change in velocity and the flow is unsteady. To analyse this problem we render the flow steady by superposing a uniform velocity equal to the celerity of the wave in the opposite direction as shown in Fig. 10.9*b*. Thus the disturbance remains stationary. The changes in velocity and pressure are given by the differential form of the Bernoulli equation $-dp = \rho V \cdot dV$ and the equation of continuity is given by $\rho V =$ constant or $\rho dV + V d\rho = 0$. Eliminating dV from these two relationships, we get $V = (dp/d\rho)^{1/2}$.

We also have the equation of state $K = \rho dp/d\rho$. The expression for V (which is also the expression of the celerity of the elastic wave, c now becomes

$$V = c = \left(\frac{K}{\rho}\right)^{1/2} \tag{10.6}$$

It is important to note that the celerity depends on both the Bulk modulus of the fluid as well as the density of the fluid. The average value of K for water is 2.075×10^6 N/m^2 and hence $c = (2.075 \times 10^6/998)^{1/2} = 1442$ m/s.

An elastic wave will be modified, reflected (or both) as it passes from one medium to another or if the geometric form of the restraining boundary is changed. Among examples of reflected waves are the familiar echo, submarine detection by sonar and more importantly for our discussion, the cyclic reflection of a water hammer wave as it is reflected alternatively and repeatedly at reservoirs, valves and at sections where boundary form changes repeatedly.

In the examples cited above, two distinctly different types of boundaries should be noted:

1. A relatively solid boundary at which the normal velocity is constant but the pressure varies such as at a valve or closed end of a pipe.
2. A boundary such as a reservoir (the pressure is dictated by the hydrostatic head at the reflection point) at which the pressure remains constant and the velocity is free to change.

In the first instance, the reflection is positive and a compression wave is reflected as a compression wave and an expansion wave is reflected as an expansion wave. The reflection is in the same sense as the arriving one.

In the second case, the reflection is negative and a compression wave is reflected as an expansion wave and an expansion wave is returned as a compression wave so that the boundary remains unchanged i.e., the reflection is in the unlike sense.

If two waves or wave systems come together, their combined effect in every case will be the algebraic sum of their individual effects: like waves being additive and unlike waves being subtractive. It may be seen therefore that if two equal compression waves travelling in opposite directions produce at the moment of coincidence, the condition of zero velocity (as required for reflection of a wave from a solid boundary). The combined pressure change at that instant is twice that of either the individual wave, and subsequently each wave proceeds in its original form in opposite directions. The case of a wave reflected from a solid boundary is similar to the meeting of two waves of the same sense. At the point of reflection, the velocity is reduced to zero and the pressures being additive are doubled in their magnitude. More explicitly, if the pressure intensity of a compression wave is p, then at the instant of reflection the pressure becomes $2p$ and an expansion wave of pressure intensity $-p$ will on reflection attain a value of $-2p$. The reflected wave may be looked upon as a wave of similar form travelling in the opposite direction.

Similarly equal but unlike waves—one being a wave of compression and another being an expansion wave—satisfy the requirement of zero change in pressure but the velocity is doubled at the point of reflection. On reaching a reservoir of constant depth, a compression wave is returned as an expansion wave. These concepts are directly applicable to the water hammer phenomena as well as surges in open channels and will be discussed in subsequent sections.

10.11 Basic Equations of Water Hammer

10.11.1 Equation of Continuity

Let us consider a pipe section of length dx (x-axis is oriented along the axis of the inclined pipe) and radius r as shown in Fig. 10.10.

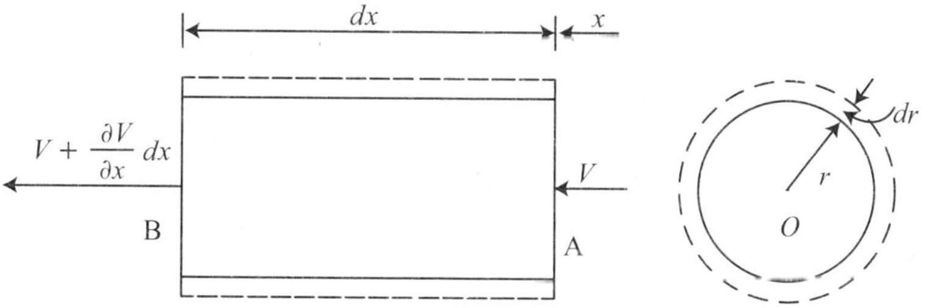

Fig. 10.10 Definition sketch

Let the velocity be V at section A and $[V + (\partial V/\partial x)dx]$ at section B. Hence a volume of water equal to $[\pi r^2 \cdot (\partial V/\partial x)dx]dt$ enters the section through A during the time interval dt which is not balanced by a corresponding flow at section B. This difference between inflow and outflow is compensated by the compression of the water column under increased pressure as well as the expansion of the pipe line under increased pressure.

Compression of the Water Column

The water column of length dx due to an increase in pressure $(\partial p/\partial t) \cdot dt$ in the time interval dt reduces the volume by $[\pi r^2 dx(\partial p/\partial t)dt/K]$, where K is the bulk modulus of the liquid.

Expansion of the Pipe

Owing to the increase in pressure $(\partial p/\partial t) \cdot dt$, the pipe radius r is also increased by dr. The circumferential strain is $2\pi dr/2\pi r = dr/r$. The circumferential stress caused by the increase in pressure is $2r(\partial p/\partial t)dt/2e = (r/e)(\partial p/\partial t)dt$, where e is the thickness of the pipe assumed small in relation to the radius of the pipe. The young's modulus E is the ratio of stress to strain and hence we have $E = (r/e)(\partial p/\partial t)dt/(dr/r)$. Solving for dr, we get

$$dr = \left(\frac{r^2}{Ee}\right)\left(\frac{\partial p}{\partial t}\right) \cdot dt$$

The corresponding change in volume is $2\pi r \cdot dr \cdot dx = (2\pi r^3 dx/Ee)(\partial p/\partial t)dt$. Continuity demands that

$$-\pi r^2\left(\frac{\partial V}{\partial x}\right) dx \cdot dt = \pi r^2 \cdot dx \cdot \left(\frac{\partial p}{\partial t}\right) \cdot \frac{dt}{K} + \left(\frac{2\pi r^3 dx}{Ee}\right) \cdot \left(\frac{\partial p}{\partial t}\right) \cdot dt$$

$$-\frac{\partial V}{\partial x} = \left(\frac{\partial p}{\partial t}\right)\left[\frac{1}{K} + \frac{2r}{Ee}\right]$$

Equating $\gamma/g[1/K + 2r/Ee] = 1/a^2$, we get ($a$ has the dimensions of velocity)

$$a = \sqrt{\frac{g/\gamma}{1/K + 2r/Ee}} = \sqrt{\frac{K/\rho}{1 + (K/E)(D/e)}} \tag{10.7}$$

a represents the velocity of propagation of an elastic pressure wave in a pipe line. This is less than the celerity of an elastic wave in an unbounded fluid medium given by equation (10.6). The expression for $-\partial V/\partial x$ can be now written as (with $p = \gamma H$)

$$-\frac{\partial V}{\partial x} = \left(\frac{g}{a^2}\right) \cdot \frac{\partial H}{\partial t}$$

$$-\frac{\partial^2 V}{\partial x^2} = \left(\frac{g}{a^2}\right)\left(\frac{\partial^2 H}{\partial x \partial t}\right) \tag{10.8}$$

$$-\frac{\partial^2 V}{\partial x \partial t} = \left(\frac{g}{a^2}\right)\left(\frac{\partial^2 H}{\partial t^2}\right)$$

10.11.2 Dynamic Equation

From the one-dimensional Euler's equation along the x-direction (assuming no losses), we have

$$\frac{\partial V}{\partial t} + V\left(\frac{\partial V}{\partial x}\right) = -\frac{\partial(p/\rho + gz)}{\partial x} = -\frac{g\partial(p/\gamma + z)}{\partial x} = -g\frac{\partial H}{\partial x}$$

where H replaces the piezometric head, $(p/\gamma + z)$. If $\partial V/\partial x$ is negligible compared to $\partial V/\partial t$, then we have

$$\frac{\partial V}{\partial t} = -g\frac{\partial H}{\partial x}$$

$$\frac{\partial^2 V}{\partial x \partial t} = -g\left(\frac{\partial^2 H}{\partial x^2}\right) \qquad (10.9)$$

$$\frac{\partial^2 V}{\partial t^2} = -g\left(\frac{\partial^2 H}{\partial x \partial t}\right)$$

Equations (10.8) and (10.9) are to be solved simultaneously for H and V.

Eliminating H we get

$$\frac{\partial^2 V}{\partial t^2} = a^2\frac{\partial^2 V}{\partial x^2}$$

Eliminating V, we get

$$\frac{\partial^2 H}{\partial t^2} = a^2\frac{\partial^2 H}{\partial x^2}$$

The general solutions of these equations are

$$H - H_0 = f\left(t - \frac{x}{a}\right) + F\left(t + \frac{x}{a}\right)$$

$$V - V_0 = \frac{g}{a}\left[f\left(t - \frac{x}{a}\right) - F\left(t + \frac{x}{a}\right)\right]$$

We have so far considered x and V as positive when they are measured from the reservoir to the valve but it is more convenient to locate the origin at the valve and measure x as positive in the direction from the valve to the reservoir as the wave travels up the pipe line when the valve is operated either closed or opened. Then these equations are rewritten as

$$H - H_0 = F\left(t - \frac{x}{a}\right) + f\left(t + \frac{x}{a}\right) \qquad (10.10)$$

$$V - V_0 = -\frac{g}{a}\left[F\left(t - \frac{x}{a}\right) - f\left(t + \frac{x}{a}\right)\right] \qquad (10.11)$$

The physical meaning of the above equations, for example equation (10.10) is that $F(t - x/a)$ represents a pressure wave which moves in the direction of positive x with a velocity a. Similarly $f(t + x/a)$ represents a pressure wave which moves in the negative x-direction with a velocity a as shown in Fig. 10.11.

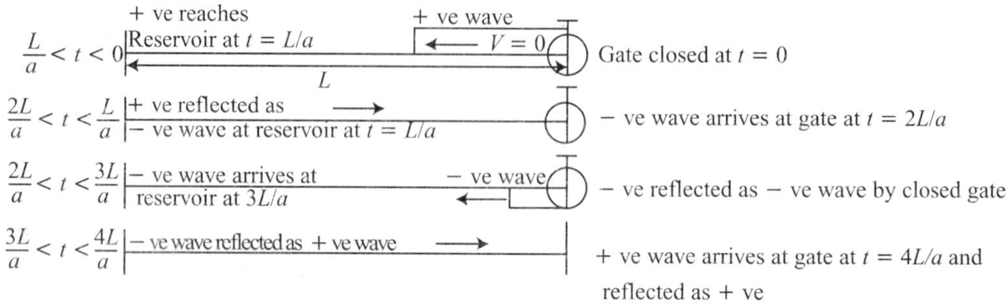

Fig. 10.11 Movement of pressure waves

If the valve at the end of the pipe is closed suddenly (or closed in a relatively short interval) we have $V = 0$, and at the instant of closure there is no return wave from the reservoir, hence $f(t - x/a) = 0$. Equations (10.10) and (10.11) give

$$H - H_0 = F\left(t - \frac{x}{a}\right) = -\frac{a}{g}(V - V_0)$$

$$\Delta H = \frac{\Delta p}{\gamma} = -\frac{a}{g}(-V_0) \text{ since } V = 0$$

$$\Delta p = \rho a V_0 \tag{10.12}$$

This pressure wave will travel upstream along the pipe with a velocity a and reach the reservoir at time $t = L/a$, where L is the length of the pipe. At the reservoir, the pressure does not change as it is controlled by the reservoir level. Hence $H - H_0 = 0$. Therefore, we have

$$F\left(t - \frac{L}{a}\right) = -f\left(t + \frac{L}{a}\right)$$

An expansion wave of equal magnitude $\Delta p = -\rho a V_0$ starts downstream at the same wave celerity a, thus returning the pressure in the pipe to the original condition as shown in Fig. 10.12.

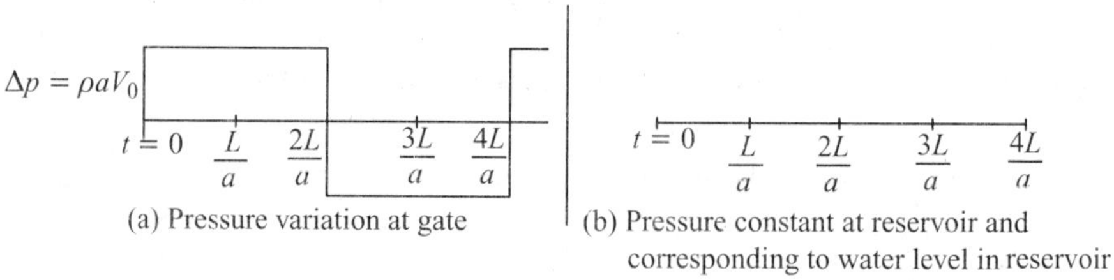

Fig. 10.12 Pressure changes at $t = L/a, 2L/a, 3L/a, 4L/a$

And by (10.11), we have

$$V - V_0 = -\left(\frac{2g}{a}\right) F\left(t - \frac{L}{a}\right)$$

This reveals that the change in velocity at the reservoir locations is twice that produced by the direct wave at other points. In fact this additional change in velocity produces the reflected or f-wave. This reflected wave reaches the gate at time $2L/a$. At the gate the reflection is always in the like sense and an expansion wave is returned as an expansion wave and the pressure now becomes negative and equal to $-\rho a V_0$ (Fig. 10.9). The expansion wave travels upstream again and reaches the reservoir at time $t = 3L/a$. At the reservoir, the expansion wave is reflected in the negative sense or a positive is propagated restoring the pressure to the normal condition. The positive reaches the gate at $t = 4L/a$ and is reflected positively at the gate and the pressure again rises to $\Delta p = \rho a V_0$. The cycle is repeated endlessly in the absence of any loss. The periodic changes in pressure at mid-point of the pipe length is shown in Fig. 10.13.

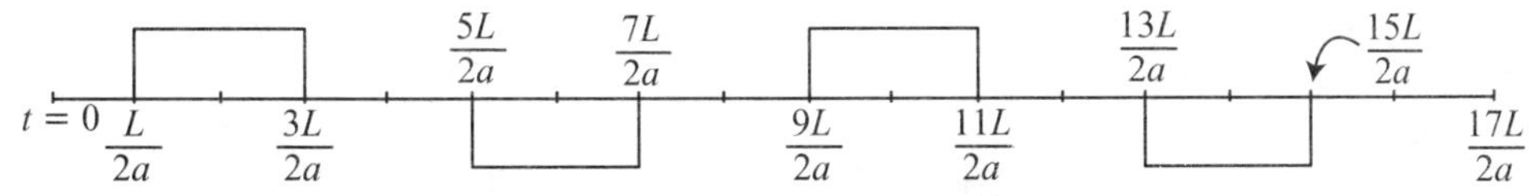

Fig. 10.13 Pressure fluctuations at mid-point

However in practice friction dampens the fluctuations as shown in Fig. 10.14 and the water hammer effects die down rapidly. The initial pressure rise at the gate will be $\rho a V_0$. During established conditions of flow in the pipe, the piezometric head rises in the upstream direction. Hence the pressure head at the gate must continue to rise proportional to the head loss in the pipe. It has been shown that the rise in pressure head is approximately $0.707h_f$.

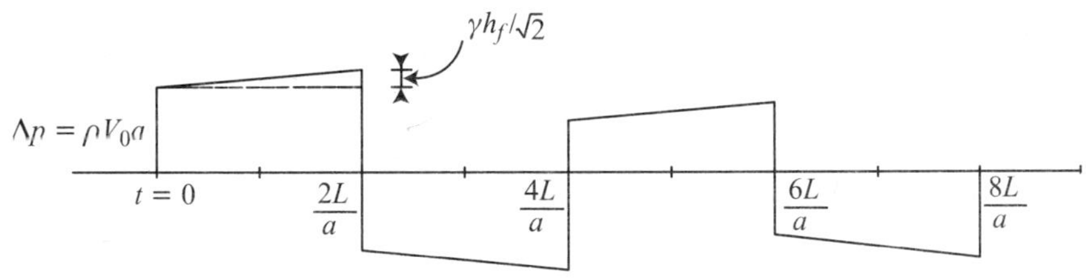

Fig. 10.14 Dampening effect due to friction

If the gate is closed in a finite but short time $t_1 < 2L/a$ (known as abrupt closure), the pressure fluctuations will occur as shown in Fig. 10.15. The maximum pressure intensity at the gate will

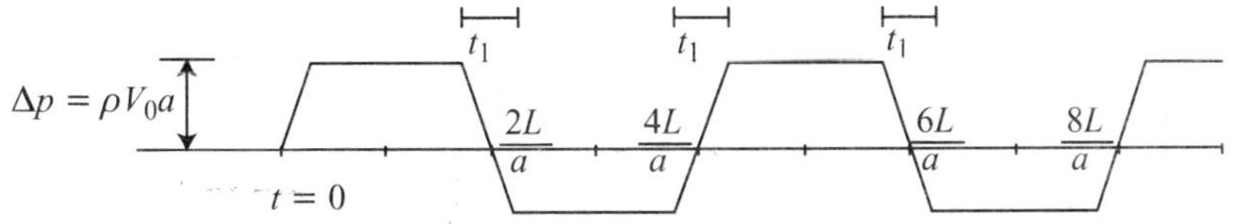

Fig. 10.15 Pressure changes due to closing of gate in a short time

remain the same as that due to an instantaneous gate closure ($\Delta p = \rho a V_0$) as long as the time of closure is smaller than the time required for the wave to reach the reservoir ($t_0 < L/a$). The greater the time required to close the valve, the smaller is the increase in pressure. As $t_0 \to 0$, the inertial effects dominate and the pressure rise will reach $\rho a V_0$.

The arguments may be extended to partial reflections at changes in the cross-sectional areas of pipes and branching of pipes or partial opening or closing of gates.

EXAMPLE 10.7 A steel pipe line 1800 m long takes off from a reservoir terminating in a control valve at its extreme end. The diameter of the pipe is 1 metre with a thickness of wall equal to 1 cm. When the valve is fully open, the discharge is 1000 litres per second. Calculate the maximum rise in pressure in the pipe line due to an abrupt closure of the valve reducing the flow to zero. E for steel $= 2.1 \times 10^{11}$ N/m^2; $K = 2.1 \times 10^9$ N/m^2

Velocity of propagation of the pressure wave a is given by

$$\frac{1}{a^2} = \frac{\gamma}{g}\left[\frac{1}{K} + \frac{d}{Ee}\right]$$

$$\frac{1}{a^2} = 998\left[\frac{1}{2.1 \times 10^9} + \frac{100}{2.1 \times 10^{11}}\right] = 998\left[\frac{2}{2.1 \times 10^9}\right]$$

$$a = 1025.72 \text{ m/s}$$

Velocity $V_0 = 4Q/\pi d^2 = 4 \times 1/\pi \times 1^2 = 4/\pi$ m/s

Maximum pressure rise will be immediately upstream of the valve $\Delta p = \rho a V_0 = 998 \times 1025.72 \times 4/\pi = 1303377$ N/m^2 or in terms of pressure head

$\Delta H = \Delta p/\gamma = 1303377/998 \times 9.81 = 133.13$ m.

Gradual closure of valve If the valve is closed gradually in time $T > 2L/a$, the rise in head is approximately determined as

$$\Delta H = \frac{2L/a}{T}\left(\frac{aV_0}{g}\right)$$

In Example 10.7, if the time of closing is 10 seconds, we have $2L/a = 2 \times 1800/1052.72 = 3.42$ s and the pressure head rise is

$$\frac{3.42}{10} \times 133.13 = 55.9 \text{ m}$$

10.12 Oscillations of Liquid Column in a U-Tube

We will consider the oscillations of a liquid column in a frictionless U-tube of uniform diameter as shown in Fig. 10.16

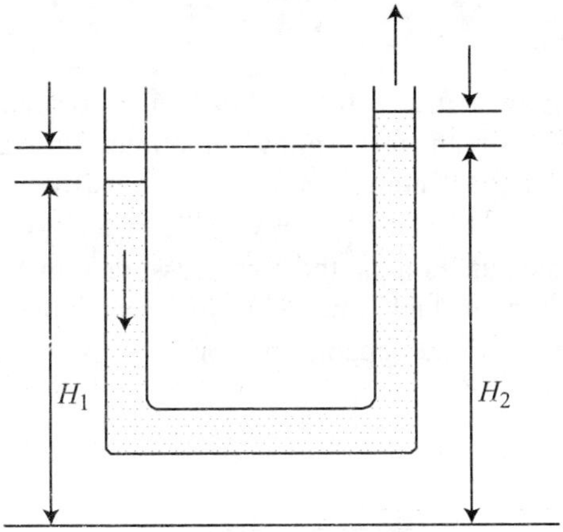

Fig. 10.16 Oscillations in a U-tube

The dynamic equation becomes

$$\left(\frac{L}{g}\right)\frac{\partial V}{\partial t} = -2z$$

which can be written as

$$\left(\frac{L}{g}\right)\left(\frac{\partial V}{\partial z}\right)\left(\frac{\partial z}{\partial t}\right) = \left(\frac{L}{g}\right)\left(\frac{\partial V}{\partial z}\right)V = \left(\frac{L}{g}\right)\frac{\partial(V^2/2)}{\partial z} = -2z$$

$$\frac{\partial V^2}{\partial z} = \frac{-4gz}{L}$$

Integration of the preceding equation yields

$$V^2 = -2gz^2/L + C$$

Since $V = 0$ when $z = z_{\max}$, the integration constant $C = 2gz_{\max}^2/L$. Hence

$$V = \frac{dz}{dt} = \sqrt{\frac{2g}{L}(z_{\max}^2 - z^2)}$$

Integrating again we get

$$t = \sqrt{\frac{L}{2g}}\sin^{-1}\frac{z}{z_{\max}}$$

and the period of oscillation is

$$T = 2\pi\sqrt{\frac{L}{2g}}$$

10.13 Surge Tanks

In many high-head hydropower installations, the dam is located in a narrow gorge and the power plant is located downstream in a valley sometimes several kilometres away from the reservoir involving a long length of conveyance system consisting of a long low-pressure tunnel followed by a high-pressure penstock of relatively short length connected to the turbines. A surge tank is located at the junction of the low-pressure tunnel and the penstock. A surge tank is a large diameter vertical tank (the shape may vary depending on the local geological conditions) with a free surface capable of storing large quantities of water as shown in Fig. 10.17.

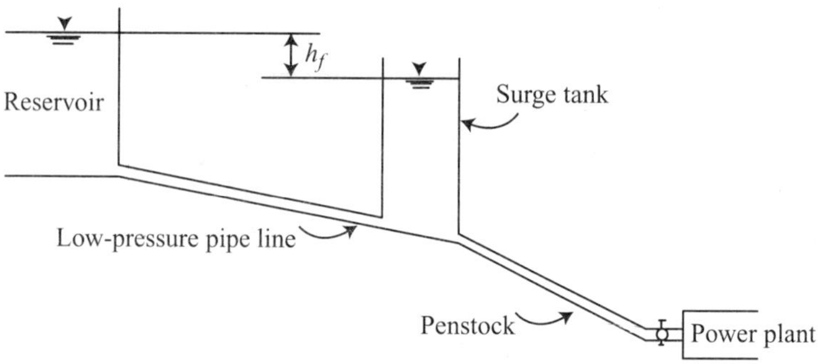

Fig. 10.17 A typical layout of a surge tank

The function of the surge tank is two-fold.

If the conduit is long, the time required to establish steady flow conditions constitutes a serious lag in time response following either an increase in demand or a rejection of demand. As explained in the earlier sections, such sudden changes in demand causes pressure waves to travel along the pipe line causing high pressures far in excess of the dynamic equilibrium pressures. In the event of a valve being opened suddenly, the surge tank can supply water to the penstock temporarily long before the water in the conduit from the reservoir has responded fully to the accelerating forces and the flow from the reservoir meets the demand. The water level in the surge tank will go down below its static level (called the draw-down). Similarly, when the valve is closed suddenly, the momentum of the flow in the low-pressure tunnel can be converted to potential energy in raising the level of the water in the surge tank above its dynamic equilibrium level. The mass of water in the low-pressure tunnel gets decelerated gradually.

The second advantage is that the pressure waves, either compression or expansion, are reflected at the surge tank itself which has a free surface. Thus it limits the influence of the pressure rise to the short length of the penstock which can be designed for these high pressures. The longer segment of the low-pressure tunnel is spared of these high-pressure transients resulting in economy in design. The low-pressure tunnel is relatively free of water hammer effects.

The flow in the surge tank will be unsteady causing oscillations of water level in the tank till they are damped out by boundary resistance.

10.14 Basic Equations of Mass Oscillations in a Simple Surge Tank

We will assume that the surge tank is a shaft of constant diameter. The area of the surge tank is very large compared to the diameter of the tunnel or penstock and we ignore the velocity of the free surface as it rises or falls. The compressibility effects are neglected.

Equation of continuity

If Q is the discharge through the turbine, A_t and A_s are the areas of the low-pressure tunnel and the surge tank as shown in Fig. 10.18, the velocities in the tunnel and the surge tank are V_t and V_s respectively then we have by continuity

$$V_t A_t = V_s A_s + Q \tag{10.13}$$

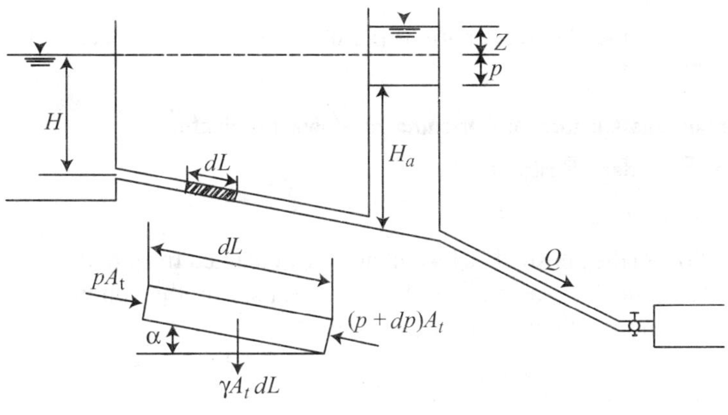

Fig. 10.18 Definition sketch

It is seen that the displacement of the free surface gives $V_s = dZ/dt$, where Z measures the water level in the surge tank above the static reservoir level.

Dynamic equation

If an element of length ds of the low-pressure tunnel is considered in isolation, the equilibrium of forces acting in the direction of the tunnel axis are

$$\underbrace{(\rho A_t ds) dV_t/dt}_{\text{change in momentum}} = \underbrace{\gamma A_t \cdot ds \cdot \sin\alpha}_{\text{weight component}} - \underbrace{A_t \cdot dp}_{\text{pressure}} - \underbrace{\gamma \cdot A_t \cdot i \cdot ds}_{\text{boundary resistance}}$$

where i stands for the energy gradient and we can write $ds \sin\alpha = dh$.

$$\frac{ds}{g}\frac{dV_t}{dt} = dh - \frac{dp}{\gamma} - i\,ds$$

Integrating the preceding equation over a finite length L of the tunnel,

$$\frac{1}{g}\frac{dV_t}{dt}\int_0^L ds = \int_{H_i}^{H_a} dh - \int_{H_i}^{H_a+Z}\frac{dp}{\gamma} - i\int_0^L ds$$

$$\left(\frac{L}{g}\right)\left(\frac{dV_t}{dt}\right) = H_a - H_i - (H_a + Z - H_i) - iL$$

$$\left(\frac{L}{g}\right)\cdot\left(\frac{dV_t}{dt}\right) + Z + iL = 0 \tag{10.14}$$

The sign of iL will be positive when the flow is from the reservoir to the surge tank as assumed in the derivation above but if the flow is in the opposite direction with V_t negative, the sign of iL will also be negative. Hence equation (10.14) can be rewritten as

$$\left(\frac{L}{g}\right)\left(\frac{dV_t}{dt}\right) + Z \pm iL = 0$$

Direct integration of the above equation is possible only in a few special cases.

10.15 Instantaneous Closure or Opening of Valve Neglecting Tunnel Boundary Resistance

Suppose that the flow in the penstock is instantaneously changed from Q_0 to Q_1 with corresponding velocities in the tunnel V_{t0} and V_{t1} in the tunnel satisfying the conditions $Q_0 = A_t V_{t0}$ and $Q_1 = A_t V_{t1}$. The equation of continuity is written as

$$V_t A_t = V_s A_s + Q = A_s\left(\frac{dZ}{dt}\right) + Q$$

Differentiating the above equation we get

$$\frac{dV_t}{dt} = \left(\frac{A_s}{A_t}\right)\left(\frac{d^2Z}{dt^2}\right)$$

The dynamic equation with $iL = 0$ becomes $(L/g)dV_t/dt + Z = 0$. Substituting for dV_t/dt, we get

$$\frac{L}{g}\frac{A_s}{A_t}\frac{d^2Z}{dt^2} + Z = 0 \tag{10.15}$$

The general solution of the foregoing equation is of the form

$$Z = C\sin(\omega t - \Phi), \text{ where } \omega = \sqrt{\frac{gA_t}{LA_s}}$$

C and Φ are constants determined from the initial conditions.

When $t = 0$, $Z = 0$, $C \sin(-\Phi) = 0$. Since C is not zero, $\Phi = 0$.
Also we have

$$A_t V_{t1} = A_s \left(\frac{dZ}{dt}\right)_{t=0} + Q_2 = A_s \left(\frac{dZ}{dt}\right)_{t=0} + A_t V_{t2}$$

Hence,

$$\left(\frac{dZ}{dt}\right)_{t=0} = \left(\frac{A_t}{A_s}\right)(V_{t1} - V_{t2})$$

$$C\omega \cos(-\Phi) = \left(\frac{A_t}{A_s}\right)(V_{t1} - V_{t2})$$

$$C = \frac{1}{\omega}\frac{A_t}{A_s}(V_{t1} - V_{t2})$$

$$Z = \frac{1}{\omega}\frac{A_t}{A_s}(V_{t1} - V_{t2}) \sin \omega t$$

The instantaneous change in flow produces sinusoidal oscillations of the water level in the surge tank with reference to the reservoir level. The period of oscillation, T is

$$\frac{2\pi}{\omega} = 2\pi\sqrt{\frac{LA_s}{gA_t}} \tag{10.16}$$

$$\text{and amplitude} = (V_{t1} - V_{t2})\sqrt{\frac{LA_t}{gA_s}} \tag{10.17}$$

$(V_{t1} - V_{t2}) > 0$ represents closure and $(V_{t1} - V_{t2}) < 0$ denotes opening. The curves representing the variation of Z with respect to time for the above conditions are shown in Fig. 10.19

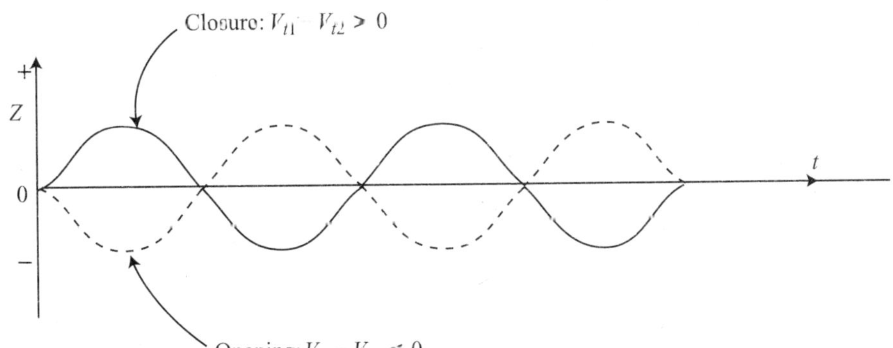

Fig. 10.19 Oscillations of water level in a surge tank

To find the maximum value of the amplitude, we can put $Q_1 = Q_{\max}$ and $Q_2 = 0$

$$Z_{\max} = \frac{Q_{\max}}{A_t}\sqrt{\frac{LA_t}{gA_s}} = Z_* \tag{10.18}$$

Z_* governs the design height of the surge tank above the reservoir level.

EXAMPLE 10.8 A cylindrical surge tank 7 metres in diameter is introduced at the end of a low-pressure tunnel 2400 m long from the supply reservoir. The area of the tunnel is 9 m^2 and $Q = 15\,\text{m}^3/\text{s}$. Find the period and maximum amplitude of oscillation following the sudden closure of the valve, ignoring all losses in the system.

Period of oscillation, $T = 2\pi(LA_s/gA_t)^{1/2} = 2\pi(2400 \times \pi \times 7^2/4 \times 9.81 \times 9)^{1/2} =$ 203.22 seconds

Maximum amplitude $Z_{\max} = (Q/A_t)(LA_t/gA_s)^{1/2} = (15/9)(4 \times 2400 \times 9/9.81 \times \pi \times 7^2) =$ 12.61 m.

10.16 Linear Closure of Valve

If the valve is closed gradually in time τ such that the discharge varies linearly with time, we have discharge at any instant of time, $t < \tau$,

$$Q = Q_0\left(1 - \frac{t}{\tau}\right)$$

then it can be shown that

$$\text{for } t < \tau,\ Z = \left(\frac{Z_*}{2\pi\theta}\right)[1 - \cos 2\pi t/T] \text{ and} \tag{10.19a}$$

$$\text{for } t > \tau,\ Z = \left(\frac{Z_*}{\pi\theta}\right)\left[\sin \pi\theta \cdot \sin 2\pi\left(\frac{t-\tau}{2}\right)/T\right] \tag{10.19b}$$

where $Z_* = V_{t0}(LA_t/gA_s)^{1/2}$ and $\theta = \tau/T$ and $T = 2\pi(LA_s/gA_t)^{1/2}$

For $t > \tau$, it is seen that Z will be a maximum when $2\pi(t - \tau/2)/T = \pi/2$ or $t = T/4 + \tau/2$ and $Z_{\max} = Z_*(\sin \pi\theta/\pi\theta)$

As $\theta = t/\tau$ tends to zero, $(\sin \pi\theta/\pi\theta)$ equals 1.0 and we get the solution for instantaneous closure $Z_{\max} = Z_*$.

EXAMPLE 10.9 For the data given in Example 10.8, find the minimum time in which the full supply of 15 m^3/s is to be shut off in order to limit the maximum height of the water level in the surge tank to 10 m above the reservoir level.

We have already found $T = 203.22$ seconds and $Z_* = 12.61$ m.

From (10.19b), $Z_{max} = Z_*(\sin \pi\theta/\pi\theta)$ and this has to be limited to 10 m. Hence $10 = 12.61(\sin \pi\theta/\pi\theta)$ or $(\sin \pi\theta/\pi\theta) = 10/12.61 = 0.7930$.

Solving for θ, we obtain $\theta = 0.366$, hence the minimum time of closure is $\tau = \theta \cdot T = 0.366 \times 203.22 = 74.38$ seconds

10.17 Instantaneous Closure of Valve Considering Tunnel Resistance

It can be shown that the maximum rise $z_m = Z_{max}/Z_*$ is given by the expression

$$(1 - 2p_0 z_m) = e^{-2p_0(2p_0 + z_m)} \tag{10.20a}$$

and the minimum level during downward surge is given by $z_n = Z_{min}/Z_*$

$$(1 + 2p_0 z_n) - (1 + 2p_0 z_m) \quad e^{-2p_0(z_m - z_n)} = 0 \tag{10.20b}$$

where $p_0 = i_0 L/Z_* =$ head loss in the tunnel for velocity V_{t0}/max. surge height for instantaneous closure.

EXAMPLE 10.10 In the previous example, the head loss in the tunnel is found to be 3.78 m corresponding to a discharge of 15 m^3/s. Compute the maximum and minimum water levels in the surge tank sequel to a total instantaneous closure of valve.

$$i_0 L = h_f = 3.78\text{ m}; \quad p_0 = 3.78/Z_* = 3.78/12.61 = 0.3$$

The first rise will be the maximum and likewise the first fall will register the minimum level in the surge tank. Subsequently the oscillations will be dampened out due to tunnel boundary resistance as shown in Fig. 10.20.

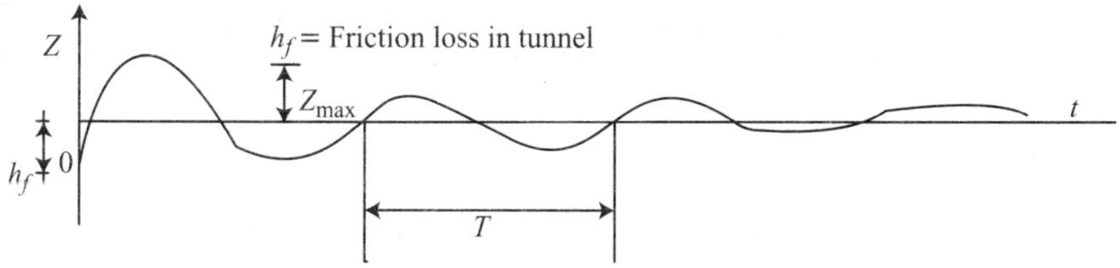

Fig. 10.20 Damped oscillations in a surge tank

Equation (10.20a) gives the maximum first rise

$$1 - 2p_0 z_m = e^{-2p_0(p_0 + z_m)} \text{ and } p_0 = 0.3$$

Hence

$$\ln(1 - 0.6z_m) = -0.6(0.3 + z_m)$$

By trial and error, $z_m = 0.811$; $Z_{max} = 0.811 \times 12.61 = 10.22$ m

Equation (10.20b) gives the first minimum fall

$$(1 + 2p_0 z_n) = (1 + 2p_0 z_m)e^{-2p_0(z_m - z_n)}; \qquad p_0 = 0.3; \qquad z_m = 0.811$$

$$\ln(1 + 0.6z_n) = \ln(1 + 0.6 \times 0.811) - 0.6(0.811 - z_n) = (-0.0901 + 0.6z_n)$$

By trial and error, $z_n = -0.61$

The minimum water level in the surge tank, $Z_{\min} = -7.69\,\text{m}$

The same procedure may be followed to evaluate the successive maximum rise and fall of the water level indicating the degree of damping of oscillations in the surge tank.

10.18 Numerical Integration

The equation of continuity when differentiated yields

$$\frac{dV_t}{dt} = \left(\frac{A}{A_t}\right)\left(\frac{dV_s}{dt}\right) + \left(\frac{1}{A_t}\right)\left(\frac{dQ}{dt}\right)$$

From the dynamic equation,

$$\frac{dV_t}{dt} = -\left(\frac{g}{L}\right)(Z \pm iL)$$

Hence,

$$\frac{dV_s}{dt} = -\left[\left(\frac{g}{L}\right)\left(\frac{A_t}{A_s}\right)(Z \pm iL) - \left(\frac{1}{A_s}\right)\left(\frac{dQ}{dt}\right)\right]$$

We also have $V_s = dZ/dt$; we replace the differentials by finite differences

$$\Delta V_s = -\left(\frac{1}{A_s}\right)\left[\Delta Q + \left\{\left(\frac{gA_t}{L}\right)(Z \pm iL)\right\}\Delta t\right] \text{ and} \tag{10.21}$$

$$\Delta Z = V_s \Delta t \tag{10.22}$$

The initial values of V_t, Q, and $I_0 L$ are known. Depending on the manoeuvre of closure or opening, the variations in Q can be known with reference to time. The initial value of V_s can be determined from the continuity equation. Choosing a suitable value for the time interval Δt, ΔZ can be calculated by equation (10.22)

The value of $Z + \Delta Z$ is used in equation (10.21) to find ΔV_s and hence V_t at the end of the time interval Δt. The procedure is repeated at successive intervals of time.

The accuracy of this method will depend on the time interval Δt chosen. Arithmetical errors are cumulative in this method and the values should be checked at frequent intervals by keeping a running plot of V_s and Z which should be smooth curves.

EXAMPLE 10.11 Assuming the data of the previous examples, calculate the minimum level of water in the surge tank sequel to an instantaneous opening of the valve admitting a flow of $15\,\text{m}^3/\text{s}$.

We have $L = 2400$ m; $A_t = 9\,m^2$; $A_s = 38.5\,m^2$; loss of head, $i_0 L = 3.78$ m

Due to an instantaneous opening of valve admitting a flow of 15 m^3/s to the turbines, the continuity equation for the system is

$$A_t V_t = A_s V_s + 15 \text{ or } 9V_t = 38.5V_s + 15$$

$$\Delta V_s = -\left(\frac{1}{38.5}\right)[0 + \left(9.81 \times \frac{9}{2400}\right)(Z + iL)\Delta t$$

$$= -0.0009554(Z + iL)\Delta t$$

Choosing a time interval of 5 seconds for each step:

Initial values $Z = 0$; $i_0 = 0$; $V_t = 0$

$$V_s = -\frac{15}{38.5} = -0.3895 \text{ m/s and } \Delta Z = -0.3895 \times 5 = -1.948 \text{ m}$$

Hence $Z + \Delta Z = 0 - 1.948 = -1.948$ m

$$\Delta V_s = 0.0009554(-1.948 + 0) \times 5 = +0.0093 \text{ m/s}$$

Step 2: $V_s = -0.3895 + 0.0093 = -0.3802$ m/s; $V_t = +0.04$ m/s

The head loss is proportional to square of the tunnel velocity. When $V_{t0} = 15/9$ m/s, the head loss is 3.78 m. Hence when $V_t = 0.04$ m/s, the head loss is $3.78(0.04/1.667)^2 = 0.0022$ m.

$$\Delta Z = -0.3802 \times 5 = -1.901 \text{ m and } Z = -1.948 - 1.901 = -3.849 \text{ m}$$

$$\Delta V_s = -0.0009554(-3.849 + 0.0022) \times 5 = 0.0184 \text{ m/s.}$$

Similarly, subsequent steps are evaluated till the final value of the tunnel velocity is reached and the results are tabulated below.

t	V_s	ΔZ	Z	V_t	h_f	ΔV_s
0	−0.3895					
5	−0.3802	−1.948	−1.948	0.04	0.0022	0.0093
10	−0.3618	−1.901	−3.849	0.12	0.0196	0.0184
15	−0.3349	−1.809	−5.658	0.23	0.0745	0.0269
20	−0.3002	−1.675	−7.333	0.38	0.1995	0.0347
25	−0.2589	−1.501	−8.834	0.56	0.4267	0.0413
30	−0.2126	−1.295	−10.129	0.76	0.7796	0.0463
35	−0.1629	−1.065	−11.192	0.97	1.2800	0.0497
40	−0.1116	−0.815	−12.007	1.20	1.9590	0.0513
45	−0.0609	−0.558	−12.565	1.41	2.6900	0.0507
50	−0.0113	−0.305	−12.870	1.62	3.5700	0.0486
51.5	−0.0002	−0.014	−12.884	1.67	3.7800	0.0111

Hence the minimum level of water in the surge tank will be − 12.884 m.

10.19 Types of Surge Tanks

Simple surge tanks are sometimes uneconomical due to the topography of the area where the surge tank is to be located. In order to limit the height of the surge tank, several designs are available.

The simple surge tank may have expansion chambers to enhance the storage capacity as shown in Fig. 10.21*a*.

A surge tank with a restricted orifice as shown in Fig. 10.21*b* provides additional resistance to flow at entry. During closure of a valve, the resistance to flow at entry into the surge tank will be a maximum when Z is at its minimum and gradually reduces as the water level in the surge tank rises. When Z reaches its maximum, the resistance becomes zero. At any height Z in between, the resistance to flow at entry will be a function of the velocity at entry and together with the height Z in the surge tank greatly helps in the retardation of flow in the tunnel resulting in a reduction of the maximum height of water level reached in the tank. In an analogous manner when the water level decreases in the surge tank due to sudden opening of the valve in the penstock, the resistance to flow in the orifice helps the acceleration of the flow in the tunnel and hence the final minimum height in the surge tank will not be as low as in the absence of the orifice.

A differential type of surge tank (Fig. 10.21*c*) with a central concentric riser pipe connected to the outer chamber through port holes located at the bottom will also reduce the amplitude of oscillations in the surge tank. The diameter of the central riser is approximately equal to the tunnel diameter. The water level rises rapidly in the smaller central chamber and spillage will occur over the brim of the central chamber into the outer chamber. Further a differential head occurs between the two chambers and water enters the outer chamber gradually through the bottom port holes. The spillage is an important feature of the differential type surge tank. During draw-down conditions, the action is similar to that of the orifice tank.

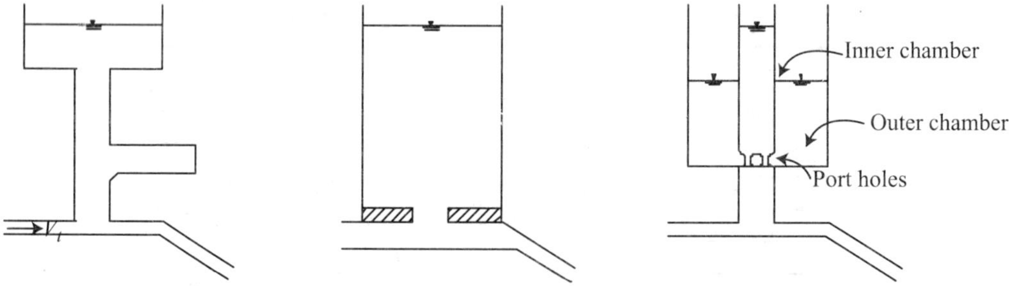

a Surge tank with expansion chambers

b Surge tank with restricted orifice

c Differential surge tank

Fig. 10.21 Different types of surge tanks

EXERCISES

In the following problems, assume $\rho = 998\ \text{kg/m}^3$ and $\nu = 1 \times 10^{-6}\ \text{m}^2/\text{s}$

10.1 A cast-iron pipe 15 cm in diameter and 15000 m long connects two reservoirs with a head difference of 20 m. There are two 90° smooth bends and four 45° elbows and a fully open gate valve in the pipe line. Calculate the discharge.

10.2 Four pipes each of length 200 m but of varying diameters are in series. The first pipe is 100 cm in diameter and the successive pipes are of diameters 75 cm, 100 cm, and 50 cm. If the last pipe discharges into the atmosphere at a point 10 m below the reservoir level, find the discharge. $C_c = 0.62$

10.3 Find the size of a supply pipe of mild steel from an overhead tank 5 km away from a town with a population of 1,50,000. The average rate of water per head is 150 litres per day. The head loss in the pipe is to be restricted to 10 m.

10.4 In the sketch of the pipe system illustrated in Example 10.2, the discharge has to be increased to $0.10\ \text{m}^3/\text{s}$. Find the required difference in water level. (use the same f-values)

10.5 A long pipe line laid between two reservoirs is to be replaced by another pipe of the same material which will increase the velocity by 10%. What is the size of the new pipe?

10.6 A pipe 5 cm in dia. is 60 m long and the velocity of flow is 2.5 m/s. What loss of head will be saved if the central 30 m length of pipe is replaced by a 7.5 cm dia. pipe., the change of section being abrupt. $C_c = 0.62$. The discharge remains unchanged.

10.7 A 50 cm dia. pipe carrying $0.16\ \text{m}^3/\text{s}$ of water divides into parallel pipes 20 cm in dia. and 25 cm in dia. which rejoin. If the 25 cm branch is 1500 m long, how long must the 20 cm pipe be for the flow to divide equally?

10.8 The total flow in three parallel pipe systems is $4\ \text{m}^3/\text{s}$. The pipe diameters and lengths of the three pipes are: $d_1 = 1.0$ m, $L_1 = 1000$ m; $d_2 = 0.5$ m, $L_2 = 800$ m; $d_3 = 0.75$ m, $L_3 = 750$ m. Find the discharge in the three pipes.

10.9 A pipe of 30 cm diameter and 1200 m long ($f = 0.01$) is carrying a discharge of $0.5\ \text{m}^3/\text{s}$ connecting two reservoirs. In order to increase the discharge, another pipe of the same diameter but 700 m long ($f = 0.008$) is connected to the old line at a point 700 m from the exit. Find the total discharge assuming that the difference in the reservoir level remains the same.

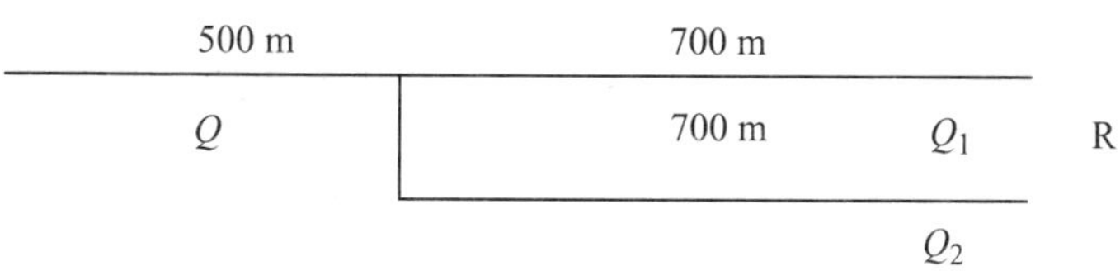

Figure for Exercise 10.9

10.10 A siphon pipe of diameter 25 cm connects two reservoirs with a difference in level of 10 m. The inlet length of the siphon is 400 m and the downward limb is 600 m (total length 1000 m). Determine the discharge and the level of the highest point of the siphon so that the pressure is not less than 2 m of water absolute. $f = 0.02$

10.11 A high-level reservoir feeds two low-level reservoirs. The pipe connections are as shown in the figure. Determine the discharge. AB = 5 km long and 75 cms in diameter; BC = 3 km long and 75 cms in diameter; BD = 4 km long and 30 cm in diameter. The difference in elevation between the reservoir level at A and the junction B is 10 m. The difference in level between junction B and reservoir level at C is 9 m. The difference in level between junction B and reservoir level at D is 6 m.

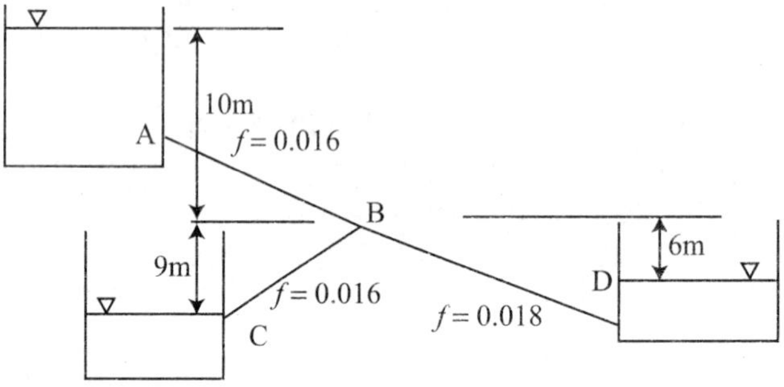

Figure for Exercise 10.11

10.12 Two reservoirs A and B discharge through pipes each 50 cm in diameter and 1.5 km long to a junction J. From junction J, the discharge is carried to a third reservoir, C by a pipe 75 cm in diameter and 0.5 km long. The surface level of reservoirs A and B are 15 m and 10 m above the reservoir level at C. Find the discharges in each of the pipes. Assume $f = 0.01$ for all pipes.

10.13 Determine the discharges in the two pipe networks shown below.

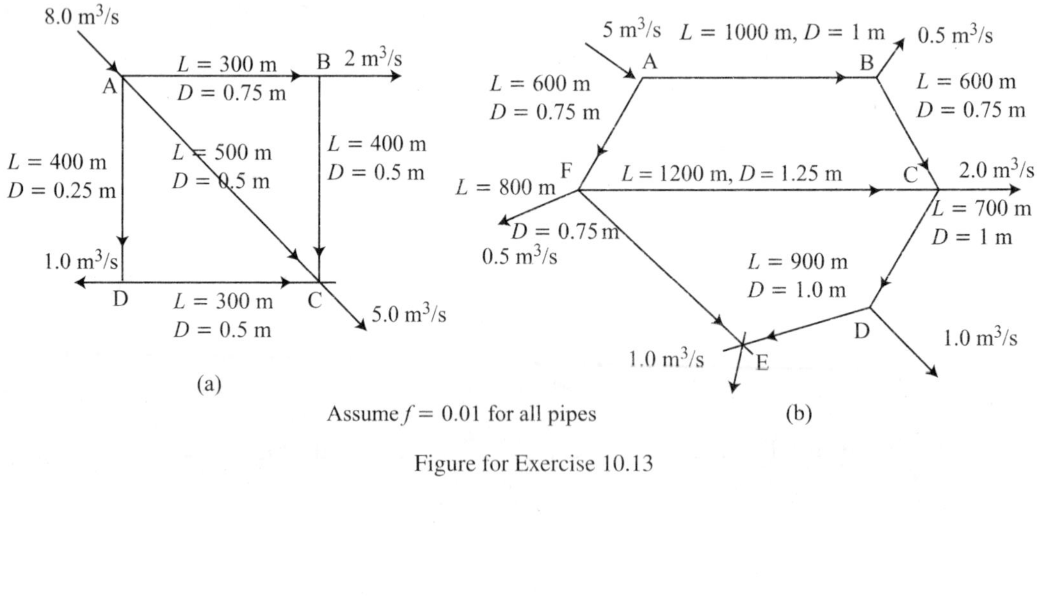

Figure for Exercise 10.13

10.14 Two sites C and D both at sea level are to be provided with water from a large storage reservoir with the water surface 25 m above sea level. Both sites should receive the same discharge. The diameters and pipe lengths are as follows
AB: 8 cm dia. and 200 m long; BC: 5 cm dia. and 200 m long; BD: 5 cm dia. and 20 m long. A pump may be necessary in pipe length BC. If so, find the power of the pump if C is to receive the same discharges as D. $f = 0.015$

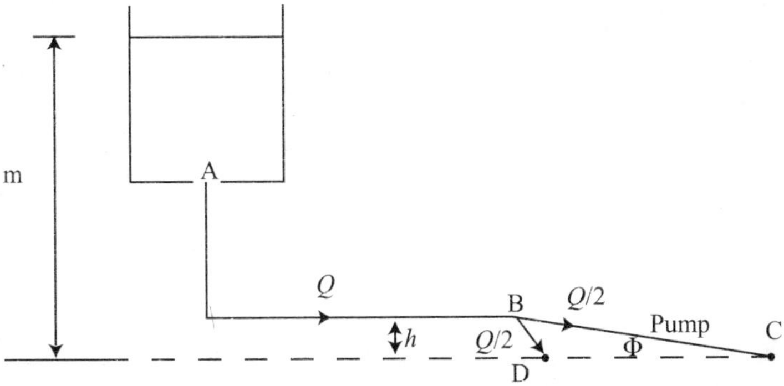

Figure for Exercise 10.14

10.15 Water flows from a reservoir through a pumping system shown in the figure. Determine the flow through the piping system through each pipe and the pressures at points 2 and 3. The elevations of the points are shown in the figure.

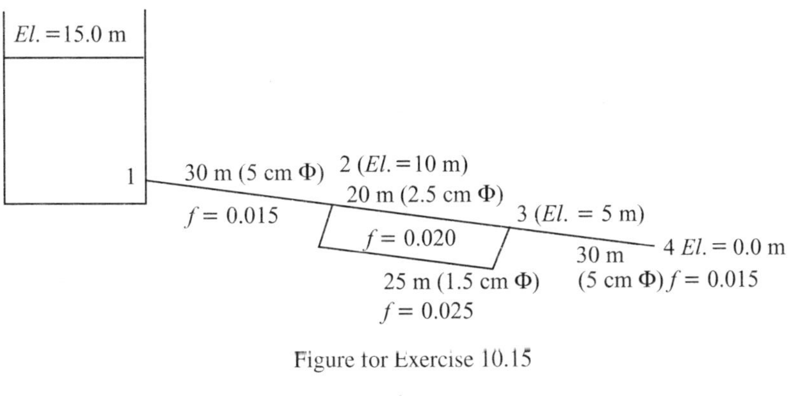

Figure for Exercise 10.15

10.16 Water is flowing at the rate of 0.5 m^3/s through a 2000 m pipe of horizontal 50 cm dia. The thickness of the pipe is 1 cm. The loss of head in the pipe is 20 m between the reservoir and the outlet. If the outlet valve is suddenly closed, determine (a) the initial change in velocity head, (b) the time required for the pressure wave to travel to the reservoir and back, (c) the maximum pressure head that will occur in the pipe and (d) changes in pressure at mid-point as a function of time.

10.17 In the preceding problem, if the valve is closed linearly, what must be the minimum duration of time for valve closure if the pressure head is not to exceed 75% of the maximum head due to sudden closure.

10.18 A simple surge tank 10 m in diameter is located at the end of a 2 m diameter pipe of length 2000 m from the reservoir. If the discharge is 25 m^3/s, find the maximum surge height and the period oscillation following sudden closure of valve ignoring all losses in the system. Find the minimum time for closure of valve in order to limit the surge height to 10 m.

11 Forces on Submerged Bodies

11.1 Introduction

We define the force on a submerged body in the direction of the flow as drag force and the force exerted on the body in a direction perpendicular to the flow direction as the lift force. We have seen in the chapter on irrotational flow that the drag force on a body immersed in an ideal fluid is zero as there are no shear forces and the pressure distribution on either side of the body on an axis normal to the flow is symmetric. However lift forces can exist if there is circulation (of strength, Γ) around the body and is given by $L = \rho U \Gamma$, where U is the uniform velocity of the flow and ρ is the density of the fluid.

Further we have seen in the chapter on laminar flow, that the force on a sphere of diameter d immersed in a fluid of viscosity μ is $F = 3\pi \mu U d$, where U is again the uniform velocity of the stream. We had assumed that the flow is very slow (creeping motion) and that the inertial terms could be neglected. There was no flow separation from the body and the forces were partly due to viscous shear on the surface of the body as well as the pressure due to their asymmetric distribution on either side of the body. At higher velocities, inertial effects cannot be ignored and in fact they play a crucial part in determining the flow pattern around the body and the forces on the body. We shall first examine the role of the inertial effects and the separation of the boundary layer from the body.

11.2 Boundary Layer Separation

For a flat plate at zero angle of incidence, the boundary layer grows in thickness with increasing distance from the leading edge. Since the pressure gradient is zero, the growth of the boundary layer may continue indefinitely. But if the plate is inclined to the flow direction, the free stream velocity outside the boundary layer is given by $U = cx^m$, where the exponent m is related to the angle of inclination of the plate θ to the x-axis by the equation $\theta = \pi m/(m+1)$. It has been found that separation of the flow on the divergent side of the plate occurs when $m = -0.09041$, corresponding to an angle of approximately 17.85°. The separation is due to adverse pressure gradient. The pressure increases in the direction of flow thus retarding the flow in the boundary layer.

Let us now consider the development of a boundary layer along a curved boundary as illustrated in Fig. 11.1

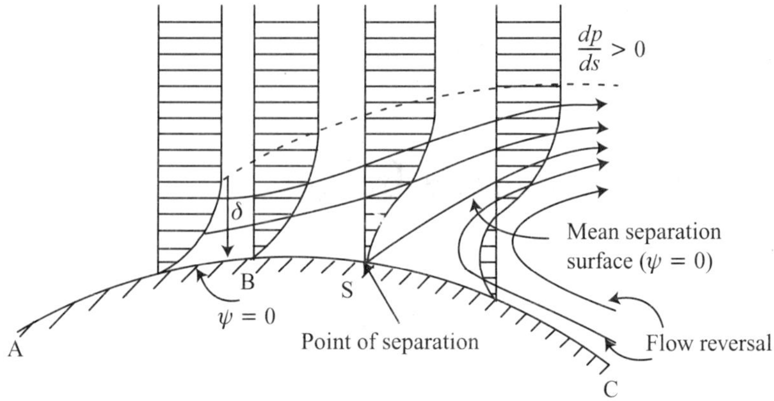

Fig. 11.1 Boundary layer separation

Along the initial curved length AB, the flow outside the boundary layer accelerates and pressure accordingly decreases in the direction of flow. We have also seen that the pressure outside the boundary layer is impressed on the boundary layer and the pressure inside the boundary layer is sensibly constant and equal to the pressure outside the boundary layer. Hence there is a favourable pressure gradient which tends to offset the retarding effects of the boundary shear. The boundary layer remains thin and its growth is not rapid. This is true for all accelerated systems. However, the conditions downstream of B are different in the sense that the expansion in flow area causes a deceleration of flow resulting in an increase of pressure in the direction of flow. The pressure gradient is referred to as an adverse one. The fluid particles in the boundary layer beyond B have now to overcome two types of forces: (a) forces due to boundary resistance and (b) forces due to adverse pressure gradient. It is evident that the available momentum in the boundary layer is low. This momentum is rapidly consumed in overcoming these two forces. The flow in the boundary is further decelerated and the boundary layer thickens rapidly by virtue of continuity in mass flow in the boundary layer. Further the velocity gradient at the boundary progressively reduces until at some point, S as shown in the figure, the gradient $\partial u/\partial n = 0$. Under such conditions of flow, the flow obviously being unable to proceed along the boundary, detaches itself from the boundary at S which is called the point of separation. The boundary shear is thus obviated.

A region of low pressure exists around the point of separation as evidenced by the curvature of the stream lines. The geometry of the detached surface must be such that the flow now occurs with a modified pressure field with minimum expenditure of energy. It would be difficult to analytically find the geometry of the separated surface owing to the complex interaction of viscous and turbulent forces in the separated region. The separated surface is not a free stream surface and its form is dependent on the Reynolds number of the primary flow and the boundary geometry downstream of the point of separation. The separated surfaces may initially move away from the boundary surface and later move again towards it and re-attach itself to the boundary enclosing a certain mass of fluid between the separated surface and the boundary. A quasi-stable eddy containing a slow rotating mass of fluid is formed in which region the pressure is low. A reversal of flow occurs within the eddy. The separated surface is a free shear surface (as distinct from the boundary shear) across which steep velocity gradients exist and along which maximum production of turbulence takes place. The turbulent shear stresses transfer momentum (across the free shear surface) from the primary flow to the eddy zone thus driving the flow in the eddy. Because the entrainment process is continuous along the free shear surface, a return flow must be established by the formation of an eddy pocket. The energy derived from the primary flow is lost in the eddy zone. The point of re-attachment of the separated surface to the boundary is a zone of relatively high pressure compared to the pressure within the eddy so that the reversal of flow in the eddy takes place from a region of high pressure to one of low pressure. The separation and the eddy pocket formed behind the body is shown in Fig. 11.2 along with a qualitative pressure distribution around the body.

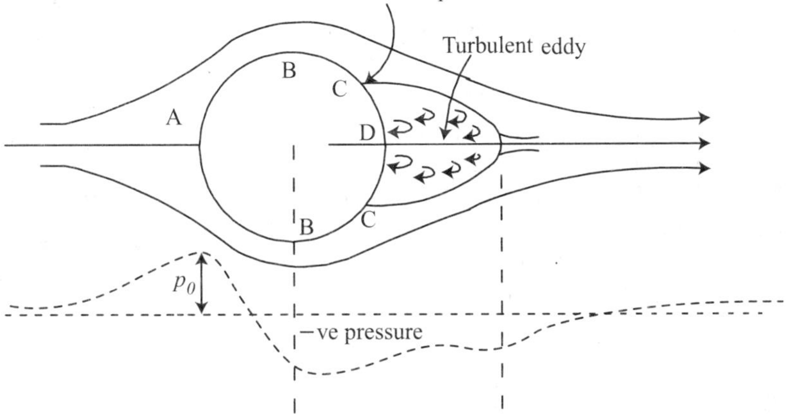

Fig. 11.2 The turbulent wake behind a body

Once the incidence of separation occurs, the pressure distribution around the object becomes modified and the point of separation moves to its final position of equilibrium. The analytical prediction of the point of separation on a smooth curved surface is in general difficult. The boundary layer equations are valid up to the point of separation beyond which the boundary layer becomes so thick that the assumptions made in the formulation of the boundary layer equations are invalidated. Flow separation displaces the external potential flow around the body to a considerable extent and the pressure distribution in the separated zone must be determined by experiments since the conditions in the primary flow and characteristics of the secondary flow are mutually dependent. It may be seen that $\partial \overline{u}/\partial n = 0$ at the point of separation. At the wall, the velocity components $\overline{u} = \overline{v} = 0$ and the boundary layer equation becomes

$$\frac{d\overline{p}}{dx} = \mu \left(\frac{\partial^2 \overline{u}}{\partial y^2}\right)_{y=0}$$

In case of an adverse pressure gradient, $dp/dx > 0$ and $(\partial^2 u/\partial y^2)_{y=0}$ is positive. However as we move into the free stream, we find that at the edge of the boundary layer, the velocity gradient decreases with increase in distance from the boundary. Hence $\partial^2 u/\partial y^2$ is negative in this region.

There must be a point of inflexion in the velocity profile between $y = 0$ and $y = \delta$. We have already indicated in an earlier chapter that the existence of a point of inflexion in the velocity profile has an important bearing on the stability of the flow in the boundary layer. An inflexion point occurs in the presence of an adverse pressure gradient and it may be said that the boundary layer separation does not occur in the absence of an adverse pressure gradient. It is not necessary, however, that separation should always occur whenever an adverse pressure gradient occurs.

A laminar boundary layer separates earlier than a turbulent one due to the fact that in the latter case, turbulence leads to a rapid exchange of momentum from the primary flow to the retarded flow in the boundary layer thus accelerating the fluid particles close to the boundary. Hence the fluid particles close to the boundary are kept in motion over a longer distance against the influence of adverse pressure gradients than is possible in the case of a laminar boundary layer. For example, the laminar boundary layer around a sphere separates near the equator. If the boundary layer is turbulent, the point of separation moves downstream. There is a distinct advantage in the shift of the point of separation downstream. If the separation is delayed by some means, the surface area behind the separated zone which is subjected to negative pressure is reduced. Thus the pressure drag on the body decreases. The reduction in pressure drag may be achieved by roughening the surface of the boundary near the nose region of the body (a golf ball is dimpled to cause roughening of the surface which delays separation thus reducing the drag so that the ball can travel longer distances) or by introducing other types of triggering devices which cause the laminar boundary layer to become turbulent. A similar shift of the separation point is noticed when the intensity of turbulence present in the approach flow increases. Other methods of reducing drag are preventing or delaying separation by either suction of the retarded fluid in the boundary layer or by injecting a stream or jet of high velocity fluid into the boundary layer as shown in Fig. 11.3.

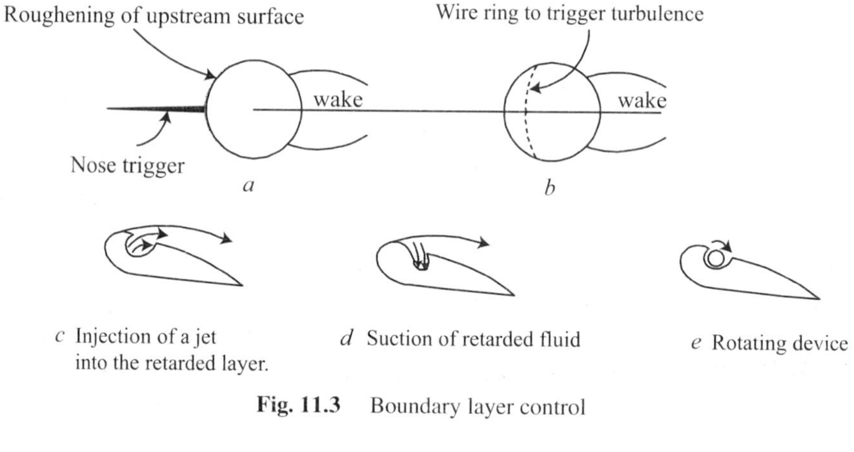

Fig. 11.3 Boundary layer control

11.3 Laminar Wake Behind a Cylinder

The wake behind bluff bodies is of considerable interest, and the particular case of past cylinders has been studied extensively ($\mathbf{R} = Ud/\nu$). Flow separation does not occur for very low Reynolds numbers ($\mathbf{R} < 1$). For Reynolds numbers less than about 40, the flow is steady and laminar. Beyond $\mathbf{R} > 1$, flow separation occurs at a position on the cylinder that moves forward as the Reynolds number increases. The flow in the separated region takes the form of two symmetric eddies attached to the cylinder on either side of the axis of symmetry as shown in Fig. 11.4.

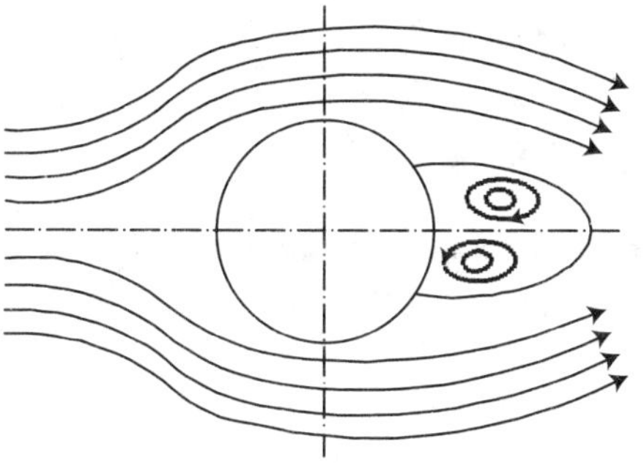

Fig. 11.4 Standing eddies at low Reynolds numbers

In the range of Reynolds numbers $40 < \mathbf{R} < 60$, the flow in the wake is essentially unstable for small disturbances. When the Reynolds number exceeds 30, the two symmetrical vortices on either side of the axis tend to stretch and become unstable. They get detached from the boundary and are washed down. Two more vortices form behind the body replacing the old ones and the process continues. For $\mathbf{R} > 60$, separation takes place alternatively on the two sides of the cylinder—the flow clinging to one side while separation takes place on the other side. The formation of these alternative vortices may be clearly seen even at large distances from the body—the vortices having a regular pattern moving in clockwise and anti-clockwise directions on either side of the axis of symmetry. This type of flow with alternate formation of vortices is called a Karman Vortex street. The formation of these vortices is shown in Fig. 11.5.

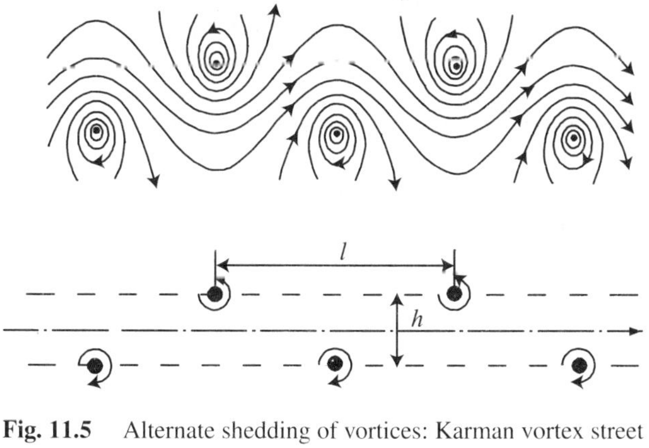

Fig. 11.5 Alternate shedding of vortices: Karman vortex street

This pattern is seen up to about $\mathbf{R} \sim 5000$ though the distance over which the vortex street is clearly seen gets reduced progressively and the vortices become indistinct after some distance.

For example, for **R** > 1000, the vortex street formation is not clearly observed beyond a distance of 10 diameters from the body. Karman showed that only a stable form of double row of vortices exists when the relative spacing $h/l = 0.281$, where h and l are the transverse and longitudinal spacing between alternate vortices as shown in Fig. 11.5. The vortex street moves with a velocity u_s which is less than that of the main stream, U. Karman obtained the average value of the drag F_D per unit length of the cylinder as

$$F_D = \rho U^2 h \left[2.83 \left(\frac{u_s}{U} \right) - 1.12 \left(\frac{u_s}{U_0^2} \right) \right] \tag{11.1}$$

The width h and u_s/U must be found experimentally. The frequency with which vortices are shed behind a circular cylinder expressed non-dimensionally by the Strouhal number nD/U which is a unique function of the Reynolds number. Taylor has given the expression for the frequency:

$$\mathbf{S} = \frac{nD}{U} = 0.198 \left[1 - \frac{19.7}{\mathbf{R}} \right] \tag{11.2}$$

The alternate shedding of vortices exerts a periodic lateral force on the body. If the frequency of vortex shedding coincides with the natural frequency of the body, resonance might occur in the lateral direction and may have a significant influence in the design of long suspension cables and the like. Since the Strouhal number is dependent on the body shape and the Reynolds number, it stands to reason that the Strouhal number should be a function of the coefficient of drag which is also dependent on the same two parameters. It has been found that

$$\mathbf{S} = \frac{0.21}{C_D^{0.75}}$$

The boundary layer around the smooth cylinder remains laminar up to $\mathbf{R} = 5 \times 10^5$ when the point of separation is approximately at an angle of 81° on the forward side of the cylinder. For $\mathbf{R} > 5 \times 10^5$, the boundary layer becomes turbulent and periodic shedding of vortices ceases since the periodic component is destroyed by turbulence.

11.4 Drag on Immersed Bodies

Whenever there is relative motion between the body and the fluid surrounding it, the body will experience a force in the direction of relative motion. If the flow is asymmetric about the body, it will also experience a force in a direction transverse to the direction of motion. The former is known as the drag force and the latter is designated as the lift force. The drag and lift forces are expressed in the Newtonian form

$$\begin{aligned} F_D &= C_D A \left(\frac{1}{2} \rho U^2 \right) \\ F_L &= C_L A \left(\frac{1}{2} \rho U^2 \right) \end{aligned} \tag{11.3}$$

where C_D and C_L are dimensionless coefficients of drag and lift respectively and A is a characteristic projected area of the body in the direction of forces measured. C_D and C_L are functions of Reynolds number and the shape of the body.

The drag force arises from two causes:

(a) boundary resistance or viscous drag which is obtained by the summation of the component of the boundary shear force acting over the entire surface area of the body in the direction of the motion;

(b) pressure drag or form drag obtained by the component of the integrated pressure force acting on the body in the direction of motion.

The sum of the viscous and form drags is known as the profile drag. The resistance experienced by a flat plate kept parallel to the free stream (zero angle of incidence) is entirely due to boundary shear. In case of bluff bodies, where flow separation occurs, there is a considerable pressure difference between the fore and aft sides of the body and the form drag predominates over the viscous drag. If the body is streamlined, the pressure drag is generally low as the geometry of the body is designed in such a way that adverse pressure gradients are kept within limits so that no flow separation takes place. However, for streamlined bodies, both surface drag and form drag are to be assessed accurately.

11.4.1 Drag on Cylinders and Spheres

In the preceding section, it has been explained that the point of separation depends on the Reynolds number and the level of the initial turbulence in the main stream. Since the pressure drag of a bluff body is dependent on the location of the points of separation, we shall discuss the variation of the coefficient of drag for circular cylinders and spheres as a function of the Reynolds number. Initially, the coefficient of drag for a sphere is inversely proportional to **R** which is evident from Stokes law (Fig. 11.6)

$$F = 3\pi\mu U d = C_D\left(\frac{\pi d^2}{4}\right)\frac{\rho U^2}{2} \quad \text{and hence} \quad C_D = \frac{24}{\mathbf{R}} \tag{11.4}$$

A logarithmic plot shows this relationship as a straight line and is valid up to $\mathbf{R} = 0.1$, though it can be extended up to $\mathbf{R} = 0.2$.

In the range $1 < \mathbf{R} < 4000$, the value of C_D continuously decreases with increase in the Reynolds number. For a sphere, $C_D = 0.4$ at $\mathbf{R} = 4000$. Thereafter a slight rise in C_D is registered as reaching a value of 0.50 in the approximate range $2 \times 10^4 < \mathbf{R} < 2 \times 10^5$. Then as the laminar boundary layer becomes turbulent, the point of separation moves forward accompanied by a sudden decrease in drag and C_D reaches a low value of 0.17. This valuc of **R** near 2×10^5 is known as the critical Reynolds number. For higher Reynolds numbers ($\mathbf{R} > 4 \times 10^5$), as the skin drag increases with increasing turbulence in the boundary layer, C_D shows a slight increasing trend. Figure. 11.6 shows the variation of C_D with **R** for bodies of revolution. The figure also shows the variation of the coefficient of drag for a sphere.

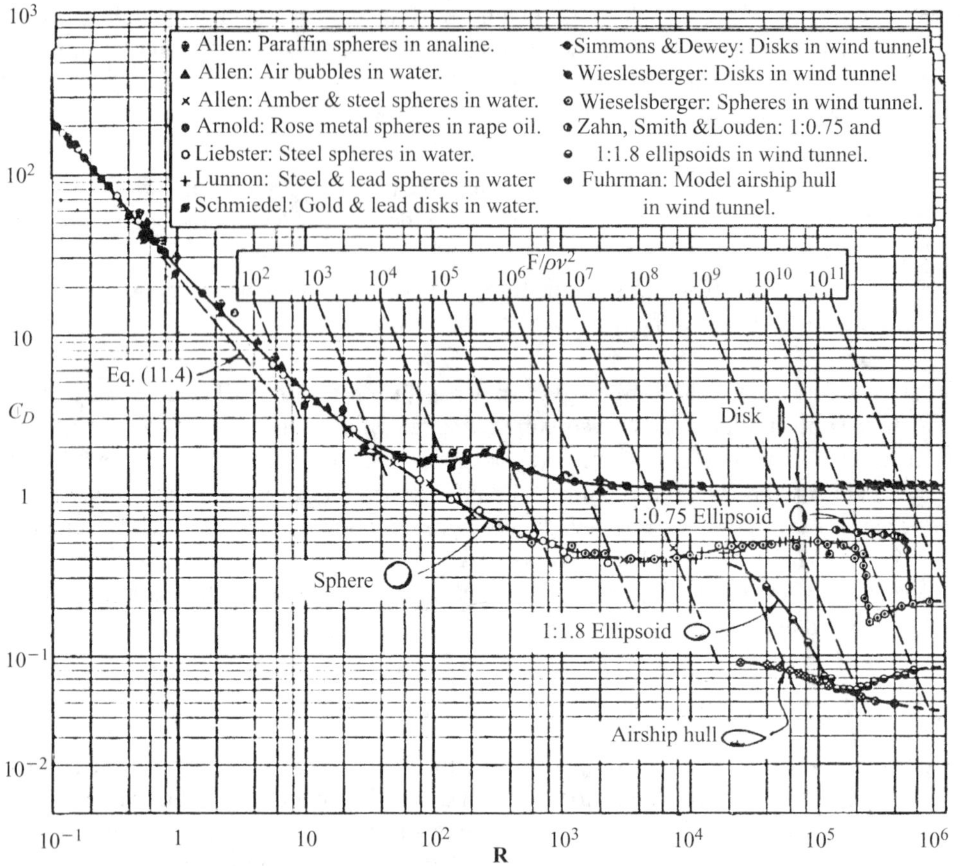

Fig. 11.6 Drag coefficient for bodies of revolution

This diagram also helps us to determine the velocity of a body of revolution directly, as it links $F/\rho\nu^2$ with C_D and $\mathbf{R}$. We have

$$C_D = \frac{F}{A}\left(\frac{\rho U^2}{2}\right) \quad \text{and} \quad \mathbf{R} = \frac{UD}{\nu}$$

Elimination of U from the above two relationships gives

$$C_D = \left(\frac{1}{\mathbf{R}^2}\right)\left(\frac{8}{\pi}\right)\left(\frac{F}{\rho\nu^2}\right) \tag{11.5}$$

In case of falling bodies, the terminal velocity can be easily determined if the intersection of the $F/\rho\nu^2$ line with the C_D: $\mathbf{R}$ curve is found for a given shape of the body. The intersection point gives the values of C_D and $\mathbf{R}$. Then velocity $U = \sqrt{2F/C_D A\rho}$ or $U = \nu\mathbf{R}/D$

A similar trend is shown in the variation of C_D with $\mathbf{R}$ for a cylinder as shown in Fig. 11.7 which shows the drag coefficient for two-dimensional bodies.

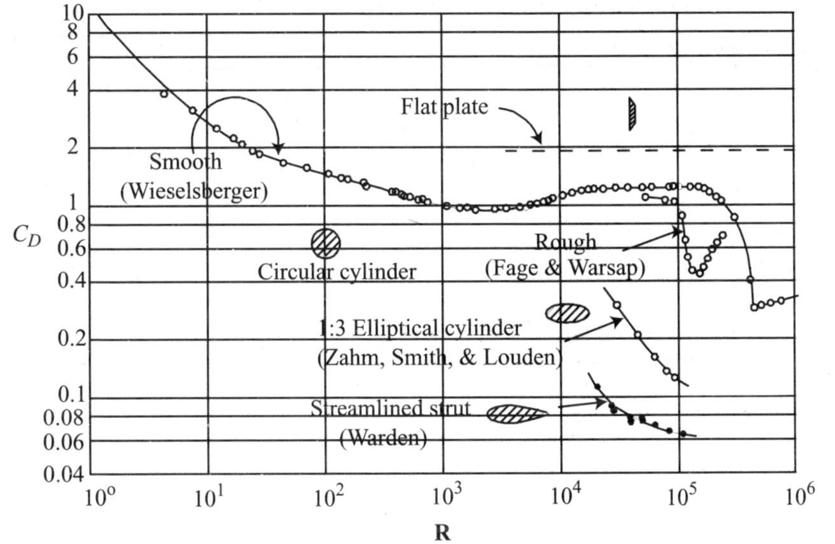

Fig. 11.7 Drag coefficient for two-dimensional bodies

Table 11.1 gives the coefficients of drag for different body shapes.
A study of the table shows certain interesting features.

(1) As the L/D ratio of a flat plate kept normal to a fluid stream increases, the drag coefficient increases as the stagnation pressure line upstream increases in length and the flow becomes more and more two-dimensional. For small ratios of L/D, the flow is more three-dimensional in character as the flow around the edges serve to improve the pressure in the wake than for large ratios of L/D. For the same reason, a rectangular plate experiences greater drag than a square plate of the same area. However, if a rectangular or square plate is cut into several rectangular strips or square strips respectively, the sum of the drags of all the cut pieces is greater than that of the uncut plate even though the total areas may be the same. Note that there is a stagnation point for each of the cut piece.

(2) For a hemisphere held with the hollow side facing the flow, not only is the region of stagnation upstream larger but the curvature of the separation surface along the rim of the hemisphere is more pronounced than the corresponding characteristics that occur when the hollow side is facing away from the flow. The drag in the former case is therefore larger than in the latter case.

(3) A thin plate held parallel to the flow experiences a surface drag and no form drag; but when the same plate is turned through 90° and held normal to the flow, the drag is primarily due to form effect. The coefficient drag does vary with the inclination of the plane to the flow direction. Generally, a different angle of attack is considered a different geometry and changes in the value of C_D are studied as a function of the Reynolds number. In a general sense, the value of C_D is a function of Reynolds number and the angle of attack, θ: $C_D = f(\mathbf{R}, \theta)$.

Table 11. 1 Drag coefficient for different shapes of bodies

Body shape	L/D	R	C_D
Flat plate kept parallel to flow		(a) Laminar range	$1.328\ \mathbf{R}_x^{-1/2}$
		(b) Turbulent $\mathbf{R} < 10^7$	$0.074\ \mathbf{R}_x^{-1/5}$
Flat plate normal to flow	1	$\mathbf{R} > 10^3$	1.16–1.18
	5		1.20
	10		1.30
	20		1.50
	30		1.60
	Infinite		1.90–1.95
Circular disk		$\mathbf{R} > 10^3$	1.12–1.17
Circular cylinder Axis perpendicular to flow	1	$10^4 < \mathbf{R} < 1.5 \times 10^5$	0.63
	5		0.74–0.80
	10		0.83
	20		0.90–0.93
	30		1.00
	Infinite		1.20
	5	$\mathbf{R} > 5 \times 10^5$	0.35
	Infinite		0.33
Sphere		$\mathbf{R} < 1$	$24/\mathbf{R}$
		$10^3 < \mathbf{R} < 2 \times 10^5$	0.47–0.50
		$\mathbf{R} > 3 \times 10^5$	0.20
Hollow hemisphere			
upstream		$10^3 < \mathbf{R} < 10^6$	1.33–1.42
downstream			0.34
Solid hemisphere			
Flat face upstream		$10^3 < \mathbf{R} < 10^6$	1.17
Flat face downstream			0.42
Square cylinder	Face $\underline{\vert r}$	3.5×10^4	2.00
Diamond shape			1.60
Elliptic cylinders (Axes ratios)	2 : 1	4×10^4	0.60
		10^5	0.46
	4 : 1	2.5×10^4	0.32
	8 : 1	2.5×10^4	0.29
		2.5×10^5	0.20
Triangular cylinder: base facing flow	120°		2.20
	60°	10^4	2.00
	30°		1.80
Vertex facing flow			
	120°		1.72
	60°	10^4	1.39
	30°		1.00

Example 11.1 Determine the terminal fall velocity of a spherical hailstone with 1 cm diameter and specific gravity 0.9. $\rho = 1.2\,\text{kg/m}^3$ and $\nu = 1.5 \times 10^{-5}\,\text{m}^2/\text{s}$.

Force = Weight of the hailstone $= 900 \times 9.81 \times (\pi/6)(0.01)^3 = 0.00462\,\text{N}$
$F/\rho\nu^2 = 0.00462/1.2 \times (1.5 \times 10^{-5})^2 = 1.712 \times 10^7$
From Fig. 11.6, we obtain $C_D \sim 0.4$ and $U = [2F/(C_D A\rho)]^{1/2}$
$U^2 = 0.00462/[0.4 \times 0.7854(0.01)^2 \times 1.2]$; $U = 11.09\,\text{m/s}$
We can also obtain the velocity by reading the Reynolds number at the intersection point and finding the velocity $U = \nu\mathbf{R}/D$

Example 11.2 A billboard has a length of 15 m and a height of 3 m and is exposed to wind blowing normally to it at a velocity of 50 km/hour. Estimate the force on the board.

$U = 50 \times 1000/60 \times 60 = 13.89\,\text{m/s}$
$\mathbf{R} = UH/\nu = [13.89 \times 3]/1.5 \times 10^{-5} = 2.78 \times 10^6$
From the table, $C_D = 1.2$ for an aspect ratio of 5. Hence the drag on the plate is $F = C_D A(0.5\rho U^2) = 1.2 \times (15 \times 3)(0.5 \times 1.2 \times 13.89^2) = 6250\,\text{N}$

Example 11.3 Find the terminal velocity of a parachute (considered as a hollow hemisphere) of diameter 10 m carrying a weight of 1000 N.

C_D for a hollow hemisphere from the table is 1.33.

$$U^2 = \frac{2F}{\left[C_D\rho(\pi D^2/4)\right]} = \frac{1000}{1.33 \times 1.2 \times 0.7854 \times 10^2} = 7.978$$

$$U = 2.82\,\text{m/s}$$

Example 11.4 A chimney 20 m in height and 2 m in dia. is subjected to a wind speed of 100 km/hour. Find the bending moment at the base due to wind load.

$U = 100 \times 1000/60 \times 60 = 27.78\,\text{m/s}$
$\mathbf{R} = UD/\nu = 27.78 \times 2/1.5 \times 10^{-5} = 3.7 \times 10^6$ and from Fig. 11.7, $C_D = 0.33$ (neglecting end effects)
$F = C_D(DL)(0.5\rho U^2) = 0.33 \times (20 \times 2)(0.5 \times 1.2 \times 27.78^2) = 6112\,\text{N}$
Bending moment at the base $= 6112 \times 10 = 6120\,\text{Nm}$

11.5 Lift Force

As we have seen before, the lift force on a body is given by $\rho U\Gamma$, where Γ is the circulation around the body. When the value of the circulation is not known and experimental results of the pressure distribution around the body are available, then the lift force is found by integrating the transverse component of the normal pressure force on the surface of the body. A symmetric body such a strut whose axis coincides with the flow direction does not experience a lift force. The pressure distribution around a symmetric strut (aerofoil) is shown in Fig. 11.8.

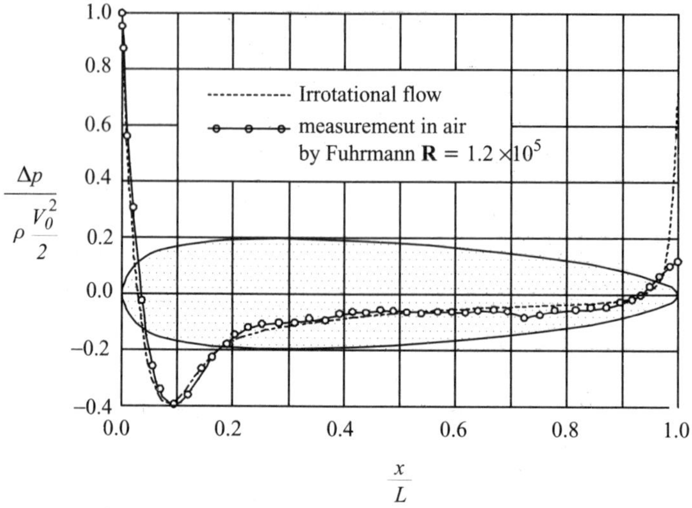

Fig. 11.8 Pressure distribution around a symmetric strut

But if the same strut is inclined to the direction of flow, a lift force acts on the body, since the flow pattern is then asymmetrical about the body and a circulation around the body is induced. It may be noticed that for a streamlined body, separation occurs almost at the tail end of the body thus minimising the form drag. However, skin friction drag becomes significant in such cases unlike for bluff bodies where the predominant drag is the form drag.

11.5.1 Development of Circulation Around an Inclined Plate

If a flat plate is held inclined to the direction of motion at a small angle, the pattern of irrotational flow will be as shown in Fig. 11.9*a* with two points of stagnation A and B. The resultant force is still zero, though a couple may be acting about the mid-point of the plate due to the eccentricity of the stagnation points.

In practice, this is not feasible as the infinite curvature at the leading and trailing edges of the plate will lead to infinite pressure gradients. This is obviated in nature by a superposition of circulation of suitable strength around the body as shown Fig. 11.9*b*. This will cause the points of stagnation to be displaced towards the trailing edge as shown in Fig. 11.9*c*. The stagnation point B coincides with the trailing edge and the stream line leaving the trailing edge will be tangential to it. It can be shown that the circulation that is required to make the stream line tangential to the trailing edge of the plate is

$$\Gamma = \pi c U \sin\theta \tag{11.6}$$

per unit length of the plate, where c is the width or chord of the plate and θ is its inclination or angle of attack. It may be noted that there will be separation at the leading edge and flow pattern

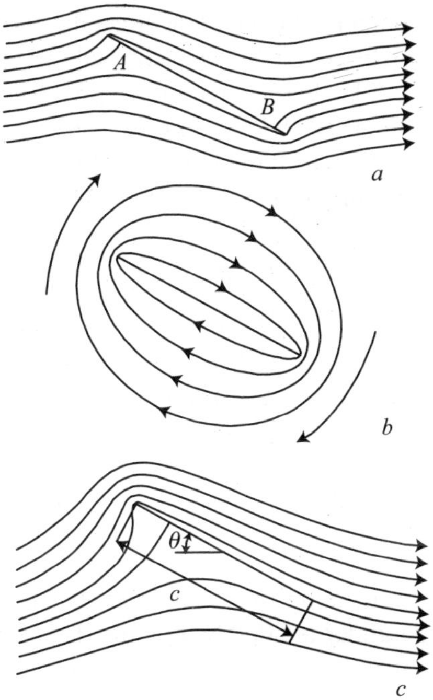

Fig. 11.9 Flow around an inclined plate with and without circulation

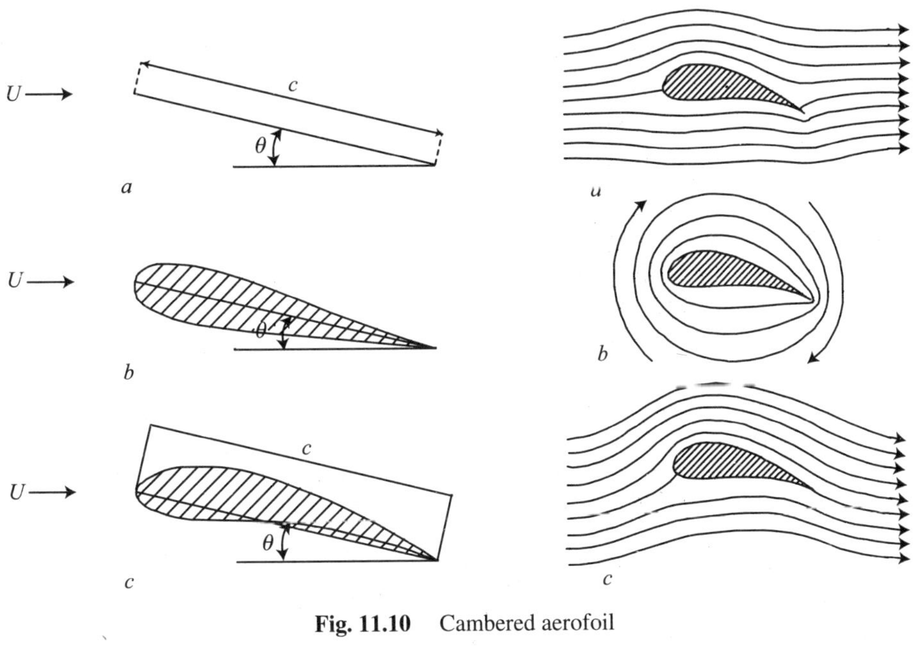

Fig. 11.10 Cambered aerofoil

will show a departure from the potential flow pattern along the entire upper surface. This may be remedied to a considerable extent by rounding the leading edge and tapering the profile gradually towards the rear as shown in Fig. 11.10*b* for a symmetric profile. Further to reduce the tendency towards separation, even for moderately high values of θ, a slight camber is given as shown in Fig 11.10*c* which also shows the final flow pattern after circulation around the aerofoil is established. This cambered profile is known as the Joukowsky profile. The lift force is now

$$L = C_L lc\left(\frac{1}{2}\rho U^2\right) = \rho U\pi lcU \sin\theta$$

$$C_L = 2\pi \sin\theta \tag{11.7}$$

EXAMPLE 11.5 A symmetric aerofoil with a chord length of 2.0 m is in an airstream of 200 m/s at an angle of 5° to the vane axis. Find the circulation around the aerofoil and the lift force.

Circulation $\Gamma = \pi cU \sin\theta = \pi 2 \times 200 \times \sin 5^\circ = 109.52\ \text{m}^2/\text{s}$
Assuming 100% efficiency, lift $L = \rho U\Gamma = 1.2 \times 200 \times 109.52 = 26285.6\ \text{N/m}$

EXAMPLE 11.6 A glider is descending at an angle of 10° to the horizontal at a speed of 200 km/hour. The weight of the glider is 3000 N. Find the lift, drag and the circulation around the wings.

$$U = 200 \times \frac{1000}{60 \times 60} = 55.56\ \text{m/s}$$

$$F_D = W \sin 10^\circ = 3000 \times 0.1736 = 520.94\ \text{N}$$

$$F_L = W \cos 10^\circ = 3000 \times 0.9848 = 2954\ \text{N}$$

$$\text{Circulation, } \Gamma = \frac{2954}{(1.2 \times 55.56)} = 44.31\ \text{m}^2/\text{s}$$

The lift force and the coefficient of lift may vary with the inclination of the body to the direction of flow. In the case of a streamlined aerofoil, the dependence of C_L on the Reynolds number may be neglected. For small angles of attack, the lift force generally increases with the increase in the angle of incidence of the flow to a reference line fixed in the aerofoil as shown in Figs. 11.11 and 11.12. However, as the angle of incidence is increased beyond a certain limit, flow separation will occur at the leading edge of the aerofoil and a condition known as stall occurs with a rapid drop not only in the value of lift force but also in the value of the lift–drag ratio (C_L/C_D). A polar diagram in which C_L is plotted as function of C_D for various angles of attack permits us to determine the maximum value of C_L/C_D and hence the most efficient angle of attack for a given aerofoil.

The design of the aerofoil is governed by the requirements of a given type of aircraft. Some require high initial lift or permit low landing speeds; others require high angles of stall; still others require only slight deviations of the lift–drag ratio with changes in the angle of attack. To meet these requirements it has become necessary to depart considerably from the cambered profiles of the Joukowsky type.

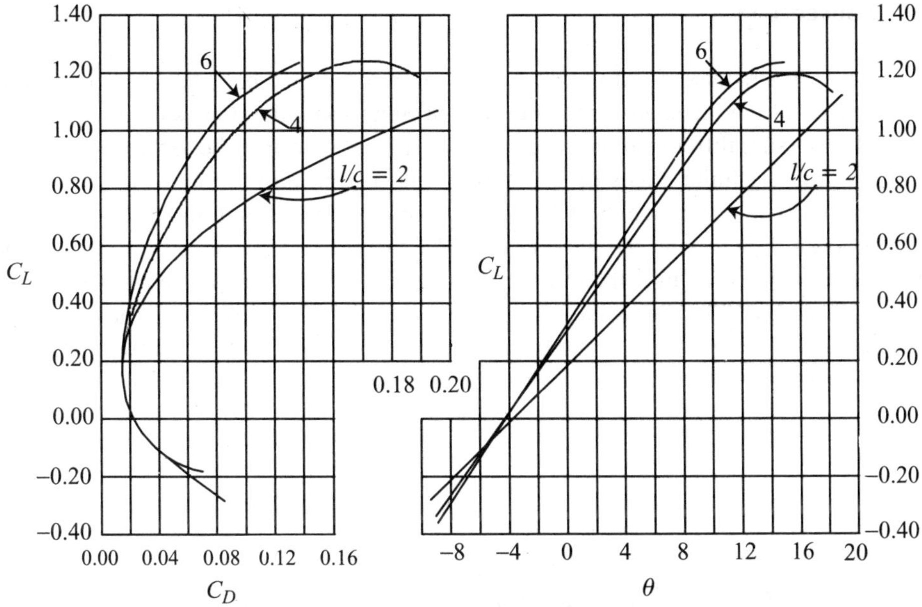

Fig. 11.11 Lift and polar diagram for three aspect ratios of a rectangular aerofoil

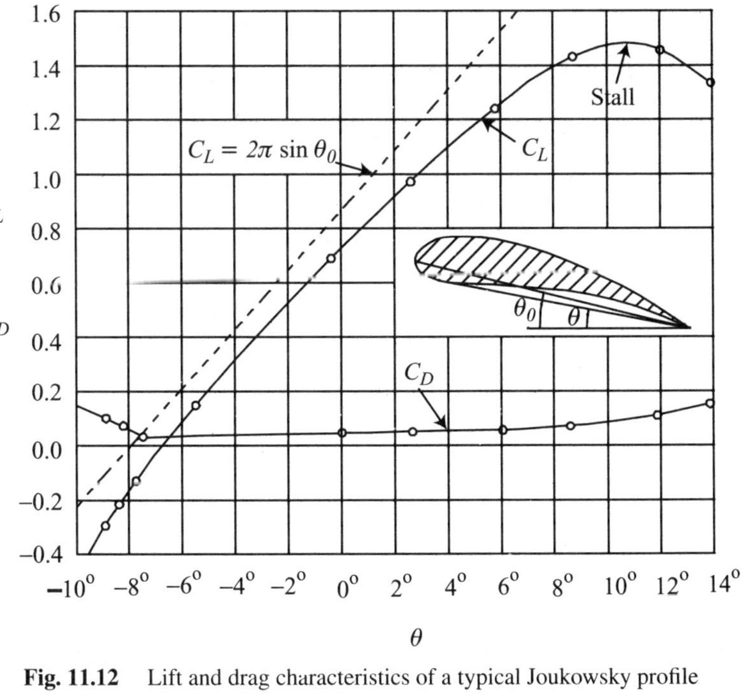

Fig. 11.12 Lift and drag characteristics of a typical Joukowsky profile

11.6 Induced Drag

For a truly two-dimensional aerofoil of infinite length, the flow characteristics are identical in all vertical planes along the span. For an aerofoil of finite span, the flow is three-dimensional in character. The pressure differences between the top and the bottom of the aerofoil must disappear gradually towards the finite edges of the aerofoil and merge into the uniform pressure of the surrounding flow. The difference in pressure between the top and bottom halves of the aerofoil causes an induced flow around the wing edges. The secondary transverse flow will occur from the high pressure at the bottom to the low pressure region at the top causing a displacement of the stream lines outward at the bottom and inward at the top. This transverse flow causes a surface of discontinuity at the edges leading to the formation of two distinct vortices known as tip vortices. A simplified picture of the formation of tip vortices is shown in Fig. 11.13.

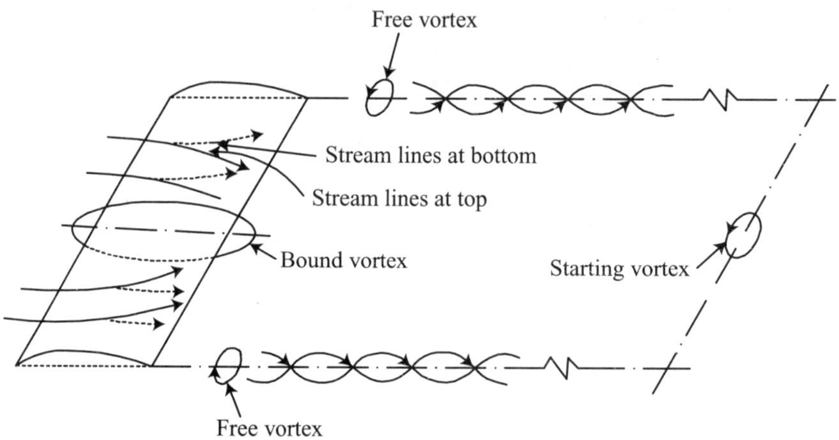

Fig. 11.13 Tip vortices for finite-span aerofoil

It may be seen that the circulation in the form of a bound vortex around the aerofoil cannot cease abruptly at the edges of the span but must continue as a free vortex. The tip vortices extend all along the flight path and must be linked with the starting vortex generated at the initial movement of the wing. These tip vortices cause an additional downward velocity in the region immediately behind the wing. The resultant relative velocity between the aerofoil and the surrounding fluid is obtained by a vectorial combination of the uniform velocity U and the downwash velocity as shown in Fig. 11.14.

Since the lift force is always visualised as a force normal to the direction of the relative velocity, this lift force is now found inclined to the vertical direction by the same angle ϕ as the resultant velocity makes with the horizontal. In other words, the effective angle of attack is reduced and the lift force has a component in the horizontal direction. This additional drag caused by the formation of the tip vortices behind a finite aerofoil is known as the induced drag, F_{Di}.

$$F_{Di} = F_L \sin\phi \simeq F_L\phi \tag{11.8}$$

for small values of ϕ.

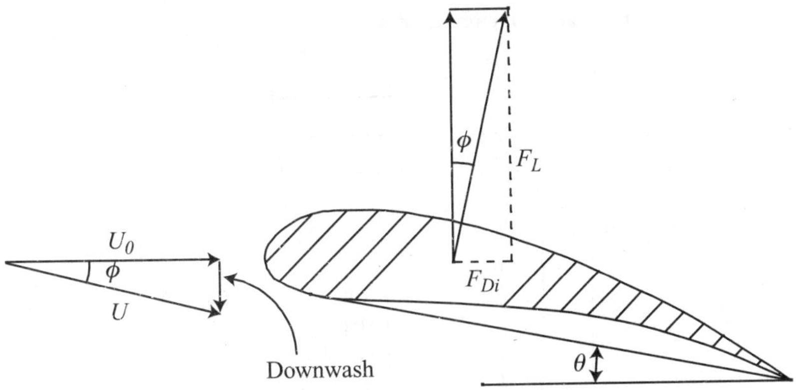

Fig. 11.14 Induced drag

It is true that the strength of circulation does not remain a constant for a finite aerofoil but varies along the span, being a maximum at the centre and decreasing towards each end. The intensity of lift force also varies along the span corresponding to the variation in circulation. In a practical problem, therefore, the continuous change in circulation from the centre towards the edges must cause the formation of vortices all along the span and not merely at the edges. This is referred to as a vortex band, and at a short distance behind the wing, these vortex sheets roll up into two distinct vortices spaced somewhat less than the wing span. These are illustrated in Figs. 11.15 and 11.16.

The lift force is no doubt proportional to the wing span l, but the energy contained in the tip vortices is not greatly altered by the variation in span. The energy supplied to these tip vortices are reflected in the induced drag on the aerofoil. This additional drag is more important for short spans than for large spans in its relative magnitude to the total surface and form drags. Based on

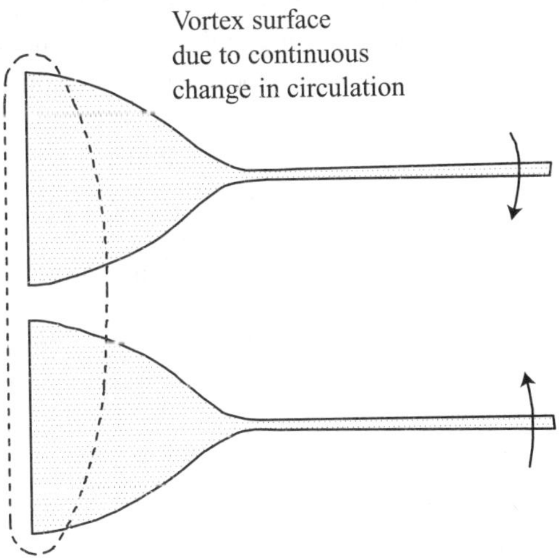

Fig. 11.15 Formation of vortices

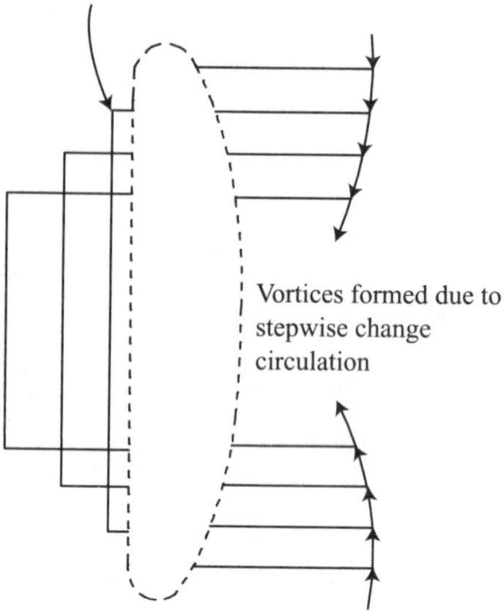

Fig. 11.16 Vortex street behind a wing

a certain arbitrary assumption of the distribution of lift on the aerofoil, it may be shown that the drag coefficient for the induced drag C_{Di} is given approximately by the formula

$$C_{Di} = \frac{C_L^2}{(\pi l/c)} \tag{11.9}$$

where l/c denotes the aspect ratio of the aerofoil. Use of this relation in conjunction with the polar diagrams permits us to isolate the induced drag from the surface and form drags.

The polar diagram for a wing of finite length ($l/c = 5$) is shown in Fig. 11.17.

Similar diagrams may be drawn from experimental measurements for different aspect ratios. Figure 11.17 also shows equation (11.5) for the aspect ratio, 5. The horizontal intercept between the two curves will represent the drag coefficient for a two-dimensional flow around a span of infinite length. Since $F_L \cos\delta \sim F_L$, the lift force is essentially the same for both cases of infinite and finite spans. The maximum lift–drag ratio is also obtained from the polar diagram. The line which is a tangent to the curve and passes through the pole evidently has a slope which is equal to the optimum value of C_L/C_D, the point of tangency indicating the most efficient angle of attack.

EXAMPLE 11.7 An aerofoil of wing span 15 m and aspect ratio $l/c = 5$ is moving at a speed of 300 km/hour at an angle of attack giving the maximum lift–drag ratio. What are the lift and drag forces?

$$U = 300 \times \frac{1000}{60 \times 60} = 83.33 \text{ m/s}$$

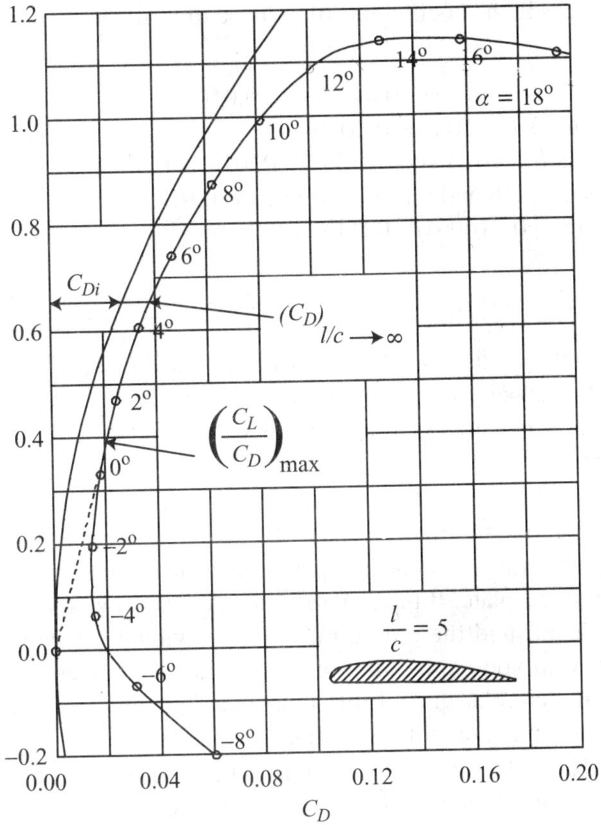

Fig. 11.17 Polar diagram for a wing of finite length

From the polar diagram of Fig. 11.17, we find that the maximum $(C_L/C_D)_{max}$ occurs at an angle of attack $\theta = 1°$.

At this angle of attack, we find $C_L = 0.4$ and C_D for infinite span = 0.011 (intercept between C_D for l/c = infinity and C_{Di})

$C_{Di} = C_L^2/(\pi l/c) = (0.4)^2/5\pi = 0.01018$

Hence C_D for $l/c = 5$ is $0.011 + 0.01018 = 0.02118$

Lift force $= C_L(lc)\rho U^2/2 = 0.4(15 \times 3) \times 1.2 \times 83.33^2/2 = 75000$ N

Drag force $= C_D(lc)\rho U^2/2 = 0.02118(15 \times 3) \times 1.2 \times 83.33^2/2 = 3971$ N

EXAMPLE 11.8 An aerofoil of infinite span has a lift coefficient of 0.75 and a drag coefficient of 0.06. What is the coefficient of induced drag for an aspect ratio of 6.

$$C_{Di} = \frac{C_L^2}{\pi(l/c)} = \frac{(0.75)^2}{6\pi} = 0.03$$

What are the changes in the coefficients of induced drag and total drag if the l/c ratio changes from 5 to 10?

C_{Di} for $l/c = 5$ is $C_L^2/\pi(l/c) = (0.4)^2/5\pi = 0.01018$
C_{Di} for $l/c = 10$ is $(0.4)^2/10\pi = 0.00509$
Hence the change is $(0.01018 - 0.00509)/0.01018 = 0.50$ or 50 %
C_D for $l/c = 5$ is 0.02118 and C_D for $l/c = 10$ is $0.011 + 0.00509 = 0.01609$
Change is $(0.02118 - 0.01609)/0.02118 = 0.24$ or 24%

Exercises

11.1. A long cylinder of dia. 10 cm with its axis perpendicular to the flow is kept in a stream of air with a velocity of 10 m/s. Find the drag force on the cylinder ($\rho = 1.2\,\text{kg/m}^3$ and $\nu = 1.5 \times 10^{-5}\,\text{m}^2/\text{s}$). If the fluid is water and the velocity is 10 m/s, find the drag ($\rho = 998\,\text{kg/m}^3$) and $\nu = 1 \times 10^{-6}\,\text{m}^2/\text{s}$.

11.2. Determine the bending moment at the base of a chimney of diameter 2.0 m and height 20 m due to a wind of 100 km/hour.

11.3. A square plate with 1.5 m side is kept normal to an air stream of velocity 20 m/s. Find the drag force on the plate. If the square plate is cut into two rectangular pieces measuring 1.5 m × 0.75 m, find the drag on the two individual pieces kept normal to the same air stream. Can you explain the difference in drag values even though the areas are identical? If a circular disk of the same area, 2.25 m^2, is kept normal to the flow, find the drag and explain the different value obtained.

11.4. A billboard has a length of 15 m and height of 3 m and is exposed to the wind blowing normal to it at a velocity of 50 km/hour. Estimate the force exerted on the board.

11.5. Determine the terminal fall velocity of a spherical hailstone of 1 cm diameter and specific gravity 0.9.

11.6. A steel sphere of diameter 0.10 m (specific gravity 8.0) is falling freely in water. Find the terminal velocity.

11.7. What is the terminal velocity of a parachute (considered as a hemisphere) of diameter 10 m carrying a weight of 1000 N.

11.8. Find the drag on a hemisphere with a hollow side is facing a stream of air at a velocity of 50 m/s. If the hollow side is facing downstream, what is the drag experienced?

11.9. An elliptical cylinder of aspect ratio 1:3 is kept in a stream of air with a velocity of 50 m/s. Find the drag if the chord length is 3 m.

11.10. What is the drag on an ellipsoid of chord length 3.6 m and aspect ratio 1:1.8 in an air stream of 50 m/s.

11.11. A kite of projected area 0.5 m^2 and weighing 2.5 N is flying in an air stream of 30 km/hour. The kite surface makes an angle of 90° to the chord and the chord makes an angle of 45° to the horizontal. If the tension in the chord is 20 N, find the coefficient of drag and lift.

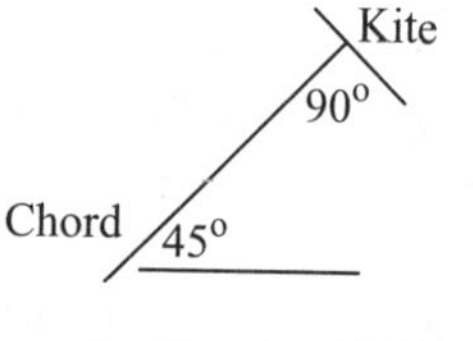

Figure for Exercise 11.11

11.12. A symmetrical strut shown in Fig. 11.8 has a chord length of 1.5 m. What is the lift per metre length on the strut if the strut is moving in water at a velocity of 4 m/s at an angle of 5° to the axis of the strut. Assume a lift efficiency of 75%.

11.13. A glider weighing 4000 N is descending with an angle of 5° to the horizontal at a speed of 250 km/hour. Find the lift, drag and circulation around the wings of chord length 2.75 m.

11.14. Determine the lift and drag of an airplane having a total wing span of 15 m and a chord length of 3.0 m ($l/c = 5.0$) when it is moving with a speed of 300 km/hour, at an angle of attack yielding a maximum lift–drag ratio, find the induced drag and the total drag and lift. (Use Fig. 11.7)

11.15. If the lift and drag coefficients for an aerofoil of infinite span at an angle of attack of 5° are 0.75 and 0.015 respectively, find the coefficients of induced drag and the total drag for a wing of $l/c = 6$. What are the changes in coefficients of induced drag and total drag if l/c changes from 5 to 10.

12 Open Channels

12.1 Introduction

An open channel is a natural or an artificial water-way in which a liquid flows with a free surface. Flow in an open channel differs from a pressure flow in a closed conduit in as much as the former has a free surface which introduces an extra boundary condition that the pressure on the free surface is subjected to ambient or atmospheric pressure. The motivating force in an open channel is the earth's gravitational attraction and hence it is also known as gravity flow.

The different types of flow that can occur in an open channel can be classified according to the changes in velocity and depth of flow with respect to space and time. If the flow conditions do not change with time, the flow is said to be steady but if they change with time, the flow is termed as unsteady. If the flow conditions do not change along the length of the channel, that is, with respect to distance or space, the flow is said to be uniform. If the flow conditions change along the length of the channel, the flow is termed as non-uniform. Thus we can have combinations of variations with both space and time. They are

steady, uniform flow;
steady, non-uniform flow;
unsteady, uniform flow (which occurs very rarely);
unsteady non-uniform flow.

Steady non-uniform flow can be either gradually varied or rapidly varied depending on the total length of the channel involved for a given change in depth and velocity. Similarly when temporal changes in flow characteristics take place, an open channel flow can be a gradually-varied unsteady flow or a rapidly-varied unsteady flow.

Further classifications are also made depending on the state of flow, namely laminar or turbulent flow, depending on the Reynolds number. If the Reynolds number, $VR/\nu < 500$, the flow is laminar and if greater than 500, the flow becomes turbulent, where $R = A/P$, A is area of flow and P is the wetted parameter. R is known as the hydraulic radius. For circular pipes flowing full, the hydraulic radius is $A/P = \text{R} = \pi D^2/4\pi D = D/4$ or $D = 4R$. The Reynolds number, $\mathbf{R} = PVD/\mu$ can be written as $\mathbf{R} = PV4R/\mu$. The critical Reynolds number at which laminar flow becomes turbulent for non-circular cross-sections is now approximately $PV\text{R}/\mu = 2000/4 = 500$.

The flow is also classified as sub-critical, critical and super-critical depending on if the value of the Froude number is less than one, equal to one or greater than one respectively. The regimes of flow are shown in Fig. 12.1.

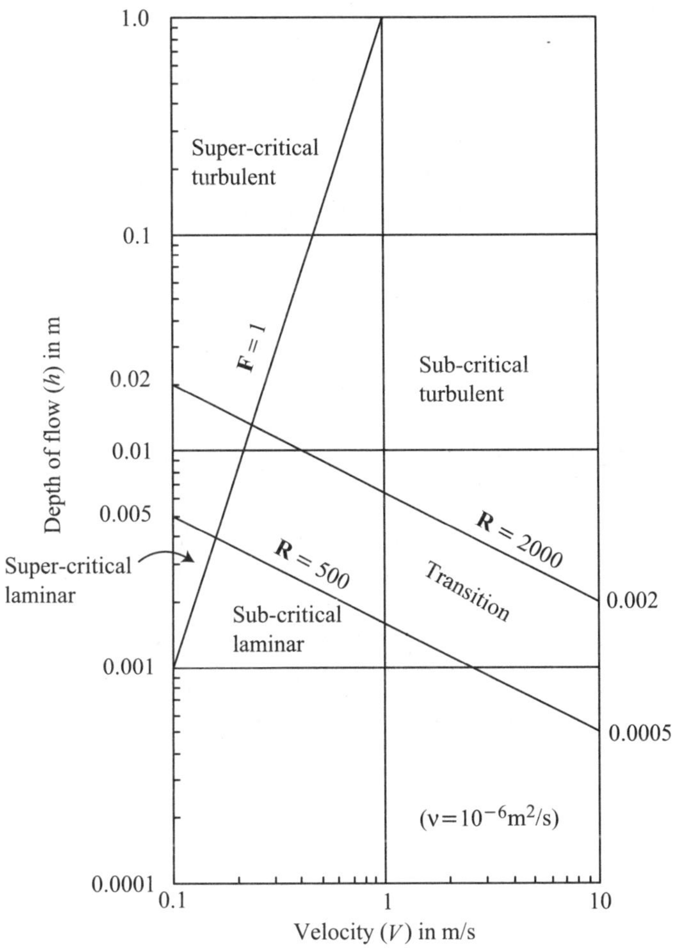

Fig. 12.1 Flow regimes in open channels

12.2 Velocity Distribution

The velocity distribution across any vertical or horizontal section is non-uniform. This is due to the fact that at the boundaries, the velocity is zero and there are steep velocity gradients in the boundary layer which may extend to the full depth or width of the channel. The velocity distribution in an open channel is extremely complex depending as it does upon various factors such as shape of the cross-section, roughness of the boundaries and the presence of bends and obstructions.

Owing to the non-uniform distribution of velocity, the velocity head and momentum of the flow at any cross-section computed by using the mean velocity of flow are less than the actual values. Therefore, correction factors are used in computing energy and momentum of the flow.

12.2.1 Energy Coefficient

The energy correction coefficient or the Coriolis coefficient designated by α can be determined as follows.

Consider a small elementary area dA in the flow cross-section and the discharge through the element $v \cdot dA$, v being the local velocity. The kinetic energy of the fluid mass passing through the element is $\rho v \cdot dA \cdot v^2/2$. The total kinetic energy for the entire cross-section of the channel can be obtained by integrating over the entire area of cross-section $\int_A \rho v^3 dA/2$. If $V = Q/A$ is the average velocity of flow, then the kinetic energy based on average velocity is $\rho V^3 A/2$ but it should be corrected by the correction factor α. Equating the two expressions for kinetic energy, we get

$$\alpha \rho V^3 A/2 = \int_A \rho v^3 dA/2$$

$$\alpha = \frac{1}{AV^3}\int_A v^3 dA \tag{12.1}$$

The local velocity v at any point can be written as $v = V + dv$, dv being the deviation from the average velocity and it can be positive or negative. Introducing this value in equation (12.1),

$$\alpha = \frac{1}{AV^3}\int_A (V + dv)^3 dA = \frac{1}{AV^3}\int_A \left[V^3 dA + (dv)^3\, dA + 3V^2 dv \cdot dA + 3V\,(dv)^2\, dA\right]$$

As dv is small, the third power of dv may be neglected and the third term disappears as the integral of $dv \cdot dA$ is zero. Hence we have

$$\alpha = 1 + \frac{3}{AV^2}\int_A (dv)^2 dA = 1 + 3\eta$$

$$\eta = \frac{1}{AV^2}\int_A (dv)^2 dA \tag{12.2}$$

12.2.2 Momentum Coefficient

The momentum coefficient or the Boussinesq coefficient designated by the symbol β is defined by considering the momentum of the flow passing through an elementary area dA and equating it to β times the momentum based on the average velocity.

$$\beta \rho A V^2 = \int_A \rho v^2 dA$$

$$\beta = \frac{1}{AV^2}\int_A (V + dv)^2 dA = \frac{1}{AV^2}\int_A \left[V^2 dA + (dv)^2 dA + 2V dv dA\right] \tag{12.3}$$

$$\beta = 1 + \frac{1}{AV^2}\int_A (dv)^2 dA = 1 + \eta$$

It can be seen from the above expressions for α and β that if the velocity distribution is uniform, $dv = 0, \eta = 0$ and $\alpha = \beta = 0$. In the case of non-uniform velocity distribution, $\eta > 0$ and $\alpha > 1, \beta > 1$. Also it is evident that for the same flow conditions and cross-sections, values of α are always larger than β.

12.2.3 Field Values of Coefficients

Experimental data indicate that the values of α vary from about 1.03 to 1.36 for fairly straight channels. The values are generally higher for small channels and lower for large streams of considerable depth. For non-uniform channels with complex cross-sections α can be as high as 1.6 and can vary from section to section. Upstream of weirs, in the vicinity of obstructions or close to sections with pronounced irregularities in alignment, values of α exceeding 2 have been observed for example, α is 3.87 at the outlet section of a draft tube and 7.4 for the spiral flow under a model turbine wheel.

Values of β vary from 1.01 to 1.12 for fairly straight prismatic channels and can be as great as 1.2 for complex sections.

Example 12.1 The velocity distribution along the depth of flow in a wide rectangular channel is approximately given by

$$\frac{v}{V_0} = \left(\frac{y}{y_0}\right)^{1/7}$$

where v is the velocity at a distance y above the bottom of the channel and V_0 is the velocity at the surface of the channel of depth y_0. Find the energy and momentum coefficients.

First we find the average velocity,

$$\overline{V} = \frac{\int_A v \cdot dA}{A} = \frac{V_0}{y_0^{1/7}}\int_0^{y_0} y^{1/7} dy = \frac{7}{8}V_0$$

$$\alpha = \frac{\int_A v^3 dA}{A\overline{V}^3} = \frac{1}{A\overline{V}^3}\int_0^{y_0} V_0^3\left(\frac{y}{y_0}\right)^{3/7} = \frac{(7/10)y_0 V_0^3}{y_0(7/8)^3 V_0^3} = 1.045$$

$$\beta = \frac{\int_A v^2 dA}{A\overline{V}^2} = \frac{(7/9)y_0 V_0^2}{(7/8)^2 y_0 V_0^2} = 1.017$$

12.3 Pressure Coefficient

In a majority of open channel flows, particularly uniform flows and gradually varied flows, where stream lines are almost parallel to each other, the pressure distribution in the vertical direction can be assumed to follow the hydrostatic law. In some cases of rapidly-varied flow such as a hydraulic jump or flow over a spillway etc., the stream lines have a pronounced curvature. This introduces an additional centrifugal force on the flow and causes a change in pressure distribution. Therefore in curvilinear flows correction factors have to be introduced in computing pressure head and pressure force.

The centrifugal force or excess pressure can be found from the relation

$$\Delta P_h = \left(\frac{\gamma h}{g}\right)\left(\frac{v^2}{r}\right) \tag{12.4}$$

where h = depth of the water column of unit cross-sectional area below the free surface and r = radius of curvature. At the channel bottom $h = y_0$ and R is the radius of curvature of the channel bottom. For practical purposes, the velocity v can be taken as the average velocity V over the section.

If the stream lines are convex downwards as in the case of the flow over a spillway, the centrifugal force acts against the gravitational force and the change in pressure $\Delta P_h < 0$. The pressure distribution is as shown in Fig. 12.2*a*. If, on the other hand, the stream lines are concave downwards as in the case of flow below a sluice gate or at the foot of a spillway (bucket region), the correction is positive as shown in Fig. 12.2*c*. For parallel flows as shown in Fig. 12.2*b*, the pressure distribution is hydrostatic ($p = \gamma h$).

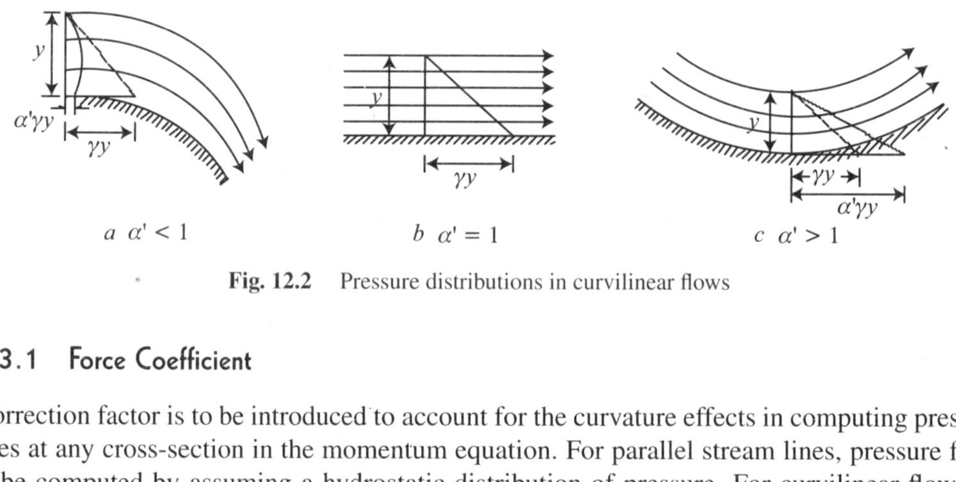

Fig. 12.2 Pressure distributions in curvilinear flows

12.3.1 Force Coefficient

A correction factor is to be introduced to account for the curvature effects in computing pressure forces at any cross-section in the momentum equation. For parallel stream lines, pressure force can be computed by assuming a hydrostatic distribution of pressure. For curvilinear flow, the pressure distribution is non-hydrostatic and a correction factor β' is introduced as follows. If the pressure distribution is hydrostatic, the total pressure force is $\gamma A z_g$, where A is the cross-section area of the flow and z_g is the depth of the centroid of the area below the free surface. If p is the pressure intensity on an elementary area dA, then the total pressure force is

$$\beta' \gamma z_g A = \int_A p dA = \int_A \gamma h \pm \Delta P_h$$

$$\beta' = 1 \pm \frac{1}{A z_g}\int_A \frac{\Delta P_h}{\gamma} dA \tag{12.5}$$

12.3.2 Effect of Slope

For parallel flow and flow down small slopes, the pressure head at any section could be represented by the depth of flow y_0 measured in the vertical plane. For this case $y_0 = d$, where d is the depth of the flow section measured in the plane of the cross-section. But when the slope angle θ becomes very large, y_0 and d are no longer equal and neither represents the true pressure at a given cross-section. The weight of the water in the column is $W = \gamma \cdot \delta a \cdot d$ (See Fig. 12.3) and acts vertically downwards. The component normal to the channel bottom is $W \cos\theta = \gamma \delta a \cdot d \cdot \cos\theta$, which exerts a pressure on the bottom of the channel equal to $W \cos\theta / \delta a = \gamma d \cos\theta$ and the corresponding pressure head is $d \cos\theta = y_0 \cos^2\theta$.

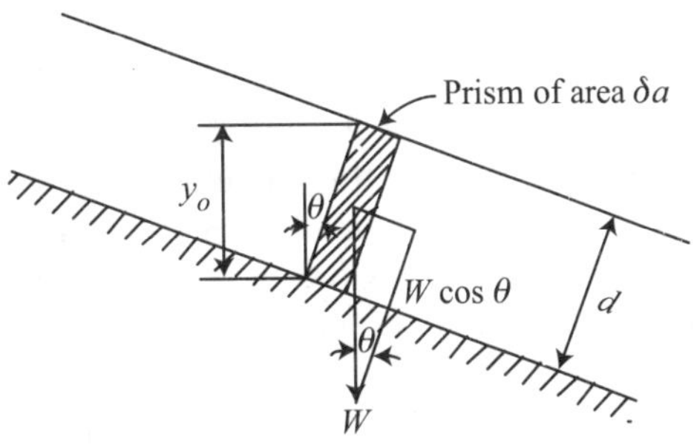

Fig. 12.3 Effect of slope

$\cos\theta$ is the correction factor which accounts for the bottom slope in computing the pressure head. Channels flatter than 1 in 10 corresponding to a 6° slope angle are considered to be mild channels for which $y_0 \sim d$ or $\cos\theta \sim 1$.

12.4 Energy Equation

The application of the Bernoulli theorem for one-dimensional flows in open channels is of great importance. Specifically Bernoulli equation is written as

$$H = V^2/2g + p/\gamma + z$$

where H is the total head with reference to the datum plane, z is the elevation above the datum, V is the velocity of the flow in the channel at the same section and p/γ represents the pressure head. When the stream lines are parallel and with a small bottom slope, the pressure distribution is hydrostatic and the sum of $(p/\gamma + z)$ remains constant along a vertical and is equal to the height of the surface above the channel bottom. If V is the average velocity over the section and α is the energy correction coefficient, we can write the Bernoulli equation for the section as

$$H = \alpha V^2/2g + y_0 + z$$

where z is the elevation of the channel bottom above the datum and y_0 is the depth of flow. For a channel with a large slope, the energy equation becomes

$$H = \alpha V^2/2g + d\cos\theta + z$$

where d is the depth of the flow section and θ is the slope of the channel.

If in addition the stream lines have an appreciable curvature in the vertical plane, the energy equation assumes the form

$$H = \alpha V^2/2g + \alpha' y_0 \cos^2\theta + z \tag{12.6}$$

where α' is the pressure correction coefficient.

The line representing the elevation of the total heads for different cross-sections along the channel is called the total energy line and its slope is known as the energy gradient denoted by the symbol S_f. The slope of the water surface is denoted by S_w and the slope of the bottom of the channel by $S_0 = \sin\theta$. If there are no losses in the system, the three lines will be parallel ($S_f = S_w = S_0 = \sin\theta$). If there are losses between any two sections 1-1 and 2-2 as shown in Fig. 12.4, the modified Bernoulli equation is written for the two sections as

$$\alpha_1 V_1^2/2g + \alpha_1' y_1 \cos^2\theta + z_1 = \alpha_2 V_2^2/2g + \alpha_2' y_2 \cos^2\theta + z_2 + h_f$$

where h_f is the energy loss between Sections 1-1 and 2-2 which includes energy expended due to boundary resistance and internal losses.

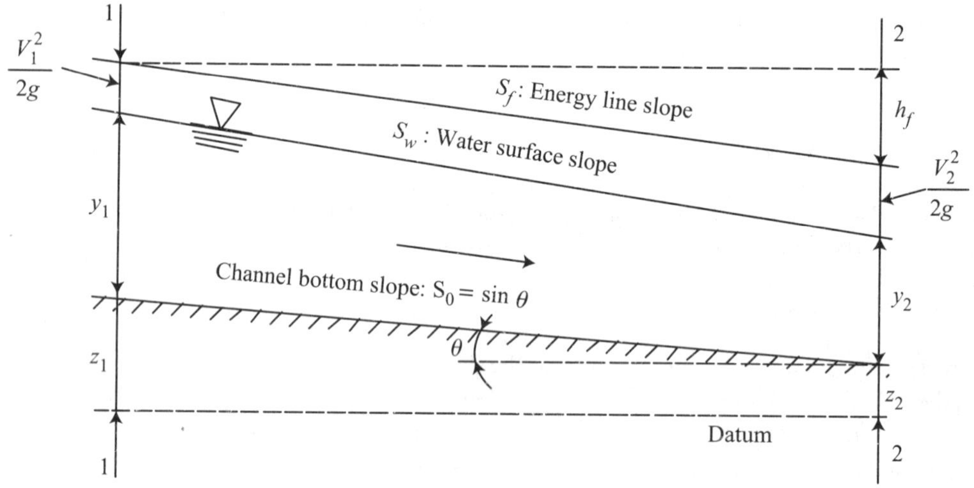

Fig. 12.4 Total head and energy loss

For cannels with small slope, zero curvature and with the energy coefficient nearly equal to 1.0, we have

$$z_1 + V_1^2/2g + y_1 = z_2 + V_2^2/2g + y_2 + h_f \tag{12.7}$$

If there are no losses (or negligible losses), we have the simplified Bernoulli theorem for open channels

$$z_1 + V_1^2/2g + y_1 = z_2 + V_2^2/2g + y_2$$

EXAMPLE 12.2 For purposes of discharge measurement, the width of a rectangular channel is reduced from 2.75 m to 2.15 m and the floor is raised 30 cm in elevation at a given section. When the depth of the approaching flow is 1.8 m, what rate of flow would be indicated by a drop of 12.5 cm at the contracted section.

Velocity at the approaching section $= V_1 = Q/2.75y_1$
Depth at the contracted section $y_2 = y_1 - 0.30 - 0.125 = y_1 - 0.425 = 1.375$
Velocity at the contracted section $V_2 = Q/2.15(y_1 - 0.425)$
Taking the bottom of the approach channel as the datum $z_1 = 0$, $z_2 = 0.30$
Applying these values in the Bernoulli equation (assuming no losses)

$$0 + \frac{Q^2}{2g(2.75y_1)^2} + y_1 = 0.30 + \frac{Q^2}{2g[2.15(y_1 - 0.425)]^2} + y_2$$

$$\frac{Q^2}{2g}\left\{\frac{1}{[2.15(y_1 - 0.425)]^2} - \frac{1}{(2.75y_1)^2}\right\} = y_1 - y_2 - 0.30 = 0.125$$

Substituting $y_1 = 1.8\,\text{m}$, $Q = 5.76\,\text{m}^3/\text{s}$

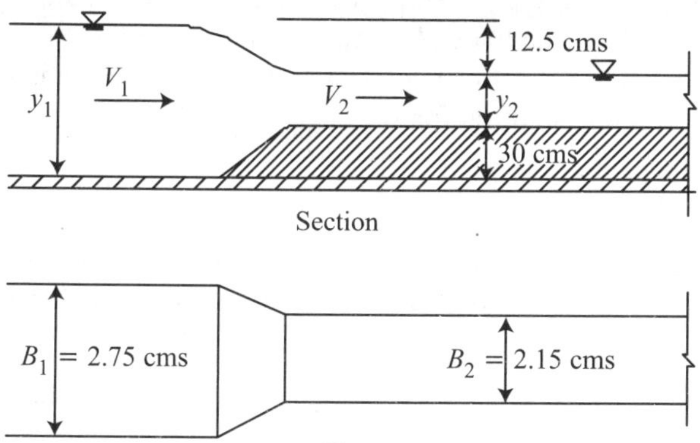

Figure for Example 12.2

12.5 Specific Energy

Specific energy is the flow energy measured with respect to the channel bottom as the datum and is denoted by E. Hence for a steeply sloping channel, we have

$$E = y_0 \cos^2\theta + \alpha V^2/2g \tag{12.8a}$$

If we assume $\cos\theta \sim 1.0$ and $\alpha \sim 1.0$, we have the expression for specific energy as

$$E = y_0 + V^2/2g \tag{12.8b}$$

Hence it is seen that specific energy is the sum of the depth of water and the velocity head. Considering the case of a rectangular channel of width B, the specific energy equation can be written as

$$E = y_0 + Q^2/2gA^2 = y_0 + Q^2/2gB^2y_0^2 = y_0 + q^2/2gy_0^2 \tag{12.8c}$$

where $q = Q/B$ = discharge per unit width.

12.6 Specific Energy Diagram

If the discharge is kept constant, it can be seen that the specific energy is a function of the depth of flow only. Equation (12.8c) can be written as

$$(E - y_0)y_0^2 = \frac{q^2}{2g} \tag{12.8d}$$

This is a cubic equation with three roots, one of which is negative and the other two positive. The plot of E vs. y_0 for a constant discharge is known as the specific diagram plotted in Fig. 12.5*a*. A dimensionless plot of y/y_c vs. E/y_c is shown in Fig. 12.5*b* for rectangular channels. The specific curve has two asymptotes along the lines $(E - y_0) = 0$ and $y_0 = 0$.

At a certain depth known as the critical depth, y_c the specific energy is a minimum, $E_{\min}$ and at this energy value there is one solution for y.

$$E_{\min} = y_c + V_c^2/2g$$

Beyond this single unique solution to equation (12.8d), for any given specific energy there are two depths y_1 and y_2 which are the two positive roots of the equation (12.8d).

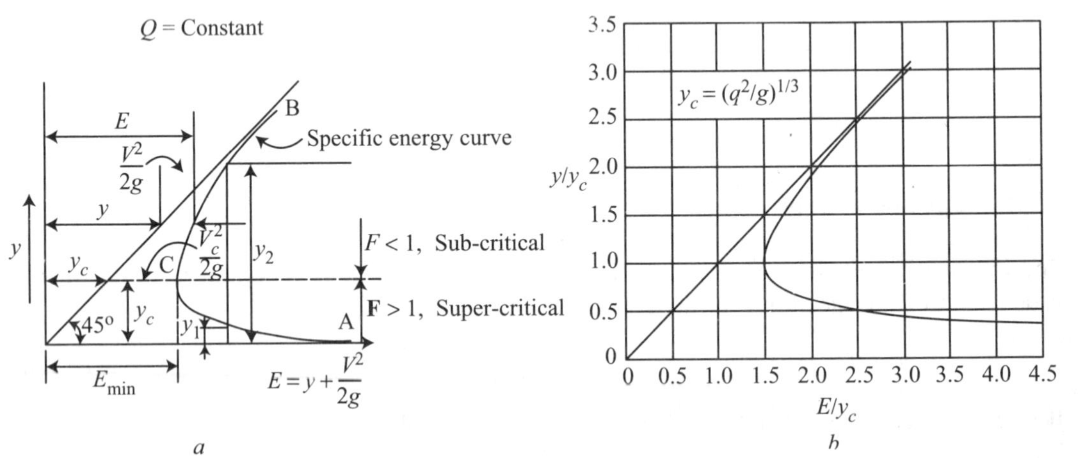

Fig. 12.5 Specific energy diagrams

The critical depth, y_c divides the flow into two distinct regimes. When the given discharge q occurs at a depth y_2 greater than y_c, the velocity $V_c > V$, and the flow is said to be tranquil or sub-critical and corresponds to the limb CB on the specific energy diagram. When the flow occurs at a depth y_1 less than y_c corresponding to the limb CA of the specific energy diagram, the velocity $V > V_c$ and the flow is said to be shooting or rapid or super-critical. Just as the Reynolds number is a criterion for distinguishing laminar and turbulent flow, the Froude number $\mathbf{F} = V/(gy)^{1/2}$ is the criterion for demarcating sub-critical and super-critical flow.

The critical depth occurs when E is a minimum or when $dE/dy = 0$. For rectangular channels of width B, we have

$$E = q^2/2gy^2 + y$$

$$\frac{dE}{dy} = 1 - q^2/gy^3 = 1 - Q^2/B^2gy^3 = 0$$

$$y_c = \left(\frac{q^2}{g}\right)^{1/3} \tag{12.9}$$

Further we have $V_c = q/y_c = (gy_c^3)^{1/2}/y_c = (gy_c)^{1/2}$

$$\frac{V_c}{(gy_c)^{1/2}} = 1 \tag{12.10}$$

at the critical depth and the specific energy is a minimum. At the critical depth, the Froude number $\mathbf{F} = 1$.

Finally the minimum specific energy at the critical depth is obtained by substituting the expression for y_c in equation (12.9) in the expression for E.

$$E = \frac{gy_c^3}{2gy_c^2} + y_c = \frac{3y_c}{2} \tag{12.11}$$

EXAMPLE 12.3 At what two depths could a flow of 11 m^3/s of water be carried by a rectangular channel 2.75 m wide? Discuss the characteristics of the flow.

The specific energy is

$$E = y + V^2/2g = y + Q^2/2gB^2y^2 = y + 11^2/(2 \times 9.81 \times 2.75^2y^2)$$

For the given values of E, Q, B, the two alternate depths are found by trial and error as $y_1 = 0.835$ m and $y_2 = 1.73$ m.

The critical depth is $y_c = (Q^2/B^2g)^{1/3} = (11^2/2.75^2 \times 9.81)^{1/3} = 1.178$ m

Comparing the two alternate depths for the given specific energy and the critical depth for the given discharge, we observe that the depth of 1.73 m being greater than the critical depth, the flow is sub-critical and $\mathbf{F} < 1$. The flow corresponding to the alternate depth of 0.835 m ($< y_c$) belongs to the super-critical regime with $\mathbf{F} > 1$.

12.7 Discharge Diagram

In the previous section, the relationship between the depth and specific energy was considered. However if the specific energy is constant and the discharge is allowed to vary with depth, we get further insight into the definition of critical depth. Considering again a rectangular channel and discharge per unit width $q = Q/B$, the specific energy equation (12.8d), we write

$$q = [2g(Ey^2 - y^3)]^{1/2} \tag{12.12}$$

A plot of equation (12.12) known as the discharge diagram is shown in Fig. 12.6 for a constant specific energy.

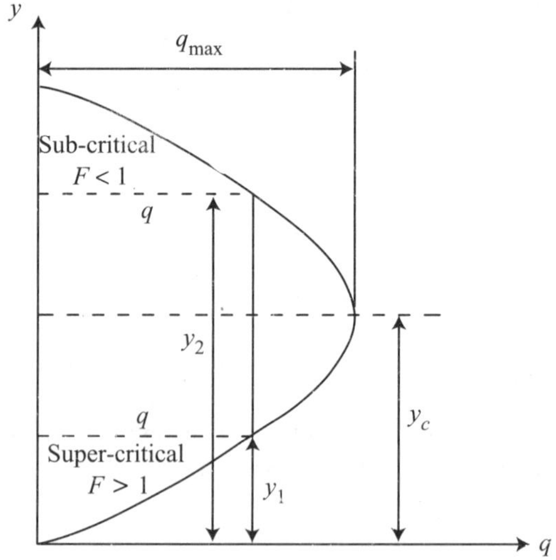

Fig. 12.6 Discharge diagram

It is evident from equation (12.12) that $q = 0$ when $y = 0$ and when $y = E$. Hence at some value of y, the discharge becomes a maximum for the given specific energy. Again to find the depth for maximum discharge we can set $dq/dy = 0$. Rewriting equation (12.12) as $q^2 = 2g(Ey^2 - y^3)$ and taking the derivative of q with respect to y and setting it to zero, we get

$$2q\left(\frac{dq}{dy}\right) = 2g(2Ey - 3y^2) = 0$$

The discharge is a maximum when $E = 3y_c/2$.

It is seen from equation (12.11) that the specific energy is also a minimum at $y = y_c$ and its value is $3y_c/2$.

Hence we conclude that at the critical depth not only is the specific energy a minimum for a given discharge but the discharge is also a maximum for a given specific energy.

12.8 Application of Specific Energy and Discharge Diagrams

Specific energy diagrams find application in problems of open channel transitions with sub-critical flows. A transition constitutes a local change in cross-section of a channel. A rise in the channel bottom elevation with constant width or a change in the width with a level floor or changes in both floor elevation and channel width constitute simple problems in open channel transitions.

A gradual rise in floor elevation is shown in Fig. 12.7*a* by an amount Δz, the width remaining constant. An increase in bottom elevation means a decrease in specific energy. If the flow is initially sub-critical, a rise in the bottom elevation would lead to a decrease in channel depth as per the specific diagram. On the other hand, if the flow is initially super-critical, a decrease in specific energy will cause the channel depth to increase. The change in depth in either case can be obtained from the specific energy diagram for a known value Δz or ΔE. It is evident that as Δz is increased to some value Δz_c (Fig. 12.7*b*), the depth of flow on the step would be critical. As this corresponds to the minimum specific energy for which flow is physically possible, further increase in floor level will increase the upstream depth which is known as the backwater effect or choking of the channel. Similar conditions will prevail if the channel width is reduced without any change in the bottom elevation.

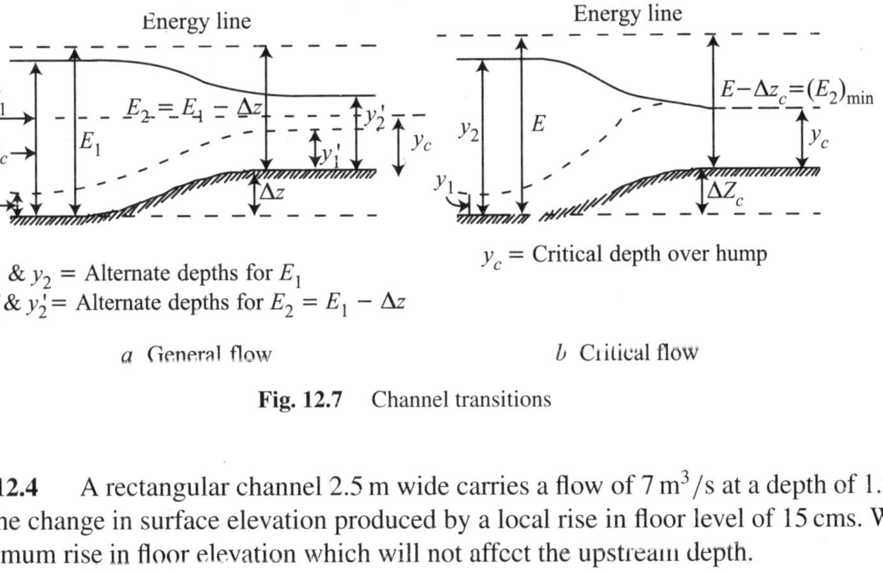

a General flow *b* Critical flow

Fig. 12.7 Channel transitions

EXAMPLE 12.4 A rectangular channel 2.5 m wide carries a flow of 7 m^3/s at a depth of 1.5 m. Compute the change in surface elevation produced by a local rise in floor level of 15 cms. What is the maximum rise in floor elevation which will not affect the upstream depth.

Applying the energy equation to Sections 1-1 and 2-2 (as shown in the sketch) and assuming no loss in energy between the sections, we have

$$y_1 + V_1^2/2g + z_1 = y_2 + V_2^2/2g + z_2$$

Substituting for $V = Q/By$, $V_1 = 7/(2.5 \times 1.5) = 1.867$ m/s; $V_2 = 7/2.5y_2$; $y_1 = 1.5$ m, $1.5 + 1.867^2/(2 \times 9.81) + 0 = 1.6776 = y_2 + 7^2/(2.5^2 \times 2 \times 9.81 y_2^2) + 0.15$;

$y_2 + 0.4/y_2^2 = 1.5276$ which when solved by trial and error gives $y_2 = 1.285$ m. Hence the change in water surface elevation is $1.5 - 0.15 - 1.285 = 6.5$ cms.

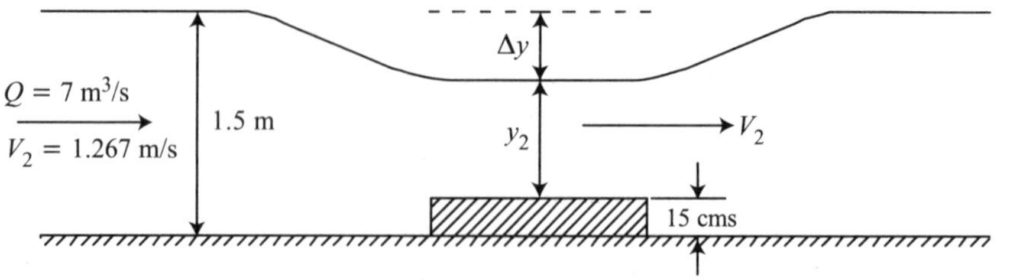

Figure for Example 12.4

For maximum floor elevation, the flow at the hump becomes critical and the critical depth is $y_c = (Q^2/gB^2)^{1/3} = (7^2/9.81 \times 2.5^2)^{1/3} = 0.928$ m and the minimum specific energy at the raised portion is $3y_c/2 = 1.392$.

The value of z_{max} is obtained by equating the total energies at the two sections.

$$z_{max} + 1.392 = 1.6776 \text{ and } z_{max} = 0.2826 \text{ m} = 28.26 \text{ cms}.$$

The problem of flow with constant specific energy but varying discharge can be exemplified by the flow under a sluice gate as shown in Fig. 12.8.

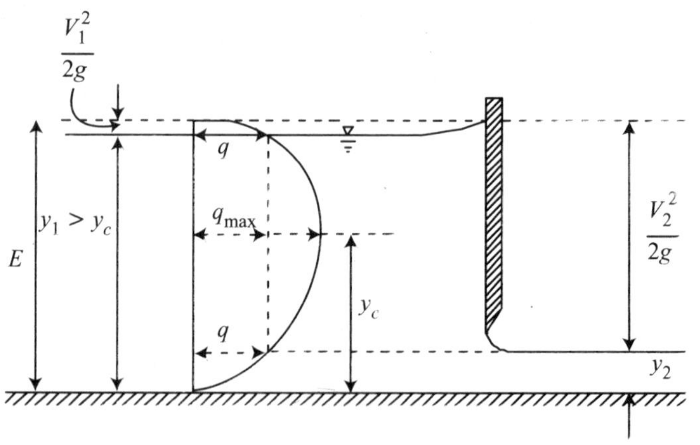

Fig. 12.8 Flow below a sluice gate

The discharge below the gate increases with increase in gate opening. For a given discharge, the depth upstream and downstream of the gate are alternate depths. When the gate is closed $q = 0$ and the upstream depth, y_1 corresponds to the total specific energy, $E (y_1 = E)$ and the downstream depth is $y_2 = 0$ (Fig. 1.29a). For a small gate opening, the flow is super-critical downstream and $q < q_{max}$ and $y_1 > y_2$. As the gate opening is increased, we have $y_2 = y_c$ and $q = q_{max}$, $y_1 = 3y_c/2$. Further increase in gate opening will not increase the discharge for a given constant specific energy, E.

12.9 Critical Depth in Non-rectangular Channels

We will now consider a general irregular cross-section as shown in Fig. 12.9. The specific energy equation for any section can be written as

$$E = y + Q^2/2gA^2 \tag{12.13}$$

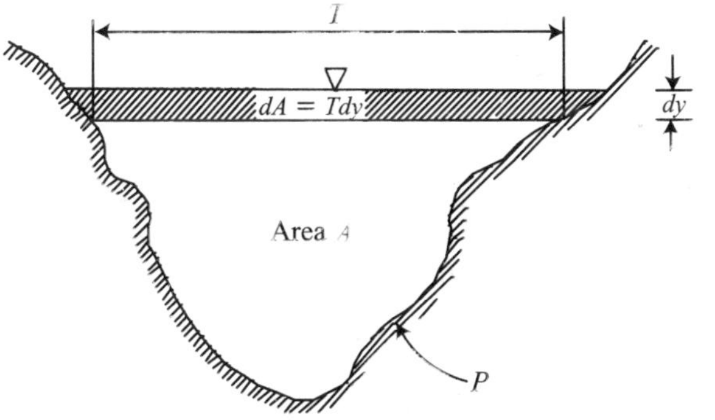

Fig. 12.9 Irregular channel cross-section

Differentiating the preceding equation with respect to y and equating to zero to get the condition for minimum specific energy

$$\frac{dE}{dy} = 1 - \left(\frac{Q^2}{gA^3}\right)\left(\frac{dA}{dy}\right) = 0$$

Referring to Fig. 12.9, for an increment dy in the depth of the flow, the increase in area $dA = T \cdot dy$, where T is the top free surface width. Thus we find that $dA/dy = T$. Substituting this in the expression for minimum specific energy, we obtain

$$1 - \left(\frac{Q^2}{gA^3}\right)T = 1 - \left(\frac{Q^2}{gA^2}\right)\left(\frac{T}{A}\right) = 0$$

A/T is the area of the channel divided by the top width and is equivalent to a depth D known as the hydraulic depth $(A/T = D)$ and $Q^2/gA^2 = V^2/g$.

Hence we have

$$\frac{V^2}{gD} = 1 \text{ or } V = (gD)^{1/2} \text{ or } \frac{V}{(gD)^{1/2}} = 1 \tag{12.14}$$

This is equivalent to the expression $V = (gy)^{1/2}$ for a rectangular section and the Froude number for an irregular cross-section is defined as

$$\mathbf{F} = \frac{V}{(gD)^{1/2}} \tag{12.15}$$

The Froude number is the ratio of the inertial force to the gravitational force. $\mathbf{F} < 1$ indicates that the gravitational forces influence the flow more than the inertial force and $\mathbf{F} > 1$ shows the predominance of inertial forces over the gravitational force.

12.9.1 Section Factor

Computation of the critical depth in non-rectangular cross-sections becomes complex since extra parameters control the area of cross-section such as the diameter of a circular channel or the side slopes and bottom width of a trapezoidal channel or the top surface width of an irregular channel.

We have seen that for a non-rectangular channel, we have obtained the criterion for critical state as $V^2/g = D$ which can also be expressed as $V^2/2g = D/2$ or in terms of discharge

$$\frac{Q^2}{2gA^2} = \frac{D}{2}$$

We now define a section factor Z_c such that

$$Z_c = AD^{1/2} = \frac{Q}{g^{1/2}} \quad \text{or} \quad Z_c^2 = \frac{A^3}{T} = \frac{Q^2}{g} \tag{12.16}$$

Z_c establishes a relation between the geometrical properties of the cross-section and the flow parameters. As Z_c is a function of the depth of flow, the critical depth can be found for a given discharge. Equation (12.16) does not contain the bed slope term S_0 which shows that the critical depth is independent of the longitudinal slope of the channel and is a function of the cross-section area and discharge. The following example shows the method of determining the critical depth for a non-rectangular section.

EXAMPLE 12.5 Compute the critical depth for a discharge of $11.36\,\text{m}^3/\text{s}$ in a trapezoidal channel of bottom width 6 m and side slope $1V : 2H$.

If y_c is the critical depth, then for the given trapezoidal cross-section, $A = y_c(B + 2y_c)$ and $T = B + 4y_c$, where T is the top width. From the property of critical section,

$$Z_c^2 = \frac{A^3}{T} = \frac{Q^2}{g} = \frac{11.36^2}{9.81} = 13.12$$

$$\frac{y_c^3(6 + 2y_c)^3}{6 + 4y_c} = 13.12$$

By a trial solution, the value of the critical depth is found to be $y_c = 0.66$ m. The critical depth can also be found graphically by drawing a plot of A^3/T vs. y for the given cross-section and reading off y_c corresponding to the value of $A^3/T = 13.12$.

12.10 Critical Depth and Wave Celerity

It has been shown that at the critical state of flow, the Froude number is unity or the velocity is given by

$$\frac{V}{(gy)^{1/2}} = 1$$

It can be shown that the celerity, c of a small disturbance wave of large wave length and small amplitude on the free surface is $(gy)^{1/2}[c = (gy)^{1/2}]$, where y is the depth of water. This implies that at critical condition, the velocity of flow is equal to the celerity of a gravity wave of small amplitude. This leads us to the important demarcation of sub-critical and super-critical flows. In still water, a disturbance moves radially outwards with velocity c as shown in 12.10a.

In sub-critical flows, $V < (gy)^{1/2}$ and a small disturbance with a celerity $c = (gy)^{1/2}$ can move upstream with a velocity $V_w = [(gy)^{1/2} - V]$ and also downstream with a velocity $[V + (gy)^{1/2}]$ as shown in Fig. 12.10b. In other words, the flow upstream is also affected by the disturbance and the sphere of influence of the disturbance is as shown in the figure. This influence of the disturbance being transmitted upstream of the location of the disturbance is known as the backwater effect and can be caused physically by a change in cross-section or an obstruction such as a dam or bridge pier etc..

In super-critical flows, however, the disturbance cannot move upstream as the velocity of the flow is greater than the celerity of the disturbance ($V > c$) and the disturbance is washed downstream with a velocity ($V - c$) as shown in Fig. 12.10d. The sphere of influence of the disturbance is also shown in the figure. It may be noted that the disturbance has no effect on the flow upstream of the location of the disturbance.

When the velocity of flow is equal to the critical velocity ($V = c$), the disturbance remains stationary as shown in Fig. 12.10c.

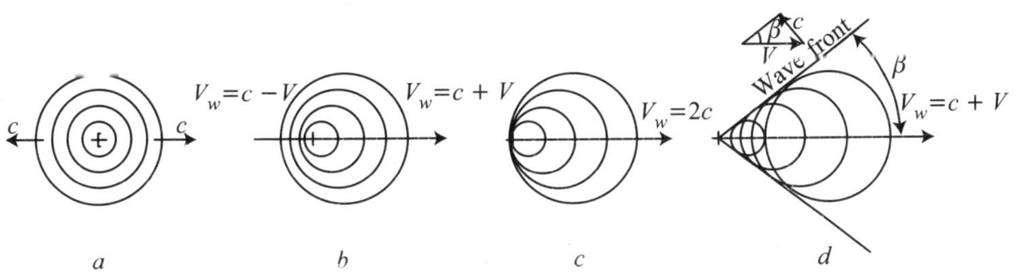

Fig. 12.10 Range of influence of a disturbance in sub-critical, super-critical and critical flows

12.11 Critical Flow and Control of Flow

A control section by virtue of either its cross-sectional area or bed elevation requires a certain minimum energy for a given rate of discharge. Hence this section controls the flow characteristics at other sections. A critical section is characterised by a minimum energy for a given discharge and hence forms a control section. From the requirement of minimum energy, one can compute the required depth of flow for a given discharge (e.g., for a rectangular channel $y_c = (q^2/g)^{1/3}$). That is, at the critical control section, the depth–discharge relationship is known and such a relationship is independent of extraneous factors such as channel roughness etc. Hence a critical section is an ideal place for the measurement of discharge in an open channel.

Control sections can be of three types. They are (1) upstream control sections, (2) downstream control sections and (3) artificial control sections.

As pointed out earlier, any surface disturbance in a sub-critical flow travels upstream affecting the flow conditions there. Hence the flow conditions in sub-critical flows are governed by conditions at the downstream end of the channel. The sub-critical flow is subject to downstream control. The critical depth can occur downstream and form a control section.

Similarly in super-critical flows, the flow is controlled by conditions upstream and is therefore subject to upstream control. The critical depth can occur upstream.

It is possible to have an artificial control section at which the depth of flow for a given discharge is either known or can be determined by such means as a sluice gate or an orifice or any other control structure. An artificial control section need not be a critical section.

Based on the principle of critical flow, devices have been developed known as critical depth metres or Parshall flumes which are used to measure discharges in open channels. A Parshall flume is similar to the venturimeter used in pipe flow measurements.

12.12 Momentum Equation

In applying the momentum equation to open channels consider the Sections 1-1 and 2-2 as shown in Fig. 12.11.

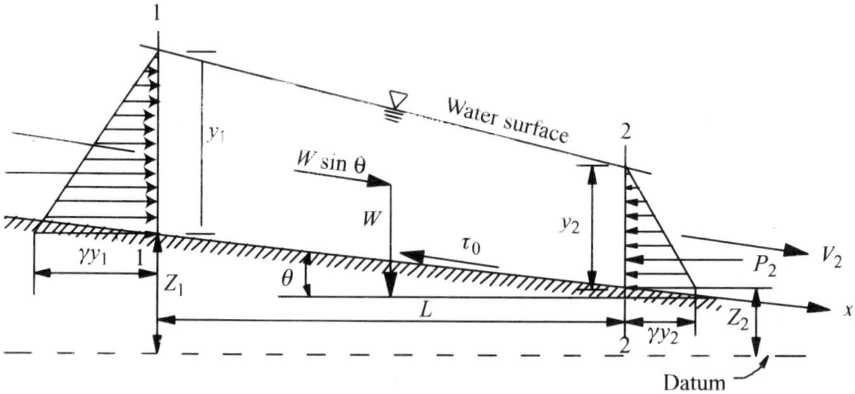

Fig. 12.11 Application of the momentum equation

Rate of change of momentum along the direction of flow = Net force along the direction of flow.

$$\int_{A_1} \rho v^2 dA - \int_{A_2} \rho v^2 dA = F_x = \int_{A_1} p dA - \int_{A_2} p dA + W \sin\theta - F_f \tag{12.17}$$

where v is the velocity through an elementary area dA at each section and p is the pressure intensity on the elementary area dA. $W \sin\theta$ is the component of the weight of water contained in between the sections along the direction of flow. θ is the inclination of the channel to the horizontal. F_f is the boundary resistance along the perimeter between the sections. The net force should include not only the above forces but also the forces on steps in the floor, forces on the sides of the channel if the channel is contracting or expanding and forces on obstructions such as bridge piers etc. as shown in Fig. 12.12.

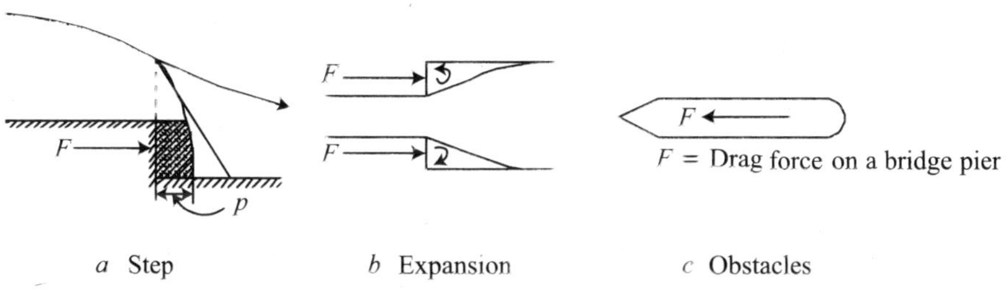

Fig. 12.12 Force components from steps, expansions, bridges

Using the correction coefficients for distribution of velocity (β) and pressure (β'), we write the above equation as

$$\rho Q(\beta_2 V_2 - \beta_1 V_1) = \beta_1' \gamma A_1 z_{1c} - \beta_2' \gamma A_2 z_{2c} - W \sin\theta - F_f \tag{12.18}$$

where z_{1c} and z_{2c} are the depths of the centroids of the areas A_1 and A_2 respectively below the free surface.

If the channel is steep, the pressure force will have to be multiplied by the slope correction factor $\cos\theta$. Rearranging the terms and dividing throughout by γ we write

$$\frac{F}{\gamma} = \left(\frac{\beta_1 Q^2}{gA_1} + \beta_1' A_1 z_{c1}\right) - \left(\frac{\beta_2 Q^2}{gA_2} + \beta_2' A_2 z_{c2}\right)$$

At any section the value of $\beta Q^2/gA + \beta' A z_c$ is known as the specific force. For rectangular channels with $\beta = \beta' = 1$ and $A = By$; $z_c = y/2$, the specific force is

$$\frac{F}{\gamma} = \frac{Q^2}{gBy} + \frac{By^2}{2} \text{ or } \frac{F}{\gamma B} = \frac{Q^2}{gy} + \frac{y^2}{2} \tag{12.19}$$

12.13 Specific Force Curve

The plot of the specific force equation (12.19) for a constant discharge is known as the specific force curve (Fig. 12.13). The curve has two asymptotes. At $y = 0$, the curve is asymptotic to the $F/\gamma B$ axis and at $y =$ infinity, the curve is asymptotic to the parabola $F/\gamma B = y^2/2$.

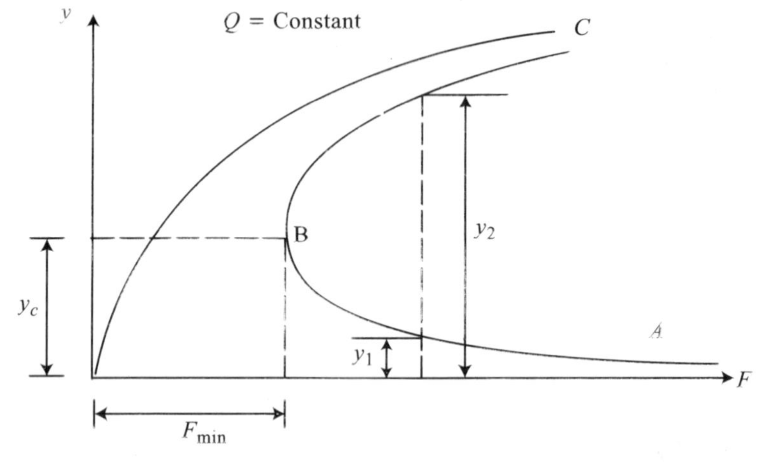

Fig. 12.13 Specific force diagram

For a given value of specific force, the specific force curve gives two depths y_1 and y_2 which are known as sequent depths. At point B, where the two depths assume a common value, the specific force is a minimum. Differentiating F with respect to y and equating to zero we get

$$\left(\frac{1}{\gamma B}\right)\left(\frac{dF}{dy}\right) = -\frac{q^2}{gy^2} + y = 0$$

Therefore we have the critical depth for minimum specific energy as

$$y_c = \left(\frac{q^2}{g}\right)^{1/3} \tag{12.20}$$

Thus we find, comparing equations (12.9) and (12.20), that both specific energy and specific force are a minimum at the same critical depth for a constant discharge.

EXAMPLE 12.6 A bridge has a pier 60 cms wide in a channel 6 m wide The upstream depth in the channel is 3 m and the velocity is 3 m/s. Assuming that the drag coefficient of the piers, C_D is 1.5. find the depth of flow downstream of the piers. Neglect bed slope and boundary resistance.

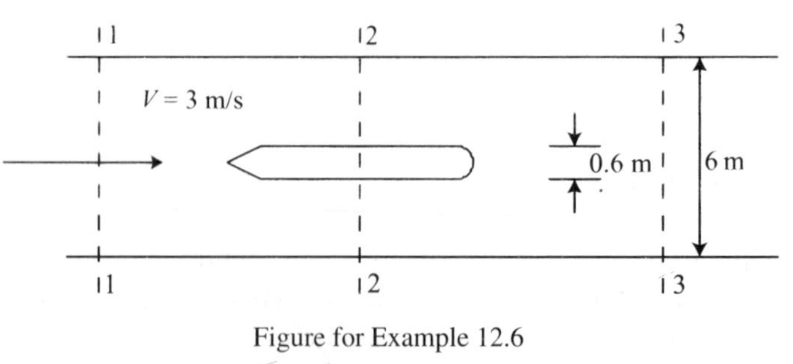

Figure for Example 12.6

$Q = 6 \times 3 \times 3 = 54\,m^3/s$ or $q = 9\,m^3/s$. Exposed area of the pier $= 0.6 \times 3 = 1.8\,m^2$ neglecting afflux at the upstream end of the pier.

Drag on pier $= C_D A(1/2\,\rho V^2) = 1.5 \times 1.8(1/2\,\rho\,3^2) = 12.15\rho$

The fluid exerts a force on the pier and the pier in turn exerts an opposite force on the fluid. Applying the momentum equation, we get

$$6(\rho g)3^2/2 + \rho \cdot 54 \times 3 = 6(\rho g)y_2^2/2 + \rho \cdot 54^2/6y_2 + 12.15\rho$$

$$426.87 - 12.15 = 414.72 = 29.43y_2^2 + 486/y_2$$

Solving by trial and error, $y_2 = 2.9\,m$.

12.14 Flow Conditions at Piers

12.14.1 Energy Approach

The flow through bridge piers may be sub-critical, critical, or super-critical. The energy equation as applied to the contracted Section 2-2 and the channel or river with a regular cross-section at 3-3 downstream of the bridge as shown in the figure for Example 12.6, assuming a level bottom is

$$\varepsilon(y_2 + V_2^2/2g) = y_3 + V_3^2/2g$$

or

$$\varepsilon y_2(2 + \mathbf{F}_2^2) = y_3(2 + \mathbf{F}_3^2)$$

where $\mathbf{F}_2$ and $\mathbf{F}_3$ are Froude numbers at Sections 2-2 and 3-3 and ε denotes the percentage recovery of energy since energy loss will occur at the two cross-sections. By continuity, we have $V_2 b_2 y_2 = V_3 B_3 y_3$. Denoting the contraction ratio b_2/B_3 as σ, the continuity equation can be written as

$$\mathbf{F}_2^2 \sigma^2 y_2^3 = \mathbf{F}_3^2 y_3^3$$

Eliminating y_2 and y_3 between the energy equation and continuity equation, we get

$$\sigma^2 = \frac{\varepsilon^3 \mathbf{F}_3^2 (2 + \mathbf{F}_2^2)^3}{\mathbf{F}_2^2 (2 + \mathbf{F}_3^2)^3}$$

If the flow is critical in between the piers at Section 2-2, then we put $\mathbf{F}_2 = 1$.

Then $\mathbf{F}_3^*$ satisfying this condition is

$$\sigma^2 = \frac{27\varepsilon^3 \mathbf{F}_3^{*2}}{(2 + \mathbf{F}_3^{*2})^3}$$

The flow between the piers is sub-critical, critical, or super-critical depending on whether $\mathbf{F}_3$ is less than or equal to or greater than $\mathbf{F}_3^*$. If the flow between the piers is super-critical, a hydraulic jump will occur downstream of the pier.

12.14.2 Momentum Approach

The energy approach suffers from the fact that an accurate estimation of energy loss (or value of ε) cannot be made. Hence it is recommended that we adopt the momentum approach in such situations. Let y_3' be the depth at the down end of the bridge. The momentum equation can be written as

$$b_2\left(\frac{\gamma y_2^2}{2}\right) + \rho b_2 y_2 V_2^2 + (B_3 - b_2)\left(\frac{\gamma y_3'^2}{2}\right) = B_3\left(\frac{\gamma y_3^2}{2}\right) + \rho B_3 y_3 V_3^2$$

y_3' is obviously between y_2 and y_3. If $y_3' = y_2$, the equation may be simplified as

$$\frac{y_2^2}{2} + \left(\frac{b_2}{B_3}\right)\mathbf{F}_2^2 y_2^2 = \frac{y_3^2}{2} + \mathbf{F}_3^2 y_3^2$$

$$\left(\frac{y_2^2}{y_3^2}\right)(1 + 2\sigma\mathbf{F}_2^2) = 1 + 2\mathbf{F}_3^2$$

But from the equation of continuity, we have $(y_2/y_3)^3 = (\mathbf{F}_3^2/\mathbf{F}_2^2\sigma^2)$. On substitution, we have

$$\frac{\mathbf{F}_2^4\sigma^4}{\mathbf{F}_3^4} = \left[\frac{(1 + 2\sigma\mathbf{F}_2^2)}{(1 + 2\mathbf{F}_3^2)}\right]^3$$

If the flow is critical between the piers, then $\mathbf{F}_2 = 1$ and the above relation reduces to

$$\sigma = \left[\frac{(2 + 1/\sigma)^3}{(1 + 2\mathbf{F}_3^2)^3}\right]\mathbf{F}_3^4$$

This may be compared with the relation obtained from the energy concept.

If $y_3' = y_3$, then it can be shown that

$$\sigma = \left[\frac{(2\mathbf{F}_3^2 + \sigma)}{(1 + 2\mathbf{F}_2^2)}\right]^3\left(\frac{\mathbf{F}_2^4}{\mathbf{F}_3^4}\right)$$

For critical condition with $\mathbf{F}_2 = 1, \sigma = (2\mathbf{F}_3^2 + \sigma)^3/27\mathbf{F}_3^4$.

12.15 Uniform Flow

Uniform flow is characterised by the fact that the total energy line and the free water surface are parallel to the channel bottom i.e., $S_f = S_w = S_0$ as illustrated in Fig. 12.14.

It is evident that uniform flow can be established only in long prismatic channels of constant bed slope wherein the flow is fully established with the boundary layer fully developed having reached the free surface. The depth of flow is known as the normal depth, y_n and the corresponding discharge is known as the normal discharge, Q_n.

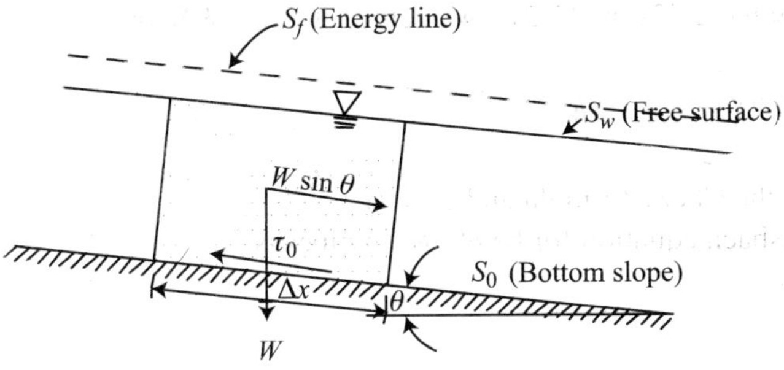

Fig. 12.14 Uniform flow

Depending on the slope of the channel, uniform flow for a given discharge can occur as sub-critical, critical or super-critical flow.

As the depths and velocities are constant along the channel in a steady uniform flow, there cannot be any acceleration or deceleration of flow. Hence in uniform flow, all forces must be in equilibrium or the gravity force must be balanced by the boundary resistance force.

12.16 Uniform Flow Formulae

Consider an element of dimensions depth y, length Δx, and area A. The hydrostatic forces on either side of the prismatic element cancel each other as the depth is uniform. The component of the weight of the element in the direction of the flow is $\gamma A \Delta x \sin\theta$. This force is resisted by the boundary shear force. If τ_0 is the mean longitudinal shear stress acting on the wetted perimeter P, the total shear stress on the element is $\tau_0 \cdot P \cdot \Delta x$. Equating the two forces, we get

$$\gamma \cdot A \cdot \Delta x \sin\theta = \tau_0 \cdot P \cdot \Delta x$$

$$\tau_0 = \gamma \left(\frac{A}{P}\right) \sin\theta = \gamma R S_0 \tag{12.21}$$

In the chapter on laminar flow, it was shown that the energy slope for uniform laminar free surface flow is

$$S_0 = \frac{3\mu V}{\gamma y^2}$$

The shear stress in a laminar flow is therefore proportional to the first power of V. This is true for small Reynolds numbers approximately up to 500. In the turbulent range τ_0 is proportional to V^2. From Newton's drag equation $F = C_D A(1/2\rho V^2)$ or dividing by area $\tau = (C_D/2)\rho V^2$ or

$$\tau_0 = c_f \rho V^2 \tag{12.22}$$

Substituting equation (12.22) in (12.21), we get $c_f \rho V^2 = \gamma R S_0$ or

$$V = \left[\left(\frac{g}{c_f}\right) RS\right]^{1/2} = C(RS)^{1/2} \tag{12.23}$$

which is known as the Chezy formula and $C = (g/c_f)^{1/2}$

The Darcy–Weisbach equation for head loss in pipes is

$$h_f = \frac{fLV^2}{2gd}$$

or

$$\frac{h_f}{L} = S = \frac{fV^2}{2gd} = \frac{fV^2}{2(4R)} = \frac{fV^2}{8R}$$

or

$$V = \left(\frac{8g}{f}\right)^{1/2} (RS)^{1/2} \tag{12.24}$$

Comparing equations (12.23) and (12.24), we find $c_f = f/8$.

Determination of the value of Chezy C is still empirical, based on the formulae developed by various investigators such as Ganguillet and Kutter, Bazin etc. It has been found that if channel boundary characteristics are similar to those found in closed conduit boundaries, the value of Chezy may be inferred by $C = (8g/f)^{1/2}$.

12.17 Manning's Formula

Another widely accepted empirical formula for an open channel design is the Manning's equation for prescribing the average velocity in the channel. In metric units, it is expressed as

$$V = \left(\frac{1}{n}\right) R^{2/3} S^{1/2} \tag{12.25}$$

where $R = A/P$ is the hydraulic radius of the channel, S is the slope of the channel and n is the Manning's roughness factor which is characteristic of the boundary material/roughness. It is evident that the equation is dimensionally non-homogeneous. We can render it dimensionally homogeneous by comparing equations (12.24) and (12.25). The correlation between Chezy C, Darcy f and Manning n is

$$C = \left(\frac{8g}{f}\right)^{1/2} = \frac{R^{1/6}}{n} \tag{12.26}$$

$$\frac{1}{(f)^{1/2}} = \left[\frac{1}{(8g)^{1/2}}\right]\left(\frac{R^{1/6}}{n}\right)$$

As n varies only with the boundary roughness, k the linear factor 1.0 appearing in Manning's equation must have the dimension of $(g)^{1/2}$. Since R has the dimension of L, it follows that n must have the dimension of $[L^{1/6}]$. As Manning's n is a function of boundary roughness which in turn depends on the roughness height, spacing of roughness elements, and their shape, all of which have linear dimensions, it follows that n must also have a linear dimension. In equation (12.26), Manning's parameter $(R^{1/6}/n)$ represents a ratio of two lengths similar to the relative roughness parameter r_0/k used in pipe flow, r_0 being the radius of the pipe. In highly turbulent pipe flow f is a function of r_0/k and according to equation (12.26), f is related to $(R^{1/6}/n)$ and hence it is possible to evaluate Manning's parameter in terms of r_0/k. Since the hydraulic radius, R of the pipe is equal to $d/4 = r_0/2 (R = r_0/2)$, we have $r_0/k = 2R/k$. The relation between r_0/k or $2R/k$ and $R^{1/6}/n$ is shown in Fig. 12.15.

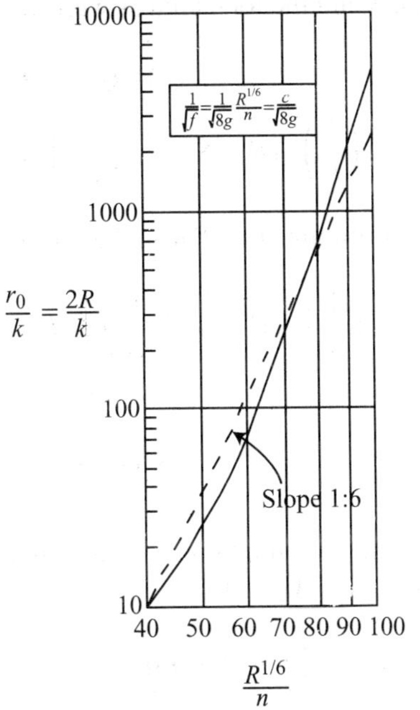

Fig. 12.15 Relation between absolute roughness and Manning's n

From Fig. 12.15, it is apparent that a small increase in the value of n requires a correspondingly large increase in the absolute roughness k. The deviation of the curve from the dotted straight line having a slope of 6:1 indicates the error involved in the empirical Manning's equation. Apart from the satisfactory results that the Manning's formula has given in field application, it has inherent defects.

1. It ignores viscosity effects and hence is least dependable at low Reynolds numbers.
2. Roughness coefficient n is assumed to remain constant at all velocities whereas it may change with different velocities.
3. It describes only the mean velocity.
4. Boundary shear is assumed constant along the perimeter.

12.18 Computation of Equivalent Manning's n for Composite Roughness

In channels where roughness along the perimeter varies, the equivalent value of n has to be found for the entire section. An example of a channel with composite roughness is the case of a stream or river with a bed made of pebbles and small boulders with the sides overgrown with grass and weeds.

There are a number of formulae for computing equivalent roughness. A popular method is based on discharge consideration. We consider the channel to be divided into N parts of areas $A_1, A_2, A_3 \ldots A_N$ with wetted perimeters $P_1, P_2, P_3 \ldots P_N$ whose roughness coefficients are $n_1, n_2, n_3 \ldots n_N$ respectively. Assuming that the total discharge is equal to the sum of the discharges through the sub-divided areas and applying the Manning's equation to each of the areas and the total area as well, we get

$$Q = Q_1 + Q_2 + Q_3 \ldots\ldots + Q_N$$

$$\left(\frac{1}{n}\right) AR^{2/3}S^{1/2} = \left(\frac{1}{n_1}\right) A_1 R_1^{2/3} S^{1/2} + \left(\frac{1}{n_2}\right) A_2 R_2^{2/3} S^{1/2} + \ldots + \left(\frac{1}{n_N}\right) A_N R_N^{2/3} S^{1/2}$$

Substituting $A = PR$ in the above equation, we write

$$\left(\frac{1}{n}\right) PR^{5/3} = \left(\frac{1}{n_1}\right) P_1 R_1^{5/3} + \left(\frac{1}{n_2}\right) P_2 R_2^{5/3} + \ldots\ldots + \left(\frac{1}{n_N}\right) P_N R_N^{5/3}$$

$$n = \frac{PR^{5/3}}{\sum_1^N (P_N R_N^{5/3})/(n_N)} \tag{12.27a}$$

Another assumption to make is that the average velocity in each of the sub-sections is the same viz., $V_1 = V_2 = V_3 = \ldots\ldots = V$. The equivalent roughness is then expressed as

$$n = \frac{\left[P_1 n_1^{1.5} + P_2 n_2^{1.5} + \ldots\ldots + P_N n_N^{1.5}\right]^{2/3}}{P^{2/3}} \tag{12.27b}$$

Yet another procedure to determine the equivalent roughness coefficient is to assume that the total force resisting the flow (KV^2P per unit length) is equal to the sum of the forces offered by resisting the flow in each of the sub-sections. The equivalent roughness is then given as

$$n = \frac{\left[P_1 n_1^2 + P_2 n_2^2 + \ldots\ldots + P_N n_N^2\right]^{1/2}}{P^{1/2}} \tag{12.27c}$$

12.19 Computation of Uniform Flow

Computation of uniform flow involves the determination of the normal depth, normal discharge and corresponding slopes for a given cross-sectional form. Two useful concepts which facilitate the computation of uniform flow are the conveyance and section factors. Conveyance K is a measure of the discharging capacity of a channel and is the ratio of discharge and the square root of the longitudinal slope.

$$K = \frac{Q}{(S_0)^{1/2}} = CAR^{1/2} \quad \text{(Chezy formula)}$$

$$K = \frac{Q}{(S_0)^{1/2}} = \left(\frac{1}{n}\right) AR^{2/3} \quad \text{(Manning's formula)}$$

$AR^{2/3}$ is known as the conveyance factor for uniform flow computations.

$$AR^{2/3} = nK = \frac{nQ}{(S_0)^{1/2}} \tag{12.28}$$

Equation (12.28) is a useful expression for the computation of normal depth given the normal discharge and bed slope. The following example shows the determination of normal depth given the discharge and the slope.

EXAMPLE 12.7 A trapezoidal channel has a base width of 9.6 metres and side slopes of $1V : 1H$ (45°). The bed slope is 0.000484. If Manning's $n = 0.0135$, find the normal depth for a discharge of 8.52 m^3/s.

If y_n is the normal depth, then for the given cross-section, we have

$$A = (9.6 + y_n)y_n, \; P = \left(9.6 + 2 \times 2^{1/2} y_n\right)$$

and the hydraulic radius

$$R = \frac{(9.6 + y_n)y_n}{(9.6 + 2.828y_n)}$$

The conveyance factor

$$AR^{2/3} = \frac{nQ}{\left(S_0^{1/2}\right)} = \frac{0.0135 \times 8.52}{(0.000484)^{1/2}} = 5.29$$

$$\frac{(9.6 + y_n)^{5/3} y_n^{5/3}}{(9.6 + 2\sqrt{2}y_n)^{2/3}} = 5.29$$

A trial and error solution gives $y_n = 0.7$ m.

Normal depth can also be found graphically by plotting $AR^{2/3}$ vs. y for the given cross-section and reading off the value of y_n for the corresponding value of $AR^{2/3} = 5.29$. For geometrically defined cross-sections, the graphical solution can be greatly facilitated by the preparation of non-dimensional charts of $AR^{2/3}$ vs. y.

12.20 Economical Cross-Section

The most efficient cross-section may be defined as that which offers least resistance to flow and hence gives maximum discharge for a given slope, area and roughness. Such a section has the least perimeter for a given area of flow section and hence maximum conveyance. The semi-circle having the least perimeter for a given cross-sectional area may be considered hydraulically the most efficient section and has been adopted in concrete aqueducts to minimise the cost. But the best hydraulic cross-section may not always be practical and modifications may have to be introduced though at the cost of efficiency. Moreover, the best hydraulic section may give a minimum cross-sectional area for a given discharge but may not give minimum excavation which depends on the relative elevation of the water surface and the ground level (free board). We will consider the most common trapezoidal cross-section and examine the conditions for best efficiency.

Let the bottom width of a trapezoidal channel be B, side slope $1 : z$ and depth of flow, y.

The area $A = y(B + zy)$ and perimeter $P = B + 2y(1 + z^2)^{1/2}$

Substituting for $B = (A/y) - zy$; $P = A/y - zy + 2y(1 + z^2)^{1/2}$

Keeping A and z constant and differentiating P with respect to y and equating it to zero, we get

$$\frac{dP}{dy} = -\frac{A}{y^2} - z + 2(1 + z^2)^{1/2} = \frac{y(B + zy)}{y^2} - z + 2(1 + z^2)^{1/2} = 0$$

On simplification we get

$$B = 2y[(1 + z^2)^{1/2} - z]$$

Substituting the value of B so obtained in the expressions for A and P and solving for P, we get

$$P = 2y[2(1 + z^2)^{1/2} - z] \quad \text{and} \quad A = y^2[2(1 + z^2)^{1/2} - z]$$

On elimination of y from the above two expressions, we get

$$P = 2\{A[2(1 + z^2)^{1/2} - z]\}^{1/2}$$

Squaring the preceding expression and differentiating with respect to z, we have

$$2P\frac{dP}{dz} = 4A\left[2 \times \frac{1}{2}(1 + z^2)^{-1/2}(2z) - 1\right]$$

Setting dP/dz to zero one obtains $2z(1 + z^2)^{-1/2} = 1$ or $z = 1/\sqrt{3}$ or

$$z = \tan 30^\circ$$

Thus the most hydraulically efficient trapezoidal section is half of a hexagon for which one gets the hydraulic radius $R = y/2$.

Similar relationships may be obtained for other cross-sectional shapes.

EXAMPLE 12.8 A trapezoidal channel with one side vertical and the other sloping at $1V : 2H$ carries a discharge of $28\,\text{m}^3/\text{s}$ at a mean velocity of 1.5 m/s. Determine the slope and cross-sectional dimensions for best hydraulic efficiency.

For the given section $A = y(B + y) = Q/V$ and $B = (A/y) - y$. The wetted perimeter $P = B + y + 5^{1/2}y = (A/y) + 5^{1/2}y$. Differentiating with respect to y and setting it equal to zero for minimum P

$$\frac{dP}{dy} = -\frac{A}{y^2} + 5^{1/2} = 0$$

or

$$y = \left(\frac{A}{5^{1/2}}\right)^{1/2} = \left(\frac{Q}{5^{1/2}V}\right)^{1/2} = \left[\frac{28}{1.5 \times 5^{1/2}}\right]^{1/2} = 2.89\,\text{m}$$

Hence we have

$$B = A/y - y = Q/Vy - y = 28/(1.5 \times 2.89) - 2.89 = 3.56\,\text{m}$$

12.21 Gradually Varied Flow

A gradually varied flow is defined as a steady flow whose depth varies gradually along the length of the channel. Gradually varied flows occur in the vicinity of controls such as dams, sluice gates, channel transitions and changes in bed slope. In all such problems, prediction of the flow profile for given flow conditions assumes importance.

The theory of gradually varied flow is based upon the following assumptions.

1. The uniform flow formula can be used to evaluate head loss in gradually varied flow and roughness coefficients applicable to uniform flow are valid even for non-uniform flow.
2. The channel slope is small.
3. The channel is prismatic.
4. The stream lines are practically parallel and curvature effects are negligible.
5. Velocity distribution coefficients are constant and in most cases may be taken as unity.

The differential equation for gradually varied flow is derived from the energy equation. The total head above a chosen datum at any cross-section is

$$H = z + y\cos^2\theta + \alpha V^2/2g$$

Choosing the channel bottom as the x-axis and differentiating the preceding equation with respect to x, we get

$$\frac{dH}{dx} = \frac{dz}{dx} + \left(\frac{dy}{dx}\right)\cos^2\theta + \frac{d}{dy}\left(\alpha\frac{V^2}{2g}\right)\cdot\left(\frac{dy}{dx}\right)$$

Solving for dy/dx and substituting $dz/dx = -S_0$, and $dH/dx = -S_f$, we have

$$\frac{dy}{dx} = \frac{S_0 - S_f}{\cos^2\theta + d/dy\left(\alpha\frac{V^2}{2g}\right)} \tag{12.29}$$

For small slopes $\cos\theta \sim 1$ and assuming α [illegible] we rewrite the above equation

$$\frac{dy}{dx} = \frac{S_0 - S_f}{1 + d/dy\left(\frac{V^2}{2g}\right)} \tag{12.30}$$

Equation (12.30) is the general differential equation of gradually varied flow. It represents the slope of the free surface with respect to the channel bottom. The term $d(V^2/2g)/dy$ represents the velocity head and has to be evaluated in terms of measurable quantities. As $V = Q/A$,

$$\frac{d}{dy}\left(\frac{V^2}{2g}\right) = \frac{d}{dy}\left(\frac{Q^2}{2gA^2}\right) = -\frac{Q^2}{gA^3}\frac{dA}{dy} = -\frac{Q^2T}{gA^3} \tag{12.31}$$

as $dA/dy = T$, the top free surface width.

Further we can introduce the section factors, $Z^2 = A^3/T$, $Z_c^2 = Q^2/g$

$$\frac{d}{dy}\left(\frac{V^2}{2g}\right) = -\frac{Z_c^2}{Z^2} \tag{12.32}$$

From the definition of conveyance factor, energy slope of gradually varied flow can be expressed as

$$S_f = \frac{Q^2}{K^2}$$

When uniform flow occurs at discharge Q, then K becomes K_n and $S_f = S_0$.

$$S_0 = \frac{Q^2}{K_n^2}$$

From the preceding two relations for the energy and bed slopes, we find

$$\frac{S_f}{S_0} = \frac{K_n^2}{K^2}$$

From the preceding relationships we have

$$\frac{dy}{dx} = S_0 \frac{(1 - K_n^2/K^2)}{(1 - Z_c^2/Z^2)} \tag{12.33}$$

This is another form of gradually varied flow equation valid for any cross-sectional form.

12.22 Classification of Flow Profiles

Equation (12.33) represents the longitudinal slope of the free surface with respect to the channel bottom. It can therefore be used to describe characteristics of various flow profiles. Flow profiles can be broadly classified into backwater curves, uniform flow profiles, and draw-down curves. When dy/dx is positive, the depth of flow increases in the direction of flow and the flow profile represents a backwater curve. When dy/dx is negative, the depth of flow decreases in the downstream direction and the flow profile represents a draw-down curve. When $dy/dx = 0$, the free surface is parallel to the channel bottom and flow is uniform.

As the bottom slope term S_0 appears in equation (12.33), it is evident that all other conditions being the same, the form of the free surface is a function of the bottom slope. Channel bottom slopes are classified as sustaining and non-sustaining. A sustaining slope is a positive slope in which the elevation of the bottom of the channel decreases in the direction of flow thus assisting the flow by gravity. It can be further classified as mild, critical and steep slopes. A mild slope is one which sustains a given discharge, Q as uniform flow at a normal depth, y_n greater than the critical depth, y_c for the given discharge $(A^3/T = Q^2/g)$ or $y_n > y_c$. Similarly a steep slope sustains a normal depth y_n less than the critical depth $(y_n < y_c)$. When the normal depth is equal to the critical depth, then the slope is described as a critical slope ($y_n = y_c$). For a given discharge and channel geometry, the critical slope described as S_c forms a dividing line between mild and steep slopes. For a mild slope, $y_n > y_c$ and $S_0 < S_c$. For a critical slope, $y_n = y_c$, $S_0 = S_c$. For steep slopes, $y_n < y_c$ and $S_0 > S_c$.

A non-sustaining slope or adverse slope is a negative slope in which the elevation of the channel bottom increases in the direction of the flow with gravitational force acting against the flow. A horizontal slope is one where $S_0 = 0$.

For a given discharge and channel conditions, the normal depth and critical depth divide the flow regime into three zones as shown in Fig. 12.16

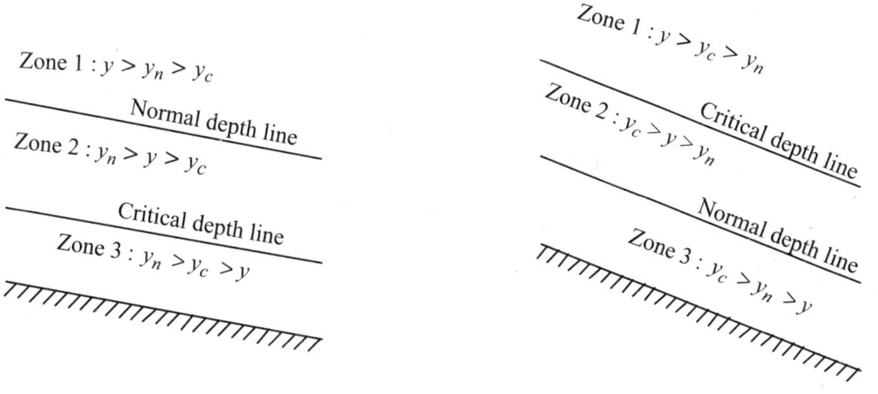

a Mild slope *b* Steep slope

Fig. 12.16 Classification of flow

Zone 1 represents the space above the normal depth line. Zone 2 lies between the normal depth and the critical depths. Finally Zone 3 lies below the critical depth.

The flow profiles are classified according to the nature of the bottom slope and the zone in which it occurs. For instance, a profile which occurs on a mild slope and in Zone 2 is denoted as an M_2 profile and so on. It will be seen in the next section that there are thirteen possible profiles, viz., M_1, M_2, M_3, S_1, S_2, S_3, C_1, C_2, C_3, H_2, H_3, A_2, A_3. Of these, twelve are varied flow profiles and one, the C_2 profile is a uniform flow profile where the normal depth and the critical depth are one and the same.

12.23 Characteristics of Flow Profiles

In the ensuing paragraphs, each of the above thirteen profiles will be individually discussed.

M_1 profile

This profile occurs on a mild slope for which $y_n > y_c$ and in Zone 1 with the actual depth $y > y_n > y_c$. Since the values of K and Z are proportional to the depth of flow, we have $K > K_n$ and $Z > Z_c$.

Therefore from equation (12.33), dy/dx is positive. Therefore the M_1 profile is a backwater curve with a depth equal to y_n at the upstream end. As depths become large on the downstream end, from equation (12.33) we infer that $dy/dx = S_0$ or the water surface becomes horizontal. Therefore the M_1 profile is asymptotic to the normal depth at the upstream end and asymptotically becomes horizontal at the downstream end. The M_1 profile shown in Fig. 12.17(a) is the most important of all profiles from the practical point of view as it occurs behind a dam or an obstruction across a sub-critical stream. It occurs also when a mild slope meets a milder slope (Fig. 12.17b).

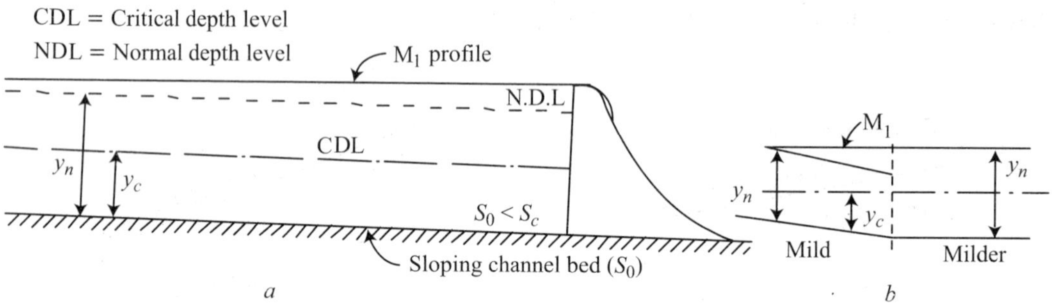

Fig. 12.17 M_1 profile

M_2 profile

This occurs in Zone 2 where we have $y_n > y > y_c$. From equation (12.33), dy/dx is negative. The M_2 profile is a draw-down curve and represents a sub-critical flow occurring on a mild slope. The upstream end is tangential to the normal depth line since $dy/dx = 0$ when $y = y_n$ which is the maximum depth possible on an M_2 profile. The downstream end of an M_2 profile ends when reaching a critical depth with a vertical slope as dy/dx becoming infinite when $y = y_c$. This profile occurs when the channel ends abruptly with a free fall as shown in Fig. 12.18a or when the channel slope changes over from a sub-critical slope to a super-critical slope as shown in Fig. 12.21 or when a milder slope is followed by a mild slope (Fig. 12.18b).

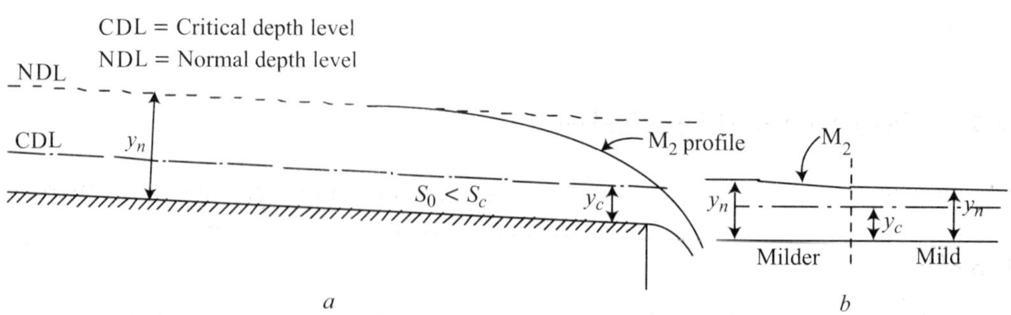

Fig. 12.18 M_2 profile

M_3 profile

On a mild slope when the initial depth, y is less than y_c and $y_n (y < y_c < y_n)$, the profile lies in Zone 3 as shown in Fig. 12.19*a*. The free surface slope dy/dx from equation (12.33) is positive. Since the starting depth is less than the critical depth, the flow is initially super-critical and the depth increases to finally reach the normal depth corresponding to the given mild slope. The minimum depth possible on an M_3 profile is theoretically zero and terminates in a hydraulic jump to reach the normal depth. This profile occurs on the downstream side of a sluice gate when the gate opening is less than the critical depth or when steep slope changes to mild slope (Fig. 12.19*b*).

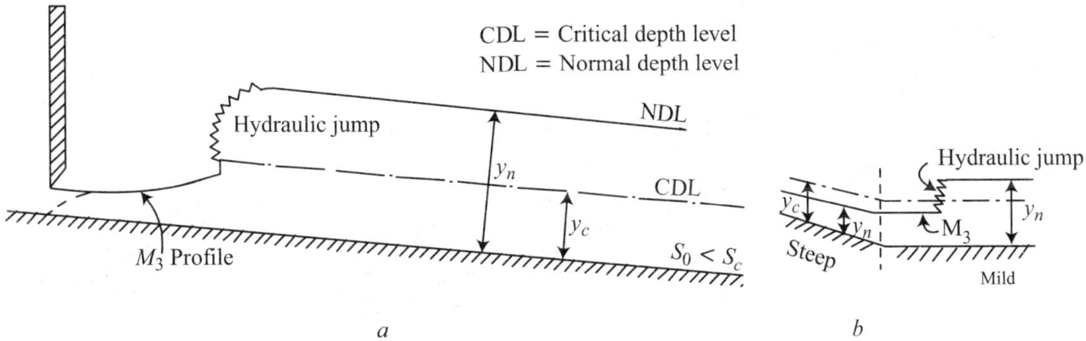

Fig. 12.19 M_3 flow profile

S_1 profile

The channel is super-critical and the initial depth is $y > y_c > y_n$. The surface slope dy/dx is positive. An S_1 profile represents a change over from super-critical flow to sub-critical flow (a backwater curve) or from a steep slope to a lesser steep slope as shown in Fig. 12.20. An S_1 profile begins with a jump as shown in Fig. 12.20 on the upstream end and becomes tangential to the depth at the downstream end. The profile behind an obstruction across a super-critical flow exemplifies the S_1 profile. It is also found when a steep channel meets a pool of high elevation.

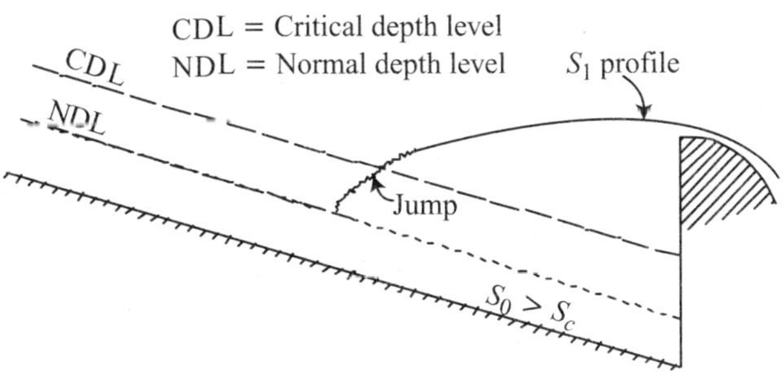

Fig. 12.20 S_1 profile

S_2 profile

The initial depth on a super-critical slope is between the normal depth and the critical depth ($y_c > y > y_n$). The free surface slope dy/dx is then negative. An S_2 profile is a draw-down curve and represents super-critical flow in a steep channel; it is generally a short profile and occurs when the channel slope changes from mild to steep as shown in Fig. 12.21. It is a change over from an M_2 to an S_2 profile. At the break in slope, critical depth occurs and $dy/dx = \tan 90°$. At the downstream end it becomes tangential to the normal depth on the steep slope. It can also occur when a steep slope is followed by a steeper slope.

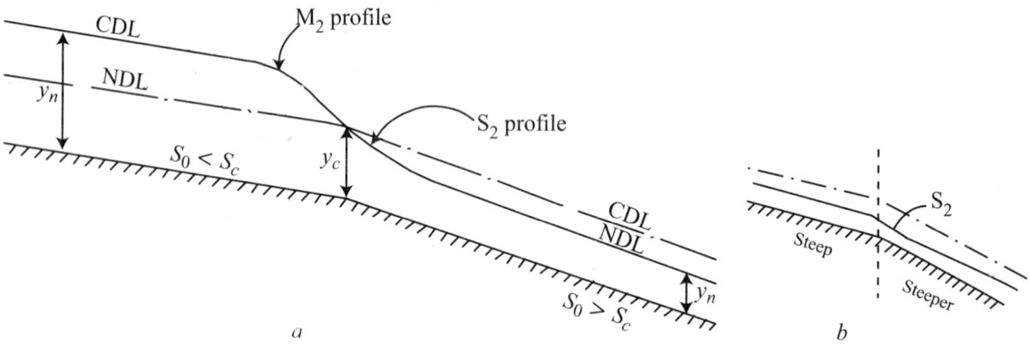

Fig. 12.21 S_2 profile

S_3 profile

It starts with a depth less than the normal depth on a steep slope and becomes tangential to the normal depth line at the downstream end as shown in Fig. 12.22. $y_c > y_n > y$ and dy/dx is positive. An example is found in the flow below a sluice gate located on a steep slope with an opening less than y_n. It is also found to form when a steeper slope is followed by a steep slope.

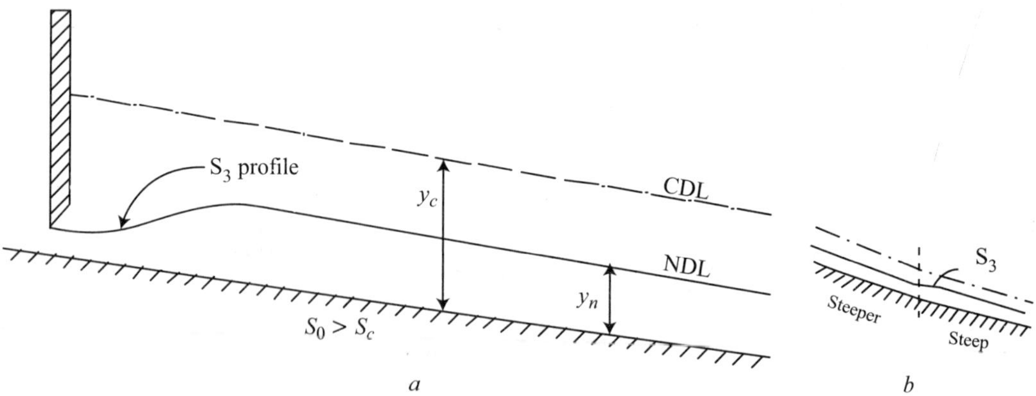

Fig. 12.22 S_3 profile

C_1 profile

This occurs on the critical slope where the normal depth is equal to the critical depth ($y_n = y_c$) and the initial depth $y > y_n$. The free surface slope dy/dx is positive and occurs when the critical slope changes over to a mild slope and the profile is formed on the critical slope as shown in Fig. 12.23. It occurs when a critical slope meets a pool of high elevation.

It resembles the backwater profile. C profiles are transition profiles between M and S profiles and possess the characteristics of both. At the upstream end, a C_1 profile starts from the critical depth which is also the normal depth in this case and becomes tangential to the normal depth of the mild slope.

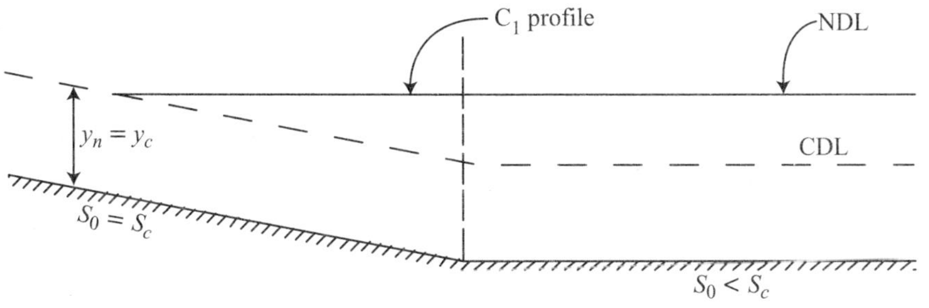

Fig. 12.23 C_1 profile

C_2 profile

The C_2 profile represents a uniform critical flow profile with $y = y_n = y_c$ and $dy/dx = 0$ as shown in Fig. 12.24.

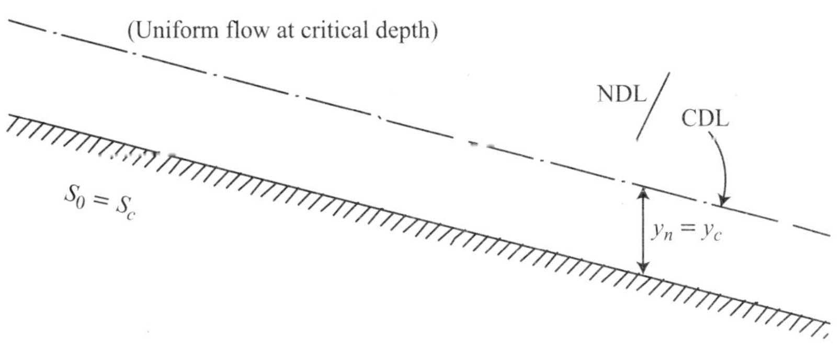

Fig. 12.24 C_2 profile

C_3 profile

In this case, the initial depth is below the critical/normal depth as illustrated by a sluice gate located on a critical slope with an opening less than the critical/normal depth as shown in Fig. 12.25 ($y < y_n = y_c$) and dy/dx is positive. The profile and increase in depth is dictated by the downstream conditions. The profile is also found when a steep slope meets a critical slope.

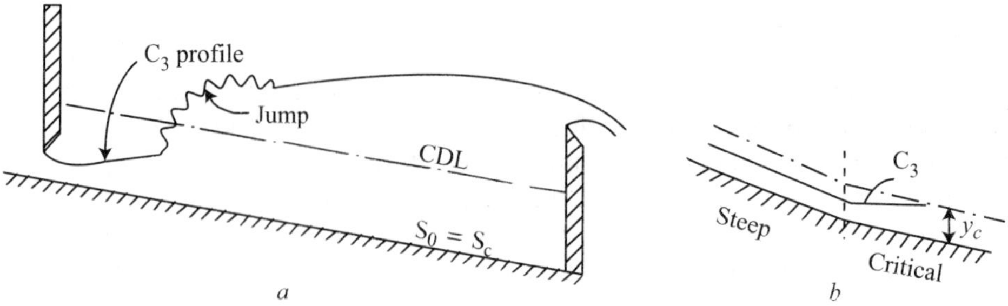

Fig. 12.25 C_3 profile

H_1 profile

As $S_0 = 0$, $K_n = Q/(S_0)^{1/2}$ becomes infinite with y_n also becoming infinite. Hence there is no flow possible with a horizontal slope and the profile is non-existent.

H_2 profile

The depth is initially greater than the critical depth ($y > y_c$) as shown in Fig. 12.26 and the profile is a draw-down curve with dy/dx being negative. The flow is sub-critical. H profiles are limiting cases of M profiles when $S_0 = 0$ with the H_2 profile corresponding to the M_2 profile.

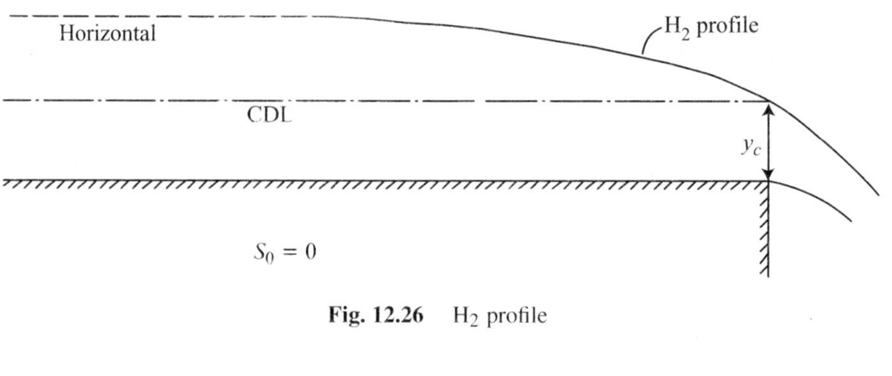

Fig. 12.26 H_2 profile

H_3 profile

The flow issues from a controlled opening such as a sluice gate located on a level floor and the initial depth $y < y_c$, the flow is super-critical. The free surface slope dy/dx is positive and the depth rises. If suitable tail-water conditions exist, a hydraulic jump may occur downstream as shown in Fig. 12.27.

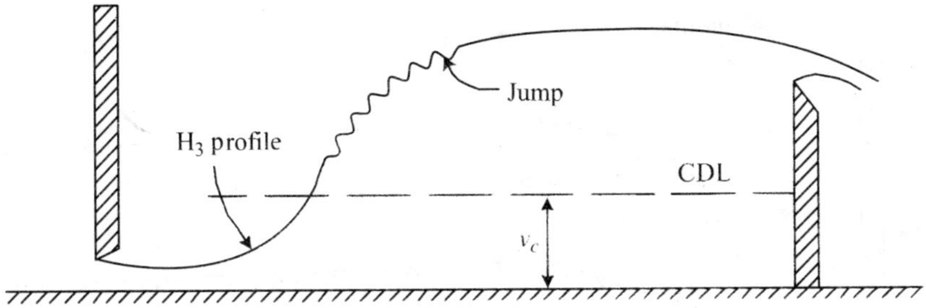

Fig. 12.27 H_3 profile

A_1 profile

If the slope is adverse, S_0 is negative and $K_n = Q/(S_0)^{1/2}$ is imaginary. Hence normal depth as well as the A_1 profile are non-existent.

A_2 profile

As in the case of the H_2 profile, $y > y_c$ and dy/dx is negative resulting in a draw-down curve. The profile upstream starts with a horizontal free surface and the draw-down curve may end in a critical depth if there is a sudden drop or if a mild or steep slope occurs immediately after the adverse slope as shown in Fig. 12.28.

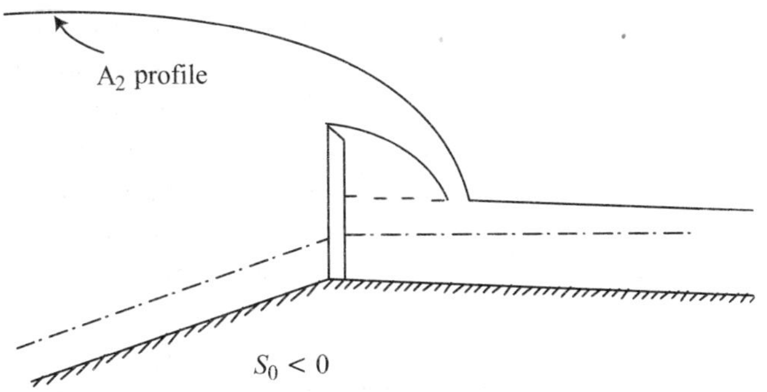

Fig. 12.28 A_2 profile

A_3 profile

Here again as in the case of a H_3 profile, the depth is initially less than the critical depth ($y < y_c$). dy/dx is positive and the profile downstream depends on the tail-water conditions and the prevailing slopes as shown in Fig. 12.29. It may also form on the adverse slope situated after a steep slope.

Table 12.1 gives a summary of the characteristics of the different flow profiles discussed above.

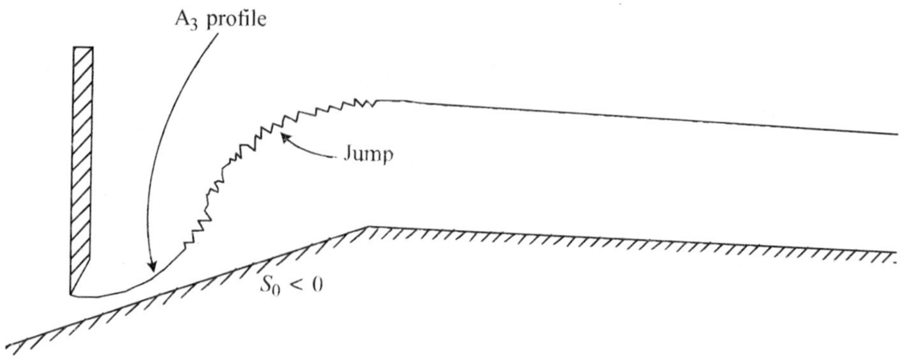

Fig. 12.29 A$_3$ profile

Table 12. 1 Summary of flow profile characteristics

Channel slope	Type of profile	Relation of y to y_n and y_c	dy/dx	Type of flow	Type of curve
Mild $S_0 < S_c$	M_1	$y > y_n > y_c$	Positive	Sub-critical	Backwater
..	M_2	$y_n > y > y_c$	Negative	Sub-critical	Draw-down
..	M_3	$y_n > y_c > y$	Positive	Super-critical	
Steep $S_0 > S_c$	S_1	$y > y_c > y_n$	Positive	Sub-critical	Backwater
..	S_2	$y_c > y > y_n$	Negative	Super-critical	Draw-down
..	S_3	$y_c > y_n > y$	Positive	Super-critical	
Critical $S_0 = S_c$	C_1	$y > y_n = y_c$	Positive	Sub-critical	
..	C_2	$y_c = y_n = y_c$	0	Critical	Uniform
..	C_3	$y_c = y_n > y$	Positive	Super-critical	
Horizontal $S_0 = 0$	H_2	$y_n > y > y_c$	Negative	Sub-critical	Draw-down
..	H_3	$y_n > y_c > y$	Positive	Super-critical	
Adverse $S_0 < 0$	A_2	$y > y_c$	Negative	Sub-critical	Draw-down
..	A_3	$y_c > y$	Positive	Super-critical	

12.24 Flow Profile Analysis

The type of profiles that occurs in a given channel layout can be determined by means of a procedure known as the flow profile analysis. If a channel consists of several reaches of different bottom slopes, the following procedure is used to analyse the flow profiles. It is assumed that the discharge and the geometry of the channel is known in each reach.

1. Plot the channel profile to an exaggerated scale.
2. Compute the normal and critical depths for each of the slopes in different reaches of the channel.
3. Locate all possible control sections such as a barrier or sluice gate.
4. Knowing the depth at the control section and the normal and critical depths, determine the type of profile for each reach.
5. Starting from the control section, sketch a continuous profile for each reach.

EXAMPLE 12.9 Sketch the free surface profile for the conditions shown in the sketch given below.

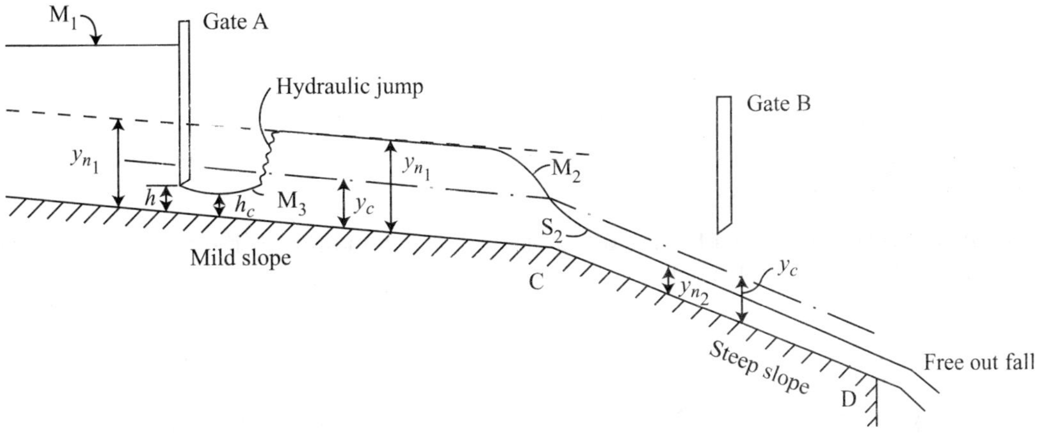

Figure for Example 12.9

y_{n1} is the normal depth on the mild slope and y_{n2} is the normal depth on the steep slope. y_c is the critical depth.

Gate A is located on the mild slope with its opening $h < y_c < y_{n1}$. The water level heads upstream of gate A resulting in a M_1 curve.

Downstream of the gate, the depth after the vena contracta (h_c) rises as an M_3 curve till the conjugate depth corresponding to y_{n1} is reached and a hydraulic jump occurs to reach the depth y_{n1}. The flow is uniform till it reaches the break in slope at C where the sub-critical flow changes to a super-critical flow. The depth at C is equal to the critical depth y_c. The transition in profile at C is marked by an M_2 profile on the mild slope changing over to an S_2 curve on the super-critical slope. The depth on the steep slope remains uniform at y_{n2} till it reaches the free out fall. Gate B has no role to play since its opening depth is above y_{n2}.

EXAMPLE 12.10 Draw the surface profile for the channel geometry as given in the sketch. Assume that the jump occurs on the horizontal slope.

Reservoir levels H_1 and H_2 are fixed and greater than y_c. The control points are at A, B and C. The depth at A has to be critical (y_c) followed by an S_2 profile and reach the normal depth on the steep slope y_n. After the location at B, the level rises and is indicated by an H_1 profile. The depth

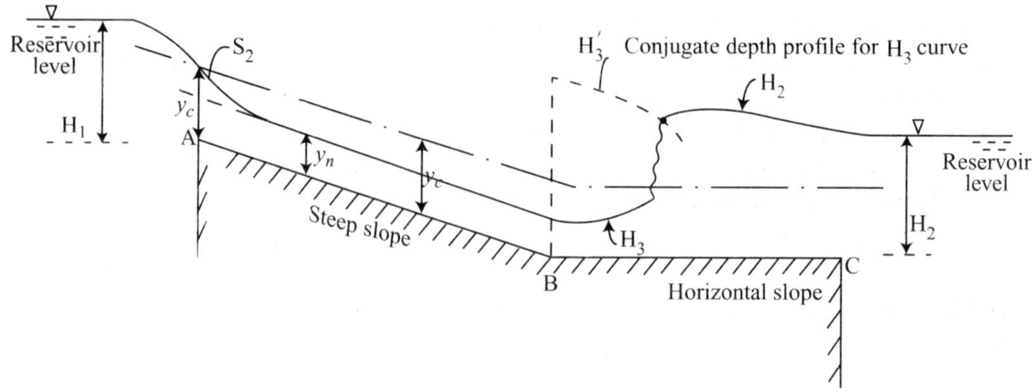

Figure for Example 12.10

at C is dictated by the reservoir level H_2. The H_2 profile has to be traced from C. The conjugate depth curve for the H_3 profile is also traced. The intersection of the conjugate profile H_3' and the H_2 profile defines the location of the hydraulic jump.

EXAMPLE 12.11 A channel has the bed profiles as shown in the sketch below. Draw the free surface profiles along the channel.

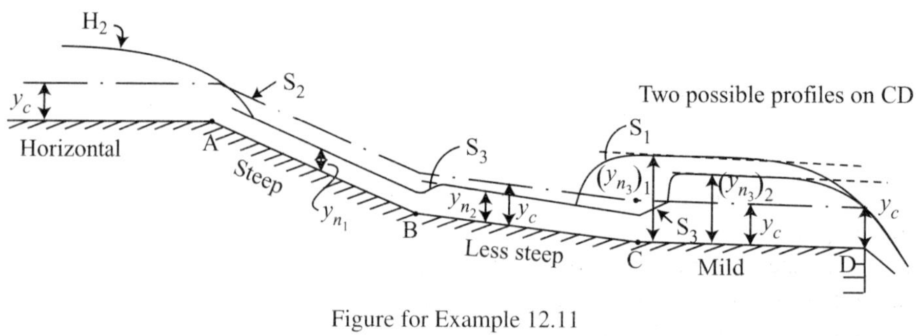

Figure for Example 12.11

At A, the slope changes from being horizontal to being steep. Hence the profile changes from H_2 to S_2, passing through critical depth at A. On the slope AB, the depth is $y_{n1} < y_c$ and on the reach BC, the depth is $y_{n2} (y_{n1} < y_{n2} < y_c)$. At B the depth increases from y_{n1} to y_{n2} with an S_3 profile. At C the slope changes from a steep slope to a mild one. The change in depth should take place via a hydraulic jump. The location of the jump depends on the normal depth $y_{n3}[(y_{n3})_1$ or $(y_{n3})_2]$ on the mild slope. If y_{n3} is greater than the conjugate depth of y_{n2}, the jump takes place on the steep slope and is accompanied by an S_1 curve till the normal depth $(y_{n3})_1$ is reached. If on the other hand, the normal depth $(y_{n3})_2$ is less than the conjugate depth of y_{n2}, then the jump takes place on the mild slope. An S_3 profile occurs after the location at C till the conjugate depth of $(y_{n3})_2$ is reached.

12.25 Integration of the Dynamic Equation

Computation of the gradually varied profiles involves the integration of the dynamic equation (12.33). Methods of integration may be broadly classified into three categories.

1. Direct integration method
2. Graphical integration method
3. Step method

There are several variations of the methods for computing gradually varied profiles. Only typical examples of the above methods of computation will be illustrated in the ensuing sections.

12.25.1 Direct Integration Method

A direct and exact integration of the gradually varied flow equation (or a closed form solution) is not available. Most methods are based on simplifying assumptions. The method described in this section was originally proposed by Bakhmeteff and vastly improved by Ven Te Chow.

According to definitions of the hydraulic exponents for uniform flow and critical flow computation, one can write

$$K_n^2 = C_1 y_n^N,\ K^2 = C_1 y^N, \quad \text{and} \quad Z_c^2 = C_2 y_c^M,\ Z^2 = C_2 y^M$$

where N = hydraulic exponent for uniform flow computation and M = hydraulic exponent for critical flow computation and C_1 and C_2 are coefficients. The values of N and M for any irregular cross-section may be obtained by the method outlined in Appendix 1 of this chapter. Substituting these expressions in equation (12.33), the dynamic equation becomes

$$\frac{dy}{dx} = S_0 \frac{1 - (y_n/y)^N}{1 - (y_c/y)^M}$$

Rearranging and solving for dx

$$dx = \frac{1}{S_0}\left[\frac{1 - (y_c/y_n)^M (y_n/y)^M}{1 - (y_n/y)^N}\right] dy$$

$$- \frac{1}{S_0}\left[\frac{1 - (y_c/y_n)^M \left(\dfrac{1}{u}\right)^M}{1 - \dfrac{1}{u^N}}\right] dy$$

Let $u = y/y_n$, then $dy = y_n du$,

$$dx = \frac{1}{S_0}\left[\frac{u^N - (y_c/y_n)^M u^{N-M}}{u^N - 1}\right] dy$$

$$dx = \frac{y_n}{S_0}\left[1 - \frac{1}{1-u^N} + \left(\frac{y_c}{y_n}\right)^M \frac{u^{N-M}}{1-u^N}\right] du \tag{12.34}$$

Assuming that M and N are constants and integrating the preceding equation

$$x = \frac{y_n}{S_0}\left[u - \int_0^u \frac{du}{1-u^N} + \left(\frac{y_c}{y_n}\right)^M \int_0^u \frac{u^{N-M}}{1-u^N} du\right] + \text{constant} \tag{12.35}$$

The second integral in the preceding equation can be transformed as follows:

Let $v = u^{N/j}$, where $j = N/(N - M + 1)$

$$dv = \frac{N}{j} u^{((N/j)-1)} du = \frac{N}{j} u^{N-M} du$$

$$\int_0^u \frac{u^{N-M}}{1-u^N} du = \frac{j}{N} \int_0^v \frac{dv}{(1-v^j)}$$

$$\int_0^u \frac{du}{1-u^N} = F(u, N) \ldots \text{ and } \ldots \int_0^v \frac{dv}{1-v^j} = F(v, j)$$

With these substitutions, the expression for x for positive slopes becomes,

$$x = \frac{y_n}{S_0}\left[u - F(u, N) + \left(\frac{y_c}{y_n}\right)^M \frac{j}{N} F(v, j)\right] + \text{constant} \tag{12.36}$$

The corresponding expressions for adverse slopes and horizontal beds are given in Appendix 2 of this chapter. The values for $F(u, N)$ for positive values are given in Appendix 3. The values of $F(u, N)$ for negative slopes are given in Appendix 4 at the end of this chapter.

To eliminate the constant of integration, we can take the difference of the distance between Sections 1 and 2 and designating the distances from an arbitrary origin as x_1 and x_2, we get the following equation (12.37).

$$x_2 - x_1 = \frac{y_n}{S_0}\left[(u_2 - u_1) - \{F(u_2, N) - F(u_1, N)\} + \left(\frac{y_c}{y_n}\right)^M \frac{j}{N} \{F(v_2, j) - F(v_1, j)\}\right] \tag{12.37}$$

The preceding equation (12.37) is the basis for computation of flow profile by the method of direct integration. In computing the flow profile, the channel is divided into a number of reaches and the length of each reach is found from equation (12.37) from known or assumed depths at the ends of each reach. The computation may follow the steps given below.

1. Compute the normal depth y_n and the critical depth y_c for the given discharge and channel slope.
2. Determine the hydraulic exponents N and M for an estimated average depth in the reach assuming that these values remain constant throughout the reach.
3. Compute $j = N/(N - M + 1)$
4. Find the values of $F(u, N)$ and $F(v, j)$ from varied function tables.
5. Compute the length of the reach from equation (12.37).

Special case of wide rectangular section (Manning's formula)

For the special case of a wide rectangular section, we have $A = By$ and $P = B + 2y$. Therefore $R = A/P = By/(B + 2y)$ and since $B \gg y$, we can write $R = y$.
$K = (1/n)By \cdot y^{2/3} = (1/n)By^{5/3}$ and $K^2 = (1/n^2)B^2y^{10/3} = C_1y^{10/3}$ and $N = 10/3$
$Z = A^3/T = B^3y^3/B = B^2y^3 = C_2y^3$; hence $M = 3$.
$J = N/(N - M + 1) = (10/3)/(10/3 - 3 + 1) = 2.5$

EXAMPLE 12.12 A wide rectangular channel laid on a slope of 0.001 carries a discharge of $3.72\ m^3/s$ of water per unit width of the channel. A weir is constructed across the channel which increases the depth just upstream of the weir to 3.05 m. Compute the surface profile by the direct integration method. Manning's $n = 0.025$.

Normal depth: $AR^{2/3} = nQ/(S_0)^{1/2}$; $y_n^{5/3} = 0.025 \times 3.72/(0.001)^{1/2}$; $y_n = 1.91$ m
Critical depth $y_c = (q^2/g)^{1/3} = (3.72^2/9.81)^{1/3} = 1.121$ m.
Since $y_n > y_c$, the slope is a mild slope and the profile will be an M_1 curve.

The computations are carried out from the downstream control created by the weir at which location, the depth is given as 3.05 m. The computations are carried out starting from the known depth of 3.05 m and proceeding upstream till a depth of 1% more than the normal depth (i.e., 1.93 m) is reached.

Table for Example 12.12 Direct integration method

y	$u = y/y_n$	$v = u^{N/j}$	$F(u, N)$	$F(v, j)$	x	L
3.050	1.60	1.871	0.158	0.287	2835	0
2.960	1.55	1.794	0.172	0.309	2720	115
2.863	1.50	1.718	0.188	0.333	2606	229
2.770	1.45	1.641	0.207	0.363	2480	355
2.675	1.40	1.566	0.229	0.397	2354	481
2.482	1.30	1.419	0.289	0.482	2073	762
2.294	1.20	1.275	0.381	0.615	1741	1094
2.100	1.10	1.135	0.558	0.864	1288	1547
1.930	1.01	1.013	1.218	1.856	145	2690

Wide rectangular section (Chezy formula)

If the Chezy formula is used for the velocity $V = C(RS_0)^{1/2}$, $Q/S_0^{1/2} = K = (1/C)AR^{1/2} = (1/C)By \cdot y^{1/2} = C_1y^{3/2}$ and K^2 is proportional to y^3. Hence $N = 3$. Also $y_n^3 = (q^2/C^2S_0)^{1/3}$, $Z = A^3/T$ and as before Z^2 is proportional to y^3. Therefore, $Z = 3$, $y_c^3 = (q^2/g)^{1/3}$.

$$J = \frac{N}{(N - M + 1)} = 3$$

Substituting $y_c^3/y_n^3 = C^2 S_0/g$ in equation (12.37), we get

$$dx = x_2 - x_1 = \frac{y_n}{S_0}\left[(u_2 - u_1) - \left(1 - \frac{C^2 S_0}{g}\right)\{F(u_2, 3) - F(u_1, 3)\}\right]$$

$F(u, 3)$ can be found from varied flow function tables. Bresse was the first to integrate $F(u, 3) = \int_0^u du/(1 - u^3)$ and prepare tables to evaluate the integral for different values of u which are known as the Bresse backwater function tables. A determination of flow profile using the backwater function tables is known as the Bresse method and is applicable to wide rectangular channels.

12.25.2 Graphical Integration Method

Consider two channel sections as shown in Fig. 12.30a at distances x_1 and x_2 from the origin and with depths y_1 and y_2 respectively. The distance x between the sections measured along the channel bottom can be expressed as

$$x = x_2 - x_1 = \int_{y_1}^{y_2} \frac{dx}{dy} dy \tag{12.38}$$

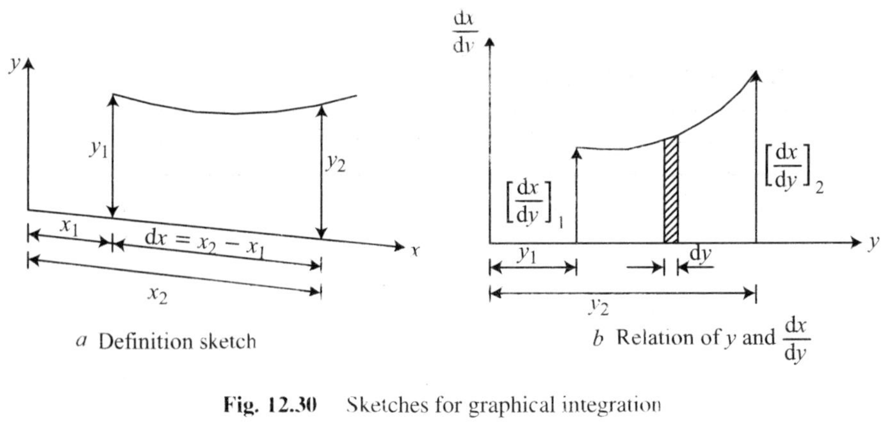

a Definition sketch

b Relation of y and $\frac{dx}{dy}$

Fig. 12.30 Sketches for graphical integration

The graphical integration method is based on the evaluation of the integral on the right-hand side of equation (12.38) by a graphical procedure. This is done by assuming several values of y and computing corresponding values of dx/dy from equation (12.33). The curve of y vs. dx/dy is then constructed as shown in Fig. 12.30b. According to equation (12.38), the distance x between the two sections under consideration is equal to the area enclosed by the curve, the x-axis and the ordinates of dx/dy corresponding to the depths y_1 and y_2 at the two sections.

This method can be applied to prismatic as well as non-prismatic channels of any shape or slope.

EXAMPLE 12.13 A trapezoidal channel with a base width of 12 m and side slopes of 1:1 and a bed slope of 0.0004 delivers water at a normal depth of 2.4 m. Determine the surface profile by the graphical integration method if the channel ends in a sudden drop. Manning's $n = 0.022$

$$A = (12 + 2.4) \times 2.4 = 34.56\,\text{m}^2;\ P = 12 + 2 \times 1.414 \times 2.4 = 18.7872;$$
$$R = A/P = 34.56/18.7872 = 1.84\,\text{m}$$

From the Manning's equation discharge, $Q = (1/n)AR^{2/3}S_0^{1/2} = 47.2\,\text{m}^3/\text{s}$.

The depth at the end will be the critical depth y_c.

$$\frac{Q^2}{g} = \frac{A^3}{T}$$
$$\frac{47.2^2}{9.81} = 227.1 = \frac{[(12 + y_c)y_c]^3}{(12 + 2y_c)}$$

and solving for $y_c = 1.129$ m.

As the slope is mild and the depth decreases from 2.4 m to 1.129 m, the profile is an M_2 profile.

Table for Example 12.13 Graphical integration method

y (metre)	T (metre)	A (m^2)	P (m)	R (m)	$R^{2/3}$	$K = AR^{2/3}/n$	$Z = A^{3/2}/T^{1/2}$	dx/dy	ΔA (m^2)	L (m)
1.129	14.256	14.80	15.185	0.975	0.983	661	15.10	0	0	0
1.20	14.40	15.84	15.39	1.030	1.019	734	16.6	45.2	4	4
1.40	14.80	18.78	15.96	1.178	1.115	951	21.1	221	45	49
1.80	15.60	24.82	17.08	1.451	1.282	1449	31.4	1138	239	288
2.00	16.00	28.00	17.66	1.588	1.362	1737	37.0	2642	364	652
2.15	16.30	30.22	18.08	1.673	1.409	1934	41.25	4280	570	1222
2.30	16.60	32.90	18.51	1.778	1.467	2195	46.2	13120	1091	2313
2.35	16.7	33.75	18.65	1.810	1.485	2278	48.0	27840	1025	3338
2.376	16.75	34.15	18.72	1.821	1.491	2315	48.6	43350	942	4280

12.25.3 Direct Step Method

In the step methods, the channel is divided into smaller reaches and the computation is carried step by step from one end of the reach to the other. The length of sub-reaches should be small enough to minimise the error arising out of averaging water surface elevation for the sub-reach from slopes at either ends. Moreover, in the step methods, computations should be carried starting from a control section in the downstream direction for steep channels and in the upstream direction for mild channels. A wrong direction of computation leads to erroneous profiles.

There are a number of step methods each suitable for a particular situation. Some are direct methods and others involve trial and error computations.

The direct step method described below is a simple method especially suitable for prismatic channels.

Consider the channel reach of length Δx (Fig. 12.31).

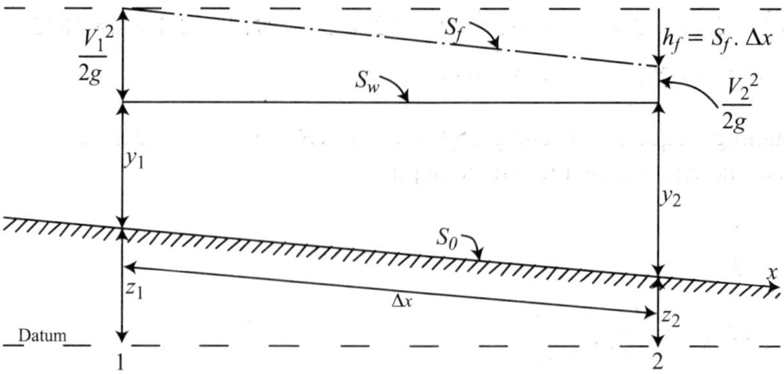

Fig. 12.31 Definition sketch for direct step method

Equating the total heads at the end of the two sections 1 and 2, we have

$$z_1 + y_1 + V_1^2/2g = z_2 + y_2 + V_2^2/2g + h_f$$

But $z_1 - z_2 = S_0 \Delta x$, and $h_f = S_f \Delta x$ and substituting in the above equation

$$S_o \Delta x + y_1 + V_1^2/2g = y_2 + V_2^2/2g + S_f \Delta x$$

$$\Delta x = \frac{[(y_2 + V_2^2/2g) - (y_1 + V_1^2/2g)]}{S_0 - \overline{S_f}} = \frac{E_2 - E_1}{S_0 - \overline{S_f}} = \frac{\Delta E}{S_0 - \overline{S_f}} \tag{12.39}$$

In equation (12.39), E_1 and E_2 are the specific energies at Sections 1 and 2 respectively and $\overline{S}_f$ is the average friction slope for the reach computed from Manning's formula using average velocity and average hydraulic radius for the reach.

EXAMPLE 12.14 A rectangular concrete flume 6 m wide takes off from a reservoir with the flume sill 2.4 m below the reservoir level. If the slope of the flume is 0.01 and Manning's $n = 0.015$, compute the profile by direct step method.

The specific energy corresponds to the reservoir depth above the sill which is 2.4 m. (Velocity in the reservoir is assumed to be zero.)

The critical depth corresponding to the specific energy of 2.4 m is $(2/3) \times 2.4 = 1.6$ m and occurs over the sill. The corresponding discharge

$$Q = Bg^{1/2} y_c^{3/2} = 6 \times 9.81^{1/2} \times 1.6^{3/2} = 38 \text{ m}^3/\text{s}$$

The normal depth in the channel is calculated by

$$Q = A.V = 6y_n \left(\frac{1}{0.015}\right) \left[\frac{6y_n}{(6 + 2y_n)}\right]^{2/3} (0.01)^{1/2}$$

Solving for y_n by trial and error, we get $y_n = 1.1$ m.

Table for Example 12.14

y	A (m^2)	P (m)	R (m)	$R^{4/3}$	V^2	$V^2/2g$ (m)	E (m)	S_f	$\overline{S}_f$	$S_0 - \overline{S}_f$	ΔE (m)	Δx (m)	L (m)
1.60	9.60	9.20	1.042	1.056	15.68	0.799	2.399	0.00334	–	–	–	–	0
1.55	9.30	9.10	1.021	1.028	16.70	0.851	2.401	0.00366	0.0035	0.0065	0.002	0.31	0.31
1.50	9.00	9.00	1.000	1.000	17.80	0.906	2.406	0.00400	0.0038	0.0062	0.005	0.81	1.12
1.45	8.70	8.90	0.979	0.972	19.10	0.974	2.424	0.00442	0.0042	0.0058	0.018	3.11	4.23
1.40	8.40	8.80	0.955	0.941	20.50	1.045	2.445	0.00490	0.0047	0.0053	0.021	3.94	8.17
1.30	7.80	8.60	0.906	0.877	23.80	1.212	2.512	0.00610	0.0055	0.0045	0.067	13.4	21.57
1.20	7.20	8.40	0.857	0.814	27.90	1.421	2.621	0.00771	0.0069	0.0031	0.109	35.2	56.77
1.11	6.66	8.22	0.851	0.756	32.50	1.658	2.768	0.00965	0.0087	0.0013	0.147	111	168

The normal depth is less than the critical depth and hence the profile is S_2 over the channel. The profile starts with the critical depth at the entrance to the channel and ends asymptotically with the normal depth. We will assume again that it is sufficient to calculate the length of the profile to a depth of 1% more than the normal depth i.e., 1.11 m. The table given exemplifies the method.

If the procedure is reduced to one single step $E_2 = 2.768$, $E_1 = 2.399$, $\Delta E = 0.369$

$$\frac{(S_{f1} + S_{f2})}{2} = \frac{(0.00334 + 0.00965)}{2} = 0.00646,\ S_0 = 0.01$$

$x = 0.369/(0.01 - 0.00646) = 104.2$ m. Hence the accuracy increases if the reach is subdivided into short reaches.

12.25.4 Standard Step Method

Flow profile computation in non-prismatic channels and particularly natural rivers becomes cumbersome and involved as the cross-section varies along the channel or river. The method most suitable for computing flow profiles in natural streams is the standard step method.

The length of the river is divided into a number of sub reaches in such a way that the hydraulic elements are sensibly constant for each sub-reach. Thus the lengths of each sub-reach is known. As far as possible, the number of sub-reaches should also be small. Next the data regarding the hydraulic elements such as A, R, P, S_0 etc., at the end section of all reaches are collected by either field surveys or with the help of available topographical maps of the river. The discharge in the river for which the profile is required is to be estimated or computed from available rating curves. Next an estimate is made of the roughness coefficient for different sub-reaches. Necessary information for the standard step method is the elevation of the water surface at a control section.

The basic equation for the standard step method is derived by considering two sections 1 and 2 and equating the total energies at the two sections.

$$z_1 + y_1 + \alpha_1 V_1^2/2g = z_2 + y_2 + \alpha_2 V_2^2/2g + h_f + h_e \tag{12.40}$$

or

$$H_1 = H_2 + h_f + h_e$$

In the above equation $h_f = (S_f)_{av}\Delta x$, where $(S_f)_{av} = (S_{f1} + S_{f2})/2$, and h_e is the eddy loss term which is computed empirically from $h_e = K(\Delta V^2/2g)$, where K is a coefficient. For gradually converging or diverging channels, K varies from 0 to 0.2 and for abrupt expansions or contractions K is usually taken as 0.5. Other eddy losses could be due to change in alignment such as bends and meanders.

The standard step method is based on the solution of equation (12.40) by a trial and error procedure and carried out step by step from the first station to the last. At the initial station where the depth is known, the value of the total energy, H_1 is computed. An assumption is made for the depth at the end of the first sub-reach and the value of the right-hand side of equation (12.40) including friction and eddy losses is calculated and compared with H_1. If the two total energies agree, then the assumed depth is correct. If the agreement is not good, a fresh value for the depth is assumed and the total energy is calculated again. The procedure is repeated for every reach till the total energy is in agreement for all the sections.

12.26 Rapidly Varied Flow

In rapidly varied flows, changes in depth and velocity take place in relatively short lengths compared to gradually varied flows. The stream lines usually have pronounced curvatures. Rapidly varied flow occurs in the vicinity of control structures and in transitions. Examples of rapidly varied flow are found in hydraulic jumps, flow over spillways, flow below sluice gates etc. Due to appreciable stream line curvature, the pressure distribution may vary from hydrostatic distribution and the velocity distribution coefficients α and β are usually greater than unity. However, since the length over which the changes are taking place is short, the boundary resistance over the short reach of the channel is generally ignored.

12.27 Hydraulic Jump

The hydraulic jump is one of the most important rapidly varied flow phenomenon and is used in many hydraulic structures as an energy dissipation device. The hydraulic jump occurs whenever the flow changes its state from super-critical to sub-critical. The depth changes from a super-critical depth, $y_1(y_1 < y_c)$ to a sub-critical conjugate depth, $y_2(y_2 > y_c)$.

The flow is turbulent and there will be energy loss but the main advantage is that the kinetic energy is converted to potential energy.

Hydraulic jump on a level bed in a rectangular channel is analysed by the application of the momentum equation. In this case, the weight component is zero and we can ignore the boundary resistance in the short reach in which the jump takes place. The specific forces before and after the jump shown in Fig. 12.32 are equal.

$$\frac{q^2}{gy_1} + \frac{y_1^2}{2} = \frac{q^2}{gy_2} + \frac{y_2^2}{2}$$

Continuity: $q = V_1 y_1 = V_2 y_2$

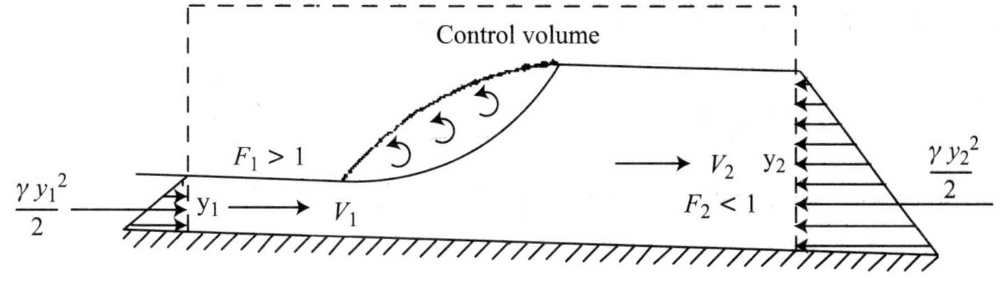

Fig. 12.32 Hydraulic jump

$$\frac{V_1^2 y_1}{g} + \frac{y_1^2}{2} = \frac{V_1^2 y_1^2}{gy_2} + \frac{y_2^2}{2}$$

$$\mathbf{F}_1^2 y_1^2 + \frac{y_1^2}{2} = \mathbf{F}_1^2 \frac{y_1^3}{y_2} + \frac{y_2^2}{2}$$

$$\mathbf{F}_1^2 + \frac{1}{2} = \mathbf{F}_1^2 \frac{y_1}{y_2} + \frac{1}{2}\frac{y_2^2}{y_1^2}$$

$$\left(\frac{y_2}{y_1}\right)^3 - (2\mathbf{F}_1^2 + 1)\left(\frac{y_2}{y_1}\right) + 2\mathbf{F}_1^2 = 0$$

$$\left[\left(\frac{y_2}{y_1}\right)^2 + \left(\frac{y_2}{y_1}\right) - 2\mathbf{F}_1^2\right]\left(\frac{y_2}{y_1} - 1\right) = 0$$

$$\left(\frac{y_2}{y_1}\right)^2 + \left(\frac{y_2}{y_1}\right) - 2\mathbf{F}_1^2 = 0$$

$$\frac{y_2}{y_1} = \frac{1}{2}\left(\sqrt{1 + 8\mathbf{F}_1^2} - 1\right)$$

where y_1 and y_2 are the depths before and after the jump and $\mathbf{F}_1 = V_1/(gy_1)^{1/2}$ is the initial Froude number.

In terms of the downstream Froude number, $\mathbf{F}_2 = V_2/(gy_2)^{1/2}$, the ratio of depths is obtained as

$$\frac{y_1}{y_2} = \frac{1}{2}\left(\sqrt{1 + 8\mathbf{F}_2^2} - 1\right)$$

12.27.1 Energy Loss in the Hydraulic Jump

The energy loss in a hydraulic jump on a level floor in a rectangular channel can be derived as the difference of total energies before and after the jump and in this case it is the same as the difference between the specific energies before and after the jump.

$$\Delta E = E_1 - E_2 = (y_1 + V_1^2/2g) - (y_2 + V_2^2/2g)$$

$$\Delta E = (y_1 - y_2) + \left(\frac{q^2}{2g}\right)\frac{(y_2^2 - y_1^2)}{y_1^2 y_2^2}$$

But from the momentum equation, $q^2/gy_1 + y_1^2/2 = q^2/gy_2 + y_2^2/2$.

Hence

$$\frac{q^2}{g}\left(\frac{1}{y_1} - \frac{1}{y_2}\right) = \frac{(y_2^2 - y_1^2)}{2}$$

$$y_c^3 = \frac{q^2}{g} = \frac{[y_1 y_2 (y_1 + y_2)]}{2}$$

On substitution for q^2/g in the expression for ΔE, we get

$$\Delta E = (y_1 - y_2) + \frac{y_1 y_2}{4}(y_1 + y_2)\frac{(y_2^2 - y_1^2)}{y_1^2 y_2^2}$$

$$\Delta E = \frac{(y_2 - y_1)^3}{4 y_1 y_2}$$

The ratio of the specific energy after the jump to that before the jump i.e., E_2/E_1 is known as the efficiency of the jump. Efficiency of the jump which is a measure of the energy dissipating capacity can be expressed in terms of the initial Froude number, $\mathbf{F}_1$.

$$\frac{E_2}{E_1} = \frac{(8\mathbf{F}_1^2 + 1)^{3/2} - 4\mathbf{F}_1^2 + 1}{8\mathbf{F}_1^2(2 + \mathbf{F}_1^2)}$$

$$\frac{h_j}{E_1} = \frac{y_2 - y_1}{E_1} = \frac{\sqrt{1 + 8\mathbf{F}_1^2} - 3}{\mathbf{F}_1^2 + 2}$$

where h_j is the height of the hydraulic jump.

12.27.2 Length of Jump

The length of the jump, L_j is defined as the distance from the toe of the jump to a point on the surface immediately downstream of the roller (Fig. 12.32). There is no theoretical expression for the length of the jump. But many investigators have experimentally determined it in terms of other jump parameters. A few empirical formulae for the length of the jump are given below.

$L_j = 5(y_2 - y_1)$ Bakhmeteff and Matzke
$L_j = (y_2 - y_1)(8 - 0.05y_2/y_1)$ Woycicki
$L_j = 10.3y_1(F_1 - 1)$ Chertoussov
$L_j = 5.2y_2$ Safranez

A plot of jump characteristics as a function of the initial Froude number results in characteristic curves which can be used conveniently for design purposes.

Example 12.15 Water flowing down a spillway at a rate of 9.5 m^3/s per unit width leaves the apron horizontally with a mean velocity of 12 m/s. Determine (a) the depth of tail-water necessary to form a hydraulic jump, (b) head lost in the jump and (c) approximate length of the jump.

$$y_1 = q/V_1 = 9.5/12 = 0.791 \text{ m}$$

$$\mathbf{F}_1 = 12/(9.81 \times 0.791)^{1/2} = 4.308$$

$$y_2/y_1 = \sqrt{[(1 + 8\mathbf{F}_1^2) - 1]/2} \text{ and } y_2 = 4.44 \text{ m}$$

The head lost in the jump $(y_2 - y_1)^3/4y_1y_2 = 3.47$ m
L_j = length of the jump $= 5(y_2 - y_1) = 18.55$ m

12.27.3 Location of the Jump

Location of the jump involves fixing the place where the jump will form for a given channel condition. A jump can form below a spillway, downstream of a sluice gate or when the slope changes from steep to mild. For a satisfactory operation of structures involving a jump it is necessary to locate the jump in the structure for the design conditions and ensure that the jump remains in a prescribed place for all operating conditions.

Theoretically, the jump will occur in a horizontal rectangular channel between two depths which satisfy the momentum equation i.e., specific force must be the same at the two sections. If the downstream depth is fixed by the tail-water conditions, then the depth at the beginning of the jump should correspond to the conjugate of the tail-water depth. A jump occurring below a sluice gate is taken for illustration of the location of the jump.

Figure 12.33 show a sluice gate on a mild channel. The issuing super-critical stream forms an M_3 profile which theoretically extends up to the critical depth. Tail-gate at the downstream end creates an M_2 profile. M_2 and M_3 profiles are drawn first.

Curve A′B represents depths conjugate to the depths on the M_3 profile. If the length of the jump is not considered, it is evident that the jump will form in the vertical section EK, K lying on the intersection of the M_2 profile and curve A′B. The sequent depth of y_2 is found from the equation $y_1/y_2 = 1/2[(1 + 8\mathbf{F}_2^2)^{1/2} - 1]$. To include the length of the jump in the analysis, the curve BGFG′ is drawn depicting the length of the jump for different conjugate depths. The length of the jump from any of the equations cited in Section 12.27.2 for each of the conjugate depth on the curve BKA′ is set off beyond the conjugate depth curve. This is shown as BGFG′. The intersection of BGFG′ and the M_2 profile DFF′ at F gives the end of the jump. The jump takes place between E and F and they satisfy both the momentum equation and the equation for length of jump. In Fig 12.33, KF is the required intercept equal to the length of the jump and the jump will form between E and F. Similar arguments are made for the jumps ocurring when the slope changes from steep to mild or an obstruction occurs on the steep slope. The jump occurs on the steep slope if the sequent depth corresponding to y_1 is smaller than the normal depth on the mild slope followed by an S_1 curve. If the sequent depth of y_1 is larger than the depth on the mild slope, then the jump occurs on the mild slope preceded by an M_3 curve.

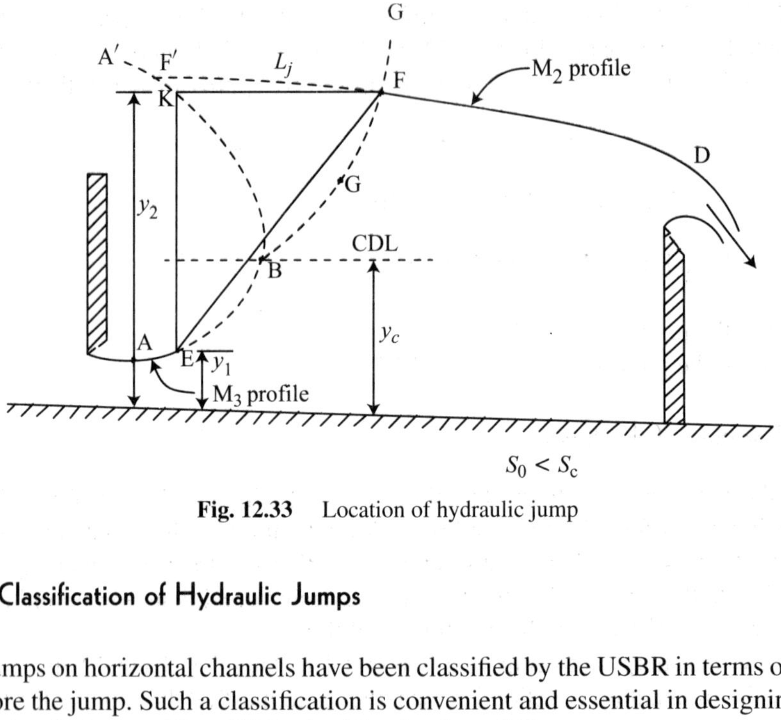

Fig. 12.33 Location of hydraulic jump

12.27.4 Classification of Hydraulic Jumps

Hydraulic jumps on horizontal channels have been classified by the USBR in terms of the Froude number before the jump. Such a classification is convenient and essential in designing structures involving hydraulic jumps. The USBR classification is as follows:

Value of F	Description of the jump
1.0	No jump
1–1.7	Undular jump with mere surface undulations
1.7–2.5	Weak jump; small roller develops; energy loss low
2.5–4.5	Oscillating jump; due to the oscillation of the high velocity stream, waves are developed downstream
4.5–9	Steady jump: jump well-balanced; energy dissipation is 45–75% of the initial energy
9 and above	Strong jump, jump very rough, energy dissipation about 85%

Since the jump is steady and energy dissipation is high at Froude numbers above 4.5, an attempt should be made to obtain an initial stream before the jump attains a high Froude number of this order. This can easily be achieved by expanding the channel since for super-critical streams an expansion would result in decrease in depth and increase in Froude number.

12.28 The Oblique Jump

A contraction in the width of the channel as shown in Fig. 12.34 causes an oblique jump if the flow is initially super-critical. The oblique jump resembles the shock wave in supersonic flow of gases. The oblique jump starts from the point of boundary discontinuity and occurs at a characteristic angle β (known as the wave angle) to the initial direction of flow. This wave angle depends on the angle of deflection of the boundary, θ for a given Froude number. The depth of flow changes abruptly along the wave front from y_1 upstream to y_2 after the front. The flow called an oblique jump is super-critical before and after the jump and in this respect differs from the normal jump where the flow changes from super-critical to the sub-critical state.

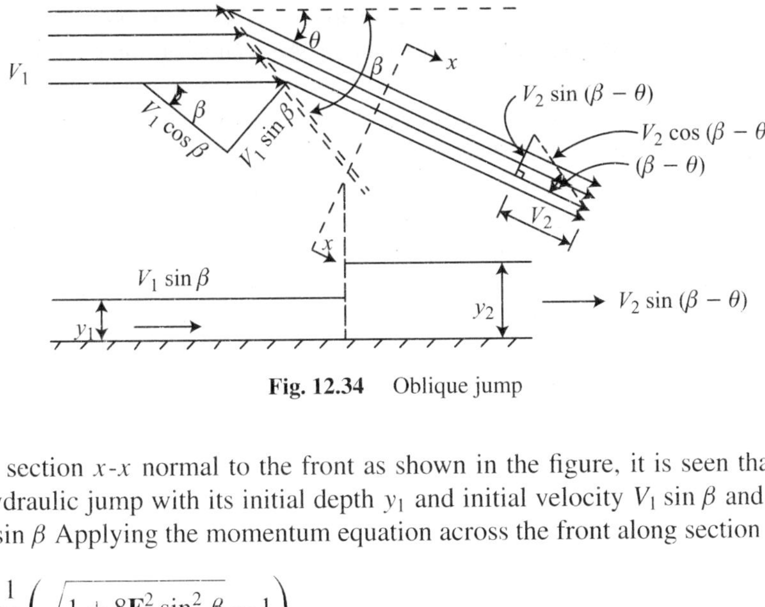

Fig. 12.34 Oblique jump

Taking a section x-x normal to the front as shown in the figure, it is seen that it resembles a normal hydraulic jump with its initial depth y_1 and initial velocity $V_1 \sin\beta$ and initial Froude number $\mathbf{F}_1 \sin\beta$ Applying the momentum equation across the front along section x-x, we get

$$\frac{y_2}{y_1} = \frac{1}{2}\left(\sqrt{1 + 8\mathbf{F}_1^2 \sin^2\beta} - 1\right)$$

The above equation has a striking similarity to the normal jump equation in a horizontal rectangular channel. When $\theta = 0$, $\beta = 90°$.

As no momentum change takes place parallel to the wave front, the tangential velocities before and after the jump must be the same. From the velocity triangles shown in the figure

$$V_1 \cos\beta = V_2 \cos(\beta - \theta)$$

From continuity,

$$y_1 V_1 \sin\beta = y_2 V_2 \sin(\beta - \theta)$$

$$\frac{y_2}{y_1} = \frac{\tan\beta}{\tan(\beta - \theta)}$$

By eliminating y_2/y_1 from the two equations, we get

$$\tan\theta = \frac{\tan\beta\left(\sqrt{1 + 8\mathbf{F}_1^2 \sin^2\beta} - 3\right)}{2\tan^2\beta + \sqrt{1 + 8\mathbf{F}_1^2 \sin^2\beta} - 1}$$

12.29 Channel Transitions

12.29.1 Contractions and Expansions in Sub-critical Flow

A channel transition may be defined as a local change in channel cross-section which produces a variation in flow from one uniform state to another. An analysis of an open channel transition involves the determination of the free surface profile for a given boundary configuration.

In Section 12.7, simple transitions involving a change in bottom elevation for constant width were described. In fact, the design of sub-critical flow transitions are based on the simple application of specific energy diagrams. For example, for a rectangular channel with a change in width and constant discharge, a series of specific energy diagrams are plotted for different sections along the transition corresponding to each value of $q = Q/B$. Apart from the specific energy curves, other information required are the geometric shape of the channel both upstream and downstream, the discharge Q, the elevation of the total energy line at the upstream section and the geometric shape of the transition.

To determine the water surface profile, a smooth bottom profile is assumed between the end sections. Thus the values of the specific energy $E = H - z$ are fixed at different sections for which E vs. y plots are available. From specific energy diagrams, corresponding values of y are determined. If the resultant water surface is discontinuous, a new bottom profile is assumed or adjustments made in the longitudinal spacing of the sections or geometry of the transition (linear or circular or elliptic) itself till a smooth surface profile is obtained. The method prescribed above is valid for both sub-critical contractions and expansions.

Channel contractions

If the width of a rectangular channel is to be reduced from B to b as shown in Fig. 12.35, one of the recommended practice (known as the Russian method) is to design the contraction as per the formula given:

$$B_x = b + 2x\tan\theta[1 - (1 - x/L)]$$

The optimum angle between the channel axis and the line joining the channel sides between the entrance and the exit sections is 12.5° or $\tan\theta_{max} = (B - b)/2L = 0.22169 = 1/4.5$

A simple cylindrical quadrant transition seems to have worked well with the radius equal to $(B - b)/2$ as shown in Fig. 12.36.

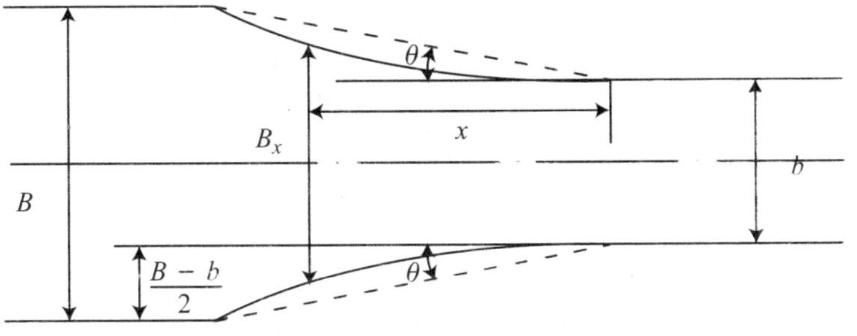

Fig. 12.35 Channel contraction (Russian method)

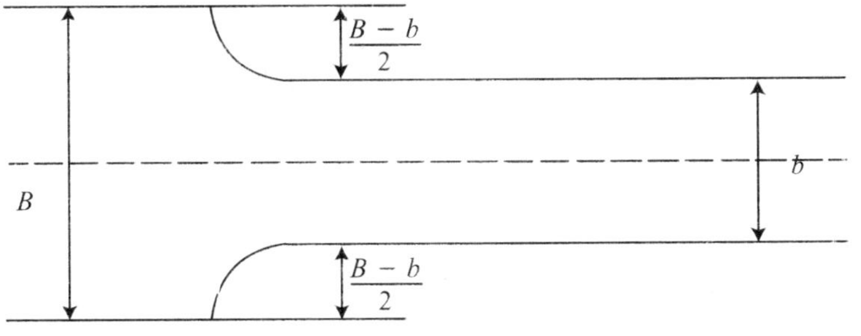

Fig. 12.36 Cylindrical quadrant transition

The energy loss in channel contractions is expressed as $C_L V_2^2/2g$, where V_2 is the velocity in the smaller channel. The maximum value of C_L for square-edged contractions is about 0.23 and 0.11 for a cylindrical quadrant transition. The Russian profile has a very low loss coefficient, less than 0.05. These values, however, are dependent on the width-to-depth ratio.

Channel Expansions

Contrary to popular belief, a straight wall flare is superior to a curved wall flare of equal length. This is because the flare angle increases continuously and separation will occur at a later stage though the stream may stick to the wall in the initial reaches. For instance Hind's S-curved warped transition has a maximum central angle of 50° and flow separation has always occurred in such transitions. It has been found that the maximum central angle should not exceed 11°26′ which corresponds to a 1 in 10 side wall flare. The recommended length of transition varies from $2\,(B_2 - B_1)$ to $2.5\,(B_2 - B_1)$, where B_1 and B_2 are the channel widths before and after the expansion. One type of transition is based on the assumption that the depth remains constant. The velocity varies linearly along the transition length and the profile is given as (B_x at any distance x from the beginning of the transition)

$$B_x = \frac{B_1 B_2 L}{L B_2 - (B_2 - B_1)x}$$

The loss in energy in expansion is usually taken as $0.3(V_1^2/2g - V_2^2/2g)$.

We will consider a sub-critical channel expansion in a rectangular channel as shown in Fig. 12.37. Let P correspond to the specific energy E_1 in the approach channel of width B_1, depth y_1, velocity, V_1, then $E_1 = y_1 + V_1^2/2g$. P lies on the curve corresponding to the unit discharge, q_2. The transition can be done in several ways. We can initially keep the width constant and move along the curve q_2 = constant and travel to point Q by increasing the bed level by Δz (the specific energy is now $E_1 - \Delta z$) and move to the point R vertically above on the q_1 curve (corresponding to the width B_2) by increasing the width. Alternately we can increase the width first to reach the point S on the q_1 curve and then reach R to account for the loss in specific energy due to increase in bed level. However these abrupt changes will cause energy losses. Hence we should gradually increase the width and the bed elevation and follow the path PR.

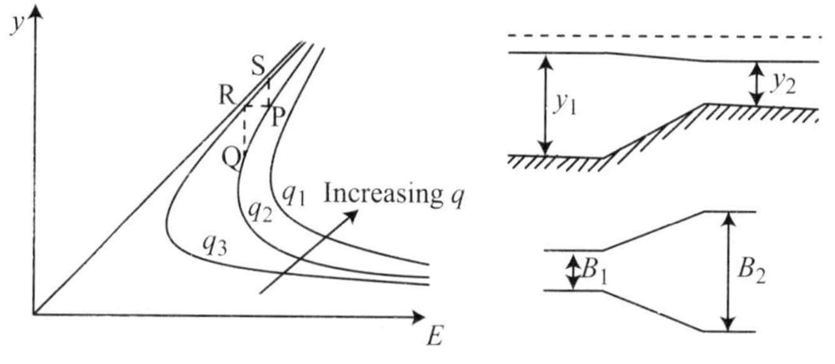

Fig. 12.37 Channel expansion

EXAMPLE 12.16 Let the approach channel be of width $B_1 = 2$ m, depth $y_1 = 1.5$ m, and velocity = 2 m/s. After expansion, the width $B_2 = 4$ m, depth $y_2 = 0.8$ m. Design the channel expansion.

Length of transition $= 2.5(B_2 - B_1) = 2.5(4 - 2) = 5$ m.

$$Q = 2 \times 2 \times 1.5 = 6\ \text{m}^3/\text{s};\ V_2 = 6/(4 \times 0.8) = 1.875\ \text{m/s}$$

Loss in energy due to expansion $= 0.3(V_1^2/2g - V_2^2/2g) = 0.3(2^2 - 1.875^2) = 0.0074$ m.

Let the elevation of the bed in the downstream channel after expansion be 10.0 m. Then the total energy in the downstream channel is $10.0 + 0.8 + 1.875^2/2g = 10.979$ m.

Hence the energy in the approach channel is $10.979 + 0.0074 = 10.9864$ m.

The bed elevation of the approach channel is $10.9864 - 1.5 - 2^2/2g = 9.283$ m. The difference in bed elevation before and after expansion is $10.000 - 9.283 = 0.717$ m. The water surface elevation in the approach channel is $9.283 + 1.5 = 10.783$ m and that in the downstream channel is 10.8 m.

12.29.2 Contractions in Super-Critical Flow

While in sub-critical transitions, minimum energy dissipation with maximum economy is the criterion for design, in super-critical transitions elimination of surface disturbances assumes primary importance. A channel contraction in a super-critical flow causes positive waves. A linear contraction results in the formation of an oblique jump as discussed earlier. Even if the contraction is gradual, positive waves generated at different points tend to converge and finally form a finite oblique jump. The free surface after the jump will be very rough and will persist over a considerable distance downstream. A good contraction design must eliminate these disturbances effectively within a short distance downstream. This is achieved by adjusting the transition walls such that a positive disturbance generated by a wall contraction at the upstream end A or B (Fig. 12.38) are cancelled out by the negative disturbances caused by channel expansions at the downstream end C and D thus preventing the disturbances propagating further into the downstream channel.

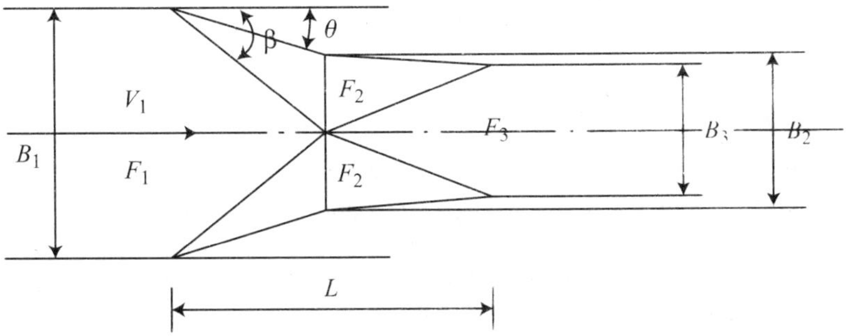

Fig. 12.38 Channel contraction in a super-critical flow

This is practical if the channel contraction is designed for a given Froude number. It has been found that the straight wall contraction shown in Fig. 12.38 gives satisfactory results in terms of reducing cross waves. The length of the transition is

$$L = \frac{(B_1 - B_3)}{2 \tan \theta}$$

From continuity, we have $B_1 y_1 V_1 = B_3 y_3 V_3$ and since $\mathbf{F} = V/(gy)^{1/2}$, we get

$$\frac{B_1}{B_3} = \left(\frac{y_3}{y_1}\right)^{3/2} \left(\frac{\mathbf{F}_3}{\mathbf{F}_1}\right)$$

The relationships among $\mathbf{F}_1$, y_2/y_1, β and θ are given in section 12.28.

12.29.3 Channel Expansions in Super-Critical Flow

Expansions in super-critical flow are required in many hydraulic structures such as downstream of a sluice gate, or at the foot of the spillway or at the toe of a steep chute. A careful design of such expansions is necessary to prevent separation of a high velocity stream from channel walls.

Such separation occurs in rapidly diverging transitions creating eddy zones between walls and surfaces of separation. A good expansion should confine a major portion of the flow within the boundaries and prevent creation of surface disturbances in the channel downstream.

Due to the limitations of theoretical solutions, the design of super-critical expansions is based on an empirical approach arising from experimental results. The results of Rousė and others given below can be usefully employed for the design of channel expansions in super-critical flows.

The general relationship among the different parameters of super-critical expansion can be expressed by

$$\frac{y}{y_1} = f\left(\frac{x}{B_1\mathbf{F}_1}, \frac{z}{B_1}\right)$$

where y = depth of flow at (x, z), y_1 = depth of approach flow, $\mathbf{F}_1$ = Froude number of approach flow, B_1. x = longitudinal coordinate measured from the outlet, z = lateral coordinate measured from the centre line of the channel.

From experiments, the boundary form for the most efficient expansion was found to be

$$\frac{z}{B_1} = \frac{1}{2}\left(\frac{x}{B_1F_1}\right)^{3/2} + \frac{1}{2}$$

The preceding equation represents a boundary form which confines 90% of the flow. If the transition from the divergent walls of the expansion and the parallel wall of the downstream of the channel is abrupt, surface disturbances may be produced. One way to prevent surface irregularities is to create a finite steady hydraulic jump at the end of the expansion. Surface disturbances can be eliminated by providing smooth transitions (by providing reverse curves) between the diverging walls and the downstream channel. Rouse and others have developed a set of curves for different values of B_2/B_1 as shown in Fig. 12.39.

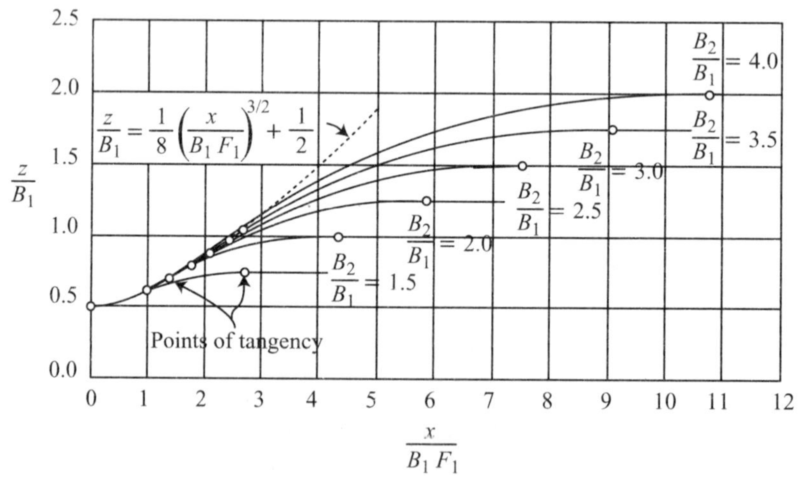

Fig. 12.39 Expansions in super-critical flows

12.30 Stilling Basins

Stilling basins form a special case of channel transitions in which the flow regime changes from super-critical to sub-critical with the formation of a hydraulic jump within a contained basin length. The hydraulic jump dissipates part of the energy of the stream and more importantly changes the kinetic energy of the flow to potential energy thus reducing the hazard of erosion of the downstream reach of the channel. A stilling basin is a paved channel of short length located at the foot of the spillway or a chute or any other source of super-critical stream. The aim is to dissipate the fast moving stream before it reaches the unprotected river bed downstream thus preventing dangerous bed degradation. Thus a good stilling basin must ensure jump formation and make it stable under flow conditions.

To stabilise the jump, stilling basins are provided with special appurtenances such as chute blocks, floor blocks, and sills. The chute blocks are used to divide the incoming stream and lift a portion of it from the floor thus reducing the length of the jump. Also, they stabilise the jump and improve its performance.

The sill which is provided at the end of the stilling basin is meant to provide the necessary depth of water as a cushion for the formation of the jump, reduce the length of the jump, control scouring of the river bed downstream by the upward deflection of the stream and diffusing the residual portion of energy of the high velocity jet that may reach the end of the basin.

Floor blocks are placed in the middle of the basin usually in a staggered pattern in two or more rows to dissipate the energy by impact. The blocks exert a force in the opposite direction on the stream and thus reduce its momentum. Floor blocks are generally used in low structures where the incoming velocities are not high. They are not very suitable for higher velocities because of the cavitation damage.

There are many designs of stilling basins suitable for different conditions. The following are a few typical examples.

1. The SAF basin: It is suitable for small structures where the incoming Froude number is $F_1 = 1.7 - 2.7$
2. The USBR basin II: This is useful on structures such as spillways, large canal structures and etc., where $\mathbf{F}_1 > 4.5$ (Fig. 12.40)
3. The USBR basin IV: This is designed for jumps with initial Froude numbers in the range of 2.5 to 4.5 which usually occur on canal structures and diversion dams.

12.31 Canal Bends

Flow in an open channel bend is similar to the flow in pipe bends. But the presence of a free surface makes the flow pattern more complex. Bends are introduced in artificial channels due to a change in linear alignment but in natural rivers, they are caused by a phenomena known as meandering in erodable soils.

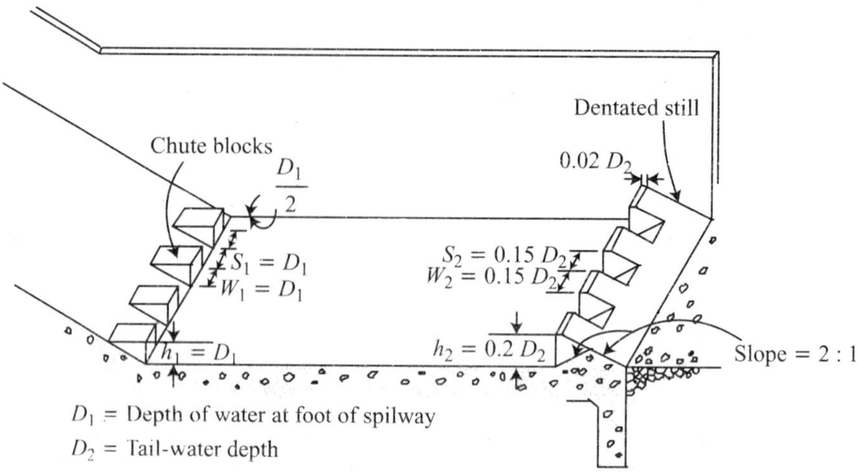

Fig. 12.40 USBR stilling basin II

Behaviour of the flow in an open channel bend depends upon whether the flow is sub-critical or super-critical. The most important aspects of sub-critical flow in an open channel bend are (a) occurrence of super-elevation (b) secondary circulation and (c) energy loss.

12.31.1 Super-elevation

Super-elevation refers to the lateral slope of the water surface in a bend; there is a rise in the water surface elevation above the mean level at the outer bank of the channel and a lowering of the water surface in the inner radius of the bend as shown in Fig. 12.41. Super-elevation is the result of the centrifugal forces acting on the flow in the bend.

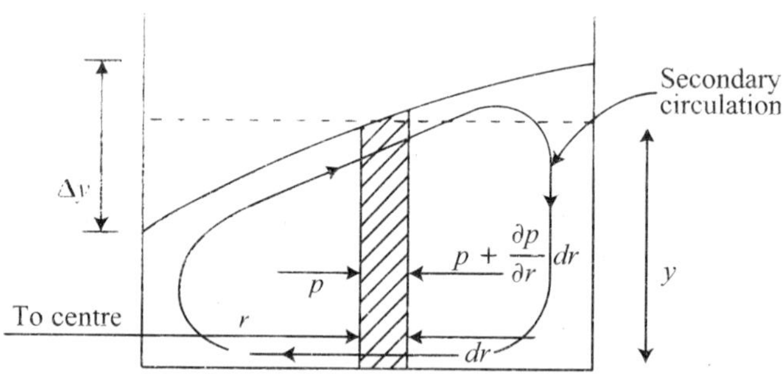

Fig. 12.41 Channel bend

Consider a vertical column of water of depth y and cross-sectional area $rd\theta dr$ at a distance r from the centre of the radius of the bend. For simplicity, we will assume that frictional effects are negligible, that the stream lines are all concentric arcs and the pressure distribution along the vertical is hydrostatic. The elementary column of water is in equilibrium under the action of the centrifugal force and the radial pressure gradient.

$$r d\theta \cdot y \left[p + \left(\frac{dp}{dr} \right) dr - p \right] = \alpha(\rho \cdot y \cdot r d\theta \cdot dr) \cdot \frac{V^2}{r}$$

where α is a factor to account for the non-uniform velocity distribution in the vertical and v is the tangential velocity. Dividing by γ and simplifying, we get $d/dr(p/\gamma) = dy/dr = \alpha V^2/gr$, assuming hydrostatic distribution of pressure along a vertical $p = \gamma y$. Note that dy/dr is the lateral slope of the free surface. The super-elevation (the difference in elevation between the inner and outer walls of the channel along a radial line) is

$$\Delta y = \int_{r_i}^{r_0} \frac{\alpha V^2}{gr} dr$$

where r_i is the inner radius and r_0 is the outer radius of the bend. It has been found that the flow in the bend can be conveniently approximated by a free vortex for which $V = C/r$, where C is the strength of the vortex. Assuming $\alpha = 1$, we have

$$\Delta y = \int_{r_i}^{r_0} \frac{C^2}{gr^3} dr = -\frac{C^2}{2g} \left(\frac{1}{r^2} \right)_{r_i}^{r_0} = \frac{C^2}{2g} \left[\frac{1}{r_i^2} - \frac{1}{r_0^2} \right] = \frac{C^2}{2gr_i^2 r_0^2} \left(r_0^2 - r_i^2 \right)$$

C, the strength of the vortex can be expressed in terms of the average depth, average velocity and discharge as follows. The average tangential velocity is

$$\overline{V} = \frac{\int_{r_i}^{r_0} V dr}{r_0 - r_i} = \frac{\int_{r_i}^{r_0} \frac{C}{r}}{r_0 - r_i} = \frac{C}{r_0 - r_i} \ln \frac{r_0}{r_i}$$

The average depth of flow is

$$\overline{y} = \frac{\int_{r_i}^{r_0} y dr}{r_0 - r_i} = \frac{\int_{r_i}^{r_0} (E - C^2/2gr^2) dr}{r_0 - r_i} = E - \frac{C^2 (r_0 - r_i)}{2gr_0 r_i}$$

where E is specific energy and $y = E - V^2/2g$. Now discharge $Q = \overline{V}\overline{y}(r_0 - r_i) = C(E - C^2/2gr_0 r_i) \ln(r_0/r_i)$.

12.31.2 Secondary Circulation

The velocity distribution along a vertical plane is not uniform, being zero at the boundary and reaching a maximum value at the free surface. Since the centrifugal force is proportional to V^2, the fluid particles at the top experience greater centrifugal force relative to the fluid particles near the boundary. Hence the fluid particles at the top move outwards due to the differential centrifugal force and to maintain continuity of flow a secondary circulation is set up in the bend transverse to the main flow as shown in Fig. 12.41. The fluid particles at the top move outwards to the outer bank and the fluid particles at the bottom move towards the inner side of the bend. In rivers with erodable sides, this has an adverse effect. The outer bank is attacked by the secondary current and causes erosion of the outer concave side of the bend and the sediment so eroded is carried by the bottom current to the convex inner side of the bend and is deposited there. This causes the bend curvature to increase and eventually is a cause for meanders in alluvial soils. The secondary current superimposed on the main flow in fact causes a spiral motion. The spiral motion develops at the bend entrance and increases in magnitude as the flow proceeds along the bend. The spiral motion decays gradually in the straight reach of the channel downstream of the bend due to viscous dissipation. The strength of the spiral motion is defined as a percentage of the mean kinetic energy of the flow at any section.

12.31.3 Loss of Energy in the Bend

Another important feature of the bend is the energy loss in the bend. The bend causes a distortion of the free surface and a portion of the energy lost may be attributed to wave drag. A portion of the energy is also lost due to boundary resistance. The secondary current also absorbs some energy which is eventually lost to the main flow when it decays. In sharp bends, there is flow separation on the inner side of the bend and eddy zones are formed which also extract energy from the main flow. The total energy loss due to the bend may be expressed as $h_e = f_c\, V_{\text{av}}^2/2g$, where f_c is the bend resistance factor and V_{av} is the average velocity in the channel. The value of f_c has been found experimentally in terms of bend parameters.

When the flow is super-critical, there are positive waves generated from the outer bank and expansion waves generated from the inner bank. They intersect each other, get modified and reach the opposite side to get reflected in the like sense. The water surface rises along the wall where positives are reflected and falls where negative waves are deflected. The water surface reaches maximum and minimum depths alternately along the bend. The process continues in the bend and even after the bend in the straight reach, till viscous dissipation of the waves takes place. The surface in the bend will be highly disturbed.

12.32 Unsteady Flow in Open Channels

Unsteady flow in open channels or rivers is caused by increase in inflow due to rainfall (floods), or due to the operation of control structures such as locks and gates. Common examples of unsteady flow in rivers are the propagation of tides in estuaries, closing or opening of turbine gates due to load fluctuations and flood waves created by sudden operation of gates.

12.32.1 Open Channel Surges

The class of waves which are produced in open channels due to the sudden operation of control structures, such as sudden opening or closing of gates, resulting in an abrupt increase or decrease in flow is known as surges. Surges are characterised by abrupt changes in curvature of the free surface with steep advancing or retreating fronts. Because of the rapid changes in the depth, surges are classified under rapidly varied unsteady flow. A surge with an advancing front moving upstream or downstream is known as a positive surge. A surge with a retreating front moving upstream or downstream is known as a negative surge.

12.32.2 Positive Surge

A positive surge has an advancing steep front but a stable profile. The analysis of the positive surge is simplified by the assumption that the wave profile will not change in shape as it moves down the channel. Such a fow is known as a uniformly progressive wave which is characterised by the following special features.

1. The wave fronts at successive positions at different times are parallel.
2. Velocity of the wave front is greater than the mean velocity of the main stream at any section.
3. The wave front moves with a constant velocity.

A monoclinal rising wave is a special wave form of uniformly progressive wave which can be conveniently approximated to surges for mathematical analysis. The monoclinal rising wave is a translatory wave of stable form that travels along the channel at an absolute velocity V_w from an upstream region of uniform flow with depth y_1, velocity V_1, discharge Q_1 to the downstream region of another uniform flow with depth y_2, velocity V_2, and discharge Q_2. The depth at the wave front changes abruptly from the upstream section to the downstream section as shown in Fig. 12.42. The velocity of the wave front V_w is greater than V_1 and V_2. We can analyse the flow by the method of superposition.

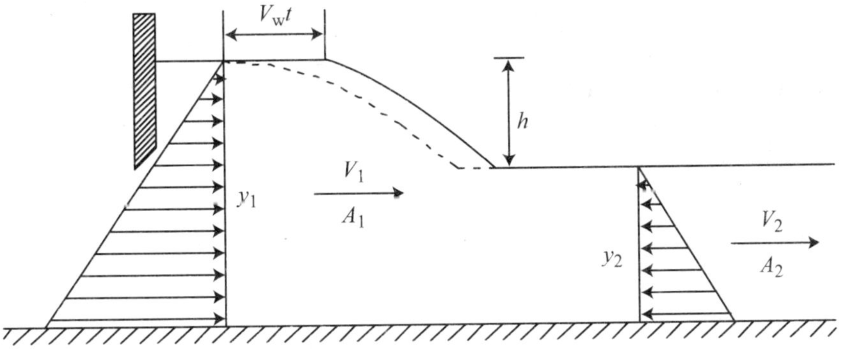

Fig. 12.42 Positive surge formation following sudden opening of gate

We superimpose on the flow an equal and opposite velocity, V_w so that the wave front becomes stationary. Now the velocity in the upstream and downstream regions will be $(V_w - V_1)$ and $(V_w - V_2)$. Therefore from continuity, we have

$$(V_w - V_1)A_1 = (V_w - V_2)A_2$$

$$V_w = \frac{V_1A_1 - V_2A_2}{A_1 - A_2} = \frac{Q_1 - Q_2}{A_1 - A_2}$$

The preceding equation gives the absolute velocity of the monoclinal rising wave. The equation was derived from the continuity principle. We can also analyse the monoclinal rising wave from the momentum principle. The mass of water between the gate and the wave front is $\rho(V_w - V_2)A_2$ and its velocity changes from V_1 to V_2. According to Newton's law, the change in momentum is equal to the force and the force is equal to the difference in hydrostatic pressure between the sections.

$$F = \gamma A_2 y_{2c} - \gamma A_1 y_{1c} = \left(\frac{\gamma}{g}\right)(V_w - V_2)(V_2 - V_1)$$

where y_{1c} and y_{2c} are the centroids of the areas A_1 and A_2 respectively. From the continuity equation, we have $V_2 = (V_1 A_1 + V_w A_2 - V_w A_1)/A_2$.

Substituting for V_2 in the momentum equation, we get

$$(V_w - V_1)^2 = \frac{(A_2 y_{2c} - A_1 y_{1c})g}{A_1(1 - A_1/A_2)}$$

$$V_w = V_1 + \sqrt{\frac{(A_2 y_{2c} - A_1 y_{1c})g}{A_1(1 - A_1/A_2)}}$$

The preceding equation gives the absolute velocity of the simplified surge. It is evident that the square-root term is the celerity of the wave.

$$c = \sqrt{\frac{A_2 y_{2c} - A_1 y_{1c}}{A_1(1 - A_1/A_2)} g}$$

For rectangular channels, $y_{1c} = y_1/2$ and $y_{2c} = y_2/2$, $A_1 = By_1$, $A_2 = By_2$, where B is the channel width. Then we obtain the celerity as

$$c = \sqrt{\frac{g y_2}{2 y_1}(y_1 + y_2)}$$

Equating the expressions for V_w obtained from momentum and continuity principles, we get

$$V_w = c + V_1 = \left[\frac{g y_2 (y_1 + y_2)}{2 y_1}\right]^{1/2} + V_1 = \frac{V_1 y_1 - V_2 y_2}{y_1 - y_2}$$

On simplification we have the relation between the initial velocities and depths of the surge.

$$(V_1 - V_2)^2 = \frac{g(y_1 - y_2)^2 (y_1 + y_2)}{2 y_1 y_2}$$

Multiplying the above equation by the square of the celerity, c^2, we get

$$V_1 - V_2 = \pm\left(\frac{gh}{c}\right)\left[\frac{y_1 + y_2}{2 y_1}\right]$$

where $h = y_1 - y_2$ = height of surge. If the height of the surge is small, the above expression is simplified as

$$V_1 - V_2 = \pm \frac{gh}{c}$$

Equations derived above can be used to compute the speed of propagation or the height of a surge given the initial conditions. The analysis was based on the assumption that the wave profile does not change as it moves along the channel. But in reality, the wave profile changes gradually due to the action of boundary resistance. The change or alteration in the wave profile is more in natural channels than in smooth prismatic channels. But in small lengths of artificial prismatic channels, the change is not noticeable. Hence the above analysis can be used in such cases with negligible error.

EXAMPLE 12.17 The depth and velocity of flow in a rectangular channel are 0.9 m and 1.5 m/s respectively. If the rate of inflow at the upstream end is suddenly doubled, what will be the height and the absolute velocity of the surge.

Let y_1, V_1 be the depth and velocity of the flow upstream of the surge and y_2, V_2 be the corresponding values downstream of the surge. Let V_w be the absolute velocity of the surge, $V_2 = 1.5$ m/s, $y_2 = 0.9$ m.

Initial discharge per unit width $= 0.9 \times 1.5 = 1.35\ \text{m}^3/\text{s/m}$

The discharge is doubled $= 2 \times 1.35 = 2.70\ \text{m}^3/\text{s/m}$ and $V_1 = 2.7/y_1$ m/s. By the principle of continuity, $y_1(2.7/y_1 - V_w) = 0.9(1.5 - V_w)$ or $V_w = 1.35/(y_1 - 0.9)$ and by the momentum principle, we write

$$\left(\frac{\gamma q}{g}\right)(V_2 - V_1) = \gamma\left(\frac{y_1^2}{2} - \frac{y_2^2}{2}\right) \text{ and } q = y_1\left(\frac{2.7}{y_1 - V_w}\right)$$

Hence,

$$y_1\left[\frac{2.7}{y_1} - \frac{1.35}{(y_1 - 0.9)}\right]\left(1.5 - \frac{2.7}{y_1}\right) = \left(\frac{g}{2}\right)(y_1^2 - 0.9^2)$$

A trial and error solution gives the positive real value of $y_1 = 1.182$ m.

The height of the surge is $1.182 - 0.9 = 0.282$ m and the velocity of the surge $V_w = 1.35/0.282 = 4.79$ m/s, $V_1 = 2.7/1.182 = 2.2$ m/s.

EXAMPLE 12.18 The depth and velocity of flow in a rectangular channel are 1.2 m and 1.8 m/s respectively. If the rate of flow at the upstream end is suddenly increased so that the depth of the flow is doubled, determine the absolute velocity of the resulting surge.

By continuity principle:

$$2.4(V_w - V_2) = 1.2(V_w - 1.8);\ V_2 = V_w/2 + 0.9$$

By momentum principle:

$$\frac{\gamma q}{g}(V_1 - V_2) = \left(\frac{\gamma}{2}\right)(y_2^2 - y_1^2)$$

$$1.2(1.8 - V_w)(1.8 - V_w/2 - 0.9) = \left(\frac{g}{2}\right)(2.4^2 - 1.2^2)$$

$$(1.8 - V_w)^2 = 3.6g \text{ and } V_w = (3.6g)^{1/2} + 1.8 = 7.74 \text{ m/s}$$

12.33 Negative Surge

Unlike positive surges, negative surges are not stable in form. The negative wave may be visualised as a wave front composed of a large number of smaller waves superimposed on each other. The higher wave in a greater depth moves faster than those in lower depths. In the case of positive surges, the higher waves will overtake the waves in smaller depths and finally assume a steep wave front with a stable form. In the case of a negative surge, the top waves will retreat faster than the bottom ones and the wave front will be stretched out, becoming flatter as the wave front advances. Examples of negative surges are seen on the downstream side of a gate when the discharge is suddenly decreased due to a lowering of a sluice gate. Similarly, the negative surge is seen on the upstream side when the gate is abruptly opened. In the case of a dam break also, a negative surge takes place and travels up the reservoir till the reservoir becomes empty.

A negative surge moving upstream is seen in Fig. 12.43.

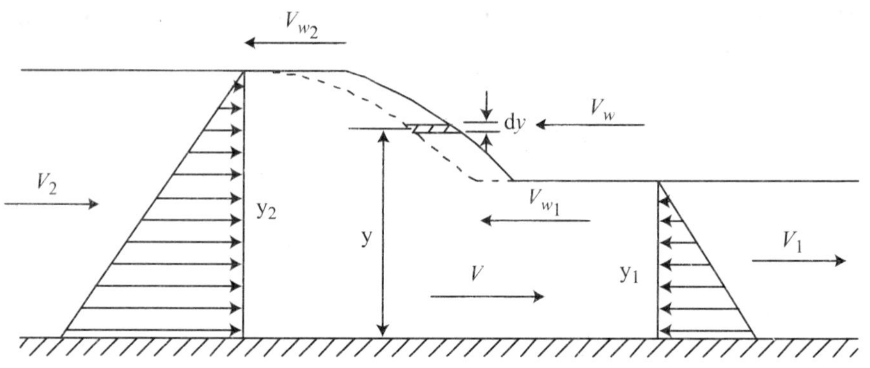

Fig. 12.43 A negative surge

The velocity V_w of the wave actually varies from point to point with change in depth. Let V_w be the velocity at a given section where the depth is y and the velocity of flow is V. After a time dt the depth changes by dy. The change in momentum of the vertical element between y and $(y + dy)$ is equal to the change in hydrostatic pressure due to change in depth. Assuming a channel of unit width, we write the momentum equation

$$\frac{\gamma y^2}{2} - \frac{\gamma(y+dy)^2}{2} = \left(\frac{\gamma}{g}\right)\left[\left(y + \frac{dy}{2}\right)(V + V_w)\right] dV$$

On simplification and omitting higher order terms, we get

$$dy = -\left[\frac{(V + V_w)}{g}\right] dV$$

Considering the surge to be made up of an infinite number of small waves of infinitesimal height, the velocity of a small wave at the point under consideration is derived from the equation for the positive wave $V_w = [(gy_2/2y_1)(y_1 + y_2)]^{1/2} + V$. With $y_1 \sim y_2$, we write

$$V_w = (gy)^{1/2} - V$$

At the wave crest where the depth is y_2. $V_{w2} = (gy_2)^{1/2} - V_2$ and at the wave trough where the depth is y_1, $V_{w1} = (gy_1)^{1/2} - V_1$.

Elimination of V_w between the equations for dy and V_w, we have

$$\frac{dy}{y^{1/2}} = -\frac{dV}{g^{1/2}}$$

Integrating the preceding equation between y_2 to y and V_2 to V and solving for V

$$V = V_2 + 2(gy_2)^{1/2} - 2(gy)^{1/2}$$

$$V_w = (gy)^{1/2} - V = 3(gy)^{1/2} - 2(gy_2)^{1/2} - V_2$$

$$V_{w1} = 3(gy_1)^{1/2} - 2(gy_2)^{1/2} - V_2$$

The expression for the wave length λ after a time t from the time the negative wave is created can be written as

$$\lambda = (V_{w2} - V_{w1})t$$

Example of a sudden break of a dam

The water profile shown in Fig. 12.44 after a sudden collapse of a dam may be obtained from the equations derived for a negative surge.

In this case V_2 (velocity in the reservoir) is zero and $y_2 = H$ and hence velocity at any depth y is

$$V = 2[(gH)^{1/2} - (gy)^{1/2}]$$

$$c = V - (gy)^{1/2} = 2(gH)^{1/2} - 3(gy)^{1/2}$$

The surface profile is given by

$$x = ct = [2(gH)^{1/2} - 3(gy)^{1/2}]t$$

The profile is a parabola with the vertex at the leading edge. The curve will get flatter and flatter till the entire reservoir is emptied.

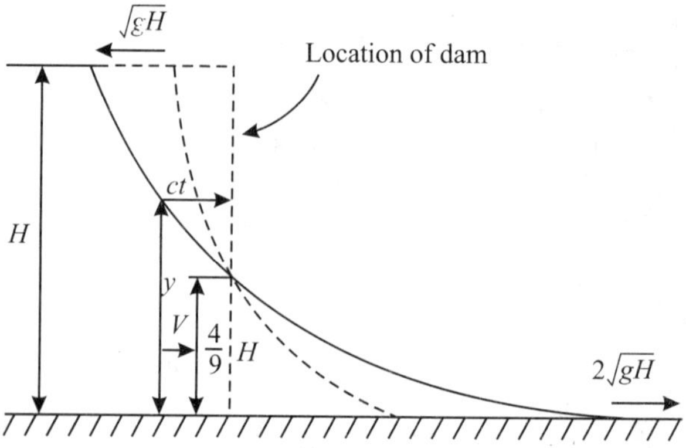

Fig. 12.44 Sudden break of dam

At $x = 0$, $y_0 = (4/9)H$ and the depth at the location of the dam remains constant and the velocity at the dam is $V = (2/3)(gH)^{1/2}$ and the discharge per unit width is $q = (8/27)g^{1/2}H^{3/2}$.

Theoretically, the front of the parabola has zero depth and moves with a velocity $(2gH)^{1/2}$. However in practice, friction of the ground causes a positive surge formation with a finite depth at the wave front and the velocity is also retarded.

EXERCISES

12.1. A channel of rectangular cross-section 3 m wide with Manning's $n = 0.014$ and slope 0.001 leads from a lake whose surface level is 3 m above the channel bed at the lake outlet. Compute the depth of uniform flow in the channel and the corresponding discharge. What will be the discharge in the channel if the slope becomes 0.002?

12.2. A venturi flume has an approach channel of 3 m width and a hump 13.5 cms high in the throat which is 1.6 m wide. Calculate the discharge when the upstream depth is 1 m and the venturi coefficient is 0.95.

12.3. A rectangular channel 3 m wide carries a water discharge of 8.5 m^3/s at a depth of 1.5 m. If the width is reduced to 2 m at a given section what change in bottom elevation will cause no change in water elevation?

12.4. What rate of discharge per unit width will cause depths of 2 m and 0.3 m to occur upstream and downstream of a sluice gate? What is the corresponding gate opening? $C_c = 0.6$

12.5. Water flows at a velocity of 2.44 m/s at a depth of 1.83 m in a rectangular channel. If the channel width is reduced from 3.05 m to 2.14 m and the bottom elevation is increased by 30.5 cms at a given section, determine the change in water surface elevation.

12.6. A sluice gate is mounted on a horizontal rectangular channel 3 m wide. A short distance upstream of the gate, the depth is 3 m. The gate opening is 1 m and the contraction coefficient is 0.61. Calculate the thrust on the gate.

12.7. Two rows of staggered floor blocks are installed in a stilling basin 6 m wide, in order to assist the formation of a stable hydraulic jump in the basin. The blocks are 60 cms in height 0.6 m wide and have an effective drag coefficient of 0.3 based on the upstream velocity and combined frontal area of ten blocks. If the discharge is 27 m^3/s and the upstream depth is 60 cms, find the downstream tail-water depth required to form a jump (a) if the blocks are installed and (b) in the absence of the blocks.

12.8. Water flows in a rectangular channel in which there is a smooth step of height Δz in the channel bed without change in width. Prove that the force on the step per unit width of the channel in the direction of the flow is equal to

$$F = \gamma y_1^2 \frac{(1-r)^3 + 4r(\Delta z/y_1)}{1+r}$$

where $r = y_2/y_1$.

12.9. Water flows in a rectangular channel in which the width changes smoothly from b_1 to $b_2 = kb_1$, with no change in bed level. Prove that the force on the side wall transition in the direction of flow is equal to

$$F = \frac{1}{2} b_1 y_1^2 \frac{1 - k^2 r^3 - 3kr(1-r)}{1+kr}$$

where $r = y_2/y_1$.

12.10. A rectangular channel 6 m wide has Manning's $n = 0.015$. Determine (a) the normal slope if the normal depth is 0.375 m for a discharge of 6 m^3/s, (b) the critical slope and the corresponding normal depth for a discharge of 6 m^3/s, (c) the critical slope at the critical normal depth of 0.375 m and compute the corresponding discharge.

12.11. Determine the critical depth in the following sections for a discharge of 3 m^3/s.

(i) rectangular section 6 m wide
(ii) triangular section with a vertex angle of 90°
(iii) trapezoidal section with bottom width of 6 m and side slopes of $1V{:}2H$
(iv) circular section of 4 m diameter
(v) a parabolic section with a width of 5 m at a depth of 1 m

12.12. Using Manning and Chezy formulae, determine the depths for maximum velocity and discharge in a circular conduit.

12.13. A rectangular flume is 0.6 m wide and laid on a slope of 0.001. When the channel surface is smooth, the measured normal depth and discharge are 0.415 m and 0.252 m^3/s. The corresponding values when the channel is roughened are 0.4 m and 0.147 m^3/s. (a) Determine the discharge for a normal depth of 0.4 m if the bed is rough and the walls are smooth. (b) Determine the discharge for a normal depth of 0.4 m if the bed is smooth and the walls are rough. (use eqn. 12.27c)

12.14. A channel is laid on a critical slope and the normal depth is equal to the critical depth ($y_n = y_c$). Show that if the wetted perimeter, P is to be a minimum, the depth of flow must be equal to three-quarters of the width.

12.15. A rectangular channel 4.5 m wide and 1.2 m deep is laid on a slope of 0.0001 with Manning's $n = 0.015$. Calculate the saving in excavation and lining in cubic metres and square metres respectively per kilometre if, hydraulically, the most efficient section is used for the same slope and discharge assuming that there is no free board.

12.16. A trapezoidal channel carries a discharge of 40 m^3/s at a mean velocity of 1 m/s. The side slopes are $1H{:}1V$ and Manning's $n = 0.015$. Calculate the section dimensions and longitudinal slope for best efficiency.

12.17. A trapezoidal channel with a longitudinal slope of 0.0001 carries a discharge of 10 m^3/s. The bottom width of the channel is 10 m and the side slopes are $2H{:}1V$. Find the depth of flow. ($n = 0.015$)

12.18. A long rectangular channel 4 m wide carries a discharge of 6 m^3/s, the bottom slope changing from 0.01 to 0.0001. Discuss the different profiles on the two slopes. Will a hydraulic jump occur? If so, will it occur on the steep slope or the mild slope? $n = 0.015$

12.19. A trapezoidal channel of 3 m bottom width and side slopes of $1H{:}1V$ is laid on a slope of 0.0001 and carries a discharge of 2.5 m^3/s. $n = 0.016$. If the canal ends in an abrupt drop, compute the surface profile by (1) graphical integration method, (2) direct integration method and (3) direct step method.

12.20. A rectangular channel 3 m wide is laid on a slope of 0.0002 and carries a discharge of 9 m^3/s. A small barrier is built across the channel which raises the depth just upstream of the barrier to 2 m. Assuming $n = 0.016$, compute the depth of flow 100 m upstream of the barrier.

12.21. A rectangular channel 6 m wide carries a discharge of 84 m^3/s from one lake to another. The lakes are 1.75 km apart and the difference in the surface level is 5 m. Choose a bed slope and bed levels so as to minimise the excavation required, assuming that there is no free board.

12.22. Determine the depths before and after a hydraulic jump in a horizontal rectangular channel 10 m wide and carrying a discharge of 10 m^3/s. The energy loss in the jump is 2 m.

12.23. A rectangular channel is 10 m wide and is laid on a slope of 0.01. If the slope changes to 0.001 at a given section and the depth in the mild reach is 1.5 m, locate the hydraulic jump. $n = 0.015$

12.24. Water issuing from a sluice gate has a depth of 0.3 m and a velocity of 24 m/s at the vena contracta. Determine (a) tail-water depth necessary to form a hydraulic jump, (b) the length of the jump and (c) the efficiency of the jump

12.25. The width of an open channel is to be reduced from 1.8 m to 0.9 m. The approach Froude number is 4.6 with a depth of 0.5 m. Compute the oblique wave angle and sketch the position of wave fronts. The depth after the oblique jump is given as 1.5 m.

12.26. Water discharges from a lake under a sluice gate into a wide rectangular channel with a slope of 0.005. At the gate, the channel bed is 3 m below the lake level. Downstream of the gate, the flow becomes uniform at a depth of 0.6 m. Now the gate is completely opened. Determine the speed and height of the surge thus formed.

12.27. Water is flowing at a depth of 1.5 m and a velocity of 0.9 m/s. If the discharge upstream is suddenly doubled, find the depth and the speed of the resulting surge.

12.28. A river is flowing at a depth of 2.5 m with a velocity of 0.9 m/s when it meets a tidal bore which abruptly increases the depth to 3.6 m. Find the speed with which the bore moves upstream and the magnitude and direction of the velocity of water behind the bore.

12.29. Sketch the possible surface profiles in the following cases. Assume all reaches to be sufficiently long enough to obtain stable conditions of flow in the reach.

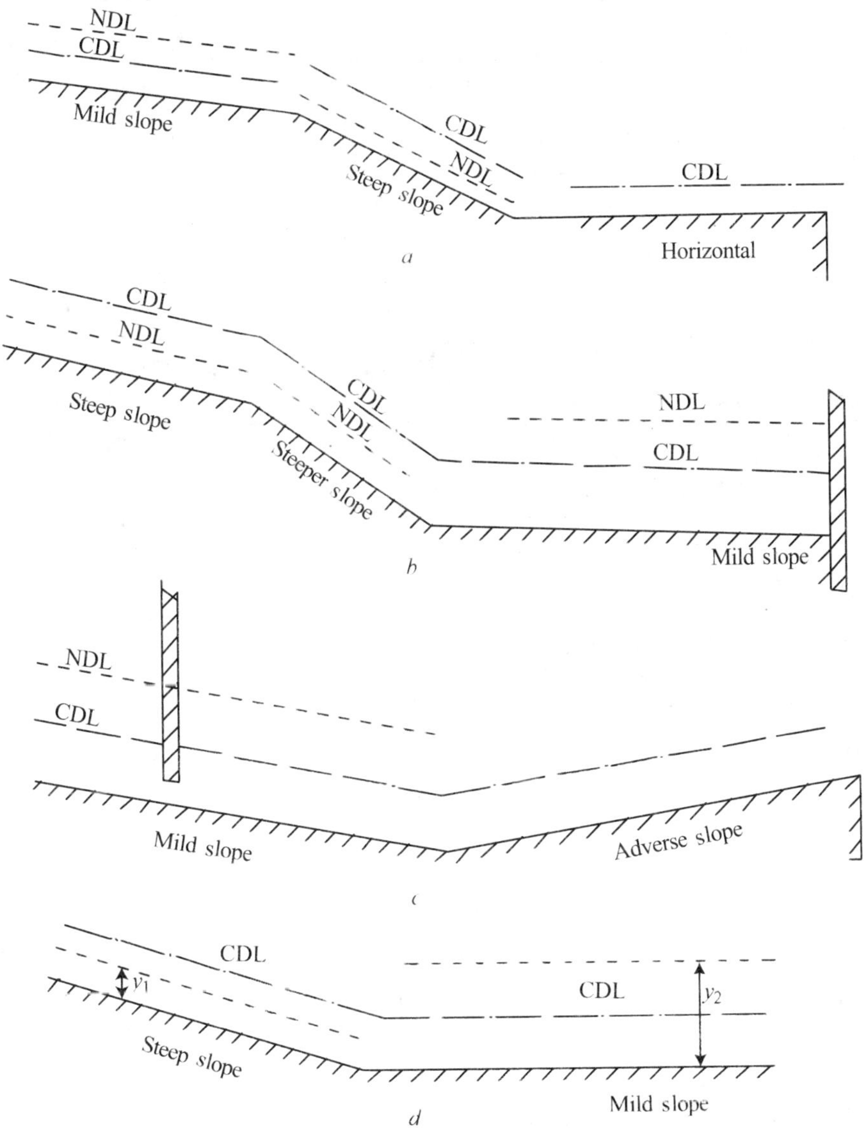

(a) if y_2 is less than conjugate depth of y_1

(b) if y_2 is greater than conjugate depth of y_1

Appendix 1

Determination of N and M values.

$$K^2 = C_1 y^N \tag{1}$$

K = conveyance factor, C_1 is a constant and N is a parameter called the hydraulic exponent for uniform flow computation.

Taking logarithms of both sides of the above equation and differentiating

$$\frac{d(\ln K)}{dy} = \frac{N}{2y} \tag{2}$$

Using Manning's equation, we have

$$K = \frac{1.49}{n} A R^{2/3} \tag{3}$$

Taking logarithms of both sides and differentiating with respect to y

$$\frac{d(\ln K)}{dy} = \frac{1}{A}\frac{dA}{dy} + \frac{2}{3}\frac{1}{R}\frac{dR}{dy} \tag{4}$$

$dA/dy = T$, surface width; $R = A/P$

$$\frac{d(\ln K)}{dy} = \frac{1}{3A}\left(5T - 2R\frac{dP}{dy}\right) \tag{5}$$

Equating the right-hand sides of (1) and (5), we get

$$N = \frac{2y}{3A}\left(5T - 2R\frac{dP}{dy}\right) \tag{6}$$

For rectangular channels,

$$N = \frac{10}{3} - \frac{8}{3}\left[\frac{y/B}{1 + 2(y/B)}\right] \tag{7}$$

For very wide rectangular channels where $B \gg y$, $N = 10/3$.

In a similar fashion we can express M in terms of channel geometry.

$$Z^2 = C_2 y^M$$

Taking logarithms of both sides and differentiating w.r.t. y

$$\frac{d}{dy}(\log Z) = \frac{M}{2y} \tag{8}$$

However,

$$Z = A\sqrt{\frac{A}{T}} \tag{9}$$

$$\frac{d}{dy}(\ln Z) = \frac{3}{2A}\frac{dA}{dy} - \frac{1}{2T}\frac{dT}{dy} = \frac{3T}{2A} - \frac{1}{2T}\frac{dT}{dy} \tag{10}$$

Equating the right-hand sides of (8) and (10)

$$M = \frac{3Ty}{A} - \frac{y}{T}\frac{dT}{dy}$$

For rectangular channels $M = 3$.

Appendix 2

Varied Flow Function

For adverse slopes (non-sustaining or negative slopes), we start with

$$\frac{dy}{dx} = \frac{-S_0 - S}{1 + d(V^2/2g)/dy}$$

The corresponding equation for flow profile can be shown to be

$$x = -\frac{y_n}{S_0}\left[u - \int_0^u \frac{du}{1+u^N} - \left(\frac{y_c}{y_n}\right)^M \int_0^u \frac{u^{N-M}}{1+u^N}du\right] + \text{constant}$$

The varied flow function are

$$F(u, N)_{-S_0} = \int_0^u \frac{du}{1+u^N}$$

$$F(v, J)_{-S_0} = \int_0^u \frac{u^{N-M}}{1+u^N}du = \frac{J}{N}\int_0^v \frac{dv}{1+v^J}$$

$$v = u^{N/J} \quad \text{where} \quad J = N/(N - M + 1)$$

$$x = -\frac{y_n}{S_0}\left[u - F(u, N)_{-S_0} + \left(\frac{y_c}{y_n}\right)^M F(v, J)_{-S_0}\right]$$

An abridged version of $F(u, N)_{-S_0}$ is given in Appendix 4.

Horizontal Channels

For horizontal channels, $S_0 = 0$. We start with

$$\frac{dy}{dx} = \frac{-(Q/K)^2}{1 - (Z_c/Z)^2}$$

$$Q = K_c\sqrt{S_c} \quad \text{and} \quad \left(\frac{K_c}{K}\right)^2 = \left(\frac{y_c}{y}\right)^N$$

$$\left(\frac{Z_c}{Z}\right)^2 = \left(\frac{y_c}{y}\right)^M \quad \text{and let} \quad p = \frac{y}{y_c}$$

$$\frac{dy}{dx} = S_c\frac{p^{M-N}}{1 - p^M}$$

Integrating and solving for x, we get

$$x = \frac{y_c}{S_c}\left(\frac{p^{N-M+1}}{N - M + 1} - \frac{p^{N+1}}{N + 1}\right) + \text{constant}$$

Appendix 3

Varied function table for positive slopes*

u \ N	2.5	3	$3\frac{1}{3}$	4	5	7	u \ N	2.5	3.0	$3\frac{1}{3}$	4	5	7
0.00	0.000	0.000	0.000	0.000	0.000	0.000	1.12	0.907	0.626	0.509	0.354	0.225	0.110
0.04	0.040	0.040	0.040	0.040	0.040	0.040	1.13	0.878	0.602	0.488	0.337	0.212	0.102
0.08	0.080	0.080	0.080	0.080	0.080	0.080	1.14	0.851	0.581	0.479	0.322	0.201	0.095
0.10	0.100	0.100	0.100	0.100	0.100	0.100	1.15	0.824	0.561	0.452	0.308	0.191	0.089
0.14	0.140	0.140	0.140	0.140	0.140	0.140	1.16	0.802	0.542	0.436	0.295	0.181	0.084
0.18	0.181	0.180	0.180	0.180	0.180	0.180	1.17	0.782	0.525	0.421	0.283	0.173	0.078
0.22	0.222	0.221	0.220	0.220	0.220	0.220	1.18	0.760	0.509	0.406	0.272	0.165	0.073
0.26	0.263	0.261	0.261	0.260	0.260	0.260	1.19	0.740	0.494	0.393	0.262	0.157	0.068
0.30	0.305	0.302	0.301	0.300	0.300	0.300	1.20	0.723	0.480	0.381	0.252	0.150	0.064
0.34	0.347	0.343	0.342	0.341	0.340	0.340	1.22	0.692	0.454	0.358	0.235	0.138	0.057
0.38	0.391	0.385	0.383	0.382	0.381	0.380	1.24	0.662	0.431	0.338	0.219	0.127	0.051
0.42	0.435	0.428	0.425	0.423	0.421	0.420	1.26	0.633	0.410	0.320	0.205	0.117	0.046
0.46	0.481	0.472	0.469	0.464	0.462	0.460	1.28	0.609	0.391	0.303	0.193	0.108	0.041
0.50	0.528	0.517	0.512	0.506	0.503	0.500	1.30	0.587	0.373	0.289	0.181	0.100	0.037
0.54	0.578	0.563	0.557	0.550	0.544	0.541	1.32	0.568	0.357	0.275	0.171	0.093	0.034
0.58	0.631	0.612	0.604	0.594	0.587	0.582	1.34	0.579	0.342	0.262	0.162	0.087	0.030
0.62	0.686	0.663	0.653	0.640	0.630	0.623	1.36	0.531	0.329	0.251	0.153	0.081	0.028
0.66	0.746	0.717	0.705	0.688	0.675	0.665	1.38	0.513	0.316	0.239	0.145	0.076	0.026
0.70	0.811	0.776	0.760	0.739	0.722	0.708	1.40	0.496	0.304	0.229	0.138	0.071	0.023
0.75	0.900	0.857	0.837	0.808	0.784	0.763	1.42	0.481	0.293	0.220	0.131	0.067	0.021
0.80	1.008	0.950	0.924	0.887	0.854	0.823	1.44	0.467	0.282	0.211	0.125	0.063	0.019
0.85	1.139	1.065	1.030	0.980	0.935	0.891	1.46	0.455	0.272	0.203	0.119	0.059	0.018
0.90	1.319	1.218	1.172	1.103	1.039	0.974	1.48	0.444	0.263	0.196	0.113	0.056	0.017
1.00	∞	∞	∞	∞	∞	∞	1.50	0.432	0.255	0.188	0.108	0.053	0.015
1.01	1.867	1.419	1.218	0.936	0.681	0.422	1.55	0.405	0.235	0.172	0.097	0.046	0.012
1.02	1.602	1.191	1.014	0.766	0.546	0.327	1.60	0.380	0.218	0.158	0.087	0.040	0.010
1.03	1.436	1.060	0.896	0.668	0.469	0.272	1.65	0.359	0.203	0.145	0.079	0.035	0.008
1.04	1.321	0.967	0.813	0.660	0.415	0.236	1.70	0.340	0.189	0.135	0.072	0.031	0.007
1.05	1.242	0.896	0.749	0.548	0.374	0.208	1.75	0.322	0.177	0.125	0.065	0.027	0.006
1.06	1.166	0.838	0.697	0.506	0.342	0.187	1.80	0.308	0.166	0.116	0.060	0.024	0.005
1.07	1.111	0.790	0.651	0.471	0.315	0.168	1.85	0.293	0.156	0.108	0.055	0.022	0.004
1.08	1.059	0.749	0.618	0.441	0.292	0.153	1.90	0.279	0.147	0.102	0.050	0.020	0.004
1.09	1.012	0.713	0.586	0.415	0.272	0.140	1.95	0.268	0.139	0.095	0.046	0.018	0.003
1.10	0.973	0.681	0.558	0.392	0.254	0.129	2.00	0.257	0.132	0.089	0.043	0.016	0.003
1.11	0.939	0.652	0.532	0.372	0.239	0.119	3.00	0.135	0.056	0.034	0.012	0.003	0.000

**Abridged version: For a full version, refer to Chow Ven Te, Open Channel Hydraulics, McGraw Hill Book Co. 1959*

Appendix 4

Varied function table for negative slopes*

u \ N	2.4	3.0	3.6	4.0	5.0	u \ N	2.4	3.0	3.6	4.0	5.0
0.00	0.00	0.000	0.000	0.00	0.00	1.11	0.00	0.006	0.907	0.907	0.935
0.04	0.040	0.040	0.040	0.040	0.040	1.12	0.864	0.891	0.910	0.921	0.939
0.08	0.080	0.080	0.080	0.080	0.080	1.13	0.868	0.895	0.914	0.925	0.943
0.10	0.100	0.100	0.100	0.100	0.100	1.14	0.872	0.899	0.918	0.928	0.947
0.14	0.140	0.140	0.140	0.140	0.140	1.15	0.876	0.903	0.922	0.932	0.950
0.18	0.179	0.180	0.180	0.180	0.180	1.16	0.880	0.907	0.926	0.936	0.953
0.22	0.217	0.220	0.220	0.220	0.220	1.17	0.884	0.911	0.930	0.939	0.957
0.26	0.256	0.259	0.260	0.260	0.260	1.18	0.888	0.915	0.933	0.943	0.960
0.30	0.294	0.297	0.299	0.300	0.300	1.19	0.892	0.918	0.937	0.947	0.963
0.34	0.331	0.337	0.339	0.339	0.340	1.20	0.896	0.921	0.940	0.950	0.966
0.38	0.368	0.374	0.377	0.378	0.380	1.22	0.900	0.929	0.944	0.956	0.972
0.42	0.405	0.411	0.414	0.417	0.419	1.24	0.908	0.935	0.950	0.962	0.977
0.46	0.440	0.449	0.453	0.455	0.478	1.26	0.919	0.942	0.960	0.968	0.982
0.50	0.475	0.485	0.491	0.493	0.497	1.28	0.926	0.948	0.965	0.974	0.987
0.54	0.507	0.520	0.527	0.531	0.536	1.30	0.933	0.955	0.981	0.979	0.991
0.58	0.539	0.554	0.563	0.567	0.574	1.32	0.940	0.961	0.976	0.985	0.995
0.62	0.571	0.578	0.598	0.603	0.611	1.34	0.948	0.967	0.982	0.990	0.999
0.66	0.600	0.618	0.630	0.637	0.647	1.36	0.954	0.973	0.987	0.994	1.002
0.70	0.629	0.649	0.663	0.670	0.682	1.38	0.960	0.979	0.993	0.998	1.006
0.75	0.664	0.686	0.701	0.709	0.724	1.40	0.966	0.984	0.997	1.001	1.009
0.80	0.695	0.720	0.737	0.746	0.762	1.42	0.972	0.989	1.001	1.005	1.012
0.85	0.725	0.752	0.770	0.780	0.798	1.44	0.979	0.995	1.006	1.009	1.016
0.90	0.754	0.781	0.801	0.811	0.834	1.46	0.983	1.000	1.010	1.014	1.018
0.95	0.783	0.809	0.829	0.840	0.861	1.48	0.989	1.005	1.014	1.016	1.020
1.00	0.808	0.834	0.856	0.867	0.887	1.50	0.996	1.009	1.017	1.020	1.022
1.01	0.812	0.840	0.861	0.873	0.893	1.55	1.007	1.020	1.026	1.029	1.028
1.02	0.818	0.845	0.866	0.877	0.898	1.60	1.020	1.030	1.035	1.035	1.032
1.03	0.822	0.850	0.871	0.882	0.902	1.65	1.032	1.039	1.042	1.041	1.036
1.04	0.829	0.855	0.877	0.888	0.907	1.70	1.044	1.048	1.049	1.047	1.039
1.05	0.831	0.859	0.881	0.892	0.911	1.75	1.054	1.057	1.055	1.052	1.042
1.06	0.837	0.864	0.885	0.896	0.915	1.80	1.064	1.065	1.060	1.057	1.045
1.07	0.841	0.869	0.889	0.901	0.919	1.85	1.074	1.071	1.066	1.061	1.047
1.08	0.846	0.873	0.893	0.905	0.923	1.90	1.084	1.079	1.071	1.065	1.049
1.09	0.851	0.877	0.898	0.909	0.927	1.95	1.092	1.085	1.075	1.068	1.051
1.10	0.855	0.881	0.903	0.913	0.931	2.00	1.100	1.090	1.078	1.071	1.053

**Abridged version: For a full version refer to Chow Ven Te, Open Channel Hydraulics, McGraw Hill Book Co, 1959.*

13 Compressible Flow

13.1 Introduction

This book is essentially devoted to incompressible flow but a small chapter is introduced here so that the compressibility effects especially at supersonic speeds may be appreciated by the student. Some of these effects are not observed at low speeds or in incompressible flows. We shall largely limit our study to frictionless flow outside the boundary layer, since the boundary layer is relatively very thin in high speed flows compared to low speed flows. Further we shall ignore the effect of gravity whose effects are minor in high speed flows of gases.

Since density changes are accompanied by temperature changes, it will be useful to recapitulate the relations for a perfect gas with constant specific heats.

$$\text{Equation of state: } p = \rho RT \tag{13.1}$$

$$\text{Internal energy: } e = C_v T \tag{13.2}$$

$$\text{Enthalpy: } h = C_p T \tag{13.3}$$

$$\text{Specific heats: } C_p = \frac{kR}{(k-1)} \tag{13.4}$$

$$C_v = \frac{R}{(k-1)} \tag{13.5}$$

$$\frac{C_p}{C_v} = k \tag{13.6}$$

$$\text{Speed of sound: } c = (kRT)^{1/2} \tag{13.7}$$

$$\text{Entropy change: } S_2 - S_1 = C_p \ln\left(\frac{T_2}{T_1}\right) - R\ln\left(\frac{p_2}{p_1}\right) \tag{13.8}$$

The following relations are also applicable to an isentropic process of a perfect gas.

$$\frac{p_2}{p_1} = \left(\frac{\rho_2}{\rho_1}\right)^k \tag{13.9}$$

$$\frac{T_2}{T_1} = \left(\frac{\rho_2}{\rho_1}\right)^{k-1} = \left(\frac{p_2}{p_1}\right)^{(k-1)/k} \tag{13.10}$$

For air at standard temperature and pressure, $R = 287\,\text{m}^2/(\text{s}^2\text{K})$; $C_p = 1005\,\text{m}^2/(\text{s}^2\text{K})$ and $C_v = 718\,\text{m}^2/(\text{s}^2\text{K})$ and $k = 1.4$. K is the temperature on the Kelvin scale.

13.2 Speed of Sound

The speed of sound is defined as the rate at which an infinitesimal disturbance such as a pressure pulse propagates in a medium with respect to the frame of reference of the medium. Actual sound waves comprised of pressure disturbances of finite amplitude travel only slightly faster than the speed of an infinitesimal disturbance (designated as the speed of sound). We will derive an equation for the speed of sound assuming that the control volume moves with the wave, thereby analysing a steady flow problem.

Let us consider a small section of a pressure wave as it propagates at velocity c through a medium as shown in Fig. 13.1. As the wave travels through the gas initially at pressure p and density ρ, it produces infinitesimal changes Δp, $\Delta \rho$, ΔV. These changes must obey laws of continuity and force–momentum. We can allow the control volume to travel with the wave or we can superpose an equal and opposite velocity equal to c to render the disturbance front stationary.

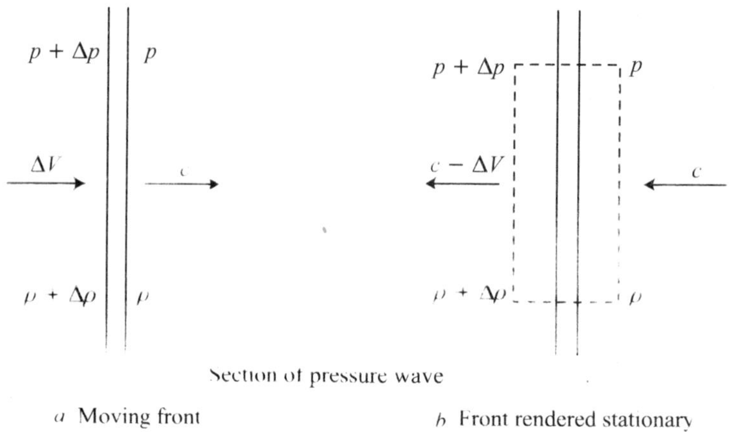

Fig. 13.1 Propagation of a disturbance

Mass balance of the control volume requires

$$A\rho c = A(\rho + \Delta\rho)(c - \Delta V)$$

Since the amplitude of the disturbance is assumed small, we can neglect the higher order of terms to obtain

$$\Delta V = c\left(\frac{\Delta\rho}{\rho}\right) \tag{13.11}$$

The momentum equation applied to the steady flow gives

$$pA - (p + \Delta p)A = A\rho c(c - \Delta V) - (\rho Ac)c$$

Simplifying the momentum equation, we get

$$\Delta p = \rho c \Delta V \tag{13.12}$$

Eliminating ΔV from the continuity and momentum equations, we have

$$\left\{c^2 = \frac{\Delta p}{\Delta \rho}\right\} \tag{13.13}$$

The speed of propagation of the disturbance is related to change in pressure and change in density across the wave. In incompressible flows, density is constant, $\Delta\rho = 0$ and the propagation speed becomes infinite.

This is not realistic and we must realise that liquids are also compressible to some extent and that the property of bulk modulus of liquids has been taken into account while determining the pressures due to water hammer.

The above equation gives an expression for the speed of a general pressure wave. The sound wave is a special type of pressure wave producing only infinitesimal changes in pressure and density and may be regarded as a reversible process. Since there is negligible heat transfer, we can regard it as an adiabatic process and a reversible adiabatic process is an isotropic process. We can replace $\Delta p/\Delta\rho$ by the partial derivatives at constant entropy

$$c^2 = \left(\frac{\partial p}{\partial \rho}\right)_s \tag{13.14}$$

Since for a perfect gas, p/ρ^k = constant and $p = \rho RT$, we have

$$c = \left(\frac{kp}{\rho}\right)^{1/2} = (kRT)^{1/2} \tag{13.15}$$

For example, the speed of sound in air at 15°C is, with R = 287 J/kg K and k = 1.4, $c = [1.4 \times 287 \text{ J/kg K} \times 288 \text{ K}]^{1/2} = 340$ m/s

Finite amplitude waves across which there is a discontinuous change of pressure is called a shock wave. Shock waves are not isentropic and they propagate faster than the sonic speed.

The speed of sound can also be expressed in terms of bulk modulus

$K = -\mathrm{d}p/(\mathrm{d}V\!\!\!/ /V\!\!\!/)$, where $V\!\!\!/$ is the volume and $\mathrm{d}V\!\!\!/$ is the change in volume due to a change in pressure $\mathrm{d}p$. The negative sign is due to the fact that an increase in pressure causes a decrease in volume.

$$\frac{\mathrm{d}V\!\!\!/}{V\!\!\!/} = \frac{\mathrm{d}v_s}{v_s} = \frac{-\mathrm{d}\rho}{\rho} \text{ and } K = \rho\left(\frac{\mathrm{d}p}{\mathrm{d}\rho}\right)$$

Hence $$c = \left(\frac{K}{\rho}\right)^{1/2} \tag{13.16}$$

13.3 Mach Number

The Mach number, **M** is defined as the ratio of the velocity of the fluid to the local velocity of sound in the medium $\mathbf{M} = V/c$. Since the square of the Mach number $V^2/c^2 = V^2/kRT$, the Mach number may be interpreted as the ratio of the kinetic energy (which is proportional to V^2) to the thermal energy (which is proportional to T). The Mach number is also a measure of compressibility and for an incompressible fluid, K = infinity and $\mathbf{M} = 0$. For perfect gases where compression is isentropic, $K = kp$.

Compressible flows can be classified on the basis of the Mach number.

(a) Incompressible flow: When $\mathbf{M} < 0.3$, density variations due to pressure changes may be neglected and the fluid flow may be treated as an incompressible flow.
(b) Subsonic flow: The Mach number lies in the range $0.3 < \mathbf{M} < 1.0$ and shock waves do not appear anywhere in the flow.
(c) Transonic flow: The Mach number lies in the range of 0.8 and 1.2. Shock waves appear and lead to an appreciable increase in drag. The transonic flow is difficult to handle analytically since non-linear terms must be retained and further inviscid and viscous aspects of the flow cannot be separated.
(d) Supersonic flow: The Mach number lies in the range of 1 to 3. Shock waves are present sometimes and supersonic flow is more adept to analytical solutions since disturbances propagate along certain lines known as characteristic lines.
(e) Hypersonic flow: When the Mach number is greater than 3, heat is generated in the boundary layers causing dissociation of molecules and other chemical effects.

13.4 One-Dimensional Flow (Basic Equations)

Let us consider a stream tube of area $A(x)$ at any section x-x as shown in Fig. 13.2. We shall assume that the curvature of the stream lines compared to the diameter of the stream tube is small.

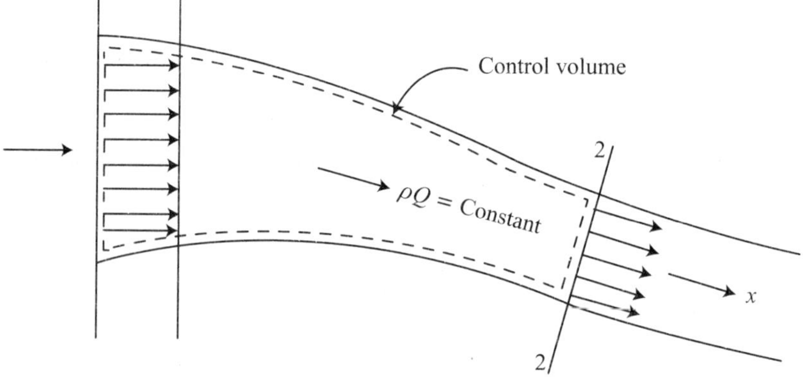

Fig. 13.2 Stream tube for one-dimensional flow

The variations of the velocity, density and pressure across the tube are ignored and the average velocity, density and pressure at any section are the representative properties at that section [$u(x)$, $\rho(x)$, $p(x)$].

Continuity equation

The conservation of mass requires $\rho V A =$ constant for steady flows. Taking logarithms of both sides and differentiating with respect to x, we have

$$\left(\frac{1}{\rho}\right)\left(\frac{\mathrm{d}\rho}{\mathrm{d}x}\right)+\left(\frac{1}{V}\right)\left(\frac{\mathrm{d}V}{\mathrm{d}x}\right)+\left(\frac{1}{A}\right)\left(\frac{\mathrm{d}A}{\mathrm{d}x}\right)=0$$

$$\frac{\mathrm{d}\rho}{\rho}+\frac{\mathrm{d}V}{V}+\frac{dA}{A}=0 \tag{13.17}$$

Energy equation

Applying the energy equation to two points 1 and 2 on a stream line,

$$-\dot{m}_1\left(h_1+\frac{V_1^2}{2}+gz_1\right)+\dot{m}_2\left(h_2+\frac{V_2^2}{2}+gz_2\right)=\dot{Q} \tag{13.18}$$

where h is the enthalpy which is the sum of internal energy e and the work done by pressure forces per unit mass pv_s or p/ρ.

The elevation terms (or the component of weight of gas or the gravity effect) between the two sections may be neglected. If the flow is adiabatic, the right-hand side disappears. The equation of continuity requires that the mass flow rate is constant ($\dot{m}_1 = \dot{m}_2$) so (13.18) becomes

$$h_1+\frac{V_1^2}{2}=h_2+\frac{V_2^2}{2}$$

Since the two points 1 and 2 are arbitrarily chosen, we conclude

$$h+\frac{V^2}{2}=\text{constant} \tag{13.19}$$

along a stream line in an adiabatic flow.

The constant in this expression is called total enthalpy or stagnation enthalpy, h_0 since this value is obtained if the flow is brought to rest and $V = 0$. If h_0 is the same for all stream lines, the flow is homenergic.

The equation (13.19) may now be compared with the Bernoulli equation. We express the specific enthalpy as the sum of specific internal energy, u and p/ρ and write the above equation as

$$\left(u+\frac{p}{\rho}\right)+\frac{V^2}{2}=\text{constant}$$

If the flow is incompressible and there is no heat transfer, the specific internal energy is constant and the equation reduces to the Bernoulli's equation.

Temperature

The enthalpy of an ideal gas is $h = C_p T$, where C_p is the specific heat at constant pressure. Substituting in equation (13.19) and dividing by $C_p T$, we get

$$1 + \frac{V^2}{(2C_p T)} = \frac{T_0}{T} \tag{13.20}$$

where T_0 is the stagnation temperature (when the flow is brought to zero isentropically). But $C_p = kR/(k-1)$ and $c^2 = kRT$.

$$T_0 = T\left[1 + \frac{(k-1)\mathbf{M}^2}{2}\right] \tag{13.21}$$

If an aircraft is flying at $\mathbf{M} = 1.6$ at a temperature of $-50°$, the total or stagnation temperature is

$$T_0 = (273 - 50)[1 + 0.2 \times 1.6^2] = 337 \text{ K or } 64°\text{C}$$

Pressure

The thermodynamic relationship between pressure and temperature in an isentropic flow is

$$\frac{p_1}{p_2} = \left(\frac{T_1}{T_2}\right)^{k/(k-1)} = \left\{\frac{1 + [(k-1)/2]\mathbf{M}_2^2}{1 + [(k-1)/2]\mathbf{M}_1^2}\right\}^{k/(k-1)} \tag{13.22}$$

To find the stagnation pressure put $p_1 = p_0$ and $p_2 = p$, $T_1 = T_0$, $T_2 = T$, then we have

$$p_0 = p\left[1 + \frac{(k-1)\mathbf{M}^2}{2}\right]^{k/(k-1)} \tag{13.23}$$

If the flow is isentropic the total pressure is constant along a stream line. However, unlike total temperature, total pressure may not be constant along stream lines in adiabatic flows. For example in shock waves, though the flow is adiabatic, it is not reversible and therefore not isentropic. The total pressure variation along a stream line can be found by

$$\frac{p_{01}}{p_{02}} = \left(\frac{p_1}{p_2}\right)\left(\frac{T_2}{T_1}\right)^{k/(k-1)} \tag{13.24}$$

Density

Since $p_2/p_1 = (\rho_2/\rho_1)^k$, the total density at the point where the velocity becomes zero is

$$\rho_0 = \rho\left[1 + \frac{(k-1)\mathbf{M}^2}{2}\right]^{1/(k-1)} \tag{13.25}$$

where ρ is the local density. If the flow is isentropic, the total density, ρ_0 is constant and we can determine the local density along a stream line as a function of the Mach number.

13.5 Area–Velocity Relation in One-Dimensional Isentropic Flow

If the flow is inviscid, Euler equation for steady flow is written as

$$\rho V\left(\frac{\mathrm{d}V}{\mathrm{d}x}\right)+\frac{\mathrm{d}p}{\mathrm{d}x}=0$$

Substituting $c^2 = \mathrm{d}p/\mathrm{d}\rho$ or $\mathrm{d}p = c^2\mathrm{d}\rho$, the preceding equation becomes

$$-\left(\frac{V}{c^2}\right)\left(\frac{\mathrm{d}V}{\mathrm{d}x}\right)=\left(\frac{1}{\rho}\right)\left(\frac{\mathrm{d}\rho}{\mathrm{d}x}\right)$$

From equation (13.17), we can substitute for the expression on the right-hand side

$$\left(\frac{1}{V}\right)\left(\frac{\mathrm{d}V}{\mathrm{d}x}\right)\left[\left(\frac{V}{c}\right)^2-1\right]=\left(\frac{1}{A}\right)\left(\frac{\mathrm{d}A}{\mathrm{d}x}\right)$$

$$\frac{1}{V}\frac{\mathrm{d}V}{\mathrm{d}x}=\frac{(1/A)(\mathrm{d}A/\mathrm{d}x)}{\mathbf{M}^2-1} \tag{13.26}$$

The above equation can be used to analyse the variation of velocity with changes in areas of cross-sections of ducts for subsonic and supersonic flow conditions.

For subsonic flow, $\mathbf{M} < 1$, $\mathbf{M}^2 - 1$ is negative. Hence a decrease in area leads to an increase in the velocity or an increase in area results in deceleration.

On the other hand, for supersonic flow since $\mathbf{M}^2 - 1$ is positive, a decrease in area or a contraction leads to a decrease in velocity and an increase in area or expansion will result in increase in velocity. Therefore, the velocity is minimum when the area is minimum. This principle is used in a diffuser of a jet engine. First the area is reduced to decrease the velocity so that sufficient time is available in the combustion chamber for gases to burn and then the divergence causes the velocity to increase providing sufficient thrust for the aircraft.

The area in a duct is a minimum or a maximum when $\mathrm{d}A/\mathrm{d}x = 0$ (See Fig. 13.3). At these sections from equation (13.26), we find $\mathrm{d}V/\mathrm{d}x = 0$. For subsonic flow, the velocity is a maximum and for supersonic flow the velocity is a minimum.

When $\mathbf{M} = 1$, the value of $\mathrm{d}V/\mathrm{d}x$ becomes infinite but this is physically impossible. Hence we must have $\mathrm{d}A/\mathrm{d}x = 0$ when $\mathbf{M} = 1$ so that realistic velocity gradients exist. Conversely, we conclude that sonic flow will occur at a minimum cross-section. If the flow is initially subsonic, a reduction in area (Fig. 13.3*a*) causes the flow to accelerate towards sonic speed and if the flow is initially supersonic the flow will decelerate towards sonic speed. It may be seen that the flow cannot reach sonic speed if the duct has a shape as shown in Fig. 13.3*b*. However it is not necessary that the flow reach sonic speed when the local area is a minimum as shown in Fig. 13.3*c* and *d*. We will examine the relations to be satisfied for Mach number to be unity at the minimum cross-section.

We will designate density, velocity, area, pressure, temperature at the sonic cross-section by asterisks viz., ρ^*, V^*, A^*, p^*, T^*

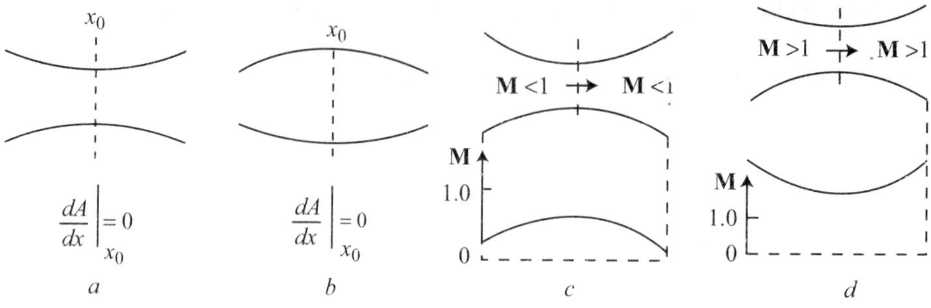

Fig. 13.3 Duct shapes ($dA/dx = 0$)

By continuity we have:

$$\rho V A = \rho^* V^* A^*$$

$$V = \mathbf{M}c = \mathbf{M}(kRT)^{1/2} \text{ and } V^* = c = (kRT^*)^{1/2}$$

$$\frac{A}{A^*} = \left(\frac{\rho^*}{\rho}\right)\left(\frac{T^*}{T}\right)^{1/2}\left(\frac{1}{\mathbf{M}}\right)$$

Since the flow is isentropic, the total pressure, total temperature, total density are all constant throughout the flow ($\rho^* = \rho_0$, $T^* = T_0$, $p^* = p_0$).

Replacing these values in equations (13.21) and (13.25) and simplifying, we get

$$\frac{A}{A^*} = \frac{1}{\mathbf{M}}\left\{\frac{1 + [(k-1)/2]\mathbf{M}^2}{(k+1)/2}\right\}^{(k+1)/2(k-1)} \tag{13.27}$$

EXAMPLE 13.1 A supersonic air tunnel is required to operate at a Mach number of 3. What is the test area if the throat area is 12 cm^2. $k = 1.4$.

$$\frac{A}{A^*} = \left(\frac{1}{3}\right)\left(\frac{1 + 0.2 \times 3^2}{1.2}\right)^3 = 4.23$$

Hence $A = 4.23 \times 12 = 51.06\,\text{cm}^2$

EXAMPLE 13.2 An air tunnel has a test-to-throat area ratio of 3.5. The absolute total pressure and temperature are 5 MPa and 400 K. Find the Mach number, pressure, temperature and velocity at the test section.

$$\frac{A}{A^*} = 3.5 = \left(\frac{1}{\mathbf{M}}\right)\left[\frac{(1 + 0.2\mathbf{M}^2)}{1.2}\right]^3$$

By trial and error, we find **M** has two values, one less than one and the other greater than one. Since the flow is supersonic, the higher value is 2.80. This can also be got easily by using the supersonic tables given in Appendix 1 (isentropic table).

We can use the tables or equations (13.21) and (13.24) to find T/T_0 and p/p_0

$$\frac{T}{T_0} = 0.389, \ T = 0.389 \times 400 = 155.6\,\text{K}$$

$$\frac{p}{p_0} = 0.037, \ p = 0.037 \times 5 = 0.185\,\text{MPa}$$

13.6 Mass–Flow Rate

The mass–flow rate at the throat is

$$\frac{dm}{dt} = \rho^* A^* V^* = \rho^* A^* (kRT^*)^{1/2}$$

From equations (13.21) and (13.25), we can express T^* and ρ^* in terms of the total temperature and density respectively (with $\mathbf{M} = 1$).

$$T^* = T_0\left[\frac{2}{(k+1)}\right] \text{ and } \rho^* = \rho_0\left[\frac{2}{(k+1)}\right]^{1/(k-1)}$$

Hence the mass flow rate $= \rho_0(kRT_0)^{1/2} A^*[2/(k+1)]^{(k+1)/2(k-1)}$

We can get the relation in terms of pressure since for ideal gas $p = \rho RT$,

$$\dot{m} = \frac{p_0 A^*}{\sqrt{RT_0}} k^{1/2} \left(\frac{2}{k+1}\right)^{(k+1)/2(k-1)} \tag{13.28}$$

$$= 0.685\, p_0 A^*/(RT_0)^{1/2} \qquad (\text{at } k = 1.4)$$

EXAMPLE 13.3 A rocket motor generates a thrust of 30,000 N at an altitude of 20 km, where the exit pressure is 5467 Pa. The pressure inside the combustion chamber (throat) is 1000 kPa and the temperature is 2500 K. Take $R = 280\,\text{m}^2/\text{s}^2\,\text{K}$ and $k = 1.4$. Assuming that the flow is isentropic, calculate the throat and exit areas.

$$p_e = 5467\,\text{Pa}, \ p_0 = 1000\,\text{kPa}, \ T_0 = 2500\,\text{K}, \ \frac{p_e}{p_0} = \frac{5467}{1000 \times 1000} = 5.467 \times 10^{-3}$$

Thrust or force = mass × velocity = $(\rho_e A_e V_e) V_e$

From (13.23), Mach number at exit is 4.15;

From (13.27), $A_e/A^* = 12.2$;

From (13.21), $T_e/T_0 = 0.225$; All the values may be obtained from Table A.1.
Therefore, the exit temperature is $T_e = 0.225 \times 2500 = 562\,\text{K}$
The exit density is $\rho_e = p_e/RT_e = 5467/(280 \times 562) = 0.0347\,\text{kg/m}^3$
The exit velocity $V_e = \mathbf{M}_e(kRT_e)^{1/2} = 4.15 \times (1.4 \times 280 \times 562)^{1/2} = 1948\,\text{m/s}$
The exit area is $A_e = \text{Thrust}/\rho_e V_e^2 = 30000/(0.0347 \times 1948)^2 = 0.228\,\text{m}^2$
The throat area is $0.228/12.2 = 0.0187\,\text{m}^2$

13.7 Normal Shock Wave

A shock wave is a wave front normal to the flow, similar to a sound wave but having a finite strength. It has some thickness. The flow properties are discontinuous across the front and isentropic relations can be used to analyse the flow across the shock wave. This is analogous to the hydraulic jump in open channels. The flow changes from a supersonic condition to a subsonic one. As a result of the deceleration, there is an increase in static temperature, pressure and density. We can derive the relationships between the properties on either side of the shock wave with reference to Fig. 13.4.

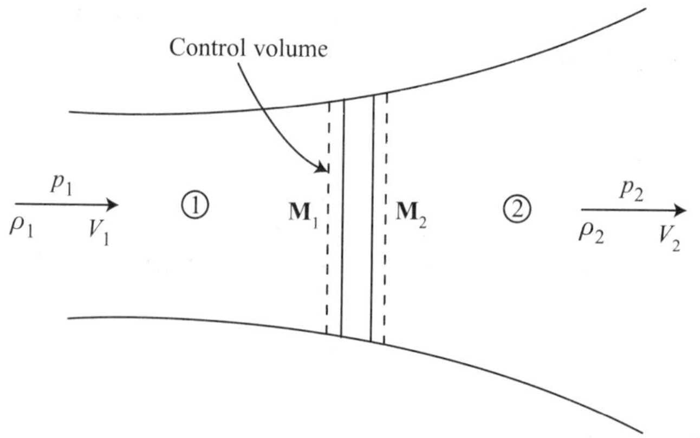

Fig. 13.4 Normal shock wave

The control volume chosen has the two sections 1 and 2 close to each other and area change is neglected.

1. Continuity gives $\rho_1 A V_1 = \rho_2 A V_2$
2. Momentum equation gives $(p_1 - p_2)A = A(\rho_2 V_2^2 - \rho_1 V_1^2)$ or since $\rho V^2 = \rho c^2 \mathbf{M}^2 = \rho(kp/\rho)\mathbf{M}^2 = kp\mathbf{M}^2$, we can rewrite

$$p_1 + kp_1\mathbf{M}_1^2 = p_2 + kp_2\mathbf{M}_2^2$$

$$\frac{p_2}{p_1} = \frac{1 + k\mathbf{M}_1^2}{1 + k\mathbf{M}_2^2} \tag{13.29}$$

3. Energy equation: $h_1 + V_1^2/2 = h_2 + V_2^2/2$

   But $h = C_pT = \left[\frac{kR}{(k-1)}\right]\left(\frac{p}{\rho R}\right)$ or $h = \frac{kp}{(k-1)\rho}$

   The energy equation finally is of the form

$$\left[\frac{k}{(k-1)}\right]\left(\frac{p_1}{\rho_1}\right) + \frac{V_1^2}{2} = \left[\frac{k}{(k-1)}\right]\left(\frac{p_2}{\rho_2}\right) + \frac{V_2^2}{2} \tag{13.30}$$

   Elimination of ρ_2 and V_2 from the three equations above yields

$$\frac{p_2}{p_1} = 1 + \left[\frac{2k(\mathbf{M}_1^2 - 1)}{k+1}\right] \tag{13.31}$$

   Using the momentum equation, the following relations may be obtained

$$\mathbf{M}_2^2 = \frac{(k-1)\mathbf{M}_1^2 + 2}{2k\mathbf{M}_1^2 + 1 - k} \tag{13.32}$$

$$\frac{\rho_2}{\rho_1} = \frac{V_1}{V_2} = \frac{(k+1)\mathbf{M}_1^2}{(k-1)\mathbf{M}_1^2 + 2} \tag{13.33}$$

$$\frac{T_2}{T_1} = 1 + \frac{2(k-1)}{(k+1)^2}\frac{k\mathbf{M}_1^2 + 1}{\mathbf{M}_1^2}(\mathbf{M}_1^2 - 1) \tag{13.34}$$

EXAMPLE 13.4 A normal shock wave occurs in air at a Mach number of 1.5. The static pressure and temperature of the air upstream of the shock wave is 100 kPa absolute and 15°C. Find the Mach number, pressure and temperature after the shock front.

$\mathbf{M}_2$ is determined by (13.32); or by Table A.2 in the Appendix.

$$\mathbf{M}_2^2 = \frac{0.4 \times 1.5^2 + 2}{2 \times 1.4 \times 1.5^2 + 1 - 1.4} = 0.49$$

$$\mathbf{M}_2 = 0.7$$

To find p_2, we can use either (13.29) or (13.31) or Table A.2

$$\frac{p_2}{p_1} = 1 + \frac{2 \times 1.4 \times (1.5^2 - 1)}{1.4 + 1} = 2.46$$

$$p_2 = 246\,\text{kPa absolute}$$

$$\frac{T_2}{T_1} = 1 + \frac{[2 \times 0.4 \times 1.4 \times 1.5^2 + 1] \times 1.25}{2.4^2 \times 1.5^2} = 1.34$$

$$T_2 = 385\,\text{K}$$

The total pressure does change across the wave front though the total temperature remains unchanged. The total pressure before and after the shock can be calculated by equation (13.23)

$$p_{01} = 100(1 + 0.2 \times 1.5^2)^{3.5} = 367\,\text{kPa}$$

and $$p_{02} = 246[1 + 0.2(0.7)^2]^{3.5} = 341\,\text{kPa}$$

The entropy increase is given by

$$\Delta s = R \ln \left(\frac{p_{01}}{p_{02}}\right) = 287 \left(\ln \frac{367}{341}\right) = 21\ \text{J/kg K}$$

EXAMPLE 13.5 A pitot tube is used to measure the Mach number of a supersonic flow in terms of pitot pressure and static pressure of the free stream. A shock exists between the incoming supersonic stream (1) and the tip of the pitot tube (3). The stream line in region (2) is brought to rest isentropically at location (3). Find an expression for p_3.

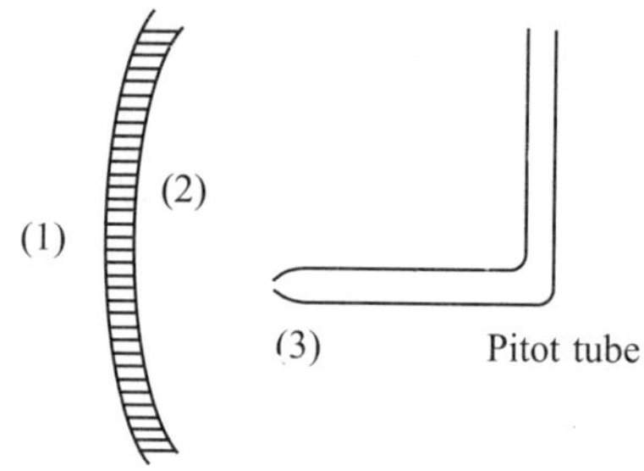

Figure for Example 13.5

$p_3 = p_{0,2}$ (stagnation pressure or total pressure)
The ratio of p_2/p_1 is obtained by the shock relationship (13.31)

$$\frac{p_2}{p_1} = 1 + \frac{2k(\mathbf{M}_1^2 - 1)}{k+1}$$

$$\frac{p_{02}}{p_2} = \left[1 + \frac{k-1}{2}\mathbf{M}_2^2\right]^{k/k-1} \quad \text{from equation (13.23)}$$

$$\mathbf{M}_2^2 = \frac{(k-1)\mathbf{M}_1^2 + 2}{2k\mathbf{M}_1^2 - (k-1)} \quad \text{from equation (13.32)}$$

$$\frac{p_{02}}{p_1} = \frac{p_3}{p_1} = \frac{p_{02}}{p_1}\frac{p_2}{p_1} = \left[1 + \frac{k-1}{2}\mathbf{M}_2^2\right]^{k/k-1}\left[1 + \frac{2k(\mathbf{M}_1^2 - 1)}{k+1}\right]$$

$$\frac{p_3}{p_1} = \frac{\left[(k+1)/2\mathbf{M}_1^2\right]^{k/k-1}}{\left[(2k/k+1)\mathbf{M}_1^2 - (k-1)/(k+1)\right]^{1/k-1}}$$

p_3 is measured and p_1 is the known free stream static pressure and hence $\mathbf{M}_1$ can be found for a known value of k. In the absence of a shock wave, the Mach number is given by

$$p_0/p_1 = \left[1 + \frac{k-1}{2}\mathbf{M}_1^2\right]^{k/k-1}$$

13.8 Fanno and Rayleigh Lines

We will examine the conditions for the occurrence of a normal shock wave. Let us assume that the normal shock illustrated in Fig. 13.4 pertains to the steady, one-dimensional, adiabatic flow of an ideal gas. We will also assume that the area remains constant across the shock (The channel area may vary but nor appreciably through the shock thickness). We will further assume that all upstream conditions are known and the unknown quantities downstream of the shock are T_2, p_2, ρ_2 and V_2. Since there are four equations available (equations of continuity, momentum, energy and state), a solution for these variables is possible. We will also introduce equation (13.8) which incorporates the equation of state for entropy change $(S_2 - S_1)$. In an adiabatic process $S_2 \geqslant S_1$.

By considering the equations of energy, continuity and state, we can relate S_2 and T_2. By assigning arbitrary values for S_2, values of T_2 can be found. The plot of T vs. S when plotted in the TS-plane gives a line known as the Fanno line shown in Fig. 13.5. Similarly, considering the equations of momentum, continuity and state, we can relate S_2 and T_2. This second relation between T and S when plotted on the same TS-plane (Fig. 13.5) gives a curve known as the Rayleigh line. The Fanno and Rayleigh lines intersect at points (1) and (2) as shown in the figure, and these two points satisfy all the four equations of energy, momentum, continuity and state.

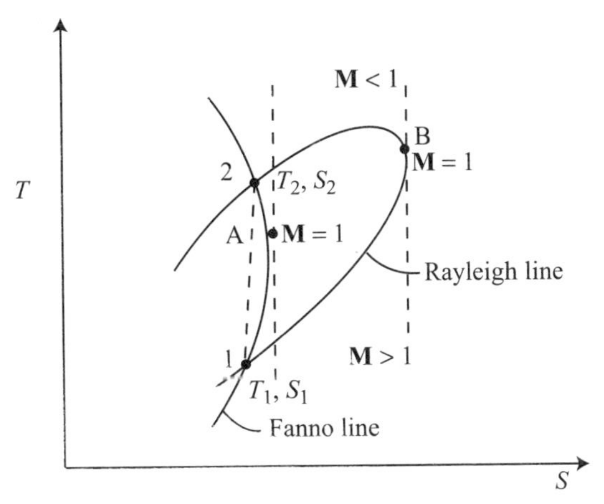

Fig. 13.5 Fanno and Rayleigh lines

The points of maximum entropy on these lines are located at A on the Fanno line and at B on the Rayleigh line. These points correspond to Mach number, $\mathbf{M} = 1$. The lower part of these curves correspond to $\mathbf{M} > 1$ (supersonic) and the upper part to $\mathbf{M} < 1$ (subsonic).

As $S_2 > S_1$, the flow proceeds from point 1 to point 2 with a rise in entropy and the supersonic flow changes to a subsonic flow across the shock. Tables are available for describing values along Fanno and Rayleigh lines for various values of Mach number for specified values of k.

13.9 Flow of a Perfect Gas in a Duct of Constant Area with Heat Transfer

Consider an elementary volume as shown in Fig. 13.6.

Let dQ/dM be the rate of heat transfer to the fluid (heat added per unit mass)

First law of thermodynamics:

$$C_pT_1 + \frac{V_1^2}{2} + \frac{dQ}{dM} = C_pT_2 + \frac{V_2^2}{2} \tag{13.35}$$

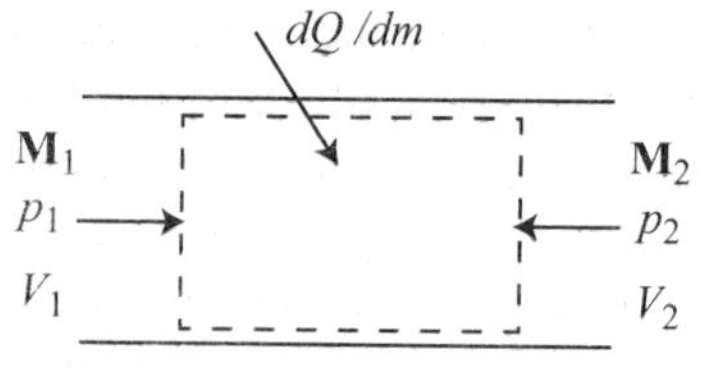

Fig. 13.6 Elementary volume of a duct

For constant area of duct, momentum equation: $p_1 - p_2 = \rho_2 V_2^2 - \rho_1 V_1^2$

Equation of continuity: $\rho_1 V_1 = \rho_2 V_2$

The ratio of flow parameters can be got by substituting $\rho = p/RT$ and $V = \mathbf{M}c = \mathbf{M}\sqrt{kRT}$

$$\frac{p_1}{p_2} = \frac{1 + k\mathbf{M}_2^2}{1 + k\mathbf{M}_1^2} \quad \text{from equation (13.29)}$$

with $p = \rho RT$;

$$\frac{T_1 V_2}{T_2 V_1} = \frac{1 + k\mathbf{M}_2^2}{1 + k\mathbf{M}_1^2} \tag{13.36}$$

Substituting $V = \mathbf{M}\sqrt{kRT}$;

$$\frac{T_1}{T_2} = \left(\frac{\mathbf{M}_1}{\mathbf{M}_2}\right)^2 \left(\frac{1 + k\mathbf{M}_2^2}{1 + k\mathbf{M}_1^2}\right)^2 \tag{13.37}$$

$$\frac{V_1}{V_2} = \left(\frac{\mathbf{M}_1}{\mathbf{M}_2}\right)^2 \left(\frac{1 + k\mathbf{M}_2^2}{1 + k\mathbf{M}_1^2}\right) \tag{13.38}$$

The ratio of stagnation pressures is obtained from equations (13.23) and (13.29)

$$\frac{(p_0)_1}{(p_0)_2} = \frac{1 + k\mathbf{M}_2^2}{1 + k\mathbf{M}_1^2} \left[\frac{1 + [(k-1)/2]\mathbf{M}_1^2}{1 + [(k-1)/2]\mathbf{M}_2^2}\right]^{k/(k-1)} \tag{13.39}$$

Further it is convenient to determine the flow parameters between any section of duct and the critical section where $\mathbf{M}_2 = 1$ (choking conditions). We set $\mathbf{M}_2 = 1$ and drop the subscript 1.

$$\frac{p}{p^*} = \frac{1 + k}{1 + k\mathbf{M}^2} \tag{13.40}$$

$$\frac{T}{T^*} = \left[\frac{\mathbf{M}(1 + k)}{1 + k\mathbf{M}^2}\right]^2 \tag{13.41}$$

$$\frac{V}{V*} = \frac{\mathbf{M}^2(1+k)}{1+k\mathbf{M}^2} \tag{13.42}$$

$$\frac{p_0}{p_0^*} = \frac{1+k}{1+k\mathbf{M}^2}\left[\frac{2(1+[(k-1)/2]\mathbf{M}^2)}{1+k}\right]^{k/k-1} \tag{13.43}$$

Combining (13.21) and (13.41)

$$\frac{T_0}{T_0^*} = \frac{2(k+1)\mathbf{M}^2[1+[(k-1)/2]\mathbf{M}^2]}{(1+k\mathbf{M}^2)^2} \tag{13.44}$$

These ratios may be found in the gas tables under the title *Rayleigh line*. Finally the amount of heat added per unit mass of fluid is

$$q = C_{p_0}(T_{0_2} - T_{0_1}) \tag{13.45}$$

Example 13.6 Air enters a duct with a pressure of 550 kPa and $T_1 = 220$ K and a Mach number of 0.4. Heat is added so that the exist temperature is 280 K. Assuming that the friction may be neglected, find the exit Mach number and the total pressure change through the duct. Find also the amount of heat added per kg mass. ($C_p = 1$ kJ/kg K)

$$\frac{T_1}{T_2} = \frac{220}{280} = 0.786 = \left[\frac{0.40}{\mathbf{M}_2}\frac{1+0.4\mathbf{M}_2^2}{(1+1.4\times 0.40^2)}\right]^2 = 0.107\left[\frac{1+1.4\mathbf{M}_2^2}{\mathbf{M}_2}\right]^2$$

$$\mathbf{M}_2 = 0.49$$

$$T_{0_1} = T_1[1+(k-1)\mathbf{M}_1^2/2] = 220[1+0.4\times 0.4^2/2] = 227.04$$

$$T_{0_2} = T_2[1+(k-1)\mathbf{M}_2^2/2] = 280[1+0.4\times 0.49^2/2] = 293.4$$

Heat added per unit mass:

$$q = C_p(T_{0_2} - T_{0_1}) = 1(293.4 - 227.04) = 66.36 \text{ kJ/kg K}$$

$$p_1/p_2 = (1+1.4\times 0.49^2)/(1+1.4\times 0.4^2) = 1.092$$

$$p_2 = 550/1.092 = 503.66 \text{ kPa}$$

$$p_{0_1} = 550(1+0.2\times 0.4^2)^{3.5} = 614.10 \text{ kPa}$$

$$p_{0_2} = 503.66(1+0.2\times 0.49^2)^{3.5} = 593.51$$

$$(p_{0_1} - p_{0_2}) = 614 - 593.51 = 20.49 \text{ kPa}$$

EXAMPLE 13.7 Oxygen at a velocity of $V_1 = 525\,\text{m/s}$ at a pressure of $80\,\text{kN/m}^2$ and $T = -10°\text{C}$ flows in a 60 mm dia. frictionless pipe with sonic conditions at exit. How much heat transfer per kg of fluid is required?

$C_p = 0.219\,\text{J/kg K};\ k = 1.4;\ R = 260\,\text{m}^2/\text{s}^2\text{K}$

$$c = \sqrt{KRT} = \sqrt{1.4 \times 260 \times 263} = 309.41\,\text{m/s}$$

$$\mathbf{M} = V/c = 525/309.41 = 1.7$$

$$T_{0_1} = T\left[1 + \frac{k-1}{2}\mathbf{M}^2\right] = 263[1 + 0.2 \times 1.7]^2 = 414.4\,\text{K}$$

For sonic conditions $\mathbf{M}_2 = 1$:

$$\frac{p_1}{p_2} = \frac{1+k}{1+k\mathbf{M}_1^2} = \frac{2.4}{1 + 1.4 \times 1.7^2} = 0.4756$$

$$T_2 = T_1\left(\frac{\mathbf{M}_2^2}{\mathbf{M}_1^2}\right)\left(\frac{p_2}{p_1}\right)^2 = 263 \times \frac{1.0}{1.7^2} \times \left(\frac{1}{0.4756}\right)^2 = 402\,\text{K}$$

$$T_{0_2} = 402[1 + 0.2 \times 1^2] = 482\,\text{K}$$

$$q = c_p(T_{0_2} - T_{0_1}) = 0.219(482 - 414.4) = 14.7\,\text{k cal/kg}$$

EXAMPLE 13.8 A fuel mixture having properties same as air enters a combustion pipe chamber at $V_1 = 120\,\text{m/s}$ and $T_1 = 500\,\text{K}$. (a) If the exit flow at the end of the duct is choked ($\mathbf{M}_2 = 1$), what is the heat addition? (b) What will be the Mach number and temperature if 1200 KJ/kg is added during combustion. $C_p = 1.005\,\text{KJ/kg K}$

(a) $$T_{0_1} = T_1 + \frac{V_1^2}{2C_p} = 500 + \frac{120^2}{2 \times 1.005} = 572\,\text{K}$$

$$\mathbf{M}_1 = \frac{V_1}{\sqrt{kRT_1}} = \frac{120}{\sqrt{1.4 \times 287 \times 500}} = 0.268$$

$$\frac{(T_0)_1}{T_0^*} = \frac{2(k+1)\mathbf{M}^2[1 + (k-1)/2\mathbf{M}^2]}{(1 + k\mathbf{M}^2)^2} = \frac{2 \times 2.4 \times 0.268^2[1 + 0.2 \times 0.268^2]}{(1 + 1.4 \times 0.268^2)^2}$$

$$= \frac{0.3497}{1.2112} = 0.289$$

$$T_0^* = \frac{572}{0.2890} = 1981\,\text{K}$$

$$q = C_p(T_0^* - T_{0_1}) = 1.005(1981 - 572) = 1663\,\text{KJ/kg}$$

(b) $(T_0)_2 = (T_0)_1 + \dfrac{1200}{1.005} = 572 + 1194 = 1409\,\text{K}$

$$\frac{(T_{02})}{T_0^*} = \frac{1409}{1981} = 0.711$$

$$\mathbf{M}_2 = 0.513$$

$$T_2 = (T_0)_2/1 + \frac{k-1}{2}\mathbf{M}^2 = \frac{1409}{1 + 0.2 \times 0.513^2} = 1396\,\text{K}$$

13.10 Oblique Shock Waves

A shock wave can be oriented obliquely as it happens at a contraction in area or at the leading edge of an inclined plate as shown in the Fig. 13.7. The velocity changes from V_1 to V_2 across the front as indicated in the figure. V_1 has components $V_1 \sin\beta$ normal to the front and $V_1 \cos\beta$ parallel to the front, β being the shock angle or wave angle.

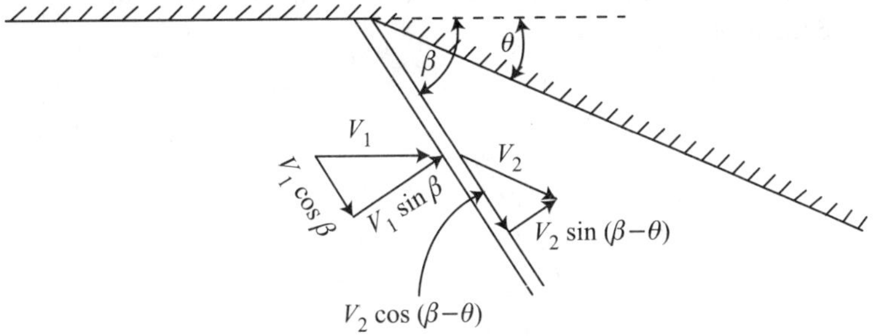

Fig. 13.7 Oblique shock wave

After crossing the front, the normal velocity will be less than $V_1 \sin\beta$ and as seen from the figure there is a sudden change in the direction of flow and the resultant velocity V_2 is deflected towards the front by an angle θ (called the deflection angle), with component $V_2 \sin(\beta - \theta)$ normal to the front and $V_2 \cos(\beta - \theta)$ parallel to the front. The Mach number normal to the front is $V_1 \sin\beta/c_1 = \mathbf{M}_1 \sin\beta$ before the shock and the Mach number normal to the front after the shock is $\mathbf{M}_2 \sin(\beta - \theta)$. The velocity components parallel to the front before and after the shock does not change. $V_1 \cos\beta = V_2 \cos(\beta - \theta)$. The mass rate of flow across the front is $\rho_1 V_1 \sin\beta = \rho_2 V_2 \sin(\beta - \theta)$. Hence, we have $\rho_2/\rho_1 = \tan\beta/\tan(\beta - \theta)$

The expressions for p_2/p_1, ρ_2/ρ_1, T_2/T_1, and $(S_2 - S_1)/C_v$ are therefore obtained by replacing $\mathbf{M}_1$ by $\mathbf{M}_1 \sin\beta$.

$$\frac{p_2}{p_1} = 1 + \frac{2k}{k+1}(\mathbf{M}_1^2 \sin^2\beta - 1) \tag{13.46}$$

$$\frac{\rho_2}{\rho_1} = \frac{(k+1)\mathbf{M}_1^2 \sin^2\beta}{(k-1)\mathbf{M}_1^2 \sin^2\beta + 2} = \frac{\tan\beta}{\tan(\beta - \theta)} \tag{13.47}$$

$$\frac{T_2}{T_1} = 1 + \frac{2(k-1)}{(k+1)^2}\frac{k\mathbf{M}_1^2 \sin^2\beta + 1}{\mathbf{M}_1^2 \sin^2\beta}(\mathbf{M}_1^2 \sin^2\beta - 1) \tag{13.48}$$

The relationship between $\mathbf{M}_2$ and $\mathbf{M}_1$ is given by

$$\mathbf{M}_2^2 \sin^2(\beta - \theta) = \frac{(k-1)\mathbf{M}_1^2 \sin^2\beta + 2}{2k\mathbf{M}_1^2 \sin^2\beta + 1 - k} \tag{13.49}$$

The relationship between the deflection angle θ and the shock angle β for a given $\mathbf{M}_1$ is

$$\tan\theta = 2\cot\beta \frac{\mathbf{M}_1^2 \sin^2\beta - 1}{\mathbf{M}_1^2(k + \cos 2\beta) + 2} \tag{13.50}$$

For a given $\mathbf{M}_1$, θ becomes zero at $\beta = \pi/2$ and at $\sin^{-1}(1/\mathbf{M}_1)$ and there is a maximum deflection angle, θ_{max}. For each $\mathbf{M}_1$, there are two possible values of deflection angle, a weak shock corresponding to a lower value of β and a strong shock corresponding to the higher value of the shock angle.

If the supersonic flow is passing over a gradually curved wall as shown in Fig. 13.8, the wave fronts are now Mach lines each inclined at an angle of $\mu = \sin^{-1}(1/\mathbf{M})$ to the local direction of the wave and $\mathbf{M}$ is the local Mach number.

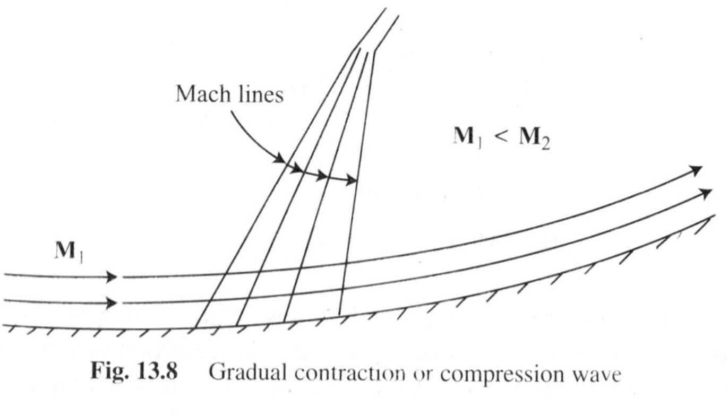

Fig. 13.8 Gradual contraction or compression wave

The Mach number decreases along the flow so that the Mach angle increases. The Mach lines generated at different points along the curve will coalesce and finally form an oblique shock.

13.10.1 Expansion Waves

In the case of gradual expansion, the Mach number increases and the Mach angle decreases causing the Mach lines to diverge as shown in Fig. 13.9*a*. If the expansion is sudden as shown in Fig. 13.9*b*, the flow expands through a *fan* of Mach lines known as the Prandtl–Meyer expansion fan.

The Mach number increases through the fan, the first Mach line making an angle $\mu_1 = \sin^{-1}(1/\mathbf{M}_1)$ to the local flow direction and the second Mach line making an angle $\mu_2 = \sin^{-1}(1/\mathbf{M}_2)$ to the local flow direction and so on till the flow has undergone a complete change in direction equal to the deflection of the wall.

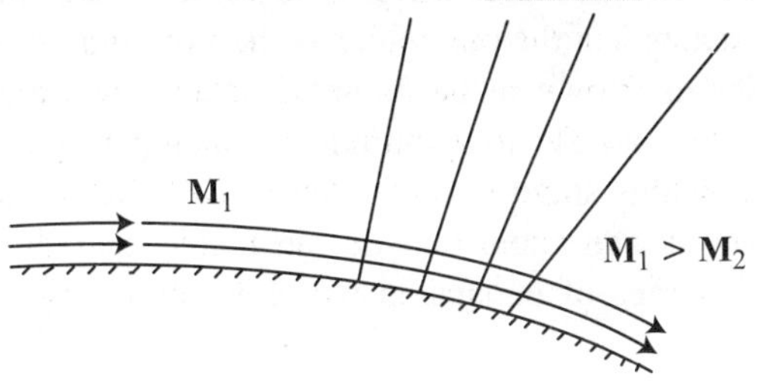

a Gradual expansion

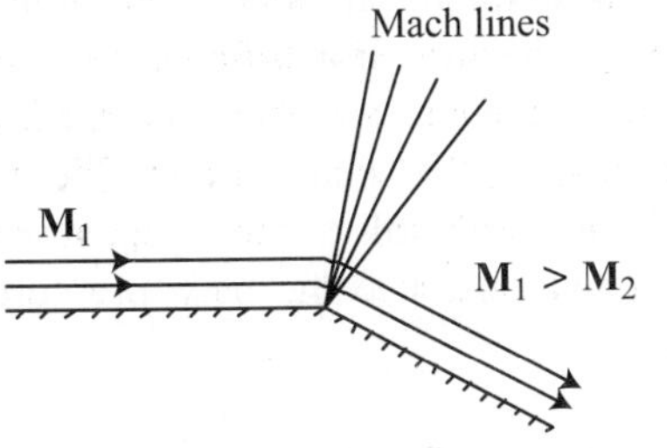

b Prandtl–Meyer abrupt expansion

Fig. 13.9 Gradual and abrupt expansion

13.11 Behaviour of Nozzle Flow as a Function of Back Pressure

It is interesting to study the behaviour of a nozzle as the exit pressure is varied. It will be assumed that the supply comes from a large reservoir where the pressure, p_0 is maintained at a constant value. In subsonic flow, the pressure at the exit of the nozzle is the same as the back pressure, p_b.

13.11.1 Convergent Nozzle

We will first consider the case of a convergent nozzle as shown in Fig. 13.10.

When the flow is subsonic throughout the nozzle, the pressure along the nozzle follows the curve (a). As the area decreases in subsonic flow, the velocity increases and pressure decreases.

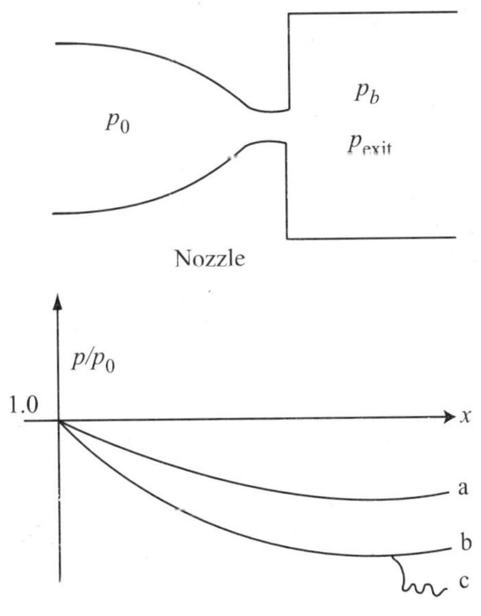

Fig. 13.10 Pressure distribution along a convergent nozzle for different back pressures

As the back pressure is decreased, the mass flux increases with increase in velocity and Mach number. This continues till sonic conditions are reached at the exit which is the minimum cross-section. The pressure drop follows the curve (b) as shown in the figure. Further lowering of the back pressure has no effect on the flow in the nozzle as disturbances cannot propagate upstream crossing the sonic barrier. The nozzle at this stage is said to be *choked*. If the back pressure is lowered still further, supersonic flow is generated outside the nozzle and the jet pressure adjusts itself to the back pressure by a series of expansion waves as shown by the curve (c).

13.11.2 Convergent–Divergent Nozzle

A typical convergent–divergent nozzle is shown in Fig. 13.11.

The curve (a) depicts subsonic flow throughout. As the back pressure is lowered, sonic conditions are reached at the throat and maximum mass rate of flow occurs for a given p_0. Further lowering of the back pressure has no effect on the flow upstream of the throat. Again we have reached the *choked* condition. However, beyond the throat, the diverging nozzle causes the flow to become supersonic. Depending on the value of the back pressure, a normal shock wave will occur in the diffuser section (inside the nozzle) as shown by curve (c). The position of the

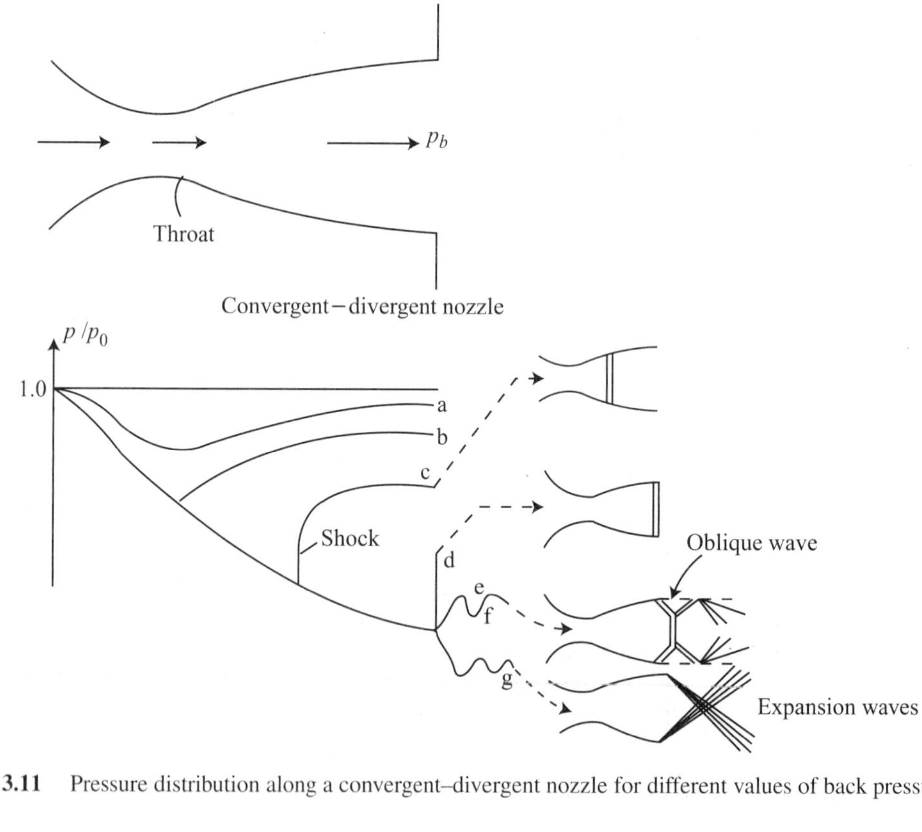

Fig. 13.11 Pressure distribution along a convergent–divergent nozzle for different values of back pressure, p_b

front will advance as the pressure is further decreased and at some point the normal shock wave will occur at the exit plane of the diffuser as shown by curve (d). The flow in the diffuser section is supersonic before the front but after the shock front the flow will be subsonic. If the back pressure is further reduced, no normal shock wave occurs either inside the nozzle or at the exit plane. The jet pressure adjusts itself to the back pressure by means of oblique shock waves outside the exit plane of the nozzle as shown by curve (e). These oblique waves vanish when $p_b = p_f$. With further reduction of back pressure, the jet pressure adjusts itself to the back pressure by means of oblique expansion waves as shown by curve (g).

EXAMPLE 13.9 A convergent–divergent nozzle has a throat area of $0.2\,\mathrm{m}^2$ and exit area of $0.7\,\mathrm{m}^2$. The absolute pressure and temperature of the supply reservoir upstream are 400 kPa and 800 K. The Mach number upstream of the shock is 2.44. Find the area at the location of the shock and the exit temperature.

We proceed in a sequential order: Reservoir to throat; throat to shock front; before shock front to after shock front; after shock front to exit section.

From the table,

$$A/A^* = 2.496 \text{ for } \mathbf{M}_1 = 2.44$$

The area at which shock takes place is $A = 2.496 \times 0.2 = 0.4992\,\mathrm{m}^2$

From the shock table,

$$\mathbf{M}_2 = 0.519; \quad \frac{p_{02}}{p_{01}} = \frac{p_{02}}{p_0} = 0.523; \quad p_{02} = 0.523 \times 400 = 209.2\,\mathrm{kPa}$$

A_2/A_2^* corresponding to $\mathbf{M} = 0.519$ is 1.3 from the isentropic table.

$A_2^* = 0.4992/1.3 = 0.3846\,\mathrm{m}^2$

The flow from the shock area to the exit area is isentropic. Hence

$A_3/A_2^* = 0.7/0.3846 = 1.82$. For this value of area ratio, in subsonic range

$T_3/T_0 = 0.977$ and $T_3 = 0.977 \times 800 = 782\,\mathrm{K}$.

EXAMPLE 13.10 The total pressure in a nozzle with an area ratio, $A/A^* = 4.0$ is 1.3 MPa. If the back pressure is 100 kPa, find the conditions at the exit.

From the tables, for $A/A^* = 4.0$, $\mathbf{M} = 2.94$ and the corresponding pressure ratio $p/p_0 = 0.0298$ and $p = 0.0298 \times 1300 = 38.7\,\mathrm{kPa}$

The back pressure is higher and the nozzle is said to be *over-expanded.*

EXAMPLE 13.11 A converging–diverging nozzle is supplied from a large constant pressure reservoir. The exist area is $250\,\mathrm{cm}^2$ and the throat area is $125\,\mathrm{cm}^2$. The reservoir pressure and temperature are 400 kPa and 350 K. Determine the range of pressure over which a normal shock pressure will occur in the nozzle.

Sonic conditions exist at the throat and supersonic flow downstream of the throat. For a shock to occur just downstream of the throat, for a throat ratio of $250/125 = 2.0$, the subsonic solution from the isotropic table is $\mathbf{M} = 0.31$; $p/p_0 = 0.936$

$$p_b = 0.936 \times 400 = 374.4\,\text{kPa}$$

For a shock to occur at the nozzle exit plane, the Mach number just before the shock can be found from the supersonic range of the isotropic table for $A/A^* = 2$ as $\mathbf{M} = 2.20$ and $p/p_0 = 0.094$. Pressure p_1 before the shock $p_1 = 0.094 \times 400 = 37.6\,\text{kPa}$.

From the normal shock tables for $\mathbf{M} = 2.20$, $\mathbf{M}_2$ after the shock is 0.547 and $p_b/p_0 = 5.48$

$$p_b = 5.48 \times 37.6 = 206.05\,\text{kPa}$$

A normal shock will therefore appear in the nozzle for $206.06\,\text{kPa} < p_b < 374.4\,\text{kPa}$. Oblique shocks will occur for pressures $37.6\,\text{kPa} < p_b < 206.05\,\text{kPa}$ and expansion waves for $p_b < 37.6\,\text{kPa}$.

EXAMPLE 13.12 A Laval nozzle has an expansion ratio (exit/throat) of 4.0. A normal shock wave occurs where the area ratio is 2.0. The total pressure upstream of the shock is 1 MPa. Find the static pressure at the exit.

From the isentropic table, we find that the Mach number corresponding to an area ratio of 2.0 is $\mathbf{M} = 2.20$, and from the shock table, the Mach number after the shock is 0.547, the ratio of the total pressures before and after the shock is $p_{02}/p_{01} = 0.6281$ and hence the total pressure downstream of the shock is $0.6281 \times 1000 = 628.1\,\text{kPa}$.

The value of A/A^* corresponding to the Mach number behind the shock wave (2.20) for the subsonic part is 1.26. Thus the ratio of the nozzle area to the throat area that would be needed to develop sonic flow downstream of the shock wave is $A^*/A_t = (A/A_t)/(A/A^*) = 2.0/1.26 = 1.59$.

Hence the exit area A_e is found by $A_e/A^* = (A_e/A_t)/(A^*/A_t) = 4.0/1.59 = 2.52$.

The Mach number corresponding to this ratio is $\mathbf{M}_e = 0.24$.

The static pressure corresponding to this Mach number is
$p_e = 628/[1 + 0.2 \times 0.24^2]^{3.5} = 603\,\text{kPa}$

13.12 Flow through a Truncated Nozzle

The truncated version of the Laval nozzle is shown in Fig. 13.12 in which the nozzle is cut off at the throat section and the diffuser section is absent.

The flow at the nozzle exit can never be supersonic as the diverging section is not present and hence at the critical stage, the flow can be sonic at the exit. The critical pressure ratio is $p^*/p_0 = [2/(k+1)]^{k/(k-1)}$ and for air with $k = 1.4$ is 0.528

If the back pressure, p_b is less than p^*, the flow at the exit is sonic and pressure equilibrium is achieved by a series of expansion waves. The mass rate of flow is calculated by equation (13.28).

If the back pressure is greater than $p*$, the flow has to exit subsonically.

Mach number at the exit,

$$\mathbf{M}_e = \sqrt{\frac{2}{k-1}\left[\left(\frac{p_0}{p_b}\right)^{(k-1)/k} - 1\right]} \quad \text{from equation (13.23)}$$

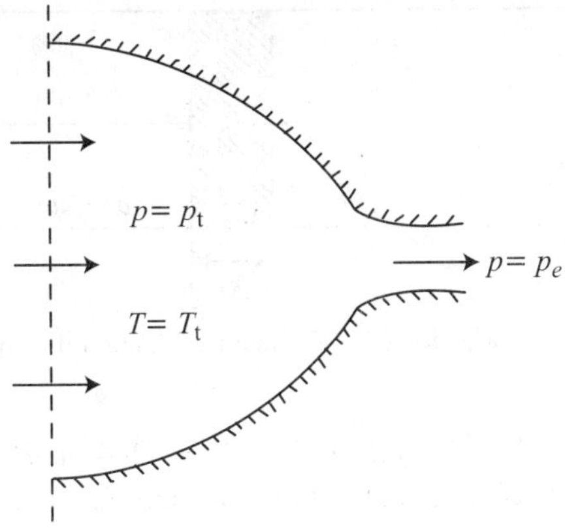

Fig. 13.12 Truncated nozzle

The static temperature and the speed of sound at the exit are found by the following expressions

$$T_e = \frac{T_0}{1 + [(k-1)/2]\mathbf{M}_e^2} \qquad \text{from equation (13.21)}$$

$$c_e = \sqrt{kRT_e}$$

The gas density at the exit is given by $\rho_e = p_b/(RT_e)$.
The mass rate of flow $= (\rho A\mathbf{M}c)_e$.

13.13 Compressible Flow in a Pipe with Friction

We widen our scope to consider some flows which are non-isentropic but truly one-dimensional. Pipe friction and heat transfer make the flow non-isentropic. We will consider two problems: one in which the pipe is insulated and the flow can be treated as adiabatic, and the second one, an isothermal flow which approximates to the flow in long pipe lines.

The equation of continuity is ρV = constant or $\mathrm{d}V/V + \mathrm{d}\rho/\rho = 0$.

Because the flow is adiabatic, the energy equation still holds and so do the ideal gas relations.

$$h + \frac{V^2}{2} = \text{constant}$$

Substituting $h = C_pT$, $C_p = kR/(k-1)$, we have the energy equation in differential form:

$$\frac{kR\mathrm{d}T}{(k-1)} + V\mathrm{d}V = 0$$

The momentum equation (Fig. 13.13) is written as

$$A[p - (p + \mathrm{d}p)] - \tau_0 P\Delta x = \rho VA[(V + \mathrm{d}V) - V]$$

where P is the circumference of the pipe on which the wall shear stress, τ_0 is acting.

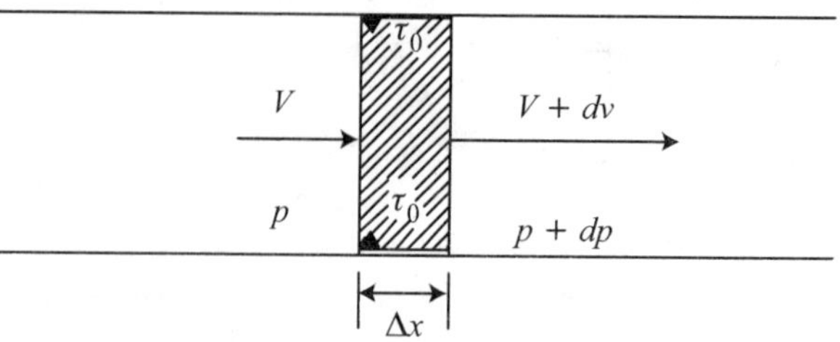

Fig. 13.13 Momentum equation in a pipe

We can express $\tau_0/\rho = fV^2/8$ and $A/P = R = D/4$ or $P = 4A/D$ for circular pipes. The momentum equation in the differential form now becomes

$$\rho V \mathrm{d}V + \mathrm{d}p + \frac{f\rho V^2 \mathrm{d}x}{2D} = 0$$

Dividing throughout by p to get:

$$\frac{\rho V \mathrm{d}V}{p} + \frac{\mathrm{d}p}{p} + \frac{f\rho V^2 \mathrm{d}x}{(2pD)}$$

Each term can be developed in terms of the Mach number.
$V^2 = (\mathbf{M}^2 kp)/\rho$; or $\rho V^2/p = k\mathbf{M}^2$ and the last term becomes $fk\mathbf{M}^2\mathrm{d}x/2D$

Use of the equation of state: $p/\rho = RT = c^2/k$ renders the first term $\rho V \mathrm{d}V/p = kV\mathrm{d}V/c^2 = k\mathbf{M}^2(\mathrm{d}V/V)$

The equation of state in differential form is: $\mathrm{d}p/p = \mathrm{d}\rho/\rho + \mathrm{d}T/T$

From the equation of continuity: $-\mathrm{d}V/V = \mathrm{d}\rho/\rho$ and from the energy equation

$$\frac{\mathrm{d}T}{T} = \frac{-(k-1)V\mathrm{d}V}{kRT} = -(k-1)\mathbf{M}^2\left(\frac{\mathrm{d}V}{V}\right)$$

Hence $\dfrac{\mathrm{d}p}{p} = \dfrac{-\mathrm{d}V}{V} - (k-1)\mathbf{M}^2\left(\dfrac{\mathrm{d}V}{V}\right)$

Substituting these expressions in the momentum equation, we get

$$(\mathbf{M}^2 - 1)\left(\frac{\mathrm{d}V}{V}\right) + \frac{kf\mathbf{M}^2\mathrm{d}x}{2D} = 0$$

The Mach number is $\mathbf{M} = V/c = V/(kRT)^{1/2}$ which in differential form is

$$\frac{\mathrm{d}\mathbf{M}}{\mathbf{M}} = \frac{\mathrm{d}V}{V} - \frac{\mathrm{d}T}{2T}$$

Again using the energy equation to eliminate T, $\mathrm{d}T/T = -(k-1)\mathbf{M}^2(\mathrm{d}V/V)$

$$\frac{\mathrm{d}\mathbf{M}}{\mathbf{M}} = \left(\frac{\mathrm{d}V}{V}\right)\left[1 + \frac{(k-1)\mathbf{M}^2}{2}\right]$$

Finally we eliminate $(\mathrm{d}V/V)$ from the momentum equation to get a relation between the Mach number and the distance x along the pipe and the equation describes the distribution of the Mach number along the pipe.

$$\frac{(1-\mathbf{M}^2)\mathrm{d}\mathbf{M}}{\mathbf{M}^3\left\{1+[(k-1)/2]\mathbf{M}^2\right\}} = \frac{kf\,\mathrm{d}x}{2D}$$

$$\left[\frac{1}{\mathbf{M}^3} - \frac{k+1}{2\mathbf{M}} + \frac{(k+1)(k-1)\mathbf{M}}{4\left\{1+[(k-1)/2]\mathbf{M}^2\right\}}\right]\mathrm{d}\mathbf{M} = \frac{kf\,\mathrm{d}x}{2D}$$

Integrating

$$\frac{-1}{2\mathbf{M}^2} - \frac{k+1}{2}\ln\mathbf{M} + \frac{k+1}{4}\ln\left(1+\frac{k-1}{2}\mathbf{M}^2\right) = \frac{kfx}{2D} + C$$

The integration constant, C can be evaluated by defining L as the distance corresponding to $\mathbf{M} = 1$. Evaluating and substituting it in the preceding equation, we finally have

$$\frac{1-\mathbf{M}^2}{k\mathbf{M}^2} + \frac{k+1}{2k}\ln\left[\frac{(k+1)\mathbf{M}^2}{2+(k-1)\mathbf{M}^2}\right] = \frac{f(L-x)}{D} \tag{13.51}$$

where x is the distance from the origin corresponding to a Mach number $\mathbf{M}$.

The above equations reveal that when the flow is subsonic, $\mathrm{d}\mathbf{M}/\mathrm{d}x > 0$ and the Mach number increases with distance along the pipe. If the flow is supersonic, then $\mathrm{d}\mathbf{M}/\mathrm{d}x < 0$ and the Mach number decreases along the pipe. We conclude that the effect of the wall friction is to cause the Mach number to approach unity. A subsonic flow can attain a maximum value of $\mathbf{M} = 1$ at the end of the pipe. A supersonic flow can change to a subsonic flow by the occurrence of shock waves in the pipe.

If $\mathbf{M}_0$ is the entrance Mach number at $x = 0$, the length of the pipe L where sonic velocity is achieved with $\mathbf{M} = 1$ is for air with $k = 1.4$

$$\frac{fL}{D} = \frac{5}{7}\left(\frac{1}{\mathbf{M}_0^2} - 1\right) + \frac{6}{7}\ln\frac{6\mathbf{M}_0^2}{\mathbf{M}_0^2 + 5} \tag{13.52}$$

EXAMPLE 13.13 If $\mathbf{M}_0 = 0.3$, $f = 0.02$, $D = 5$ cms, find L.

Substituting the value of $\mathbf{M}_0 = 0.3$ in the above equation, we get

$$\frac{fL}{D} = 5.3 \text{ or } L = \frac{5.3 \times 0.05}{0.02} = 13.25\,\text{m}$$

$$\text{If } \mathbf{M}_0 = 3, \quad \frac{fL}{D} = 0.52, \; L = \frac{0.52 \times 0.05}{0.02} = 1.3\,\text{m}$$

$$\text{If } \mathbf{M}_0 = 2, \quad \frac{fL}{D} = 0.3042, \; L = \frac{0.3042 \times 0.05}{0.02} = 0.855\,\text{m}$$

The distance between the two sections at which the Mach numbers are 3 and 2 respectively in the same pipe is $1.3 - 0.855 = 0.445$ m.

The pressure, velocity and temperature are given in terms of the upstream conditions at entrance p_e, V_e, T_e and p^*, V^*, T^* which are values at $\mathbf{M} = 1$

$$\frac{p^*}{p_e} = \mathbf{M}_0\sqrt{\frac{(k-1)\mathbf{M}_0^2 + 2}{k+1}} \tag{13.53}$$

$$\frac{V^*}{V_e} = \frac{1}{\mathbf{M}_0}\sqrt{\frac{(k-1)\mathbf{M}_0^2 + 2}{k+1}} \tag{13.54}$$

$$\frac{T^*}{T} = \frac{(k-1)\mathbf{M}_0^2 + 2}{k+1} \tag{13.55}$$

13.14 Isothermal Flow

In an isothermal flow in a constant area pipe, the energy equation is merely

$$T = \text{constant}$$

The speed of sound, $c = (kRT)^{1/2}$ is also constant along the pipe. Further, pressure, p is proportional to density and $\mathrm{d}p/p = \mathrm{d}\rho/\rho$ which in turn is equal to $-(\mathrm{d}V/V)$.

The momentum equation derived earlier is

$$k\mathbf{M}^2\left(\frac{\mathrm{d}V}{V}\right) + \frac{\mathrm{d}p}{p} + \frac{kf\mathbf{M}^2\mathrm{d}x}{2D} = 0$$

With $\mathrm{d}p/p = -(\mathrm{d}V/V)$ and $\mathrm{d}V/V = \mathrm{d}\mathbf{M}/\mathbf{M}$, the momentum equation becomes

$$\left(\frac{\mathrm{d}\mathbf{M}}{\mathbf{M}}\right)(k\mathbf{M}^2 - 1) + \frac{kf\mathbf{M}^2\mathrm{d}x}{2D} = 0$$

$$\frac{\mathrm{d}\mathbf{M}}{\mathrm{d}x} = \frac{f}{2D}\frac{k\mathbf{M}^3}{1 - k\mathbf{M}^2}$$

If $\mathbf{M} < (1/k^{1/2})$, then the Mach number increases with increase in x but if $\mathbf{M} > 1$, $\mathrm{d}\mathbf{M}/\mathrm{d}x$ is negative and the Mach number decreases with distance. The preceding equation is rewritten as a sum of partial fractions

$$\frac{\mathrm{d}\mathbf{M}}{k\mathbf{M}^3} - \frac{\mathrm{d}\mathbf{M}}{\mathbf{M}} = \frac{f\,\mathrm{d}x}{2D}$$

Integrating

$$\frac{-1}{2k\mathbf{M}^2} - \ln \mathbf{M} = \frac{fx}{2D} + C$$

We evaluate the integration constant by designating L as the maximum distance at which $\mathbf{M} = 1/k^{1/2}$

$$\frac{fL}{D} = \ln(k\mathbf{M}^2) + \frac{1 - k\mathbf{M}^2}{k\mathbf{M}^2} \tag{13.56}$$

For lengths greater than L, choking occurs and the mass rate of flow decreases.

In an isothermal flow, since the temperature is constant, c is constant and velocity is directly proportional to velocity. The pressure is directly proportional to density. Hence we can write the relationship between pressure and Mach number as

$$\frac{\mathrm{d}p}{p} = -\frac{\mathrm{d}\mathbf{M}}{\mathbf{M}}$$

which on integration yields $\ln p = -\ln \mathbf{M} + C$

The integration constant can be evaluated by designating the pressure at the maximum distance, L as p_T, where the value of $\mathbf{M} = 1/k^{1/2}$ and substituting the value of the constant, we get pressure p at any section where the value of the Mach number is $\mathbf{M}$

$$\frac{p}{p_T} = \frac{1}{\mathbf{M}k^{1/2}} \tag{13.57}$$

EXAMPLE 13.14 Helium is being transported in a pipe 25 cms in dia. and 500 m long at 20°C. The pressure at the exit is 60 kPa. Determine the flow rate through the pipe and the inlet pressure. $k = 1.66$, gas constant $= 2077$ J/kg K, $f = 0.01$

Mach number at exit : $\mathbf{M} = 1/k^{1/2} = 1/1.66^{1/2} = 0.7761$
Velocity $= \mathbf{M}c = 0.7761 \times (1.66 \times 2077 \times 293)^{1/2} = 780$ m/s; $\rho = p/RT$
Mass rate of flow $= \rho AV = (60 \times 1000/2077 \times 293) \times (\pi/4)0.25^2 \times 780 = 37.75$ kg/s
$fL/D = 0.01 \times 500/0.25 = 20 = \ln(k\mathbf{M}^2) + (1 - k\mathbf{M}^2)/k\mathbf{M}^2$
Solving for $k\mathbf{M}^2 = 0.04$
Pressure $p = 60/(0.04)^{1/2} = 120$ kPa

EXERCISES

13.1. Calculate the speed of sound in helium and hydrogen at 50°C. For helium, $\rho = 0.168\,\text{kg/m}^3$, $R = 2077\,\text{m}^2/\text{s}^2\text{K}$, $C_p = 5187\,\text{J/kg K}$, $k = C_p/C_v = 1.66$. For hydrogen, $\rho = 0.0851\,\text{kg/m}^3$, $R = 4127$, $C_p = 14223$, $k = 1.41$

13.2. What is the temperature at the nose of a supersonic fighter flying at $\mathbf{M} = 2$ at 273 K?

13.3. An airplane travels at 850 km/hour at sea level at $T = 15°\text{C}$. How fast would the airplane be flying at the same Mach number at an altitude where the temperature is $-50°\text{C}$. $R = 287\text{m}^2/\text{s}^2\text{K}$, $K = 1.4$.

13.4. An object is placed in an air flow with a static pressure of 250 kPa absolute at a static temperature of 25°C and a velocity of 300 m/s. What are the pressure and temperature at the stagnation point?

13.5. A normal shock pressure occurs in a nitrogen stream having a velocity of 600 m/s at a static temperature of −30°C and a static pressure of 75 kPa. Calculate the Mach number, pressure and temperature downstream of the shock wave and entropy increase across the wave. $R = 296\,\text{m}^2/\text{s}^2\text{K}$, $K = 1.404$

13.6. A normal shock wave in air occurs with initial conditions before the shock given as $p_1 = 70\,\text{N/m}^2$, $T_1 = 5°\text{C}$, $V_1 = 425\,\text{m/s}$. Find these parameters after the shock.

13.7. A truncated nozzle is used for measuring the mass rate of flow of methane. The data: area of nozzle $= 2.5\,\text{cm}^2$; area of pipe $= 10\,\text{cm}^2$; upstream total pressure and temperature are 150 kPa and 40°C. Back pressure = 100 kPa. Calculate the mass rate of flow of methane. Also assuming that the Bernoulli theorem is valid, calculate the mass rate of flow, the density being the density at exit. ($K = 1.31$; $R = 518\,\text{m}^2/\text{s}^2\text{K}$)

13.8. At the throat area of a convergent–divergent nozzle, the total pressure $p_0 = 220\,\text{kPa}$ and $T_0 = 450\,\text{K}$. The mass rate of flow is 2.7 kg/s and at the exit, the Mach number is 2.5. Determine the throat and exit areas, exit pressure, temperature and velocity. The flow is isotropic. $k = 1.4$ and $R = 287\,\text{m}^2/\text{s}^2\text{K}$

13.9. Consider an air flow in a convergent–divergent diffuser with the data given as follows. At entrance: area $A_1 = 100\,\text{cm}^2$, $\mathbf{M}_1 = 2.1$, $p_1 = 60\,\text{kPa}$; area at throat $A_2 = 75\,\text{cm}^2$. Determine the static pressure, stagnation pressure, and the Mach number at the exit. Assume isentropic flow except for the normal shock wave at the throat. Exit area = $125\,\text{cm}^2$

13.10. Determine the atmospheric pressure necessary for the shock wave to position itself at an allocation where the area is $120\,\text{cm}^2$, in the divergent portion of a convergent–divergent diffuser whose values at entrance are $A_1 = 200\,\text{cm}^2$, $\mathbf{M}_1 = 0.3$ and $p_1 = 400\,\text{kPa}$; Area at the exit is $140\,\text{cm}^2$.

13.11. The Mach number of air flowing out of the exit of a 2.5 cm brass tube is 0.75. The temperature is 100°C and the atmospheric pressure is 100 kPa. At what distance upstream of the exit is the Mach number equal to 0.25.

13.12. Oxygen flows through a pipe 3 cm in diameter and 10 m long. It discharges to the atmosphere ($p_a = 100\,\text{kPa}$). If the absolute static pressure at the beginning of the pipe is 350 kPa and the total temperature is 300 K, calculate the mass rate of flow in the pipe.

13.13. Methane is pumped into a 15 cm steel pipe at a pressure of 1 kPa and a temperature of 350 K and with a velocity of 25 m/s. What is the pressure 2500 m downstream? $K = 1.31$, $R = 518\, m^2/s^2K$; $f\,(\text{average}) = 0.015$. The flow is isothermal. What is the maximum Mach number at exit and corresponding length of pipe?

13.14. Mach number in a frictionless duct flow reduces from 2.00 to 1.75. Determine the ratios of pressure, temperature, velocity and density. $k = 1.4$.

13.15. What is the maximum length of 30 mm dia. pipe for air flow to take place given the initial Mach number as 0.25. $f = 0.015$.

Appendix

Table A. 1. One-dimensional isentropic relations*

M	A/A^*	p/p_0	ρ/ρ_0	T/T_0	M	A/A^*	p/p_0	ρ/ρ_0	T/T_0
0.00	...	1.000	1.000	1.000	0.54	1.27	0.820	0.868	0.945
0.01	57.87	0.9999	0.9999	0.9999	0.56	1.24	0.808	0.859	0.941
0.02	28.94	0.9997	0.9999	0.9999	0.58	1.21	0.796	0.850	0.937
0.04	14.48	0.999	0.999	0.9996	0.60	1.19	0.784	0.840	0.933
0.06	9.67	0.997	0.998	0.999	0.62	1.17	0.772	0.831	0.929
0.08	7.26	0.996	0.997	0.999	0.64	1.16	0.759	0.821	0.924
0.10	5.82	0.993	0.995	0.998	0.66	1.13	0.747	0.812	0.920
0.12	4.86	0.990	0.993	0.997	0.68	1.12	0.734	0.802	0.915
0.14	4.18	0.986	0.990	0.996	0.70	1.09	0.721	0.792	0.911
0.16	3.67	0.982	0.987	0.995	0.72	1.08	0.708	0.781	0.906
0.18	3.28	0.978	0.984	0.994	0.74	1.07	0.695	0.771	0.901
0.20	2.96	0.973	0.980	0.992	0.76	1.06	0.682	0.761	0.896
0.22	2.71	0.967	0.976	0.990	0.78	1.05	0.669	0.750	0.891
0.24	2.50	0.961	0.972	0.989	0.80	1.04	0.656	0.740	0.886
0.26	2.32	0.954	0.967	0.987	0.82	1.03	0.643	0.729	0.881
0.28	2.17	0.947	0.962	0.985	0.84	1.02	0.630	0.719	0.876
0.30	2.04	0.939	0.956	0.982	0.86	1.02	0.617	0.708	0.871
0.32	1.92	0.932	0.951	0.980	0.88	1.01	0.604	0.698	0.865
0.34	1.82	0.923	0.944	0.977	0.90	1.01	0.591	0.687	0.860
0.36	1.74	0.914	0.938	0.975	0.92	1.01	0.578	0.676	0.855
0.38	1.66	0.905	0.931	0.972	0.94	1.00	0.566	0.666	0.850
0.40	1.59	0.896	0.924	0.969	0.96	1.00	0.553	0.655	0.844
0.42	1.53	0.886	0.917	0.966	0.98	1.00	0.541	0.645	0.839
0.44	1.47	0.876	0.909	0.963	1.00	1.00	0.528	0.632	0.833
0.46	1.42	0.865	0.902	0.959	1.02	1.00	0.516	0.623	0.828
0.48	1.38	0.854	0.893	0.956	1.04	1.00	0.504	0.613	0.822
0.50	1.34	0.843	0.885	0.952	1.06	1.00	0.492	0.602	0.817
0.52	1.30	0.832	0.877	0.949	1.08	1.01	0.480	0.592	0.810

**For a perfect gas with constant specific heat, $k = 1.4$.*

continued next page

Table A. 1. One-dimensional isentropic relations (continued)

M	A/A^*	p/p_0	ρ/ρ_0	T/T_0	M	A/A^*	p/p_0	ρ/ρ_0	T/T_0
1.10	1.01	0.468	0.582	0.805	1.74	1.38	0.191	0.306	0.623
1.12	1.01	0.457	0.571	0.799	1.76	1.40	0.185	0.300	0.617
1.14	1.02	0.445	0.561	0.794	1.78	1.42	0.179	0.293	0.612
1.16	1.02	0.434	0.551	0.788	1.80	1.44	0.174	0.287	0.607
1.18	1.02	0.423	0.541	0.782	1.82	1.46	0.169	0.281	0.602
1.20	1.03	0.412	0.531	0.776	1.84	1.48	0.164	0.275	0.596
1.22	1.04	0.402	0.521	0.771	1.86	1.51	0.159	0.269	0.591
1.24	1.04	0.391	0.512	0.765	1.88	1.53	0.154	0.263	0.586
1.26	1.05	0.381	0.502	0.759	1.90	1.56	0.149	0.257	0.581
1.28	1.06	0.371	0.492	0.753	1.92	1.58	0.145	0.251	0.576
1.30	1.07	0.361	0.483	0.747	1.94	1.61	0.140	0.246	0.571
1.32	1.08	0.351	0.474	0.742	1.96	1.63	0.136	0.240	0.566
1.34	1.08	0.342	0.464	0.736	1.98	1.66	0.132	0.235	0.561
1.36	1.09	0.332	0.455	0.730	2.00	1.69	0.128	0.230	0.556
1.38	1.10	0.323	0.446	0.724	2.02	1.72	0.124	0.225	0.551
1.40	1.11	0.314	0.437	0.718	2.04	1.75	0.120	0.220	0.546
1.42	1.13	0.305	0.429	0.713	2.06	1.78	0.116	0.215	0.541
1.44	1.14	0.297	0.420	0.707	2.08	1.81	0.113	0.210	0.536
1.46	1.15	0.289	0.412	0.701	2.10	1.84	0.109	0.206	0.531
1.48	1.16	0.280	0.403	0.695	2.12	1.87	0.106	0.201	0.526
1.50	1.18	0.272	0.395	0.690	2.14	1.90	0.103	0.197	0.522
1.52	1.19	0.265	0.387	0.684	2.16	1.94	0.100	0.192	0.517
1.54	1.20	0.257	0.379	0.678	2.18	1.97	0.097	0.188	0.513
1.56	1.22	0.250	0.371	0.672	2.20	2.01	0.094	0.184	0.508
1.58	1.23	0.242	0.363	0.667	2.22	2.04	0.091	0.180	0.504
1.60	1.25	0.235	0.356	0.661	2.24	2.08	0.088	0.176	0.499
1.62	1.27	0.228	0.348	0.656	2.26	2.12	0.085	0.172	0.495
1.64	1.28	0.222	0.341	0.650	2.28	2.15	0.083	0.168	0.490
1.66	1.30	0.215	0.334	0.645	2.30	2.19	0.080	0.165	0.486
1.68	1.32	0.209	0.327	0.639	2.32	2.23	0.078	0.161	0.482
1.70	1.34	0.203	0.320	0.634	2.34	2.27	0.075	0.157	0.477
1.72	1.36	0.197	0.313	0.628	2.36	2.32	0.073	0.154	0.473

continued next page

Table A. 1. One-dimensional isentropic relations (continued)

M	A/A^*	p/p_0	ρ/ρ_0	T/T_0	M	A/A^*	p/p_0	ρ/ρ_0	T/T_0
2.38	2.36	0.071	0.150	0.469	2.78	3.43	0.038	0.097	0.393
2.40	2.40	0.068	0.147	0.465	2.80	3.50	0.037	0.095	0.389
2.42	2.45	0.066	0.144	0.461	2.82	3.57	0.036	0.093	0.386
2.44	2.49	0.064	0.141	0.456	2.84	3.64	0.035	0.091	0.383
2.46	2.54	0.062	0.138	0.452	2.86	3.71	0.034	0.089	0.379
2.48	2.59	0.060	0.135	0.448	2.88	3.78	0.033	0.087	0.376
2.50	2.64	0.059	0.132	0.444	2.90	3.85	0.032	0.085	0.373
2.52	2.69	0.057	0.129	0.441	2.92	3.92	0.031	0.083	0.370
2.54	2.74	0.055	0.126	0.437	2.94	4.00	0.030	0.081	0.366
2.56	2.79	0.053	0.123	0.433	2.96	4.08	0.029	0.080	0.363
2.58	2.84	0.052	0.121	0.429	2.98	4.15	0.028	0.078	0.360
2.60	2.90	0.050	0.118	0.425	3.00	4.23	0.027	0.076	0.357
2.62	2.95	0.049	0.115	0.421	3.50	6.79	0.013	0.045	0.290
2.64	3.01	0.047	0.113	0.418	4.00	10.72	0.0066	0.028	0.238
2.66	3.06	0.046	0.110	0.414	4.50	16.56	0.0035	0.017	0.198
2.68	3.12	0.044	0.108	0.410	5.00	25.00	189×10^{-5}	0.011	0.167
2.70	3.18	0.043	0.106	0.407	6.00	53.19	623×10^{-5}	0.0052	0.122
2.72	3.24	0.042	0.103	0.403	7.00	104.14	242×10^{-6}	0.0026	0.093
2.74	3.31	0.040	0.101	0.400	8.00	190.11	102×10^{-6}	0.0014	0.072
2.76	3.37	0.039	0.099	0.396	9.00	327.19	474×10^{-7}	0.0008	0.058
					10.00	535.94	236×10^{-7}	0.0005	0.048

Table A. 2. One-dimensional normal-shock relations*

M_1	M_2	$\frac{p_2}{p_1}$	$\frac{T_2}{T_1}$	$\frac{(p_0)_2}{(p_0)_1}$	M_1	M_2	$\frac{p_2}{p_1}$	$\frac{T_2}{T_1}$	$\frac{(p_0)_2}{(p_0)_1}$
1.00	1.000	1.000	1.000	1.000	1.64	0.657	2.971	1.416	0.880
1.02	0.980	1.047	1.013	1.000	1.66	0.651	3.048	1.430	0.872
1.04	0.962	1.095	1.026	1.000	1.68	0.646	3.126	1.444	0.864
1.06	0.944	1.144	1.039	1.000	1.70	0.641	3.205	1.458	0.856
1.08	0.928	1.194	1.052	0.999	1.72	0.635	3.285	1.473	0.847
1.10	0.912	1.245	1.065	0.999	1.74	0.631	3.366	1.487	0.839
1.12	0.896	1.297	1.078	0.998	1.76	0.626	3.447	1.502	0.830
1.14	0.882	1.350	1.090	0.997	1.78	0.621	3.530	1.517	0.821
1.16	0.868	1.403	1.103	0.996	1.80	0.617	3.613	1.532	0.813
1.18	0.855	1.458	1.115	0.995	1.82	0.612	3.698	1.547	0.804
1.20	0.842	1.513	1.128	0.993	1.84	0.608	3.783	1.562	0.795
1.22	0.830	1.570	1.140	0.991	1.86	0.604	3.869	1.577	0.786
1.24	0.818	1.627	1.153	0.988	1.88	0.600	3.957	1.592	0.777
1.26	0.807	1.686	1.166	0.986	1.90	0.596	4.045	1.608	0.767
1.28	0.796	1.745	1.178	0.983	1.92	0.592	4.134	1.624	0.758
1.30	0.786	1.805	1.191	0.979	1.94	0.588	4.224	1.639	0.749
1.32	0.776	1.866	1.204	0.976	1.96	0.584	4.315	1.655	0.740
1.34	0.766	1.928	1.216	0.972	1.98	0.581	4.407	1.671	0.730
1.36	0.757	1.991	1.229	0.968	2.00	0.577	4.500	1.688	0.721
1.38	0.748	2.055	1.242	0.963	2.02	0.574	4.594	1.704	0.711
1.40	0.740	2.120	1.255	0.958	2.04	0.571	4.689	1.720	0.702
1.42	0.731	2.186	1.268	0.953	2.06	0.567	4.784	1.737	0.693
1.44	0.723	2.253	1.281	0.948	2.08	0.564	4.881	1.754	0.683
1.46	0.716	2.320	1.294	0.942	2.10	0.561	4.978	1.770	0.674
1.48	0.708	2.389	1.307	0.936	2.12	0.558	5.077	1.787	0.665
1.50	0.701	2.458	1.320	0.930	2.14	0.555	5.176	1.805	0.656
1.52	0.694	2.529	1.334	0.923	2.16	0.553	5.277	1.822	0.646
1.54	0.687	2.600	1.347	0.917	2.18	0.550	5.378	1.839	0.637
1.56	0.681	2.673	1.361	0.910	2.20	0.547	5.480	1.857	0.628
1.58	0.675	2.746	1.374	0.903	2.22	0.544	5.583	1.875	0.619
1.60	0.668	2.820	1.388	0.895	2.24	0.542	5.687	1.892	0.610
1.62	0.663	2.895	1.402	0.888	2.26	0.539	5.792	1.910	0.601

**For a perfect gas with $k = 1.4$.*

continued next page

Table A. 2. One-dimensional normal-shock relations (continued)

M_1	M_2	$\frac{p_2}{p_1}$	$\frac{T_2}{T_1}$	$\frac{(p_0)_2}{(p_0)_1}$	M_1	M_2	$\frac{p_2}{p_1}$	$\frac{T_2}{T_1}$	$\frac{(p_0)_2}{(p_0)_1}$
2.28	0.537	5.898	1.929	0.592	2.68	0.497	8.213	2.322	0.431
2.30	0.534	6.005	1.947	0.583	2.70	0.496	8.338	2.343	0.424
2.32	0.532	6.113	1.965	0.575	2.72	0.494	8.465	2.364	0.417
2.34	0.530	6.222	1.984	0.566	2.74	0.493	8.592	2.396	0.410
2.36	0.527	6.331	2.003	0.557	2.76	0.491	8.721	2.407	0.403
2.38	0.525	6.442	2.021	0.549	2.78	0.490	8.850	2.429	0.396
2.40	0.523	6.553	2.040	0.540	2.80	0.488	8.980	2.451	0.389
2.42	0.521	6.666	2.060	0.532	2.82	0.487	9.111	2.473	0.383
2.44	0.519	6.779	2.079	0.523	2.84	0.485	9.243	2.496	0.376
2.46	0.517	6.894	2.098	0.515	2.86	0.484	9.376	2.518	0.370
2.48	0.515	7.009	2.118	0.507	2.88	0.483	9.510	2.541	0.364
2.50	0.513	7.125	2.138	0.499	2.90	0.481	9.645	2.563	0.358
2.52	0.511	7.242	2.157	0.491	2.92	0.480	9.781	2.586	0.352
2.54	0.509	7.360	2.177	0.483	2.94	0.479	9.918	2.609	0.346
2.56	0.507	7.479	2.198	0.475	2.96	0.478	10.055	2.632	0.340
2.58	0.506	7.599	2.218	0.468	2.98	0.476	10.194	2.656	0.334
2.60	0.504	7.720	2.238	0.460	3.00	0.475	10.333	2.679	0.328
2.62	0.502	7.842	2.260	0.453					
2.64	0.500	7.965	2.280	0.445					
2.66	0.499	8.088	2.301	0.438					

14 Dimensional Analysis and Similitude

Part I Dimensional Analysis

14.1 Introduction

The general equations of motion together with the equation of state and continuity presented in the earlier chapters form the basis for investigation of all flow phenomena. However, the application of these basic principles to solve specific problems has not always been successful due to mathematical difficulties encountered in finding solutions to differential equations satisfying complex boundary conditions. Only the simplest of problems have been amenable to rigorous analytical solutions. However in recent years, with the advent of computers, it has been possible to obtain numerical solutions for a large number of problems using finite difference or finite element techniques. Yet another class of solutions arise out of physical concepts guided by past experience and supported by experimental evidence. Finally, an exhaustive experimental program guided by dimensional analysis often provides sufficient insight into many complex flow problems to establish functional relationships among the variables governing the flow phenomena. Empirical constants linking the various significant dimensionless parameters can be determined both by laboratory experiments and field observations. In this chapter, the principles of dimensional analysis and model similitude are described.

14.2 Fundamental Dimensions

In Newtonian mechanics, the law relating force to change of momentum is expressed as $F = m \cdot a$, where the constant of proportionality has been universally accepted as unity. This in itself has defined the relationships between dimensional magnitude contained in force, mass and acceleration. For example weight of unit mass must be equal to g units of force. As such, in Newtonian mechanics the fundamental dimensions generally found convenient for use are mass [M], length [L], and time [T]. The dimensions of force can then be expressed as $[\mathrm{MLT}^{-2}]$. However, we could have chosen force, length, and time as fundamental dimensions in which case the dimension of mass becomes $[\mathrm{FLT}^{-2}]$. In problems related to thermodynamics, another dimension for temperature $[\theta]$ is required from which the dimension for heat becomes $[\mathrm{ML}^{-2}\theta^{-2}]$. Alternatively, it is sometimes more convenient to use heat [H] as the fundamental dimension and then the corresponding dimension for temperature θ is $[H^{1/2}M^{-1/2}L^{-1}]$. In problems relating to flow of electricity through conductors, a fresh dimension of electric charge may be used. These dimensions mentioned above in defining physical entities are not the only ones employed and they may vary. Basically the dimensions chosen must be independent of each other. Whatever be the dimensions chosen, they tell us the relative nature of physical quantities. They also serve a very useful purpose in determining the variation of numerical values of physical quantities when basic units of measurement are changed.

The following table illustrates the dimensions contained in several quantities commonly met with in problems related to fluid mechanics. The M-L-T system is generally employed in all flow problems.

Quantity	Dimensions in the M-L-T system	Dimensions in the F-L-T system
Mass	$[M]$	$[FL^{-1}T^{-2}]$
Mass density, ρ	$[ML^{-3}]$	$[FL^{-4}T^{2}]$
Dynamic viscosity, μ	$[ML^{-1}T^{-1}]$	$[FL^{-2}T]$
Kinematic viscosity, ν	$[L^2T^{-1}]$	$[L^2T^{-1}]$
Surface tension, σ	$[MT^{-2}]$	$[FL^{-1}]$
Bulk modulus, K and Modulus of elasticity	$[ML^{-1}T^{-2}]$	$[FL^{-2}]$
Force	$[MLT^{-2}]$	$[F]$
Pressure intensity or stress	$[ML^{-1}T^{-2}]$	$[FL^{-2}]$
Rate of angular deformation Angular velocity, ω	$[T^{-1}]$	$[T^{-1}]$
Velocity	$[LT^{-1}]$	$[LT^{-1}]$
Discharge	$[L^3T^{-1}]$	$[L^3T^{-1}]$
Acceleration	$[LT^{-2}]$	$[LT^{-2}]$
Momentum (angular)	$[ML^2T^{-1}]$	$[FLT]$
Momentum (linear)	$[MLT^{-1}]$	$[FT]$
Moment or torque	$[ML^2T^{-2}]$	$[FL]$
Work, energy, heat	$[ML^2T^{-2}]$	$[FL]$
Power	$[ML^2T^{-3}]$	$[FLT^{-1}]$
Enthalpy, quantity of heat	$[ML^2T^{-2}]$	$[FL]$ or H
Thermal conductivity	$[MLT^{-3}\theta^{-1}]$	$[FT^{-1}\theta^{-1}][HL^{-1}T^{-1}\theta^{-1}]$
Specific heat	$[L^2T^{-2}\theta]$	$[L^2T^{-2}\theta]$
Entropy	$[ML^2T^{-2}\theta^{-1}]$	$[FL\theta^{-1}]$ or $[H\theta^{-1}]$

14.3 Physical Quantity and Dimensions

We start with a general statement: *Every physical quantity can be expressed as products of powers of fundamental dimensions*. Suppose a physical quantity designated as u has dimensions M, L, T, θ contained in it, then the above statement would mean

$$u = f(\mathrm{M, L, T}, \theta) = [\mathrm{M}^a \mathrm{L}^b \,\mathrm{T}^c \,\theta^d] \tag{14.1}$$

For example, if the physical quantity involved entropy, then $a = 1, b = 2, c = -2, d = -1$ and $u = [\mathrm{M}^1\mathrm{L}^2\mathrm{T}^{-2}\theta^{-1}]$.

The proof for the validity of the statement can be found in advanced textbooks on dimensional analysis. The application of the rule will be highlighted in the ensuing section.

One of the important applications is found in the establishment of the general law for the transformation of a physical quantity from one unit of measurement to another. The physical quantity u is given by equation (14.1) in any standard set of units of measurement such as the CGS or MKS or foot–pound–second system. Let the new unit of measurement of M be α times the old unit, that of L be β times the corresponding previous unit and so on. The magnitude of u^* in the new system of measurement is given by

$$u^* = \left(\frac{\mathrm{M}}{\alpha}\right)^a \left(\frac{\mathrm{L}}{\beta}\right)^b \left(\frac{\mathrm{T}}{\gamma}\right)^c \left(\frac{\theta}{\delta}\right)^d = \left[\frac{1}{\alpha^a \beta^b \gamma^c \delta^d}\right] u \tag{14.2}$$

As an illustration, if the dynamic viscosity of a fluid at a certain temperature of 15°C is given as 0.766×10^{-3} lb/ft s in British units, we have

$$1\ \mathrm{lb} = 453.6\ \mathrm{gms},\ \alpha = \frac{1}{453.6};\ 1\ \mathrm{ft} = 30.48\ \mathrm{cms},\ \beta = \frac{1}{30.48}$$

and $\gamma = 1$ as the time unit is the same in both systems.

Since the dynamic viscosity has the dimension $[\mathrm{ML}^{-1}\mathrm{T}^{-1}]$, $a = 1, b = -1, c = -1$. The new value of μ in CGS units will be

$$\mu^* = \frac{1}{\left(\frac{1}{453.6}\right)^1 \left(\frac{1}{30.48}\right)^{-1} (1)^1} (0.766 \times 10^{-3}) = \left(\frac{453.6}{30.48}\right) \times 0.766 \times 10^{-3}$$

$$= 1.14 \times 10^{-2}\ \mathrm{gms/cm\ s}$$

14.4 Dimensional Homogeneity

It has been seen that every physical quantity has dimensions and magnitudes expressed in certain units of measurement. Hence, any functional relationship between two or more of these physical quantities must necessarily show equivalence in both dimension and magnitude. More explicitly, we may state that the two sides of an equation must not only balance numerically but also in dimension. As a corollary, it follows that if certain numerical constants appear in a dimensionally homogeneous equation, the value of the constant should remain unaltered even if the system of units of measurement is changed.

It has been seen from the Bernoulli theorem that the stagnation pressure, p_0 at the nose of the cylinder placed in a uniform flow of velocity, U_0 is given by the expression

$$p_0 = \frac{1}{2}\rho U_0^2$$

ρ being the density of the fluid. It is seen that this is a homogeneous equation. The dimensions of the two sides of the equation are identical.

$$[\mathrm{ML^{-1}T^{-2}}] = [\mathrm{ML^{-3}}][\mathrm{LT^{-1}}]^2$$

The numerical constant $1/2$ that appears on the right-hand side of the equation remains unaltered regardless of the units of measurement employed.

Very often, scientists and engineers formulate empirical relationships among variables governing a phenomena based on experiments and field observations. We may discuss one such example in detail. In the chapter on open channels, we have seen that the velocity in the open channel is given by the Manning equation in metric units

$$V = \left(\frac{1}{n}\right) R^{2/3} S^{1/2}$$

where n is a resistance factor dependent on the type of wetted surface of the bottom and sides (perimeter) of the channel. R is the hydraulic radius equal to the ratio of the area to the perimeter, A/P having the dimension [L]. S, the slope of the channel is dimensionless. Thus if n is treated as a numerical constant, then the equation does not satisfy dimensional homogeneity.

[L/T] on the left-hand side is not equal to the dimension $[\mathrm{L}^{2/3}]$ on the right-hand side. The factor n being dependent on height, spacing and shape of roughness elements of the perimeter can at best have the dimension of length only. Hence we conclude that the numerical constant 1.0 on the right-hand side of the equation must also have a dimension of length and time to satisfy dimensional homogeneity of the two sides of the equation. A critical appraisal of the equation reveals that the flow in an open channel occurs by the force of gravity. Hence the factor $g^{1/2}$ is required on the right-hand side to balance the dimension of time. Hence we modify the Manning's equation as (with $1 = Cg^{1/2}$)

$$V = \left(\frac{Cg^{1/2}}{n}\right) R^{2/3} S^{1/2}$$

whose dimensions are $[\mathrm{LT^{-1}}] = [\mathrm{L^{7/6}T^{-1}}]/[n]$. Now we conclude that $[n]$ must have the dimension of $[\mathrm{L}^{1/6}]$ to balance the equation dimensionally. Thus the empirical equation has been made dimensionally homogeneous. Similarly many other empirical non-homogeneous equations can be appropriately interpreted or evaluated afresh and rendered dimensionally homogeneous. Further it enables us to evaluate the magnitude of empirical constants when units of measurement are changed. For example, Manning's equation in metric system can be converted to the British system as follows.

We have seen that the numerical constant of 1.0 has the dimension of $g^{1/2}/\mathrm{L}^{1/6} = [\mathrm{L^{1/3}T^{-1}}]$. Since the unit of time is the same in both systems of units of measurement and 1 metre = 3.28 ft., the corresponding constant in the British system is $(3.28)^{1/3} = 1.486$. Hence in the foot–pound–second system, the Manning's equation becomes

$$V = \left(\frac{1.486}{n}\right) R^{2/3} S^{1/2}$$

The value of n is considered to have the same value in both metric and British systems and depends on the type of boundary roughness. Further discussion on the value of n has been presented in the chapter on open channels.

14.5 Summation of Quantities

Let the functional relationship between certain quantities $u_1, u_2, u_3, \ldots$ be expressed in the manner

$$u = f(u_1, u_2, u_3, \ldots u_n)$$

Then by the general law stated in Section 14.3, the quantities may be expressed as powers of the dimensions contained in them.

$$u = \mathrm{M}^a \mathrm{L}^b \mathrm{T}^c \theta^d \ldots$$

$$u_1 = \mathrm{M}^{a_1} \mathrm{L}^{b_1} \mathrm{T}^{c_1} \theta^{d_1} \ldots, u_2 = \mathrm{M}^{a_2} \mathrm{L}^{b_2} \mathrm{T}^{c_2} \theta^{d_2} \ldots, u_n = \mathrm{M}^{a_n} \mathrm{L}^{b_n} \mathrm{T}^{c_n} \theta^{d_n}$$

If the functional relationship is of the summation type

$$u = u_1 + u_2 + u_3 + \cdots + u_n$$

then dimensionally, we have

$$\mathrm{M}^a \mathrm{L}^b \mathrm{T}^c \theta^d \cdots = \mathrm{M}^{a_1} \mathrm{L}^{b_1} \mathrm{T}^{c_1} \theta^{d_1} \cdots + \mathrm{M}^{a_2} \mathrm{L}^{b_2} T^{c_2} \theta^{d_2} \cdots + \cdots + \mathrm{M}^{a_n} \mathrm{L}^{b_n} \mathrm{T}^{c_n} \theta^{d_n}$$

By dimensional homogeneity, we readily see that

$$a = a_1 + a_2 + a_3 + \cdots + a_n$$

$$b = b_1 + b_2 + b_3 + \cdots + b_n$$

$$c = c_1 + c_2 + c_3 + \cdots + c_n$$

$$d = d_1 + d_2 + d_3 + \cdots + d_n$$

Therefore, each term on the right-hand side must have the same dimension as the left-hand side.

14.6 Product of Quantities

If the functional relationship is in the form of products such as

$$u = u_1^{m_1} u_2^{m_2} u_3^{m_3} \ldots\ldots u_n^{m_n}$$

then dimensionally we have

$$\mathrm{M}^a\mathrm{L}^b\mathrm{T}^c\theta^d \cdots = M^{a_1m_1}L^{b_1m_1}T^{c_1m_1}\theta^{d_1m_1} \cdots + M^{a_2m_2}L^{b_2m_2}T^{c_2m_2}\theta^{d_2m_2} + \cdots$$
$$+ M^{a_nm_n}L^{b_nm_n}T^{c_nm_n}\theta^{d_nm_n}$$

Equating the powers of like dimensions, we have

$$\mathrm{M} : a_1m_1 + a_2m_2 + a_3m_3 + \cdots + a_nm_n = a$$

$$\mathrm{L} : b_1m_1 + b_2m_2 + b_3m_3 + \cdots + b_nm_n = b$$

$$\mathrm{T} : c_1m_1 + c_2m_2 + c_3m_3 + \cdots + c_nm_n = c$$

$$\theta : d_1m_1 + d_2m_2 + d_3m_3 + \cdots + d_nm_n = d$$

and etc.

14.7 Non-Dimensional Parameters

A special case of interest arises when u is a non-dimensional product of the variables $u_1, u_2, u_3, \ldots u_n$.

$$\mathrm{M}^0 : a_1m_1 + a_2m_2 + a_3m_3 + \cdots + a_nm_n = 0$$

$$\mathrm{L}^0 : b_1m_1 + b_2m_2 + b_3m_3 + \cdots + b_nm_n = 0$$

$$\mathrm{T}^0 : c_1m_1 + c_2m_2 + c_3b_3 + \cdots + c_nm_n = 0$$

$$\theta^0 : d_1m_1 + d_2m_2 + d_3m_3 + \cdots + d_nm_n = 0$$

and etc.

While there are n unknowns $(m_1, m_2, m_3, \ldots m_n)$, the number of equations available will be equal to the number of dimensions (say r) contained in the physical quantities $u_1, u_2, u_3 \ldots u_n$. For example if there are only three dimensions, M, L, T (as in the case of most of the problems in fluid mechanics), there will be only three equations. Thus if the number of unknowns is more than the number of linearly independent equations, a unique solution for the unknowns is not possible. Then the solution to any r number of unknowns can be expressed in terms of the other $(n - r)$ unknowns. The number of linearly independent equations is equal to the rank of the dimensional matrix formed by the coefficients of $m_1, m_2 \ldots m_n$. The matrix is

$$a_1 \; a_2 \; a_3 \ldots\ldots\ldots\ldots\ldots a_n$$

$$b_1 \; b_2 \; b_3 \ldots\ldots\ldots\ldots\ldots b_n$$

$$c_1 \; c_2 \; c_3 \ldots\ldots\ldots\ldots\ldots c_n$$

$d_1 \ d_2 \ d_3 \ldots\ldots\ldots\ldots\ldots d_n$

$\ldots\ldots\ldots\ldots\ldots\ldots\ldots\ldots$

$\ldots\ldots\ldots\ldots\ldots\ldots\ldots\ldots$

The rank of the matrix is defined as the highest value of the non-vanishing determinant which can be formed out of the preceding dimensional matrix. We can assign arbitrary but linearly independent values to $(n - r)$ unknowns and obtain solutions for the remaining r unknowns. The set of dimensionless products so obtained and which are dependent of each other is called a complete set. The procedure can be best illustrated by an example.

Let it be required to find the complete set of dimensionless products of variables: force F, velocity V, length L, density ρ, viscosity μ, gravitational acceleration g, celerity of sound wave c, and surface tension σ.

$$u = \mathrm{F}^{m_1}\mathrm{L}^{m_2}\mathrm{V}^{m_3}\rho^{m_4}\mu^{m_5}g^{m_6}c^{m_7}\sigma^{m_8}$$

Substituting the dimensions contained in the physical quantities

$$u = (\mathrm{MLT}^{-2})^{m_1}(\mathrm{L})^{m_2}(\mathrm{LT}^{-1})^{m_3}(\mathrm{ML}^{-3})^{m_4}(\mathrm{ML}^{-1}\mathrm{T}^{-1})^{m_5}(\mathrm{LT}^{-2})^{m_6}(\mathrm{LT}^{-1})^{m_7}(\mathrm{MT}^{-2})^{m_8}$$

Should the u be dimensionless $[\mathrm{M}^0\mathrm{L}^0\mathrm{T}^0]$ we get, on equating the like powers of the dimensions on both sides of the preceding dimensional equation,

$$\mathrm{M}^0 : m_1 + m_4 + m_5 + m_8 = 0$$

$$\mathrm{L}^0 : m_1 + m_2 + m_3 - 3m_4 - m_5 + m_6 + m_7 = 0$$

$$\mathrm{T}^0 : -2m_1 - m_3 - m_5 - 2m_6 - m_7 - 2m_8 = 0$$

Hence we have eight unknowns and only three equations. The dimensional matrix formed by the coefficients of m are

$$\begin{matrix} 1 & 0 & 0 & 1 & 1 & 0 & 0 & 1 \\ 1 & 1 & 1 & -3 & -1 & 1 & 1 & 0 \\ -2 & 0 & -1 & 0 & -1 & -2 & -1 & -2 \end{matrix}$$

The rank of the matrix is found to be 3 (same as the number of dimensions or number of rows in the matrix). Hence the solution to any three of the unknowns can be expressed in terms of $8 - 3 = 5$ sets of assigned values. Let us arbitrarily chose the values of m_2, m_3, m_4 and solve them in terms of m_1, m_5, m_6, m_7, m_8.

$$m_2 = -2m_1 - m_5 + m_6 - m_8$$

$$m_3 = -2m_1 - m_5 - 2m_6 - m_7 - 2m_8$$

$$m_4 = -m_1 - m_5 - m_8$$

Values of	m_1	m_5	m_6	m_7	m_8
1st set	1	0	0	0	0
2nd set	0	1	0	0	0
3rd set	0	0	1	0	0
4th set	0	0	0	1	0
5th set	0	0	0	0	1

We now assign $(n - r) = 5$ values to m_1, m_5, m_6, m_7, m_8.

We can choose the following convenient values to reduce arithmetical work.

With the first set of values, we obtain the values of $m_2 = -2, m_3 = -2, m_4 = -1$. The non-dimensional parameter will become $FL^{-2}V^{-2}\rho^{-1} = F/\rho V^2L^2$.

With the second set of values, we get $m_2 = -1, m_3 = -1, m_4 = -1$ and the non-dimensional parameter is $\mu/\rho VL$.

With the third set of values, we have $m_2 = 1, m_3 = -2, m_4 = 0$ and the parameter is c/V.

With the fourth set of values, we obtain $m_2 = 0, m_3 = -1, m_4 = 0$. The non-dimensional parameter is gL/V^2.

With the fifth set of values, we get $m_2 = -1, m_3 = -2, m_4 = -1$, yielding the parameter $\sigma/\rho LV^2$.

The complete set of non-dimensional parameters is

$$\frac{F}{\rho V^2 L^2}, \frac{\mu}{\rho VL}, \frac{gL}{V^2}, \frac{c}{V}, \frac{\sigma}{\rho LV^2}$$

If we had assigned the values of $m_5 = -1$ in the second set, $m_6 = -1$ in the third set, $m_7 = -1$ in the fourth set and $m_8 = -1$ in the fifth set, the complete set would have been

$$\frac{F}{\rho V^2L^2}, \frac{\mu}{\rho VL}, \frac{V^2}{gL}, \frac{c}{V}, \frac{\sigma}{\rho LV^2}$$

The last four parameters are seen to be reciprocals of the corresponding parameters of the earlier set. These parameters are known respectively as the Euler number, Reynolds number, Froude number (actually the square of it), Mach number and the Weber number. If we assign different values to m_1, m_5, m_6, m_7, m_8, we will get combinations of powers of the 5 non-dimensional parameters already found.

14.8 π-Theorem

The statements made in the previous section form the basis for an important theorem known as the π-theorem (sometimes known as the Buckingham π-theorem) in dimensional analysis. The general statement of the π-theorem is: *If n physical quantities are required to describe a physical phenomenon and if these quantities have m dimensions contained in them, the relationship can be reduced to one comprising of $(n - r)$ non-dimensional products, r being (equal to or less than m) the rank of the $n \times m$ dimensional matrix.*

From the previous example in Section 14.7, if a physical phenomenon is controlled by 8 variables such as $F, L, V, \rho, \mu, \sigma, g, c$, the relationship among the variables may be described by the functional form

$$f(F, L, V, \rho, \mu, \sigma, g, c) = 0$$

These variables have three dimensions (M, L, T) among them and the rank of the matrix is also three. Generally, the rank of the matrix is equal to the number of dimensions contained in them though there may be exceptions. Hence the above functional form can be reduced to a relationship among $8 - 3 = 5$ non-dimensional variables.

$$\phi(\pi_1, \pi_2, \pi_3, \pi_4, \pi_5) = 0$$

or more generally, $$\phi(\pi_1, \pi_2, \pi_3, \ldots, \pi_{n-r}) = 0$$

where $\pi_1, \pi_2, \ldots, \pi_{n-r}$ are non-dimensional parameters formed by grouping of variables as illustrated in Section 14.7. The preceding functional relationship can be written as

$$\phi\left(\frac{F}{\rho L^2 V^2}, \frac{\rho V L}{\mu}, \frac{V}{(gL)^{1/2}}, \frac{V}{c}, \frac{\rho L V^2}{\sigma}\right) = 0$$

The proof of the π-theorem is beyond the scope of the present treatment and can be found in other books[1,2]. We shall confine ourselves to the application of the theorem to practical problems.

14.9 Significance of the π-Theorem

The importance of the requirements of dimensional homogeneity and the application of the π-theorem to find a reduced number of dimensionless parameters should not be underestimated.

Dimensional homogeneity helps us to balance the dimensions of empirical equations and check analytically derived ones for homogeneity.

It helps to determine the dimensions, if any, of empirical constants.

It also helps to determine the dimensions of empirical constants when units of measurements are changed.

The role of the π-theorem is more extensive.

When the number of variables is too large, it is often difficult to correlate the effect of each one of these variables on the dependent variable. Hence the π-theorem is of great importance in as much as it groups these variables into non-dimensional parameters and reduces the number of variables by the extent of the rank of the dimensional matrix. The correlations among the reduced number of dimensionless parameters becomes easier and more meaningful. Some of the important dimensionless parameters are generally a ratio of forces such as

---

[1]Bridgman, P.W., *Dimensional Analysis*, New Haven: Yale University Press, 1937.

[2]Langhaar, H.L., *Dimensional Analysis and Theory of Models*, New York: John Wiley and Sons, 1951.

Reynolds number : Ratio of inertial to viscous forces

Froude number : Ratio of inertial forces to gravitational forces

Weber number : Ratio of inertial to surface tension forces

The flow pattern remains unchanged if these dimensionless parameters remain the same despite the fact that the individual variable in them may change their values. For example, for a fluid having different viscosity and density, the velocity, and the linear dimension involved in the value of the Reynolds number, $\rho VL/\mu$ may be so chosen such that the value of the Reynolds number remains the same. Thus the two different fluids will have the same flow pattern if the Reynolds number is the governing parameter.

These π-parameters are therefore capable of wider and more general interpretation than is possible with numerous individual variables. Further since these π-parameters are non-dimensional, they give us the opportunity of using the same graphical or analytical relationships regardless of units of measurement.

The grouping of these variables into a reduced number of non-dimensional parameters acts as a basis to an experimental program, the aim of which is to establish a functional relationship among the significant parameters and supplement it by experimentally determined coefficients. Yet another distinctive advantage of the use of the π-theorem is found in model studies. Experiments on the model are carried out at identical values of the governing or significant π-parameters as obtained on the prototype. Since the flow processes are similar between the model and the prototype at identical π-parameters, results obtained on the model can be scaled up to the prototype values. The laws of model similitude will be amplified in a later section.

14.10 Dimensional Analysis: Choice of Variables

The first step in dimensional analysis is the choice of variables governing the flow phenomena under investigation. This task is not a simple one and is wrought with pitfalls. The choice of the dependent variable (that is, the value of the variable that we wish to find out, for example, the discharge over a weir) is not a difficult one as it comes specified with the problem and how it is posed. Having made the choice of the dependent variable, we proceed to list out all other variables that have a significant effect on the dependent variable. Care should be taken to make a complete list of the variables. These variables may be classified as (1) kinematic variables such as velocity, (2) dynamic variables such as forces, accelerations, (3) geometric variables such as the diameter of an immersed body or width of a weir etc., (4) fluid properties such as density, viscosity, surface tension, vapour pressure, etc. We should not omit any important variables by oversight or wrong interpretation. For example, elastic properties of materials, gravitational acceleration though they appear to be numerical constants, are dynamic variables governing a particular phenomenon. Omission of an important variable will exclude a significant π-parameter which might have an important bearing on the problem.

At the same time, we should not include unnecessary variables unrelated to the problem on hand which will only serve to confuse the issue. We should avoid including two or more dependent variables. For example, if discharge Q and area A are chosen as variables, V should not be included as another independent variable (as $Q = A \cdot V$ and V is dependent on Q and

A). Extraneous variables which have no bearing on the problem should be avoided (for example surface tension has no role to play in a pressure flow in a closed conduit). The choice of the independent variables comes with experience and intuitive insight into the problem.

14.11 Criticisms of the Dimensional Analysis

The main criticism against dimensional analysis is that it does not give us any indication of the choice of variables.

It does not explicitly give us the functional relationships amongst the different π-parameters.

It does not supply the numerical constants that appear in functional relationships.

Dimensional analysis must be supported by experimental investigation to obtain graphical or analytical relationships among the π-parameters.

Physical understanding of the problem and experience of the investigator weighs heavily in the application of dimensional analysis and interpretation of results

We conclude that dimensional analysis is a powerful tool in the hands of an experienced investigator.

14.12 Determination of Dimensionless Parameters

Once the variables have been chosen, the grouping of the variables into $(n - r)$ dimensionless parameters can be done either by the Rayleigh method or by the method of repeating variables. The Rayleigh method has already been illustrated in Section 14.7. We will take another example to show its application.

Let it be required to find the pressure loss in a circular conduit (flowing full) in which the flow is uniform and steady. The pressure loss Δp which is required over a length L of the pipe is the unknown and hence the dependent dynamic variable. It is dependent on the average velocity (kinematic variable) V in the pipe. The geometric variables are the diameter d of the pipe and length L over which the pressure drop is to be determined and roughness of the boundary symbolised by the roughness height, k. The acceleration due to gravity g does not come into the picture as the flow is caused by a pressure gradient and gravitational effect in sloping pipes is reflected in the determination of pressure intensity at any point along the pipe. The fluid properties are density, ρ and viscosity, μ. The functional relationship among the variables is written as

$$\Delta p = f(V, d, L, k, \rho, \mu)$$

The total number of variables is 7 and the dimensional matrix formed between them and the dimensions contained in them is

Variable	Δp	V	d	L	k	ρ	μ
Dimension							
M	1	0	0	0	0	1	1
L	–1	1	1	1	1	–3	–1
T	–2	–1	0	0	0	0	–1

The rank of the matrix (highest order of the non-vanishing determinant) can easily be verified to be 3, same as the number of dimensions in this case. There are thus $7 - 3 = 4$ dimensionless parameters. The dimensionless parameters are

$$(\Delta p)^{m_1}(V)^{m_2}(d)^{m_3}(L)^{m_4}(k)^{m_5}(\rho)^{m_6}(\mu)^{m_7}$$

or in terms of the dimensions contained in them

$$(\mathrm{ML}^{-1}\mathrm{T}^{-2})^{m_1}(\mathrm{LT}^{-1})^{m_2}(L)^{m_3}(\mathrm{L})^{m_4}(\mathrm{L})^{m_5}(\mathrm{ML}^{-3})^{m_6}(\mathrm{ML}^{-1}T^{-1})^{m_7}$$

Since the parameters are dimensionless, we have each parameter with dimensions $\mathrm{M}^0\mathrm{L}^0\mathrm{T}^0$. Hence we have

$$\mathrm{M}^0 : m_1 + m_6 + m_7 = 0$$

$$\mathrm{L}^0 : -m_1 + m_2 + m_3 + m_4 + m_5 - 3m_6 - m_7 = 0$$

$$\mathrm{T}^0 : -2m_1 - m_2 - m_7 = 0$$

There are seven unknowns and 3 equations. We can solve for three chosen unknowns in terms of arbitrarily assigned values to the other four. Let us solve for m_2, m_3 and m_6 in terms of arbitrarily assigned values for m_1, m_4, m_5, m_7 as shown below, being the simplest of the four independent sets of values.

	m_1	m_4	m_5	m_7
First set	1	0	0	0
Second set	0	1	0	0
Third set	0	0	1	0
Fourth set	0	0	0	1

With the first set, we get $m_2 = -2, m_3 = 0, m_6 = -1$, the parameter is $\pi_1 = \Delta p/\rho V^2$; with the second set, $m_2 = m_6 = 0, m_3 = -1, \pi_2 = L/d$; with the third set, $m_2 = m_6 = 0, m_3 = -1, \pi_3 = k/d$; with the fourth set, $m_2 = m_3 = m_6 = -1, \pi_4 = \mu/\rho V d$. Finally we have the functional relationship among the non-dimensional parameters

$$\frac{\Delta p}{\rho V^2} = \phi\left(\frac{L}{d}, \frac{k}{d}, \frac{\mu}{\rho V d}\right)$$

It is possible to derive more information from the preceding relationship. The pressure gradient is constant or the pressure falls linearly with distance. Hence pressure drop is proportional to L. Hence we can write

$$\frac{\Delta p}{\rho V^2} = \frac{L}{d}\phi_1\left(\frac{k}{d}, \frac{1}{\mathbf{R}}\right)$$

Dividing throughout by g, we obtain after transposing certain terms

$$\frac{\Delta p}{\gamma} = h_f = \frac{LV^2}{gd}\phi_2\left(\frac{k}{d}, \mathbf{R}\right)$$

Comparing the above expression with the Darcy–Weisbach equation for head loss for turbulent flow in pipes $h_f = fLV^2/2g$, we find that the pipe resistance factor $f = \phi_3(k/d, \mathbf{R})$

We observe that the dimensional analysis has given the equation to head loss in pipes but also an insight into the dependence of friction factor f on the relative roughness, k/d and the Reynolds number.

14.13 Method of Repeating Variables

In this method, a set of repeating variables is chosen so that these variables are dimensionally independent of each other and have among them all the dimensions contained in the variables. The number of repeating variables chosen is equal to the rank of the dimensional matrix. The dependent variable should not be chosen as one of the repeating variables. Very often, the most common repeating variables chosen are the density ρ, velocity V and a linear dimension, L. These r number of repeating variables are then combined with each of the remaining variables to form $(n - r)$ number of π-parameters. The method is illustrated using the same example chosen for illustration of the Raleigh method.

As before the pressure loss is a function of the average velocity, V in the pipe, diameter d of the pipe, roughness k, length of the pipe L, fluid properties ρ and μ.

$$\Delta p = f(V, d, k, L, \rho, \mu)$$

The number of variables is 7 and the rank of the matrix is the same as the number of dimensions viz., 3. Let us choose the repeating variables as $\rho[\mathrm{ML}^{-3}]$, $V[LT^{-1}]$, $d[L]$. It is seen that these variables are independent of each other and contain among them all the three dimensions M, L, and T. Now we combine ρ, V, d with each of the remaining variables to get $7 - 3 = 4\pi$-parameters. The correct exponent to each variable is found by inspection so that the resulting parameter is dimensionless. If Δp is chosen to be in the numerator, ρ must be placed in the denominator to cancel the dimension of [M]. Since Δp contains $[\mathrm{T}^{-2}]$, V^2 must be in the denominator to cancel the dimension of [T]. We find the length has balanced itself in the numerator and denominator $[\mathrm{ML}^{-1}\mathrm{T}^{-2}]/[\mathrm{ML}^{-3}][\mathrm{L}^2\mathrm{T}^{-2}]$. Hence, the simplest combination of ρ, V, d with Δp gives $\Delta p/\rho V^2$

Similarly combination with L gives L/d

Combination with k gives k/d

Combination with μ gives $\rho V d/\mu$

The non-dimensional parameter relationship is

$$\Delta p/\rho V^2 = \phi\left[\frac{L}{d}, \frac{k}{d}, \frac{\rho V d}{\mu}\right]$$

14.14 Further Examples in Dimensional Analysis

Rise in Capillary Tube

The rise h in a capillary tube of radius R is to be determined.

It is a function of surface tension σ and the specific weight γ of the liquid besides the radius of the tube. Note that γ contains ρ and g. The functional relationship is

$$h = f(R, \gamma, \sigma)$$

The dimensional matrix is

Variable	h	R	γ	σ
M	0	0	1	1
L	1	1	−2	0
T	0	0	−2	−2

It may be found that all 3×3 determinants of the above matrix have zero value whereas some of the 2×2 matrix determinants are non-zero. The rank r of the matrix is thus 2 and not 3 (number of dimensions). As stated earlier, it is important to find the rank of the matrix, though normally r is equal to the number of dimensions. The dimensionless parameters are thus $4-2=2$. Choosing the repeating variables as h and σ and combining them with each of the remaining variables R and γ, we get

$$h, \sigma, R \text{ gives } \pi_1 = \frac{h}{R}$$

$$h, \sigma\gamma \text{ gives } \pi_2 = \frac{\sigma}{\gamma R^2}$$

$$\text{Hence we have } \frac{h}{R} = f\left(\frac{\sigma}{\gamma R^2}\right)$$

Compare the result so obtained with the theoretical expression

$$\frac{h}{R} = \frac{2\sigma}{\gamma R^2}$$

Delivery Head of a Pump

The head H to which a pump may deliver depends on the discharge Q, rotational speed N of the pump, pump diameter D, the acceleration due to gravity and the fluid properties ρ, μ.The functional relationship is

$$H = f(Q, N, D, g, \rho, \mu)$$

There are 7 variables and the rank of the matrix is 3 giving us $7 - 3 = 4$ non-dimensional parameters. With ρ, Q, D as repeating variables,

ρ, Q, D and H, we get $\dfrac{H}{D}$

ρ, Q, D and N, we get $\dfrac{Q}{ND^3}$

ρ, Q, D and g, we get $\dfrac{Q^2}{gD^5}$ and finally

ρ, Q, D, and μ, we get $\dfrac{\rho Q}{\mu D}$

$$\frac{H}{D} = \phi\left(\frac{Q}{ND^3}, \frac{Q^2}{gD^5}, \frac{\rho Q}{\mu D}\right)$$

$\rho Q/\mu D$ is the Reynolds number which may not be important in highly turbulent flow. It is possible to modify the π-parameters by obtaining the product of $(H/D)(Q/ND^3)^2(gD^5/Q^2) = gH/N^2D^2$ and thus retain only two significant parameters

$$\frac{gH}{N^2D^2} = \phi_1\left(\frac{Q}{ND^3}\right)$$

Drag on a ship

The ship being partly submerged experiences a surface drag due to viscous resistance of water and a wave drag due to free surface distortion due to the movement of the ship The drag force F is thus dependent on the speed of the ship V, acceleration due to gravity, a characteristic length of the ship L and fluid properties ρ, μ

$$F = f(V, g, L, \rho, \mu)$$

With ρ, V, L, as repeating variables, the π-parameters are

ρ, V, L, and F we get $\pi_1 = \dfrac{F}{\rho V^2 L^2}$

ρ, V, L, and g we get $\pi_2 = \dfrac{V^2}{gL}$

ρ, V, L, and μ we get $\pi_3 = \dfrac{\rho VL}{\mu}$

$$\frac{F}{\rho V^2 L^2} = \phi\left(\frac{\rho VL}{\mu}, \frac{V^2}{gL}\right)$$

Thus we find that the Euler number is a function of both Reynolds number and the Froude number.

$$E = \phi(\mathbf{R}, \mathbf{F})$$

Fall velocity of a sphere

The terminal velocity, U of a sphere of diameter, D and density, ρ_s falling freely in a fluid of density ρ and viscosity, μ may be functionally described as

$$U = f(g, D, \rho_s, \rho, \mu)$$

If we choose ρ, g, D as the repeating variables, we get the π-parameters as

$$\frac{U}{(gD)^{1/2}} = \phi\left(\frac{\rho_s}{\rho}, \frac{\rho D^{3/2} g^{1/2}}{\mu}\right)$$

If we choose ρ, U, D as repeating variables, we get

$$\frac{U}{(gD)^{1/2}} = \phi\left(\frac{\rho U D}{\mu}, \frac{\rho_s}{\rho}\right)$$

We may reduce the number of variables since we know that the motivating force, F is the submerged weight of the sphere; $F = (\pi/6)[\rho_s - \rho)gD^3]$. Hence we can replace the variables ρ_s and g by F.

$$F = (U, D, \rho, \mu)$$

Again with ρ, U, D as repeating variables, we get

$$\frac{F}{\rho U^2 D^2} = \phi\left(\frac{\rho U D}{\mu}\right)$$

This is again a relation between the Euler number and the Reynolds number. The Newtonian drag equation is given by $F = C_D(\pi D^2/4)(\rho U^2/2)$. Comparing the two equations, we observe that $C_D = \phi_1(\mathbf{R})$.

If we had chosen the repeating variables as F, D, μ, an alternative form of the π-parameters would have resulted.

$$\frac{F}{\mu U D} = \phi\left(\frac{\rho F}{\mu^2}\right)$$

The term on the left-hand side is known as the Stokes number. If the sphere is falling very slowly, then the acceleration effects around the sphere may be neglected as a close approximation. Whenever there is no acceleration (inertia forces are absent), mass density has no role to play in fluid motion. Hence we can write for very slow motion of the sphere

$$f\left(\frac{F}{\mu U D}\right) = 0 \text{ or } \frac{F}{\mu U D} = \text{constant}$$

which is the Stokes law for slow motion of spheres and the value of the constant determined analytically is 3π.

Discharge over a weir

The discharge $q = Q/B$ per unit width over a weir is dependent on the canal width B_0, width B of the weir, w the height of the weir above the canal bed, head H over weir, acceleration due to gravity, g and fluid properties ρ, μ.

$$Q = f(B_0, B, w, H, g, \rho, \mu)$$

We choose ρ, g, H as the repeating variables

Combined with $q = Q/B$ we get $\pi_1 = Q^2/B^2gH^3$

Combined with B_0, we get B_0/H

Combined with B, we get B/H

Combined with w, we get w/H

Combined with μ, we get $\rho qH/\mu$ (Reynolds number)

$$\frac{Q^2}{B^2gH^3} = \phi\left(\frac{B}{B_0}, \frac{w}{H}\right)$$

After cross multiplying the parameters B/H and B_0/H and ignoring the effects of the Reynolds number and B/H. the expression for Q becomes $Q = C_dB(g)^{1/2}(H)^{3/2}$, where C_d is a function of B/B_0 (lateral contraction effect) and w/H (approach conditions and nappe geometry).

Pipe orifice meter

Discharge Q through an orifice located in a pipe is a function of pressure difference Δp across the orifice, diameter d of the orifice, pipe diameter D, and fluid properties ρ, μ.

$$Q = f\ (\Delta p, D, d, \rho, \mu)$$

Choice of repeating variables ρ, Δp, D and combination with d gives us d/D and combination with μ gives us $D(\rho\Delta p)^{1/2}/\mu$ which is a form of Reynolds number. Finally combination with Q gives us $Q/[D^2(\Delta p/\rho)^{1/2}]$.

$$\frac{Q}{[D^2\,(\Delta p/\rho)^{1/2}]} = \phi\left(\frac{d}{D}, \mathbf{R}\right)$$

$$Q = D^2\left(\frac{\Delta p}{\rho}\right)^{1/2}\phi\left(\frac{d}{D}, \mathbf{R}\right)$$

The coefficient of discharge of an orifice meter C_d is now seen to be a function of d/D and **R**.

$$Q = C_dD^2\left(\frac{\Delta p}{\rho}\right)^{1/2}$$

If d had been chosen instead of D, we get a similar result

$$Q = C_{d1} d^2 \left(\frac{\Delta p}{\rho}\right)^{1/2}$$

Part II Model Similitude

14.15 Physical Models

Model studies of prototypes are frequently undertaken to obtain both qualitative and quantitative information about flow patterns, velocity distributions, discharges, pressures and forces, head losses, efficiencies in the proposed hydraulic structures or hydraulic machines. The experimental data may be used to calibrate weirs, orifices and generally obtain numerical constants in the correlation among π-parameters obtained by dimensional analysis.

Similitude in fluid mechanics would mean the relations between full-scale (prototype) flow and the flow in a smaller but geometrically similar model. However in many of the hydraulic models, the models are not geometrically similar and all the π-parameters may not be reproduced. Even in such cases similitude laws have been evolved though we will largely restrict our attention to geometrically similar models. If reliable quantitative information is to be obtained from model studies, one should have geometric, kinematic and dynamic similarities between the model and the prototype. Kinematic similitude would mean that the flow patterns in the prototype and model (stream lines) would be geometrically similar with similar velocity distributions. Dynamic similitude would imply that the shear stress and pressure intensity ratios at corresponding points must be a constant. By the same token, integrated values of pressure or shear stress on geometrically similar areas or forces in the model and the prototype at corresponding locations must be in similar direction and bear a constant ratio.

The criteria of creating geometric, kinematic and dynamic similarities between the prototype and the model are fully met if the values of π-parameters governing the flow phenomenon are identical both in the prototype and the model. However, simultaneous satisfaction of the requirement of identical values of all the π-parameters creates practical difficulties in modelling. For example, let us consider determining a free surface flow by Reynolds, Froude, and Weber numbers. Should we prescribe that these non-dimensional numbers be identical, the following relationships must be simultaneously satisfied between the model and the prototype.

$$\text{Reynolds criterion: } \rho_m V_m L_m / \mu_m = \rho_p V_p L_p / \mu_p$$

$$\text{Froude criterion: } V_m^2 / g_m L_m = V_p^2 / g_p L_p$$

$$\text{Weber criterion: } \sigma_m / \rho_m V_m^2 L_m = \sigma_p / \rho_p V_p^2 L_p$$

From the preceding conditions, it may be easily derived (eliminating V_m/V_p) in the process that the linear scale ratio L_m/L_p is linked with the fluid properties and gravitational forces in the model and prototype

$$\frac{L_m}{L_p} = \left(\frac{\mu_m \rho_m}{\mu_p \rho_p}\right)^{2/3} \left(\frac{g_p}{g_m}\right)^{1/3} = \left(\frac{\sigma_m \rho_m g_m}{\sigma_p \rho_p g_p}\right)^{1/2}$$

The choice of available fluids for use in the model is very limited—generally air and water are used. Force fields other than the earth's gravitational field are difficult to create. If for convenience, we use the same fluid and work the model in the same gravitational field, we obtain a trivial solution whose linear scale ratio is 1:1.

Hence we observe that the physical and economical restrictions on available space, limited choice of available fluids, and difficulties in creating force fields other than the gravitational force field compel the investigator to make certain sacrifices. Whenever it is impossible to simulate all dimensionless parameters in the model, recourse is made to select only the most significant parameter(s) and disregard the less important ones. Fortunately, in most cases only two forces are of equal magnitude and other forces are of lesser significance. However, non-satisfaction of all π-parameters does introduce some errors in model behaviour and these differences between model and prototype are referred to as scale effects. Correct interpretation of model results and its application to the prototype depends on the experience and intuition of the investigator. The use of the scale model techniques in different situations will be demonstrated by illustrative examples in the following sections.

14.16 Scale Models in Submerged Flows

In submerged incompressible flows, provided boundary geometry is similar, the pressure distribution expressed by the non-dimensional Euler number $(p/1/2\rho V^2)$ is a function of the Reynolds number $(\rho V L/\mu)$.

$$E = f(\mathbf{R})$$

In liquids, cavitation may occur at locations where the pressures are in the vicinity or lower than the vapour pressure, p_v of the liquid. Then the pressure distribution is no longer uniquely governed by the Reynolds number but also by the cavitation number, $\mathbf{K} = (p - p_v)/1/2\rho\ V^2$. Then the relationship becomes

$$E = f(\mathbf{R}, \mathbf{K})$$

In transonic and supersonic flows, since compressibility of gases plays an important role, we have to consider the influence of the Mach number. But in subsonic flows (i.e., at low Mach numbers) where density changes may be ignored without appreciable error, the Euler number remains mainly a function of the Reynolds number. This feature provides us the added facility of using air in the model as the fluid medium to replace liquids in prototypes (provided as stated earlier, cavitation does not take place in the prototype). A variety of problems belong to this category such as pressure flow in conduits, pipe fittings, orifices, mouthpieces, nozzles, pipe inlets, drag on submerged bodies etc.

EXAMPLE 14.1 A1:10 scale torpedo model was tested in a compressed air tunnel (mass density of air = 2.6 kg/m^3 and kinematic viscosity of air = 0.65×10^{-5}m^2/s). The resistance was measured as 0.85 kN when the air velocity was 35 m/s. Find the corresponding velocity and force on the prototype moving under dynamically similar conditions in the ocean (mass density of sea water 1025 kg/m^3 and kinematic viscosity of water = 1.08×10^{-6}m^2/s.

Since this is a submerged model, the Reynolds number of the model and the prototype must be identical.

$$V_m L_m/\nu_m = V_p L_p/\nu_p$$

With

$$V_m = 35 \text{ m/s}, \nu_m = 0.65 \times 10^{-5}\text{m}^2/\text{s}, \nu_p = 1.08 \times 10^{-6}\text{m}^2/\text{s}, L_m/L_p = 10,$$

we have

$$V_p = \left(\frac{L_m}{L_p}\right)(\nu_m/\nu_p)V_m = \left(\frac{1}{10}\right)\left(\frac{1.08 \times 10^{-6}}{0.65 \times 10^{-5}}\right) \times 35 = 0.378 \text{ m/s}$$

Since the Reynolds number is the same both in the model and the prototype, the Euler number will also be identical.

$$p_m/1/2\rho_m V_m^2 = p_p/1/2\rho_p V_p^2$$

Since pressure can be expressed as $p = F/A$, we write

$$F_p/A_p\rho_p V_p^2 = F_m/A_m\rho_m V_m^2$$

$$F_p = \left(\frac{\rho_p}{\rho_m}\right)\left(\frac{V_p}{V_m}\right)^2\left(\frac{A_p}{A_m}\right)F_m = \left(\frac{1025}{2.6}\right)\left(\frac{0.378}{35}\right)(10)^2 0.85 = 3.91\text{kN}$$

EXAMPLE 14.2 The performance of a gate valve in a 3 m dia. water supply pipe is to be tested on a model valve of 15 cm diameter. The velocity of flow in a prototype is 4 m/s. Suggest a suitable model discharge. Kinematic viscosity of water in the model and prototype is 1.0×10^{-2}cm^2/s.

The Reynolds number of the prototype is

$$\mathbf{R}_p = \frac{V_p D_p}{\nu_p} = \frac{4 \times 3}{1 \times 10^{-6}} = 12 \times 10^6$$

and the model velocity with $\mathbf{R}_p = \mathbf{R}_m$ is

$$V_m = \left(\frac{\nu_m}{D_m}\right) 12 \times 10^{-6} = \left(\frac{1 \times 10^{-6}}{0.15}\right) \times 12 \times 10^6 = 80\,\text{m/s}$$

and the discharge is $80 \times \pi \times (0.15)^2/4 = 1.414\,\text{m/s}$.

The velocity and the discharge in the model is too high to be obtained in an experimental laboratory. We know from experience that the pipe resistance coefficients and drag coefficients attain nearly constant values at sufficiently high Reynolds numbers. Hence it is sufficient to test the model at a lower Reynolds number ($< 12 \times 10^6$) making sure however that the flow in the model is turbulent. Choosing a value of $\mathbf{R}_m = 10^6$, we get

$$V_m = \left(\frac{\nu_m}{D_m}\right) \times 10^6 = \frac{80}{12} = 6.67\text{m/s}$$

and the model discharge is

$$Q_m = 6.67 \times \left(\frac{\pi}{4}\right) (0.15)^2 = 0.118\text{m}^3/\text{s}$$

EXAMPLE 14.3 An airfoil of surface area $0.1\,\text{m}^2$ is tested for drag and lift in an air tunnel. At an angle of attack of $5°$, the lift force measured in the model was 32.4 kN at a wind speed of 30 m/s. For a prototype of area $10\,\text{m}^2$, what are the drag and lift forces at a speed of 100 m/s?

The drag and lift coefficients (C_D and C_L) of an airfoil are functions of the shape of the airfoil and angle of attack. If the shape of the aerofoil and angle of attack are kept the same between the model and the prototype, then C_D and C_L are functions of the Reynolds number. However the lift coefficient is not very sensitive to change in the Reynolds number and we can assume that C_L for the model and the prototype are the same in the turbulent range of the Reynolds numbers. Assuming a standard air density of $1.2\,\text{kg/m}^3$, the lift coefficient in the model is

$$(C_L)_m = \frac{L_m}{A_m(1/2\rho_m V_m^2)} = 2 \times 32.4/0.1 \times 1.2 \times 30^2 = 0.6$$

The prototype lift coefficient is also the same and the lift force on the actual wing is

$$L_p = (C_L)_p A_p(1/2\rho_p V_p^2) = 0.6 \times 10 \times 0.5 \times 1.2 \times 100^2 = 36000\,\text{kN}$$

EXAMPLE 14.4 A venture-type contraction has to be tested on a model with a linear scale of 1:10. What pressure should be maintained in the model water tunnel to maintain the same cavitation index between the model and the prototype. Assume that the pressure head at the critical section in the prototype is $(h_c)_p$ and vapour pressure of water corresponds to a head of (h_v) metres of water.

In terms of head of water, the identity in the cavitation parameter can be expressed as

$$[(h_c)_p - h_v]/(V_c)_p^2/2g = [(h_c)_m - h_v]/(V_c)_m^2/2g$$

The Reynolds criterion has also to be satisfied. Assuming the kinematic coefficients of viscosity to be the same

$$\frac{(V_c)_m}{(V_c)_p} = \frac{D_p}{D_m} = 10$$

$$(h_c)_m - h_v = [(h_c)_p - h_v](V_c)_m^2/(V_c)_p^2 = 100[(h_c)_p - h_v]$$

$$(h_c)_m = 100(h_c)_p - 99h_v$$

Free surface models

In turbulent free surface problems commonly met in a large number of practical problems such as rivers, estuaries, open channels, gravity waves in oceans, weirs, hydraulic jumps, spillways etc., gravitational forces are predominant in relation to viscous and surface tension forces. In such free surface problems, Euler number is mainly a function of the Froude number and it is sufficient to ensure that the flow is turbulent in both model and prototype and simulate only the Froude number in the model. Euler numbers will be identical if Froude numbers are equal.

$$\mathbf{E} = f(\mathbf{F}) \text{ or } \frac{p}{(1/2\rho V^2)} = f\left[\frac{V}{(gL)^{1/2}}\right]$$

EXAMPLE 14.5 The discharge over a weir 150 m long and 15 m in height is to be estimated by means of a model built to a scale of 1:25. What are the scales for velocity and discharge? If the pressure measured at a given point in the model on the face of the spillway is 0.01 m, to what pressure does this correspond on the prototype?

The model length is $150/25 = 6$ m and height of the weir $15/25 = 0.6$ m.
The Froude number should be the same for the model and prototype.

$$V_m/(g_m L_m)^{1/2} = V_p/(g_p L_p)^{1/2}$$

Since the model and the prototype act under the same gravitational field, $g_m = g_p$. Hence we have

$$\frac{V_m}{V_p} = \left(\frac{L_m}{L_p}\right)^{1/2} = \left(\frac{1}{25}\right)^{1/2} = \frac{1}{5}$$

The discharge ratio is

$$\frac{Q_m}{Q_p} = \frac{V_m A_m}{V_p A_p} = \frac{V_m L_m^2}{V_p L_p^2} = \left(\frac{1}{5}\right)\left(\frac{1}{25}\right)^2 = \frac{1}{3125}$$

Equating the Euler numbers we get on substituting, $p = \gamma h$ and $g_m = g_p$

$$\frac{h_m}{V_m^2} = \frac{h_p}{V_p^2} \text{ or } h_p = \left(\frac{V_p}{V_m}\right)^2 h_m = 5^2 \times 0.01 = 0.25\,\text{m}$$

EXAMPLE 14.6 A breakwater model is built to a scale of 1:36. The prototype breakwater is to withstand waves of height 8 m and wave period of 9 seconds. What are the corresponding wave height and wave period in the model. If an armour stone of weight W_m is stable in the model, what size of stones should be used in the prototype? Mass densities of fresh water in the model and prototype may be assumed as 998 kg/m^3 and 1025 kg/m^3.

The shape of the model breakwater should be geometrically similar.
Following the Froude criteria, we have for celerity, c of wave

$$\frac{c_m}{c_p} = \left(\frac{L_m}{L_p}\right)^{1/2} = \left(\frac{1}{36}\right)^{1/2} = \frac{1}{6}$$

Since velocity can be written in terms of length and time $c = L/T$ or $T = L/c$. The time scale ratio is

$$\frac{T_m}{T_p} = \left(\frac{L_m}{L_p}\right)\left(\frac{c_p}{c_m}\right) = \left(\frac{1}{36}\right)(6) = \frac{1}{6} \text{ or } T_m = \frac{9}{6} = 1.5\,\text{seconds}$$

The force scale is obtained by equating the Euler numbers.

$$\frac{F_m}{[A_m(1/2\rho_m V_m^2)]} = \frac{F_p}{[A_p(1/2\rho_p V_p^2)]}$$

$$F_p = \left(\frac{\rho_p}{\rho_m}\right)\left(\frac{A_p}{A_m}\right)\left(\frac{V_p}{V_m}\right)^2 F_m$$

$$W_p = \left(\frac{1025}{998}\right)(36)^2(6)^2 F_m = 47918\,W_m$$

Surface tension force

EXAMPLE 14.7 A jet of water emerges from a nozzle 1 cm in diameter at a velocity of 15 m/s. The jet is found to break into a spray at a distance of about 15 cms from the nozzle. The surface tension of the fluid is 50 dynes/cm. Another fluid having a surface tension of 75 dynes/cm, mass density $\rho = 0.9\rho_1$, kinematic viscosity $\nu_2 = 1.1\nu_1$ is forced through a geometrically similar nozzle. If the two nozzle flows are to be similar, determine the scale factors for length, velocity, force and time.

Both Reynolds and Weber numbers must be identical.

Reynolds criterion: $V_1 L_1/\nu_1 = V_2 L_2/\nu_2$

$$\frac{V_1}{V_2} = \left(\frac{L_2}{L_1}\right)\left(\frac{\nu_1}{\nu_2}\right) = 0.9091\left(\frac{L_2}{L_1}\right)$$

Weber criterion: $\rho_1 V_1^2 L_1/\sigma_1 = \rho_2 V_2^2 L_2/\sigma_2$

$$\left(\frac{V_1}{V_2}\right)^2 = \left[\left(\frac{\rho_2}{\rho_1}\right)\left(\frac{L_2}{L_1}\right)\left(\frac{\sigma_1}{\sigma_2}\right)\right] = \left[0.9\left(\frac{L_2}{L_1}\right)\left(\frac{50}{75}\right)\right] = \left[0.6\left(\frac{L_2}{L_1}\right)\right]$$

Equating the values for V_1/V_2, we get

$$(0.9091)^2\left(\frac{L_2}{L_1}\right)^2 = 0.6\left(\frac{L_2}{L_1}\right)$$

$$\frac{L_2}{L_1} = \frac{0.6}{(0.9091)^2} = 0.726$$

The diameter of the second nozzle has to be 0.726 cm.

$$\frac{V_2}{V_1} = \frac{1}{0.9091 \times 0.726} = 1.515;\ V_2 = 1.515 V_1 = 1.515 \times 15 = 22.73 \text{ m/s.}$$

Force scale:

$$\frac{F_2}{F_1} = \frac{A_2 \rho_2 V_2^2}{A_1 \rho_1 V_1^2} = \left(\frac{L_2}{L_1}\right)^2\left(\frac{\rho_2}{\rho_1}\right)\left(\frac{V_2}{V_1}\right)^2 = 0.726^2 \times 0.9 \times 1.515^2 = 1.089$$

Time scale:

$$\frac{T_2}{T_1} = \frac{(L_2/V_2)}{(L_1/V_1)} = \left(\frac{L_2}{L_1}\right)\left(\frac{V_1}{V_2}\right) = \frac{0.726}{1.515} = 0.479$$

EXAMPLE 14.8 A jet of water emerges from a nozzle 1 cm in diameter at a velocity of 15 m/sec. The jet is found to break into a spray at a distance of about 15 cms from the nozzle. The surface tension of the fluid is 50 dynes/cm. Another fluid having a surface tension of 75 dynes/cm, mass density $\rho_2 = 0.9\,\rho_1$, and kinematic viscosity $\nu_2 = 1.1\,\nu_1$ is forced through a geometrically similar nozzle. If the two nozzle flows are to be similar determine the scale factors for length, velocity, force and time.

Both Reynolds and Weber numbers must be identical.

Reynolds criterion: $V_1 L_1/\nu_1 = V_2 L_2/\nu_2 : V_1/V_2 = (L_2/L_1)(\nu_1/\nu_2) = 0.9091(L_2/L_1)$

$$\text{Weber criterion: } \rho_1 V_1^2 L_1/\sigma_1 = \rho_2 V_2^2 L_2/\sigma_2,$$

$$(V_1/V_2)^2 = [(\rho_2/\rho_1)(L_2/L_1)(\sigma_1/\sigma_2)] = [0.9(L_2/L_1)(50/75)] = [0.6(L_2/L_1)]$$

Equating the values for V_1/V_2, we get $(0.9091)^2(L_2/L_1)^2 = 0.6(L_2/L_1)L_2/L_1 = 0.6/(0.9091)^2 = 0.726$

The diameter of the second nozzle has to be 0.726 cm.

$V_2/V_1 = 1/0.9091 \times 0.726 = 1.515$; $V_2 = 1.515\ V_1 = 1.515 \times 15 = 22.73$ m/sec.

Force scale: $F_2/F_1 = A_2\rho_2 V_2^2/A_1\rho_1 V_1^2 = (L_2/L_1)^2(\rho_2/\rho_1)(V_2/V_1)^2 = 0.76^2 \times 0.9 \times 1.515^2 = 1.089$

Time scale: $T_2/T_1 = (L_2/V_2)/(L_1/V_1) = (L_2/L_1)(V_1/V_2) = 0.726/1.515 = 0.479$

EXERCISES

14.1. A mass m is suspended from a weightless and frictionless spring having a spring constant k. If the mass undergoes a vertical oscillation with a period T and amplitude a_o, find a relationship for period T as a function of m, a_o, and k.

14.2. To measure the discharge Q in a pipe of diameter D_1, a venturimeter with a throat diameter D_2 is provided. The pressure difference between two points at the inlet and at the throat of the venturimeter is measured as Δp. The density and dynamic viscosity of the fluid are ρ and μ. Form the dimensionless parameters governing the discharge.

14.3. In Section 14.12, if we reframe the problem for a smooth pipe ($k = 0$) and state that the friction force exerted on the interior surface area A is dependent on the average velocity, V in the pipe, diameter D, fluid properties, ρ and μ, find the resulting dimensionless parameters. Choose ρ, V, A as repeating variables. If ρ, V, D are chosen as the repeating variables what is the alternate form of the dimensionless parameters.

14.4. When a drop of liquid falls into a pool, a small column of liquid splashes to a height h above the pool level. Assuming that the height h depends on the mass of the drop m, its velocity of fall V, surface tension σ, viscosity μ, density ρ and acceleration due to gravity g, find the relevant π-parameters.

14.5. The power P required to pump a liquid of density ρ to a height H is also a function of diameter D of the impeller and rotational speed N. Find the dimensionless groups. Should we also include g ?

14.6 A thin film of thickness b of a viscous liquid (density ρ and kinematic viscosity, ν) flows down a flat plate inclined at an angle θ to the horizontal (slope, S; motivating force is $g \sin\theta$ or gS). Find the dimensionless relationship for the velocity V of the flow.

14.7 A thin film of viscous liquid (density ρ and viscosity μ) occupies the space b between a flat stationary plate and a rotating circular plate above. Find an expression for the torque τ exerted on the circular plate of radius a rotating at speed, N.

14.8. A moveable bed consists of sand grains of uniform diameter d_s. The difference in the specific weights of sand grain and water is $\Delta\gamma_s$. The fluid density and viscosity are ρ and μ respectively. If the grains just begin to move when the bed shear stress is τ_{oc}, find its relationships with other relevant variables.

14.9. A drop of liquid of volume $\forall$ forms at the end of a tube of diameter D before it frees itself from the surface tension forces and falls under the action of gravity. Derive an expression for $\forall$ as a function of D, σ and γ.

14.10. A prototype aircraft has a cruising speed of 480 km/hour at an altitude where the pressure is 0.75 atmospheres. A model of the aircraft is made to a 1:10 scale and tested in a wind tunnel under a pressure of 12 atmospheres. If the model experiences a drag force of 500 N, what is the corresponding drag force on the prototype? (Use the relation $p/\rho = \text{constant}$).

14.11. Oil of kinematic viscosity $1.2 \times 10^{-4}\,\text{m}^2/\text{s}$ and density $800\,\text{kg/m}^3$ is flowing in a pipe of diameter 1 metre. The discharge is $1\,\text{m}^3/\text{s}$. What should the velocity and discharge in a model pipe using water as the test fluid ($\rho = 998\,\text{kg/m}^3$, $\nu = 1 \times 10^{-6}\,\text{m}^2/\text{s}$) to a scale of 1:20 be? Find the model velocity and discharge. If the pressure loss across an orifice diameter installed in the model is $50\,\text{N/m}^2$, find the corresponding pressure drop across a similar orifice in the prototype.

14.12. Cavitation characteristics in a siphon is to be modelled in a pipe of 1:10 scale. If the temperature of water in the model and the prototype are the same, find the pressure to be maintained in the model to simulate the cavitation characteristics. Vapour pressure of water at the given temperature is 4.20 kPa.

14.13. A thick viscous oil ($\nu = 2.4 \times 10^{-4}\,\text{m}^2/\text{s}$) flows through a 2.5 cm dia. pipe line at an average velocity of 1 m/s. The pressure drop observed is 30 Pascals per metre length of pipe. Kerosene ($\nu = 0.732 \times 10^{-4}\,\text{m}^2/\text{s}$) flows through the same pipe under dynamically similar conditions. What is the average velocity of the kerosene and the corresponding pressure drop? Under dynamically similar conditions, the Reynolds number and pipe roughness will be the same.

14.14. Air at atmospheric pressure ($\nu = 1.5 \times 10^{-5}\,\text{m}^2/\text{s}$ and $\rho = 1.2\,\text{kg/m}^3$) flows in a pipe of diameter 20 cms with an average velocity of 8 m/s. In the model having a scale ratio of 1:5, water ($\nu = 1 \times 10^{-6}\,\text{m}^2/\text{s}$ and $\rho = 998\,\text{kg/m}^3$) is used as the fluid medium. What is the velocity of water required for dynamic similarity and what is the pressure drop ratio between the model and the prototype?

14.15. A tank with a water depth H is emptied by an orifice at the bottom of the tank. The time required to empty the tank is a function of the cross-sectional area of tank, A, the area of orifice and g. Find the dimensionless parameter for the time of emptying, T.

In a geometrically similar model of 1:100 scale, the time taken to empty the tank is 10 minutes. What is the corresponding emptying time in the prototype?

14.16. The lift and drag force of an aerofoil of a given shape is a function of chord length C, span L, velocity V_o, speed of sound c, angle of attack, α and fluid properties ρ and μ. Form the dimensionless groups.

14.17. If both the Reynolds number and the Froude number are significant, determine the scale for kinematic viscosity in terms of the length scale if the gravitational field remains the same both for the model and the prototype.

14.18. A tidal estuary model has been built to a scale of 1:900. Determine the scale ratios for velocity. If the tidal period is 12 hours, find the tidal period in the model.

14.19. A cable 1.5 cm in dia. is found to resonate and produce a singing sound when subjected to a cross wind of 6 m/s. The resonant frequency is 160 Hz. If this cable is modelled by using a wire of 0.5 cm dia. in water, determine the velocity of flow in the model which will cause resonance, the natural frequency of the wire being 96 Hz. Since the Strouhal number is a function of the Reynolds number, check for the similarity of Reynolds numbers between the model and the prototype.

14.20. An ogee-shaped spillway 25 m long and 10 m high is to be tested on a model scale of 1:25. When the head over the weir is 0.2 m, the measured discharge is 1.55 m^3/s. What is the corresponding full-scale discharge? If the force measure on the full length of the model spillway is 3900 N, what is the force on the prototype per unit length? What is the velocity scale?

14.21. A scale model of a solar dome beneath a submarine is made to a scale of 1:5. If laboratory tests reveal that the drag on the dome is 15 N at a speed of 18 m/s, find the corresponding velocity and drag of the prototype.

14.22. Determine the head and discharge of a 1:4 model of a centrifugal pump that supplies at the rate of 0.8 m^3/s at 3 m head when running at 240 rpm. The model operates at 1200 rpm.

Solutions

Chapter 1

1.1 (a) 0.8016, (b) 7845 N/m^3, (c) 0.00125 m^3/kg

1.2 (a) 11.76 kg/m^3, (b) 13.33, (c) 0.0012

1.3 $y = 0 : \tau = 0, u = 0.148$ m/s, $y = 5$ mm $: \tau = 7.5$ N/m^2, $u = 0.326$ m/s

1.4 $8\pi^3 \mu a^3 L N^2 / 3600e$

1.5 $\pi^2 \mu a^4 / 60e$

1.6 0.382 m/s

1.7 9.05 N

1.8 $\rho = 1002.97$, Error $= 0.2927\%$

1.9 $\rho = 1076$ kg/m^3, $p = 105878$ kN/m^2

1.10 $p = 0.30$ kN/m^2, $\rho = 0.34$ kg/m^3

1.11 $p = 0.396$ kN/m^2, $\rho = 0.34$ kg/m^3

1.12 $p = 2905.4$ kN/m^2

1.13 $K = 1.266 \times 10^9$ N/m^2, $K_T = 1/K$

1.14 (a) 0.606 m^3 at 3 bars; 0.227 m^3 at 8 bars, (b) 0.829 m^3 at 3 bars; 0.411 m^3 at 8 bars

1.15 9923.8 N/m^2

1.16 3650 N m/s

1.17 $\theta = 60^\circ$

1.18 $\Delta h = 5.96$ mm

1.19 $d = 0.0149$ mm

1.20 0.0311 N/m

1.21 1.46 mm

1.22 0.00219 N

1.23 10.71 mm

Chapter 2

2.1 $p_A = -44903\,N/m^2$, $p_B = 14709\,N/m^2$, $p_C = +4903\,N/m^2$, $p_D = 18631.4\,N/m^2$

2.2 Level in A is 0.5 m above orifice, level in B is 1.5 m above orifice, pressure at C is atmospheric and the head over the orifice is constant at 0.5 m, pressure at D is –1 m.

2.3 $F_{H_1} = 110.95\,kN$, $F_{H_2} = 1202.76\,kN$, $F_{V_1} = 28.05\,kN$, $F_{V_2} = 3.92\,kN$, $M = 11.75\,kN\,m$, anti-clockwise

2.4 Tension at bottom tie is 120 kN, top tie 60 kN

2.5 $P = 299.70\,kN$

2.6 $P = \gamma HD/2(H^2 + D^2/4)^{1/2}$ acting at $5D/8$ from A along AB

2.7 $F_H = 919.26\,kN$ acting at 1.443 from B, $F_V = 752.9\,kN$, $R = 1188\,kN$ acts in a radial direction inclined at 39.32° to the horizontal

2.8 $F_H = 19.612\,kN$ at 0.33 m from base, $F_V = 30.81\,kN$ at 0.42 m from centre, $W = 60.22\,kN$

2.9 $F_{H_1} = 220.64\,kN$ at 1 m above base from left, $F_{H_2} = 55.16\,kN$ at $1/3$ m above base, $F_{V_1} = 173.29\,kN$ and $F_{V_2} = 86.6\,kN$ both at 0.637 m from centre.

2.10 $F = 498.64\,kN$ at 7.076 m along y-axis

2.11 $F = 11.56\,kN$ and $y_p = 1.768\,m$ along y-axis

2.12 (a) $\tan\theta = 0.3058$, Depth at A = 2.4675 m and at B = 0.5825 m, $p_A = 4.57\,kN/m^2$, $p_B = 18.695\,kN/m^2$, (b) $a_x = 4.905\,m/s^2$, $y_B = 3.0\,m$

2.13 Depth at A = 1.5636 m and at B = 1.4364 m

2.14 0.125 m and 1.875 m, $\theta = 15.31°$

2.15 (a) 0.994 m, 3.006 m, (b) $N = 84.6$ rpm, 0.0 m, (c) 59.82 rpm, 3.0 m (d) 103.6 rpm

2.16 $F = 6.58\,N$

2.17 $g\omega = 3.62$ radians/s

2.18 $p_B = 1.473$ bar

2.19 $p_B - p_A = 490.3\,N/m^2$

2.20 (a) BG = 0.161 m, BM = 0.175 m; therefore stable, (b) No, BM = 0.0766 m

2.21 BG = 1.98 m, BM = 1.471 m; unstable.

2.22 Unstable

2.23 (a) 72.7 kN m, (b) Reduction in restoring moment 132 kN m, (c) Division of liquid cargo hold improves stability

2.24 Stable, OG = 1.125 m, OB = 1.054 m, OM = 1.159 m

Chapter 3

3.1 No

3.2 No

3.3 $a = (x^2 - y^2)/(x^2 + y^2)^2$, $b = (y^2 - x^2)/(x^2 + y^2)^2$, $h = 4xy/(x^2 + y^2)^2$, $\omega_{xy} = 0$

3.4 $\Psi = c_1 y^2 + c_3 x^2 - c_2 xy + \text{constant}$

3.6 $\Gamma = 0$

3.7 (a) $a_1 + c_3 = b_2$, (b) $w = xz(d - 2a) + ez^2/2 + f(x, y)$

3.8 All satisfy equation of continuity

Chapter 4

4.1 $du/dt = 21.6\,\text{m/s}^2$, $dv/dt = 18\,\text{m/s}^2$, $dw/dt = 11\,\text{m/s}^2$

4.2 $p = -31\rho$

4.3 $a_r = V^2 D/2h^2$

4.4 $\Psi_B - \Psi_A = 128\,\text{m}^3/\text{s}$, $\Psi_C - \Psi_B = 25\,\text{m}^3/\text{s}$

4.5 $\ln[(x - 3)/(2 - y)] = 4z^{1/2}$

4.6 $dV/dt = (2V_o^2/L)(1 + 2x/L)$

4.7 (a) $\Psi = 2xy - x$, $\Psi = -(C \sin 2\theta)/r^2$

4.8 (a) $\Psi = Cr^2 \cos 2\theta$, (b) $(r + 2/r) \cos\theta$

Chapter 5

5.2 (b) $\Psi = m \tan^{-1}[2xy/(x^2 - y^2 - a^2)]$, $\Phi = m \ln(r_1 r_2)$

5.3 (a) $(y^2 - x^2)/2$, (b) $-2xy$, (c) $Ar^{\pi/\alpha} \cos(\pi\theta/\alpha)$

5.4 (a) $3x^2 y - y^3$, (b) $2xy$

5.6 $x = -3.7$ m, and $x = +2.7$ m

5.7 $F = -\rho\pi m^2/a$

5.8 $\mu = 33.75\,\text{m}^3/\text{s}$, $L = 370.8$ N/m, $\theta = -4.06°$ and $184.06°$

5.9 (a) 36.6 kN/m, (b) $\theta = 6°$ and $174°$, (c) 2865 rpm

5.10 (a) $\theta = 198.5°$ and $341.5°$, $L = 61.80$ N/m, (b) $\Gamma = 31.93\,\text{m}^2/\text{s}$, (c) $\Gamma = 1.62\,\text{m}^2/\text{s}$

5.12 $L = 2x$, Half width $r_b = a \cot\,(Ur_b/2\,\text{m})$

5.13 $x = \pm[(\Gamma a/\pi U) - a^2]^{1/2}$.

Chapter 6

6.1 $x = 42.65\,\text{m}, \theta = 65.09^\circ$

6.2 $r = 14.42\,\text{m}, r = 5.097\,\text{m}, y_{\text{max}} = 2.548\,\text{m}$

6.3 $V = 11.437\,\text{m/s}, x = 5.77\,\text{m}$

6.4 $\theta = 39.4^\circ$ and 59.6°

6.5 $H = H_o/2, x_{\text{max}} = H_o$

6.6 $Q = 3.19\,\text{l/s}$

6.7 $C_d = 0.7545, \Delta p = 45\,\text{kN/m}^2$

6.8 $H = 0.82\,\text{m}$

6.9 $Q = 0.0539\,\text{m}^3/\text{s}, d = 15.18\,\text{cm}, 6.71\,\text{m}, 0.178\,\text{m}^3/\text{s}$

6.10 $p_1 = 315\,\text{kN/m}^2, p_2 = 289.7\,\text{kN/m}^2$

6.11 $d = 2.945\,\text{cm}$

6.12 $\Delta p = 8.75\,\text{kN/m}^2$

6.13 2.36 cms of water, loss = 0.47 cm of water

6.14 58.2 l/s

6.15 123.15 s

6.16 19.1 s; 35.18 s

6.17 $Q = 41,75\,\text{m}^3/\text{s}$

6.18 $Q = 20.3\,\text{m}^3/\text{s}$

6.19 (a) $7.52\,\text{m}^3/\text{s}$, (b) $5.15\,\text{m}^3/\text{s}$

6.20 $Q = 2.316\,\text{m}^3/\text{s}$

6.21 Vertex to be 1.141 m above base, depth = 1.487 m

6.22 $Q = 0.388\,\text{m}^3/\text{s}$

6.23 (a) $Q = 1.786\,\text{m}^3/\text{s}$, (b) $Q_m = 2.020\,\text{m}^3/\text{s}, V = 1.68\,\text{m/s}$

6.24 Height of weir = 1.108 m

Chapter 7

7.1 (a) 6.28 N, (b) 402 N, (c) 904.8 N

7.2 $Q_1 = 0.0221\,\text{m}^3/\text{s}, Q_2 = 0.0074\,\text{m}^3/\text{s}, F = 382.6\,\text{N}$ moving plate: $F = 680.16\,\text{N}$, $P = 3400.8\,\text{N m}$

7.3 Stationary: $F_x = 938\,\text{N}, F_y = 251.3\,\text{N}$, Moving : $F_x = 527.6\,\text{N}, F_y = 141.37\,\text{N}, V_{2x} = 15.98\,\text{m/s}, V_{2y} = 15\,\text{m/s}, V_2 = 21.92\,\text{m/s}, V_2 = 0$ when $\theta = 109.47^\circ$, when V_2 makes 120°, vane angle $\theta = 137^\circ$

7.4 10.12 kN

7.5 (a) $F_x = F_y = 8.335$ kN; $R = 11.787$ kN; (b) $F_x = 8.335$ kN, $F_y = 7.96$ kN, $R = 11.53$ kN

7.6 (a) Horizontal bend: $F_x = 5.41$ kN, $F_y = 2.89$ kN, $R = 6.13$ kN, vertical bend : $F_x = 7.07$ kN, $F_y = 3.313$ kN, $R = 7.81$ kN

7.7 5200 kN

7.8 $T = 221.36$ s, $V_T = 1690$ m/s, $h = 99.5$ km, $h_{\max} = 245.16$ km

7.9 (a) 1.140 N m, (b) 275.6 rpm, (c) No, (d) No, (e) yes, anti-clockwise

7.10 Expansion: 0.717 m, Contraction: 0.566 m

7.11 $T = 65466$ N m, $P = 2056.7$ kW

7.12 $D = 2.0$ m, $d = 0.2$ m, $Q = 2.16\ \text{m}^3/\text{s}$, number of buckets = 20

7.13 $F = 48.48$ kN, $V_j = 3.46$ m/s, $V = 7.16$ m/s

7.14 $V = 28.0$ m/s

Chapter 8

8.1 $V_{av} = 1.25$ m/s, $Q = 0.6136$ l/s, pressure gradient = 25600 N/m

8.2 $R = 0.884$ cm

8.3 19.635 l/s

8.4 Loss = 4.53 m, $\tau_o = 20\ \text{N/m}^2$, $Q = 2.62$ l/s

8.5 $Q = Ut - gt^3/3\nu$

8.6 $T = 1.11$ N m

8.7 $\mu = 0.00514\ \text{N s/m}^2$

8.8 16.61 m, 42.59 m

8.9 0.0216 N

8.10 **R** = 1698 (laminar), $f = 00377$, $\Delta h = 0.00111$ m

8.11 18.39 mm/s

8.12 $w = [U/\ln(b/a)]\ln(r/a)$

8.13 $h_f = 13.05$ mm/m

8.14 86.4 rpm, $P = 0.0157$ W

8.15

$$u_1 = \left(\frac{Uy}{a}\frac{\mu_2}{\mu_1 + \mu_2}\right), u_2 = \frac{U}{a}\frac{\mu_1}{\mu_1 + \mu_2}(y - 2a) + U$$

8.16
$$u_1 = -\frac{K}{2\mu_1}y^2 + \frac{Ka(\mu_2 - \mu_1)}{2\mu_1(\mu_2 + \mu_1)} + \frac{Ka^2}{\mu_2 + \mu_1}$$
$$u_2 = -\frac{K}{2\mu_2}y^2 + \frac{Ka(\mu_2 - \mu_1)}{2\mu_2(\mu_2 + \mu_1)} + \frac{Ka^2}{\mu_2 + \mu_1}$$

8.18 $\delta_{x=0.375} = 0.53$ cm, $\delta_{x=0.75} = 0.75$ cm. Drag $= 0.12$ N, $\tau_o = 0.0282$ N/m^2

8.19 $x = 1$ m, $\delta = 1.00$ cm, $\tau_o = 0.0525$ N/m^2, $D = 1.408$ N

8.20 3.146

Chapter 9

9.1 $\delta = 5.16$ cms, $\delta_1 = 0.395$ cm, $D = 2.186$ kN

9.2 $D_{\text{lam}} = 2.34$ N, total drag $= 20.085$ N

9.3 Drag $= 139$ N, $\delta = 0.175$ m, $\tau_o = 0.0821$ N/m^2

9.4 $\Delta x = 15.52$ m, $F = 115$ N

9.5 Laminar: $L_e/D = 0.0707(U_{av}D/\nu)^{1/2}$, turbulent: $L_e/D = 1.31(UD/\nu)^{0.167}$

9.6 Smooth: $U_{\max} = 3.56$ m/s, $V_{av} = 3.10$ m/s, $Q = 0.0974$ m^3/s

9.7 $V = 0.062$ m/s, $V = 1.51$ m/s

9.8 $F = 0.0103$, $u_{\max} = 4.538$ m/s, $u_{av} = 3.961, 4.2, 4.458$ m/s

9.9 $F = 0.0115$, $h_f = 4.86$ m, $P = 19.1$ kW

9.10 $V = 2.0$ m/s, $f = 0.0119$, $u_* = 0.0771$ m/s, $\tau_o = 5.94$ N/m^2

9.11 $Y = 0.223\,a$

9.12 $f = 0.0304$, $u_{\max} = 2.35$ m/s

Chapter 10

10.1 $f = 0.023$, $V = 1.74$ m/s, $Q = 30.75$ l/s

10.2 $Q = 0.85$ m^3/s

10.3 $D = 0.56$ m

10.4 27.5 m

10.5 $d_{\text{new}} = 1.21 d_{\text{old}}$

10.6 2.81 m

10.7 518.8 m

10.8 $Q_1 = 2.34$ m^3/s, $Q_2 = 0.42$ m, $Q_3 = 1.24$ m^3/s

10.9 $Q = 0.676$ m^3/s

10.10 4.36 m

10.11 $Q_{AB} = 0.49\,m^3/s$, $Q_{BC} = 0.44\,m^3/s$, $Q_{CD} = 0.05\,m^3/s$

10.12 $Q_{AJ} = 0.58\,m^3/s$, $Q_{BJ} = 0.46\,m^3/s$, $Q_{JC} = 1.04\,m^3/s$

10.13 (a) AB = 2.54, BC = 0.54, AC = 2.93, DC = 1.53, AD = 2.53: all in m^3/s,

(b) AB = 2.51, BC = 2.01, CD = 1.26, DE = 0.26, AF = 2.49, FE = 0.74, FC = 1.25, all discharges in m^3/s

10.14 $H = 46.73$ m, $P = 3.71$ kN m/s

10.15 $Q = 2.45$ l/s, 2.0 l/s, 0.45 l/s

10.16 $\Delta H = 307.39$ m, $T = 3.377$ s, $\Delta H_{max} = 327.21$ m

10.17 $T = 5.013$ m/s

10.18 $T = 448.57$ s, $Z_{max} = 22.72$ m, $\tau = 290.28$ s

Chapter 11

11.1 (a) $C_D = 1.2$, $D = 7.2$ N/m; (b) $C_D = 0.32$, $D = 1596.8$ N/m

11.2 $D = 6482.5$ N$(C_D = 0.35)$, $B.M. = 64825$ N m

11.3 Square plate $(C_D = 1.16)$, $D = 652.5$ N, 5 rect.pieces $(C_D = 1.2) D = 675$ N

11.4 $C_D = 1.18$, $D = 6145$ N

11.5 17.13 m/s $(C_D = 0.4)$

11.6 6.76 m/s $(C_d = 0.2)$

11.7 3.89 m/s $(C_D = 1.4)$

11.8 (a) 164.93 kN $(C_D = 1.4)$, (b) 40.55 kN $(C_D = 0.34)$

11.9 150 N/m, $(C_D = 0.1)$.

11.10 240 N $(C_D = 0.08)$

11.11 $D = 14.14$ N $(C_D = 0.68)$; $L = 16.64$ N $(C_L = 0.8)$

11.12 5.91 N/m

11.13 $\Gamma = 47.82\,m^2/s$, efficiency 91.5%

11.14 $L = 75000$ N, $C_{Di} = 0.012$, $C_D = 0.0212$, $D = 39, 72$ N

11.15 C_{Di} (for $l/c = 6$) = 0.0298, $C_D = 0.0498$: C_{Di} is reduced by 50% for a change of l/c from 5 to 10 $[(1/5 - 1/10)/1/5]$; $(C_D)_{l/c=10} = 0.015 + 0.025/2 = 0.0275$

Chapter 12

12.1 (a) $y = 2.75$ m, $V = 2.215$ m/s, $Q = 18.28\,m^3/s$ (b) $S_o = 0.002$: $y_n = 2.25$ m/s, $V = 3.067$ m/s, $Q = 23.23\,m^3/s$

12.2 $Q = 2.21\,\text{m}^3/\text{s}$

12.3 $\Delta h = -0.156\,\text{m}$

12.4 $q = 1.75\,\text{m}^3/\text{s/m}$; gate opening $= 0.5\,\text{m}$

12.5 u/s depth increases by 0.42 m

12.6 55.6 kN

12.7 (a) 2.05 m, $\Delta E = 1.17\,\text{m}$ (b) 2.34 m, $\Delta E = 0.94\,\text{m}$

12.10 (a) $S_o = 0.00693$, (b) $S_c = 0.00345$, (c) $S_c = 0.00358$, $Q = 4.32\,\text{m}^3/\text{s}$

12.11 (i) 0.294 m, (ii) 1.13 m, (iii) 0.285 m, (iv) 0.640 m, (v) 0.593 m

12.12 Q_{max} at $y/D = 0.94$; V_{max} at $y/D = 0.81$

12.13 (a) $n_{\text{smooth}} = 0.00974$, (b) $n_{\text{rough}} = 0.0159$; (i) $n = 0.128$, $Q = 0.183$ m/s, (ii) $n = 0.0135$, $Q = 0.173\,\text{m}^3/\text{s}$

12.15 Saving in excavation $= 131.7\,\text{m}^3/\text{km}$ length of channel, saving in lining $= 408\,\text{m}^2/\text{km}$

12.16 $y = 4.68\,\text{m}$, $B = 3.88\,\text{m}$, $S_o = 0.000073$

12.17 $y = 1.22\,\text{m}$

12.18 $y_{\text{steep}} = 0.443\,\text{m}$, jump height $= 0.82\,\text{m}$, $y_{\text{mild}} = 0.93\,\text{m}$, jump occurs on steep slope followed by S_1 curve

12.19 M_2 curve, accuracy of L depends on the closeness of intervals chosen. Approximate distance $= 2150\,\text{m}$ from $y_c = 0.4\,\text{m}$ to $y = 1.0\,\text{m}$

12.20 $y = 1.68\,\text{m}$

12.21 $y_n = 2.37\,\text{m}$, $S_o = 0.00256$

12.22 $y_1 = 0.128\,\text{m}$, $y_2 = 1.204\,\text{m}$

12.23 $y_{n1} = 0.715\,\text{m}$, sequent depth $= 1.535\,\text{m}$, jump occurs on mild slope preceded by a short M_3 curve

12.24 (a) $y_2 - 5.79\,\text{m}$, (b) $L_j = 30\,\text{m}$ approx, (c) 80.3%, (d) strong jump

12.25 $\theta = 20.15^\circ$, $\beta = 32^\circ$, $\mathbf{F} = 1.283$

12.26 $V_w = 12.75$ m/s

12.27 $h = 1.758\,\text{m}$, $V_w = 5.23$ m/s, $\Delta h = 0.258\,\text{m}$

12.28 $c = 6.56$ m/s, $V_2 = 2.0$ m/s

Chapter 13

13.1 Helium: $c = 1055.29$ m/s; Hydrogen: $c = 1371$ m/s

13.2 218.4°C

13.3 747.86 km/hr

13.4 $p_o = 408$ kPa, $T_o = 320.4$°K

13.5 $\mathbf{M}_2 = 0.598$, $p_2 = 299.7$ kPa, $T_2 = 395$ K, $-\Delta s = 76$ J/kg K

13.6 $p_2 = 120.47$ N/m^2, $T_2 = 56.71$°C, $V_2 = 289.72$ m/s

13.7 (a) $dm/dt = 0.0602$ kg/s (b) $dm/dt = 0.0651$ kg/s

13.8 $A^* = 2.313$ m^2, $A/A^* = 2.64$, $p/p_o = 0.059$, $\rho/\rho_o = 0.132$, $T/T_o = 0.444$, $V =$ 708 m/s

13.9 $\mathbf{M}_{exit} = 0.38$, $p_{exit} = 424$ kPa, $(p_o)_{exit} = 469$ kPa, Exit area = 125 cm^2

13.10 $\mathbf{M}$(ahead of shock) = 1.507, $\mathbf{M}$(after shock) = 0.7, $\mathbf{M}_{exit} = 0.53$, $p_{exit} = 322$ kPa

13.11 14.12 m($f = 0.0148$)

13.12 0.00028 kg/s

13.13 $\mathbf{M}_{exit} = 0.874$, $L_{max} = 2991$ m, $p/p_T = 4.34$

13.14 $T_2/T_1 = 1.116$, $p_2/p_1 = 1.41$, $\rho_2/\rho_1 = 1.317$, $V_2/V_1 = 0.76$

13.15 16.96 m

Chapter 14

14.1 478.8 N/m^2

14.3 0.473 N/m and 0.394 N/m

14.4 419957 N m

14.5 (a) C has the dimension of $[L^{1/2}/T]$ which is the dimension of $g^{1/2}$. We can rewrite the equation as $C_1(gRS)^{1/2}$: C_1 is now dimensionless. (But FPS system, $V =$ $135.83(RS)^{1/2}$

14.6 $a/(1 + m/R^{1/2}]$ has the dimension of $[L^{1/2}/T]$. If m has the dimension of $[L^{1/2}]$, then hs the dimension of $[L^{1/2}/T]$

14.7 The coefficient 0.84 has the dimension $[L^{0.36}/T]$, $V = 0.548d^{0.64}$ m/s

14.8 $P = 4.83Q^{1/2}$ since the coefficient has the dimension of $[T^{1/2}/L^{1/2}]$

14.9 (a) $F/\rho V^2L^2$, (b) $\rho V^2/E$, (c) $F/\rho g^4T^6$, (d) V/nD, $\rho VD/\mu$, (e) $T/(gL)^{1/2}$ (Note that the rank of the matrix is 2. Only ρ contains and cannot combine with any other variable and from this, show that the mass of the pendulum has no part to play in determining the period of oscillation of the pendulum), (f) $C_p\mu/k$, (g) $\tau/\gamma R$, S

14.10 $F/\rho U^2D^2 = f(l/D, \rho UD/\mu)$

14.11 With ρ, ω, c as repeating variables: $F_T/\rho\omega^y D^4 = f_1(V_o/\omega D, \rho\omega D^2/\mu, c/\omega D)$

With ρ, V_o, D as repeating variables: $F_T/\rho V_o^2 D^2 = f_2(V_o D/\omega, \rho V_o D/\mu, V_o/c)$

14.12 $p/\rho\omega^2r^2 =$ Constant

14.13 $h/x =$ slope of water surface $= f(a_x/g)$

14.14 $Q/g^{1/2}H^{5/2} = f(\theta)$, neglecting velocity of approach; $Q/g^{1/2}H^{5/2} = f(V_o/(gH)^{1/2}, \theta)$ including velocity of approach

14.15 (d) The first parameter cannot be obtained by a combination of other parameters, and (e) the number of π-parameters have not remained the same

14.16 $gH/U^2 = f(gL/U^2, gt/U)$

14.17 $c = 4$ constant $(\sigma/\rho\lambda)^{1/2}$

14.18 $T(Km)^{1/2} =$ constant

14.19 With ρ, $g\Delta p$, and D_2 as repeating variables: $Q/D_1^2(\Delta p/\rho)^{1/2}$, D_1/D_2, $\rho\Delta p D_2^2/\mu_2^2$. The influence of the last parameter may be neglected. $Q = D_1^2(\Delta p/\rho)^{1/2} f(D_1/D_2)$

14.20 $F/\rho V^2 A = f(\rho V D/\mu, D^2/A)$ Alternate form: $F/\rho V^2 D^2 = f(\rho V D/\mu, D^2/A)$: note $D^2/A = (D/L)(D/P)$

14.21 $gh/V_o^2 = f(\rho V^6/mg^3, \rho V^3/g\mu, \rho V^4/g\sigma)$; convert to recognisable parameters

14.22 $P/\rho N^3 D^5 = f((Q/ND^3, H/D)$: note P is proportional to $\rho g Q H$

14.23 $V/(bgS)^{1/2} = f(\rho(bgS)^{1/2}b/\mu)$: simplify to show that $V = \text{const.}(\gamma b^2 S/\mu)$

14.24 $T/\mu N a^3 = f(b/a, \rho N a^2/\mu)$; LHS may be replaced by $T/\rho N^2 a^5$ (how?)

14.25 $\tau_o/\Delta\gamma_s d_s = f(\Delta\gamma_s d_s^3 \rho/\mu^2)$

14.26 $\forall/D^3 = f(\sigma/\rho g D^2)$

14.27 $V_m = 300$ km/hr; $F_p = 7500$ N

14.28 $V_m = 0.212$ m/s, $Q_m = 0.416$ l/s, (Pressure loss)$_m = 1445$ N/m^2

14.29 $V_m/V_p = 1/10^{1/2}$, pressure in model $= 13.92$ kPa

14.30 $V_m = 0.3$ m/s, $\Delta p = 4.465$ Pa/m

14.31 $V_m = 2.667$ m/s, $(\Delta p)_m/(\Delta p)_p = 92.43$

14.32 $T_p = 100$ minutes

14.33 $F_L/\rho V^2 c^2$ or $F_D/\rho V c^2 = f_{1,2}(\rho V c/\mu, V_o/c, L/c, \alpha)$

14.34 $\nu_m/\nu_p = (L_m/L_p)^{3/2}$

14.35 $V_m/V_p = 1/30$, $T = 0.4$ hours

14.36 1.2 m/s

14.37 4843.75 m^3/s, 2437.5 kN/m, $V_m/V_p = 1/5$

14.38 $V_p = 3.6$ m/s, $F_p = 15$ N

14.39 $Q_m = 2.5$ l/s, $H_m = 0.08$ m

Index